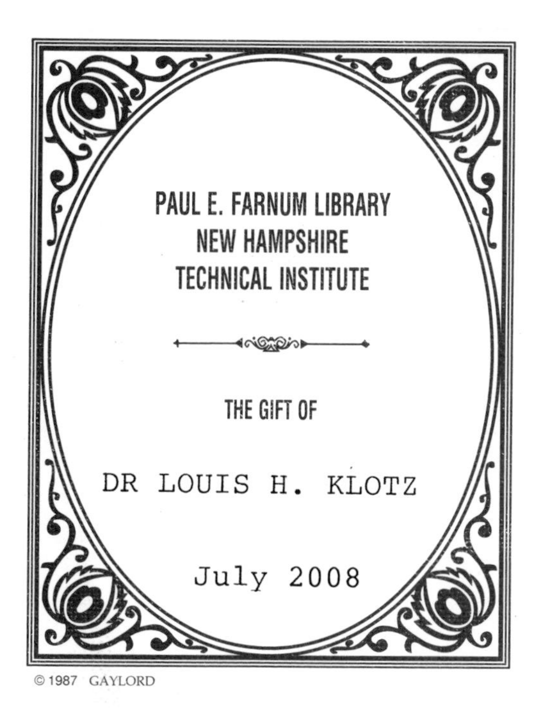

Structural Design Guide to AISC Specifications for Buildings

Structural Design Guide to AISC Specifications for Buildings

Paul F. Rice
Edward S. Hoffman

VNR
VAN NOSTRAND REINHOLD COMPANY
NEW YORK CINCINNATI ATLANTA DALLAS SAN FRANCISCO
LONDON TORONTO MELBOURNE

Van Nostrand Reinhold Company Regional Offices:
New York Cincinnati Atlanta Dallas San Francisco

Van Nostrand Reinhold Company International Offices:
London Toronto Melbourne

Library of Congress Catalog Card Number: 75-40491
ISBN: 0-442-26904-8

Manufactured in the United States of America

Published by Van Nostrand Reinhold Company
450 West 33rd Street, New York, N.Y. 10001

Published simultaneously in Canada by Van Nostrand Reinhold Ltd.

15 14 13 12 11 10 9 8 7 6 5 4 3 2

Library of Congress Cataloging in Publication Data

Rice, Paul F 1921-
Structural design guide to AISC specifications for buildings.

Includes bibliographical references and index.
1. Structural design–Handbooks, manuals, etc.
2. Building–Contracts and specifications. I. Hoffman, Edward S., 1920- joint author. II. Title.
TA658.3.R52 690 75-40491
ISBN 0-442-26904-8

PREFACE

This book is intended to guide practicing structural engineers into more profitable routine designs with the specifications promulgated or endorsed by the AISC for structural steel and joist construction in buildings.

Each new AISC Specification expresses the latest knowledge of structural steel in legal language for safe structural applications. Beginning in 1963 with the introduction of plastic design, each new edition offered better utilization of high strength steel. More economy of material was permitted, but achieved through more detailed and complicated design calculations.

The increasing complexity of the Specifications has encouraged the use of computers for design. Computer time and development of computer design programs are costly, however, and computers are not available to all engineers.

This Guide does not duplicate nor replace the AISC Specifications, the Commentary or the Manual of Steel Construction. It complements the AISC Specifications and Commentary, shows how to take full advantage of available design aids and tables in the AISC Manual. It converts some specification formulas from the review form (or trial designs) to direct design. It presents shortcut formulas, tables and charts for longhand direct design.

Tables 3-1 and 3-2 present flexural resisting moment solutions of the (9) applicable AISC equations for all doubly symmetrical rolled shapes and channels. Selection of a beam section, or solution for the flexural terms in column design, is accomplished in about three minutes.

Specification requirements scattered through various sections of the different specifications have been assembled for the analysis and design of beams, welded plate girders, rigid frames, composite construction, columns, trusses and joists. Numerical examples are based on structural steels with $F_y = 36$ or 50 ksi. Specifications for materials are discussed to aid the structural engineer in avoiding difficulties with use of obsolete specifications.

It is assumed that users of the guide are familiar with structural steel design and analysis as well as the terms and symbols in common use. The Guide indicates the Specification requirements in the order that a designer would normally require their use for design of a particular building element. The Engineer will find applicable Specifications and sections indicated in parentheses thus: AISC Steel (Joist) Spec., Section (00.00.00), following the explanations of their application. For problems outside the scope of this guide, other references are cited. References to the AISC Specification Commentary are indicated thus: "(Com. 00.00.00)."

No attempt has been made to explain each individual Section of the various Specifications. A large number of these sections have provoked little or no question in the past, and have been repeated essentially without change. Architectural (exposed) steel and bridges are not included in the scope of this book.

Two indexes are provided, a subject index and a Specification section reference index. The user wishing to locate all Specification references to a particular subject, as well as the user interested in the interpretation of a particular Specification section, should find this arrangement most convenient.

PAUL F. RICE

EDWARD S. HOFFMAN

CONTENTS

1

STRUCTURAL MATERIALS AND ECONOMICS, —SPECIFICATIONS, TESTING, AND INSPECTION

MATERIALS AND SPECIFICATIONS

General

Specifications. Structural steel shapes are now available under a variety of specifications and a variety of strength grades within the specifications. Strength grades (minimum specified yield stress in ksi) vary from 36 to 100. Table 1-1 provides a convenient summary of these specifications and grades. In addition the various specifications provide a considerable range in all other properties affecting structural design: weldability, ductility, corrosion resistance, fatigue resistance, behavior under different temperatures, etc. For welding materials, see Table 1-2.

Availability. The first decision facing the practical Engineer responsible for design of any steel structure with economy for himself and his client becomes the selection of the basic material itself. This decision is an important one, and the sophisticated design flexibility offered by the possible permutations of the expanded list of material properties, connections, and design methods now available make the choice difficult. Practical economics of design time required preclude exhaustive cost comparisons of all such permutations except for unusual projects. Practical economics of construction costs and local immediate availability simplify this decision for the usual projects. Availability in this sense is not synonymous with the term as applied in Table 1-1. Table 1-1 should be regarded merely as the inherent metallurgical limitations upon availability imposed by the manufacturing process. Economic availability implies material available from several sources (bidders) from stock for small projects, or with minimum delivery time and dependable schedules for delivery on larger projects.

Choice of Material

Grade. The most commonly available type and grade of structural steel shapes in stock is ASTM A 36 (Grade 36). (See Table 1-1.) This type comprises approximately 75 percent of the U.S. production of structural shapes and is considered to be an all-purpose type. It is suitable for all standard connection methods, including field welding. For

TABLE 1-1 Availability of Shapes, Plates, and Bars According to ASTM Structural Steel Specifications.

ASTM Specification Number	Title	Grades
	Material: SHAPES, PLATES, AND BARS (Section 1.4.1)	
A 36	Structural Carbon Steel	32 (over 8") 36 (All other)
A 529	Structural Carbon Steel	42
A 441	High-Strength Low-Alloy Structural Manganese Vanadium Steel	40, 42, 46 & 50
A 572	High-Strength Low-Alloy Columbium–Vanadium Steel	42, 45, 50, 55, 60, 65
A 242	High-Strength Low-Alloy Structural Steel	42, 46, 50
A 588	High-Strength Low-Alloy Structural Steel	42, 46, 50
	Material: PLATES AND BARS (Section 1.4.1.1)	
A 514	High-Yield Strength Quenched and Tempered Alloy Steel Plate Suitable for Welding	90 & 100
	Material: WELDED AND SEAMLESS STEEL PIPE (Section 1.4.1.1)	
A 53	Welded and Seamless Steel Pipe	Grade B
	Material: WELDED AND SEAMLESS STRUCTURAL STEEL TUBING (Section 1.4.1.1)	
A 500	Cold-Formed Welded and Seamless Carbon Steel Structural Tubing in Rounds and Shapes	
A 501	Hot-Formed Welded and Seamless Carbon Steel Structural Tubing	
A 618	Hot-Formed Welded and Seamless High-Strength Low-Alloy Structural Tubing	
	Material: STEEL SHEET AND STRIP (Section 1.4.1.1)	
A 375	High-Strength Low-Alloy Hot-Rolled Steel Sheet and Strip	
A 570	Hot-Rolled Carbon Steel Sheets and Strip, Structural Quality	Grades D, E
	Material: STEEL CASTINGS (Section 1.4.2)	
A 27	Mild-to-Medium Strength Carbon Steel Castings for General Application	65, 35
A 148	High-Strength Steel Castings for Structural Purposes	80, 50
	Materials: STEEL FORGINGS (Section 1.4.2)	
A 235	Carbon Steel Forgings for General Industrial Use	Class C1, F & G
A 237	Alloy Steel Forgings for General Industrial Use	Class A
	Materials: RIVETS (Section 1.4.3)	
A 502	Specification for Structural Rivets	Grade 1, 2
	Material: BOLTS (Section 1.4.4)	
A 307	Low-Carbon Steel Externally and Internally Threaded Threaded Standard Fasteners	
A 325	High-Strength Bolts for Structural Joints Including Suitable Nuts and Plain Hardened Washers	
A 490	Quenched and Tempered Alloy Steel Bolts for Structural Steel Joints	

TABLE 1-2 Welding Materials—AWS Specifications and Structural Welding Code, AWS D1.1-72; Rev. 1-73

AWS Specification Number	Title	Use–Grade
Material: FILLER METAL FOR WELDING (Section 1.4.5)		
A 5.1	Mild Steel Covered Arc-Welding Electrodes	Manual Shielded Metal Arc-Welding
A 5.5	Low-Alloy Steel Covered Arc-Welding Electrodes	
A 5.17	Bare Mild Steel Electrodes and Fluxes for Submerged Arc Welding (AISC Section 1.17.3)	Submerged-arc Process–F60 or F70 AWS Flux Classifications
A 5.18	Mild Steel Electrodes for Gas Metal-Arc Welding (AISC Section 1.17.3)	Gas Metal-Arc Welding
A 5.20	Mild Steel Electrodes for Flux-Cored-Arc Welding	Flux-Cored Welding
Material: STUD SHEAR CONNECTORS (Section 1.4.6)		
D 1.1-72 Rev. 1-73	Requirements of Articles 4.22 and 4.27 for Steel Stud Shear Connectors	

routine or usual projects A 36 will be the economical choice considering availability for early delivery, maximum competition among bidders, and minimum inspection required. Errors in supply of other types of steel are of little concern as they can only be of higher grade, and so field inspection for identification is minimized. Mill tests are considered adequate evidence of quality for most applications, and so quality control testing for the usual project can be avoided or minimized.* Unless otherwise noted for design examples in this book, ASTM A 36 (F_y = 36 ksi) steel will be used.

The most common choice where a higher strength grade is desired is ASTM A 572, Grade 50. This high strength, low alloy steel is weldable by the usual field methods. It can be obtained in all shapes except W 14 X 605 to 730 inclusive, and in plates and bars up to 1½ in. thick. In the usual case where a mill order is placed, the use of A 572, Grade 50, will be economical if the lighter members possible with the higher yield stress are not penalized by reductions in the allowable stress due to local buckling criteria, instability, deflection, or vibration.[1]

In some special projects such as those involving built-up members (hybrid girders), built-up box sections, or composite members, the use of Grade 50 or higher strength grades for purely tensile application can achieve significant economy through reduced tonnage, connections, or dimensions.

Special Properties

Corrosion Resistance. The selection of the higher cost, corrosion-resistant steels such as ASTM A 588 for special applications such as architecturally exposed steel or location in corrosive environments is dictated by the architectural requirements. Economies in these cases are usually achieved by the omission of otherwise required protection by other materials.

Fatigue Resistance. Fatigue resistance is seldom a matter of concern in the usual building design. Special consideration for fatigue effects is not required for either wind or earthquake loading for buildings (1.7.1). When fatigue effects must be considered, fatigue

*Under Supplement No. 3 (effective June 12, 1974), quality control requirements for traceability include material specification designation, heat number, and the mill test reports (1.26.5).

[1] "*1973 Selection Guide for Construction*," Bethlehem Steel Corp., *Modern Steels*, June, 1973.

resistance is provided in design by limiting the allowable stress range for the repetitive (live) loads. The choice of type and grade of steel is involved since the allowable stress range is limited only by the number of repetitions expected, type of stress, and the type of connections used, and is thus nearly independent of the grade of steel used. (See Appendix B, Tables B1, B2, and B3.) Except for ASTM A 514 steel (Plates and Bars, $F_y = 90$ and 100), the allowable stress ranges for various fatigue loading and connection conditions are the same regardless of grade (Table B3). Since this only exception permits an increase in the allowable stress range only for "Category A" (tension stress or reversals of stress upon base metal as rolled, as net sections in friction-bolted connections, or with specially ground and inspected full penetration groove-welded splices) the use of higher stength grades of steel under fatigue conditions offers less opportunity for tonnage savings.

For loading repetitions less than 20,000 cycles, equivalent to two applications daily for 25 years, fatigue effects need not be considered (Appendix B, Table B1). It is suggested for these conditions that fatigue be considered only for large ranges of stress, and that the allowable stress range for "Loading Condition 1" (20,000 to 100,000 cycles) be increased 50 percent (Commentary, 1.7). See also Chapter 7, "Shakedown," and Chapter 3, "Beams." For the effect of fatigue loading upon design and detailing of connections, see Chapter 5, "Connections."

Unusual Service Conditions. Other special properties of steel, for temperature effects upon ductility or the effects of long exposure to various types of radiation upon strength and ductility are usually of no concern for building structures. These effects are not considered in the AISC Specifications.[1] For projects in which these effects will occur, the user is advised to consult metallurgists for the latest available research findings.

Material Properties

General. The basic objective in the design of any structure is the selection of economically sized members to resist all expected or prescribed forces applied thereon within allowable limits on vertical and lateral displacement. The basic data required are, therefore, the complete strength-deformation relationships as forces are applied. Steels are produced to meet: (1) minimum tensile strength requirements at yielding, which is itself defined in varying terms, and (2) minimum ductility requirements, usually measured as the total deformation at rupture. Figure 1-1 shows the typical strength-deformation relationships for steels with (a) definite *yield point*, and (b) arbitrarily defined *yield stress.*

Yield Strength and Yield Point. Figure 1-1 shows curves for two steels with a definite yield point, one at $F_y = 36$ ksi and one at $F_y = 50$ ksi. The latter is marked "1." The yield point can be identified similarly by the "drop of the beam" method, the autographic diagram method, the divider method, or at any prescribed total extension under load. The rounded curves for steels "2" and "3" represent yield strengths, $F_y = 50$ ksi, measured by the offset method (0.2 percent strain beyond the elastic strain line) and the total extension under load method (0.5 percent total strain). "ASTM A 370-71 Standard Methods and Definitions for Mechanical Testing of Steel Products" provides that the yield strength for steels which have no (physical) yield point be measured by the extension under load or the offset method. The offset method is preferred unless the stress-strain relation is well known so that the total extension will correspond closely to

[1] "Proposed Standard Code for Concrete Reactor Vessels and Containments," ACI-ASME Technical Committee on Concrete Pressure Components for Nuclear Service, 1973, ASCE.

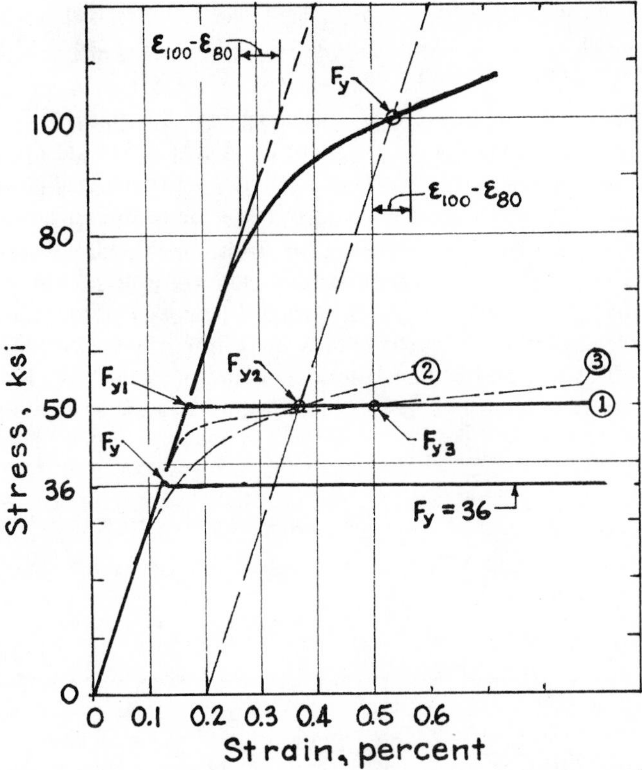

FIG. 1-1 Typical Stress-Strain Curves over the Elastic Range and Specified F_y Measurements.

that at which the specified offset occurs. A total extension of 0.5 percent for steels with $F_y \leqslant 80$ ksi is suggested. For $F_y > 80$ ksi, it is suggested that the total extension be increased by the additional elastic E above 80 ksi. For a steel with $F_y = 100$ ksi, this increase $(\epsilon_{100} - \epsilon_{80})$ has been plotted in Fig. 1-1 and corresponds closely with the 0.2 percent offset point. The reader will have discerned that the arbitrarily defined methods for measurement of yield strength do not give exactly identical results, except by chance. Similarly, the total extension under load method cannot show lower, but may well show a higher yield strength than the other prescribed methods (for example, curves "2" and "3" at $F_y = 50$ ksi in Fig. 1-1). It should be noted that ASTM A 370 does *not* specify the offset for the offset method. *It does require that the offset used shall be reported.* The values of yield strength must be reported, for example, as:

"Yield strength (0.2 percent offset) = 50,000 psi"

Particularly where the design method to be used requires calculations or knowledge of adequate inelastic rotation capacity at connections (Type 2 construction with wind connections, Type 3, and plastic or strength design, for stress reversals, "shakedown," etc.), the designer should know the shape of the stress-strain curve for the material to be used. The following brief summary may serve as a guide. ASTM Specifications A 36, A 441, A 572, and A 588 specify "yield point, min.;" A 514 specifies "yield strength, min.*" with the footnote "*Measured at 0.2 percent offset or 0.5 percent extension under load." All of the above specifications require steel to be delivered in conformance with ASTM A 6 "General Requirements" which, in turn, merely provide that all tests shall be conducted in conformance with ASTM A 370. It will also be noted that the AISC Specifications define F_y, specified minimum yield stress, as either the specified yield point or specified yield strength for steels that do not have a yield point.

Elongations. Figure 1-2 shows typical stress-strain curves through the full range of strain to rupture. Note that the elongation of steels with a definite yield point can be expected to possess a plastic range several times larger than the elastic range, approximately twelve times larger for the lower strength grades. In this range, even for most of the steels with rounded stress-strain curves, the stress does not diverge greatly from the measured yield stress. For gross deflections, beyond those utilized in structural behavior under normal loads or overloads, all steels possess a large reserve elongation capacity and some work-hardening increases in strength capacity from the yield strength to the ultimate strength. This property, although not used in design calculations, is thus of interest to the Engineer as a reassurance as well as a comparative measure of ductility.

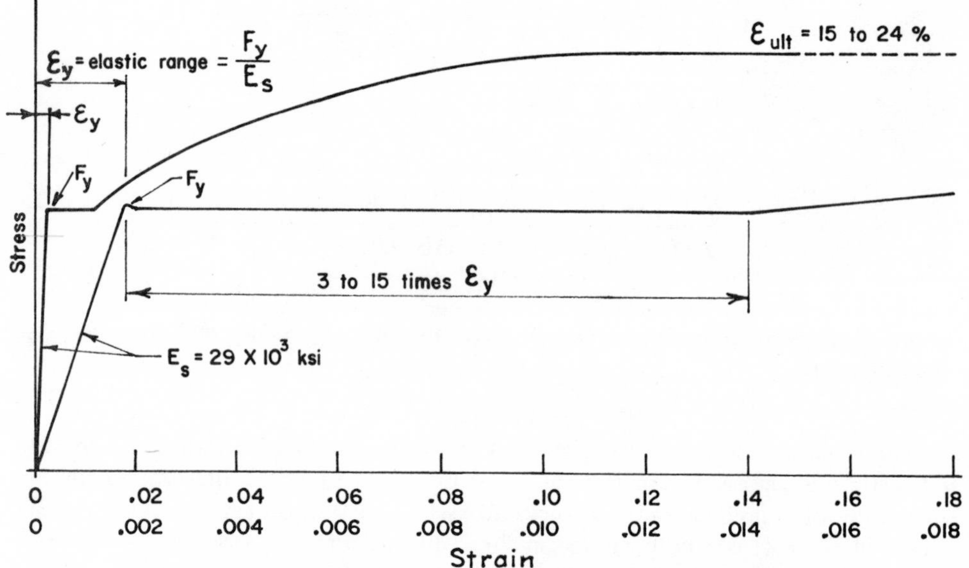

FIG. 1-2 Idealized Stress—Strain Curves over the Entire Range.

Design Assumptions

In practical applications of structural design, these actual stress-strain relationships are not available and would require calculations too complex to justify their use even if available. Practical design, therefore, is based upon the simplified relation shown in Fig. 1-3 and utilizes only the minimum specified properties of yield strength and total strain in tension. The behavior in compression is inconvenient to test and is not tested in routine control, but has been determined to be almost identical to that in tension in the elastic and early stages of plastic strain.

TESTING AND INSPECTION

General

Mill Tests. From the designers' (and clients') interest, the least inspection and testing necessary to ensure the specified quality of materials and construction to develop the

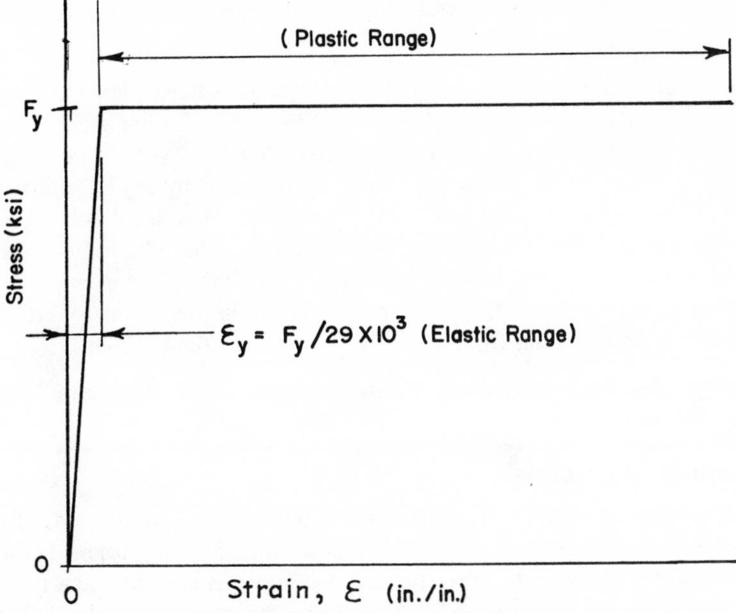

FIG. 1-3 Stress–Strain Assumed in Design

expected behavior under load means economy. Use of ASTM A 36 has the inherent advantage of requiring a minimum of testing and inspection. As previously noted, an inadvertent substitution of grade is of major concern only when higher strength grades are specified. For material quality the mill test reports are usually accepted as the sole evidence of compliance with the specifications.

Field Inspection. Field inspection of fabricated shapes and connection material is required for five principal purposes: (1) identification of the proper strength grade, (2) shop fabrication of connections as specified or shown, (3) dimensional accuracy (length) within allowable tolerances for proper erection, (4) correct sizes, and (5) workmanship, including straightness, finish, and surface condition. This aspect of inspection is so straightforward and least time consuming as to be routine. Field inspection of erection is by far more time consuming and therefore costly; it is also the more difficult and therefore susceptible to costly errors. "Field adjustments" for poorly fitted connections can change design assumptions. Improper lateral bracing, temporary or permanent, unsuited to or ineffective with the erection sequence can result in damage or collapse.

Inspections During Erection

Initial Inspection. The first inspection for erection of a structural steel frame should be performed before the actual erection begins; ideally, this inspection should be completed before the last concreting operation to ensure proper embedment of the anchor bolts. (See Section 7 (b), (d), and (e).)[1] For all but the simplest base connections, templates to fix the bolts in proper plan location, plumb, and at accurate levels are required. The practice of embedding loose anchor bolts after casting concrete usually requires too much inspection in too short a time to be reliable even for bases designed as hinged. Finally, anchor bolts must be protected against damage or displacement from any ex-

[1] AISC "Code of Standard Practice" for Steel Buildings and Bridges.

pected construction operations between the time of placing and the time of using such bolts.

Starting Erection. At this point, the approved lateral bracing plan must be initiated correctly. Lateral stability during erection is very important for immediate safety and to avoid any need for later costly corrections or delays, (Section 7 (i)).[1] Lateral bracing can be furnished by either permanent or temporary members or by any combination of the two. Anchors for any temporary guys must be in place. Any fixed column bases to be considered part of the lateral bracing during erection must be fully bolted even if (temporarily) supported on shims awaiting grouting. Bending, burning off, or displacing anchor bolts for any field adjustment of errors in location, field damage, etc. should be permitted only upon directions from, and after a review by, the Engineer.

Inspection of Field Erection—Beams, Joints, Columns. See "Inspection" paragraphs in Chapters 3, 4, and 6.

Final Approval of Erection

General. The "final approval" of erection most often will be completed, of necessity, in stages since the installation of materials by other trades usually begins before the entire steel framing is erected. In this sense, final approval is granted when the structural steel erection is deemed acceptable for attachment of other materials.[1] (See "Tolerances," Section 7 (h).)[1]

Column Base Connections. All grouting should be complete and cured. All bolts should be given final tightening if needed.

Bracing. All permanent steel lateral bracing should be in place with connections completed. Temporary bracing not required should be removed. Arrangements for removal of any temporary bracing intended to remain till bracing by other materials is in place should be made (Section 7 (i)).[1]

Beams, Joists, Columns, Girders, Trusses. Any deficiencies noted in inspections during erection should be corrected, (Section 7 (j)).[1]

Connections. All connections should be completed.

Painting. Any specified field painting or "touch-up" should be completed (1.25.5). (Also see Section 7(n).)[1]

Final Clean Up. Arrangements for clean-up vary, but any clean-up for which the steel erector is responsible should be completed (1.25.5).

ECONOMICS AND THE CHOICE OF MATERIAL

General

The function of a structural engineer is to provide a satisfactory structure with the most practicable overall economy to the client (Owner). For overall economy, the cost of the steel frame must be considered in context with, and is often subordinate to, costs of alternate designs for architectural and mechanical requirements. Without increasing these other costs more, the value of the structural analysis and design refinements and the selection of the type and grade of steel can be measured only by savings in the cost of the steel frame itself.

[1] AISC "Code of Standard Practice" for Steel Buildings and Bridges.

Variables in the Cost of Steel

The cost of the steel itself varies from time to time according to the supply and demand, domestic and imported, for the entire industry. At any one point in time, the cost of the steel frame will vary with the total tonnage, local supply and demand conditions establishing competitive prices, type and grade of steel, and the cost of the connections (fabrication and erection).

Variables Affecting the Tonnage (psf)

For a given grade of steel the unit weight of steel (psf) of necessity varies with the total height, the height/width ratio, and the number of stories or story height for the columns. Local code requirements for lateral load and specified live loads will affect the weight of steel required for both columns and floor systems. Architectural requirements for span are also major factors influencing the weight of the floor systems.

Index for Comparison

The only common index for comparison of the design efficiency, psf of steel, leaves much to be desired as an index of either design efficiency or even the cost of the steel frame. This index would be strictly applicable only when *all* of the variables of cost and tonnage are identical. In practice such ideal conditions for comparison are very unlikely. In spite of this widely recognized deficiency, the weight of steel in pounds per square foot of gross floor area is an index of great interest, probably because it is the only convenient one we have. It is easily computed, and involves fewer uncertain and unequal factors than dollars per square foot.

Within the framework of these uncertain factors, one general conclusion from comparisons of the steel index (psf) is valid. Improvements in materials (higher strength grades) and refinements in design are reducing this index steadily. Figures 1-4 (a) and (b) show this general trend from 1930 to date as taken from a recent compilation of prominent buildings.[1] As might be expected with so many uncontrolled variables, there are notable exceptions to the trend even in this small listing. The exceptions that seem to violate the trend with high values are usually explainable by unusual architectural requirements, and so these are of lesser interest than the exceptions that are ahead of the trend, those with low values. The design features by which these low values were achieved are of great interest as potential guides to greater design efficiency.

Cost Differentials, Grade 36 versus Grade 50

The practical choices of grade for the ordinary structural steel building frame as earlier noted, usually narrow down to Grade 36 versus Grade 50. Calculation of the tonnage required for alternate designs in the preliminary stage is not difficult. For the preliminary purpose of selecting the more economical grade, the tonnage differential as a percentage of the base (Grade 36) is usually sufficient. If other factors (size of members, availability, possible reduction in story height with Grade 50, etc.) are separately evaluated or not involved in the cost difference, the only additional data required for an economical choice of grade is the cost differential per ton for Grade 50/Grade 36 erected. Some recent Chicago area prices (1973 bids) per ton for alternate designs in Grade 50 and Grade 36 are shown in Table 1-3. It will be noted that the overall average premium for Grade 50 versus Grade 36 in place is approximately 10 percent for this project.

---

[1] *MODERN STEEL CONSTRUCTION*, First Quarter, 1972.

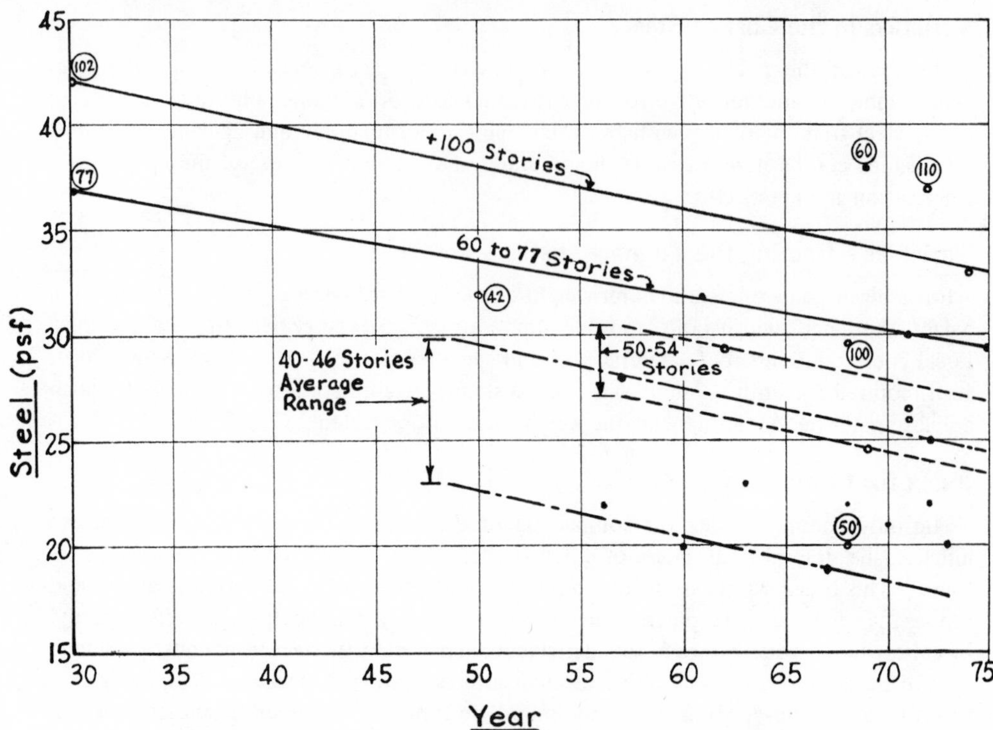

FIG. 1-4 (a) Comparisons of Buildings—Steel (psf)

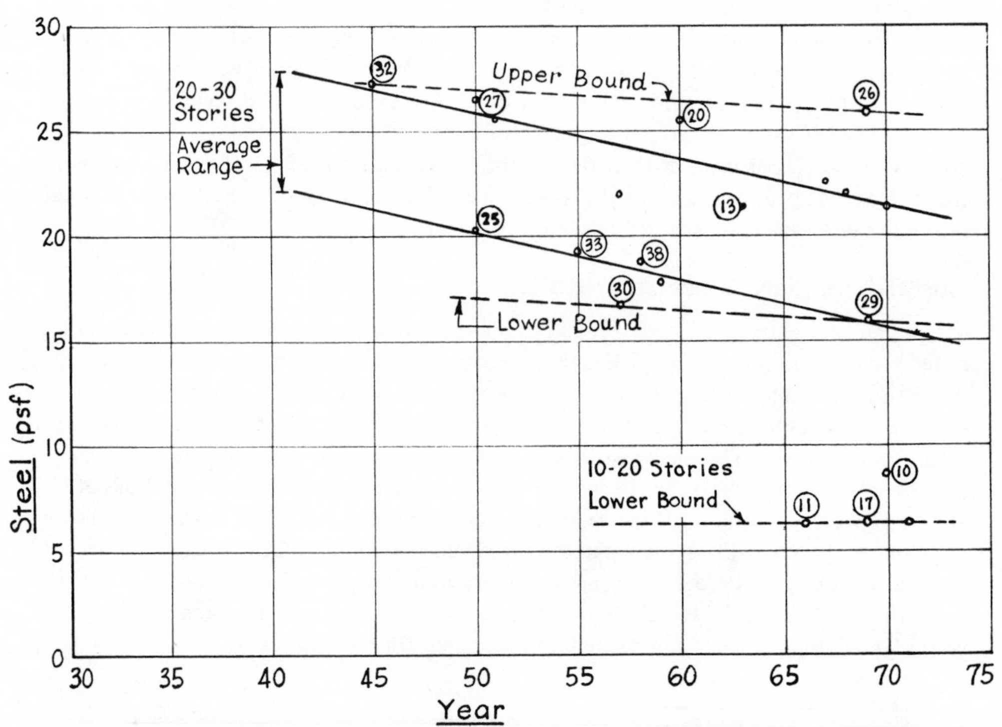

FIG. 1-4 (b) Comparisons of Buildings—Steel (psf)

TABLE 1-3 1973 Price Differentials (Bids) for Grade 50 versus Grade 36 Steel

ITEM	Amount in Order	Grade 36 Material $	Grade 36 Erection $	Grade 50 Material $	Grade 50 Erection $	Average Premium for Grade 50
Rolled shapes, framing, all sizes	+1,000 T	400	95	455	105	13.1%
Rolled shapes, framing, over 20 plf	+1,000 T	385	63	410	63	5.6%
Rolled shapes, framing, under 20 plf	+200 T	555	63	580	63	4.0%
Trusses	+600 T	523	63	548	63	
		553	87	638	110	10.8%
Plate girders	N.A.	538	63	563	63	
		563	107	648	120	9.7%
Built-up columns	+100 T	433	63	458	63	
		554	96	626	108	9.5%
Angles	N.A.	492	225	537	250	9.8%
Bracing rods	N.A.	873	431	945	475	8.9%
Pipe columns, Std. A 501	N.A.	586	102	–	–	N.A.
Shear connectors	5,000 ea.	0.24	0.70			
		0.20	0.40	–	–	–
Open web joists	+750 T	–	–	348	117	N.A.
Long span joists	+100 T	–	–	450	123	N.A.

Due to the wide fluctuation in unit costs at different times, places, and types of buildings as previously discussed, the authors recommend that a simple comparison such as that in Table 1-3 be made for current local conditions and, if possible, for alternate preliminary designs for a particular project to select the economical grade for the final design.

2

STRUCTURAL ANALYSIS AND DESIGN—ELASTIC

ELASTIC DESIGN OF SECTIONS

General

The AISC Specifications have traditionally been based upon the elastic analysis of sections and structures for so long that elastic analysis is implied throughout, except where specific references to plastic analysis, plastic design, or strength design are provided. Although untitled, "PART 1" contains design requirements based upon allowable stresses and elastic analyses with minimum exceptions to recognize plastic behavior necessary in certain joints (1.2). Under the allowable stress concept, the structure is subjected to an elastic analysis for all actual or prescribed service loads and forces (1.3), and proportioned to resist these effects without exceeding allowable stresses (1.5). The allowable stresses have been established as fractions of the specified yield strength, F_y, for the grade of steel to be used. The magnitude of the allowable stresses has been determined so as to provide safety factors approximately equivalent to the load magnification factors used in plastic (strength) design (PART 2).

Allowable Stress (Elastic Theory) Design Method (PART 1)

Pure Flexure. For determination of the resistance to pure bending (no axial load) in most flexural members where the following conditions exist, a single calculation will suffice:

$$S = 12\,M/F_b$$

where

S = the required section modulus (in.^{3}),
M = the maximum bending moment (k-ft.), and
F_b = the allowable bending stress (ksi).

The principal assumptions for the justification of this procedure include the following conditions:

1. The material is homogeneous.
2. Sections plane before bending remain plane.

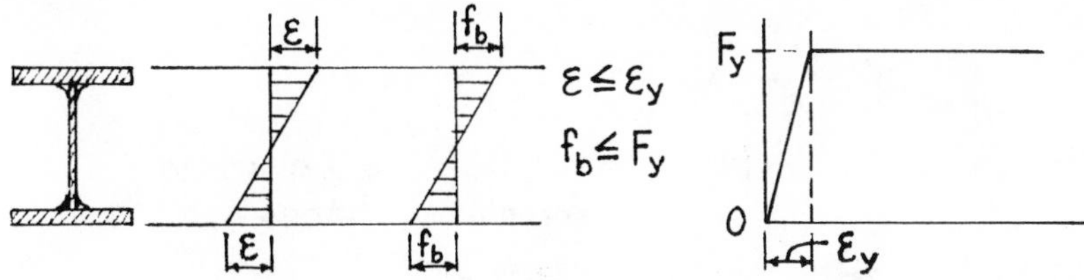

FIG. 2-1 Stress and Strain in the Elastic Range.

3. The cross-section has an axis of symmetry.
4. Stress is proportional to strain ($F_b < F_y$). (See Fig. 2-1.)
5. The bending moment is applied in the plane of symmetry. (See Fig. 2-2.)
6. The member is straight in the plane of bending.
7. The member is stable laterally to the plane of bending.

Shear. The elastic formula for the analysis of shear stress on a section is too complex for routine use with the variety of shapes available or possible for steel members. For members that possess an axis of symmetry (Fig. 2-2) in the plane of loading, two simplifying

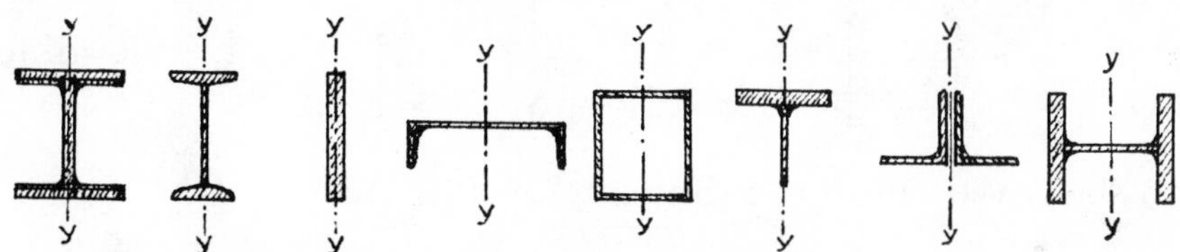

FIG. 2-2 Typical Members with Axis of Symmetry in the Plane of Loading.

assumptions that result in a negligible loss of (theoretical) accuracy are permitted (1.5.1.2):

1. The contribution of the flanges to shear capacity may be neglected.
2. The use of an average value of shear stress on the gross area of the web (overall depth times web thickness). This assumption has been justified by an appropriate reduction of the allowable shear stress.

With these assumptions, derivation and application of the accepted shear capacity formulas becomes simple. Neglecting the flanges, all symmetrical rolled shapes, box shapes, and built-up sections reduce to an equivalent rectangular section with dimensions $\Sigma t_w d$, as shown in Fig. 2-3, and shear stress becomes

$$f_v = V/\Sigma\, t_w d \leqslant F_v = 0.4\, F_y \;\ldots\ldots (1.5.1.2)$$

For special sections not included in Fig. 2-2 and without torsion (loaded in a plane through the center of twist), the theoretical shearing stress at any depth for an elastic material may be computed as:

$$f_v = VQ/It_w$$

where

f_v = the shear stress at any point in the depth, d,
Q = the static moment of the area above the point at which f_v is to be determined, about that point,
I = the moment of inertia of the section, and
V = the external shear force on the axis of symmetry.

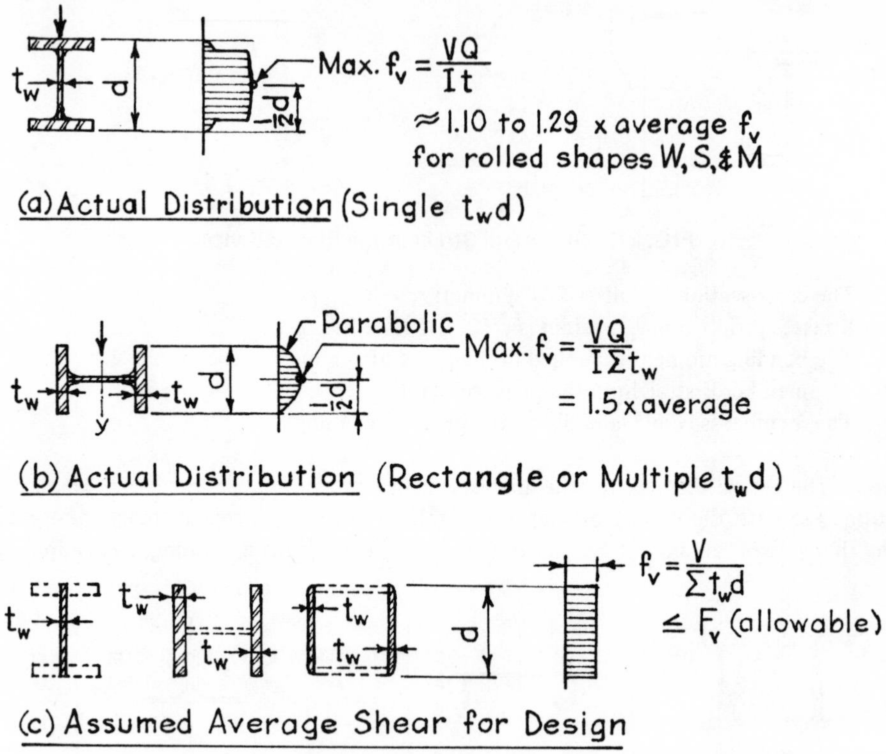

Max. $f_v = \frac{VQ}{It}$ $\approx$ 1.10 to 1.29 x average f_v for rolled shapes W, S, & M

(a) Actual Distribution (Single $t_w d$)

Parabolic — Max. $f_v = \frac{VQ}{I \Sigma t_w}$ = 1.5 x average

(b) Actual Distribution (Rectangle or Multiple $t_w d$)

$f_v = \frac{V}{\Sigma t_w d} \leq F_v$ (allowable)

(c) Assumed Average Shear for Design

FIG. 2-3 Effective Section and Distribution for Shear Stress.

Special Design Investigations

Shear Center. The shear center of a section, *c.s.* (sometimes called "center of twist") is defined as the point through which loads must pass to avoid twisting. For doubly symmetric (W, S, M) and antisymmetric (Z) sections, the *c.s.* is at the centroid. (See sketch (a).)

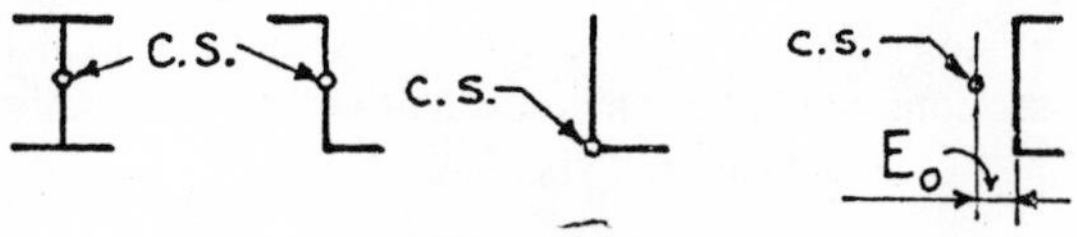

Sketch (a)

For sections unsymmetric about one or both axes (**C, L**) the location of the *c.s.* must be known so that torsion (force times distance from the *c.s.*) can be calculated. The commonly encountered problem is a channel. The *c.s.* is located on the x-x axis (an axis of symmetry) some distance outside the web on the side away from the flanges. For loads in the plane of the x-x axis no twist occurs. Loads causing bending about the x-x axis are usually applied in the plane of the web and cause torsion proportional to the load times the distance from the *c.s.* to the center-line of the web (E_0). (E_0 is tabulated for rolled channels-AISC Manual.) This torsion effect occurs especially in channel sections for

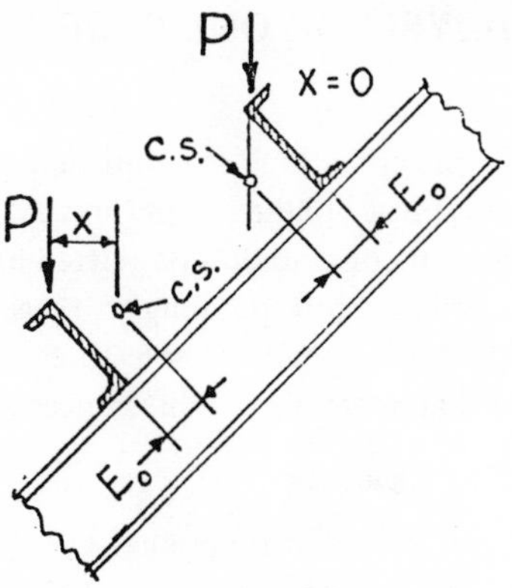

Sketch (b)

purlins on a sloping roof. The effect may be aggravated or reduced simply by the orientation of the channel to bring the force further from or closer to the *c.s.* (See sketch (b).)

Combined Stresses. For unsymmetrical sections, the bending stresses, shear stresses, and torsional warping stresses combine. The combined stress at each extremity of such sections is usually determined by algebraic combination of separately computed values of stresses. "Exact" solutions for stresses and twist angles are very laborious and seldom practicable. The problem is outside the scope of this book, and the reader is referred to available approximate solutions and design aids.*

Connections

For design of connections, see Chapter 5.

Allowable Stress and Safety Factor. For allowable stress (elastic) design (1.2), the term "safety factor" is usually expressed and accepted as the ratio of the specified minimum yield stress to the allowable stress. Though conveniently simple, this definition is a loose term particularly for continuously connected construction since it neglects the reserve strength available in the entire range of post-elastic adjustments in connections. It ignores the overload capacity of the structure between first yielding to the final overall instability. (See Chapter 7, "Plastic Design.") It also disregards the entire approach of evaluating probabilities of overload, the ratio of live load to dead load, actual strength versus minimum specified yield strength, and variations of actual section strength from computed strength, etc.

Nevertheless, the traditional use of this simple expression has considerable merit other than its simplicity. The values for the allowable stresses have been established through long experience with actual structures of all kinds, and with the benefit of numerous research and theoretical studies. The load factors for the newer plastic design (strength) method (PART 2) were established to give safety comparable to that achieved and proven through experience by the allowable stress design method (PART 1). It will be noted that the elastic "safety factor" is in general directly comparable to the "load factors" for strength design. See Chapter 7, "Load Factors."

*"Torsion Analysis of Rolled Steel Sections," Bethlehem Steel Corporation (AIA File No. 13-A-1).

ELASTIC ANALYSIS OF ONE STORY STRUCTURES

General

The application of the AISC Specifications for both analysis and design is best shown by examples. In this section typical building structures are presented with the results of analyses for the different types of construction permitted under the AISC Specifications. The designs of the various members corresponding to these analyses will be developed in the following chapters so that the user will be able to compare the resulting overall designs for the structures rather than mere weight differences for single elements.

One Story Building—Design Example 1

Types of Construction. Design Example 1 represents a very common structure, one story with three bays in the short direction and a large number in the long direction. (See Fig. 2-4 for the loads and dimensions typical for an office building.) This simple structure

TABLE 2-1 Summary of Design Variations and Preliminary Member Sizes—Design Example 1

Fig. 2-5	Type of Constr.	Joists (F_y = 50)	Columns		Beams	
			Section	F_y =	Section	F_y =
(a)	Type 1	20 *H* 6	*W* 8 × 24	36	*W* 18 × 40	50
(b)	Type 2	20 *H* 6	*W* 6 × 15.5	36	*W* 21 × 49	36
(c)	Type 2	20 *H* 6	*W* 8 × 24	36	*W* 21 × 49	36
(d)	Type 2	20 *H* 6	*W* 8 × 24	36	*W* 18 × 40	50
(e)	Type 2	20 *H* 6	*W* 6 × 15.5	36	B4-*W* 14 × 34	
					B5-*W* 14 × 26	
					B6-*W* 14 × 34	50
(f)	Type 3	20 *H* 6	*W* 8 × 24	36	*W* 18 × 40	50

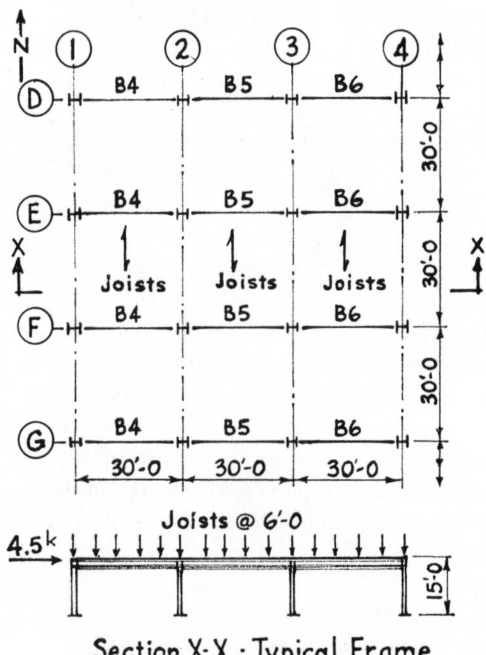

LOADS

Wind (specified) . . . . . . = 20 psf

Roofing ..... = 6.5 psf

Insulation .. = 1.5 psf

Metal Deck .. = 3.0 psf

Ceiling ..... = 7.0 psf

Joists ...... = 2.0 psf

Dead Load . . . . = 20 psf

Snow (specified) . . . . . = 30 psf

Total (D + L) . . . . = 50 psf

Joist Loads = (6)(50) = 300 plf

Beam Loads (at interior fifth points)

Allow for self-weight of beams 2.0 psf

P = (30)(6)(50 + 2) = 9,360 lbs.

P = 9,360/2 = 4,680 lbs. at edge cols.

Panel Load = (30)(30)(0.052) = 46.8 k

Wind Load = (0.020)(15)(30)/2 = 4.5 k

FIG. 2-4 Loads and Dimensions—One Story Office Building—Design Example 1.

TABLE 2-2 (a) Cantilever—Suspended Span System—Two Equal Spans.

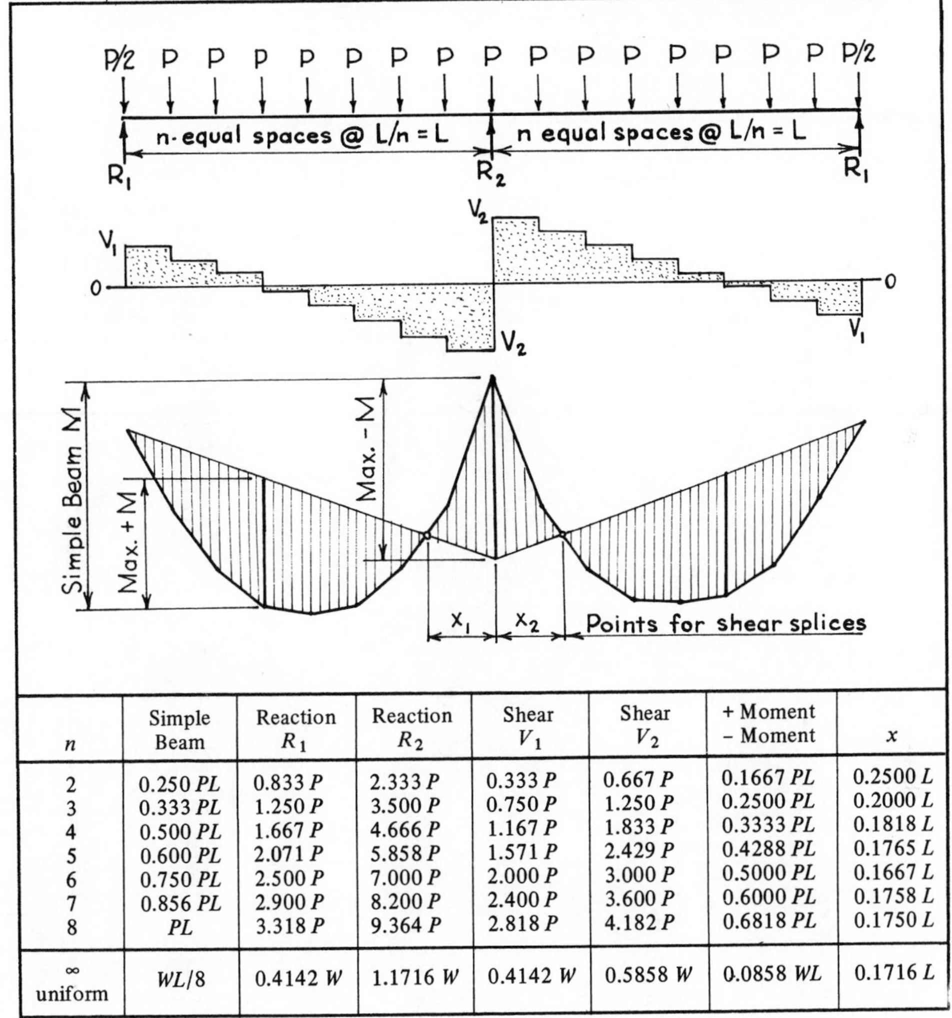

n	Simple Beam	Reaction R_1	Reaction R_2	Shear V_1	Shear V_2	+ Moment - Moment	x
2	0.250 *PL*	0.833 *P*	2.333 *P*	0.333 *P*	0.667 *P*	0.1667 *PL*	0.2500 *L*
3	0.333 *PL*	1.250 *P*	3.500 *P*	0.750 *P*	1.250 *P*	0.2500 *PL*	0.2000 *L*
4	0.500 *PL*	1.667 *P*	4.666 *P*	1.167 *P*	1.833 *P*	0.3333 *PL*	0.1818 *L*
5	0.600 *PL*	2.071 *P*	5.858 *P*	1.571 *P*	2.429 *P*	0.4288 *PL*	0.1765 *L*
6	0.750 *PL*	2.500 *P*	7.000 *P*	2.000 *P*	3.000 *P*	0.5000 *PL*	0.1667 *L*
7	0.856 *PL*	2.900 *P*	8.200 *P*	2.400 *P*	3.600 *P*	0.6000 *PL*	0.1758 *L*
8	*PL*	3.318 *P*	9.364 *P*	2.818 *P*	4.182 *P*	0.6818 *PL*	0.1750 *L*
∞ uniform	*WL*/8	0.4142 *W*	1.1716 *W*	0.4142 *W*	0.5858 *W*	0.0858 *WL*	0.1716 *L*

will be employed for comparisons of a number of variations in design permitted under the AISC Specifications. Preliminary sizes were selected for each of these variations as shown in Table 2-1. Elastic analyses were carried out for each of the resulting structural designs manually or by computer. The results of these elastic analyses are presented in Fig. 2-5 as follows:

(a) Type 1 construction. Rigid connections, including fixed column bases.
(b) Type 2 construction. Braced against sidesway; hinged column bases; connections for shear only.
(c) Same as Case (b), but not braced. Wind resistance provided by beam-column connections only.
(d) Same as Case (c), but wind connections at both base and beam-column connections.
(e) Type 2 construction. Braced against sidesway; hinged bases; shear connected "suspended spans" with the intermediate beam-column connections for "cantilever" spans. (See Tables 2-2 (a) and (b).)

TABLE 2-2 (b) Cantilever—Suspended Span System—Three or More Equal Spans.

n	Simple Beam Moment	Exterior Span							Interior Span			
		Reaction R_1	Reaction R_2	Shear V_1	Shear V_2	+ Moment	− Moment	X_1	R_3	V_3	+ Moment & − Moment	X_2
2	0.250 *PL*	0.875 *P*	2.125 *P*	0.375 *P*	0.625 *P*	0.188 *PL*	0.125 *PL*	0.2000 *L*	2*P*	0.5 *P*	0.125 *PL*	0.2500 *L*
3	0.333 *PL*	1.333 *P*	3.167 *P*	0.833 *P*	1.167 *P*	0.278 *PL*	0.167 *PL*	0.1423 *L*	3*P*	*P*	0.167 *PL*	0.1667 *L*
4	0.500 *PL*	1.750 *P*	4.250 *P*	1.250 *P*	1.750 *P*	0.375 *PL*	0.250 *PL*	0.1429 *L*	4*P*	1.5 *P*	0.250 *PL*	0.1667 *L*
5	0.600 *PL*	2.200 *P*	5.300 *P*	1.700 *P*	2.300 *P*	0.480 *PL*	0.300 *PL*	0.1304 *L*	5*P*	2 *P*	0.300 *PL*	0.1500 *L*
6	0.750 *PL*	2.625 *P*	6.375 *P*	2.125 *P*	2.875 *P*	0.562 *PL*	0.375 *PL*	0.1304 *L*	6*P*	2.5 *P*	0.375 *PL*	0.1500 *L*
7	0.856 *PL*	3.071 *P*	7.429 *P*	2.575 *P*	3.430 *P*	0.674 *PL*	0.429 *PL*	0.1250 *L*	7*P*	3 *P*	0.429 *PL*	0.1429 *L*
8	*PL*	3.500 *P*	8.500 *P*	3.000 *P*	4.000 *P*	0.750 *PL*	0.500 *PL*	0.1250 *L*	8*P*	3.5 *P*	0.500 *PL*	0.1500 *L*
9 ↕ ∞	$\frac{WL}{8}$	0.4375 *W*	1.0625 *W*	0.4375 *W*	0.5625 *W*	$\frac{43}{512}WL$	$\frac{WL}{16}$	0.1250 *L*	*WL*	0.5 *WL*	$\frac{WL}{16}$	0.1464 *L*

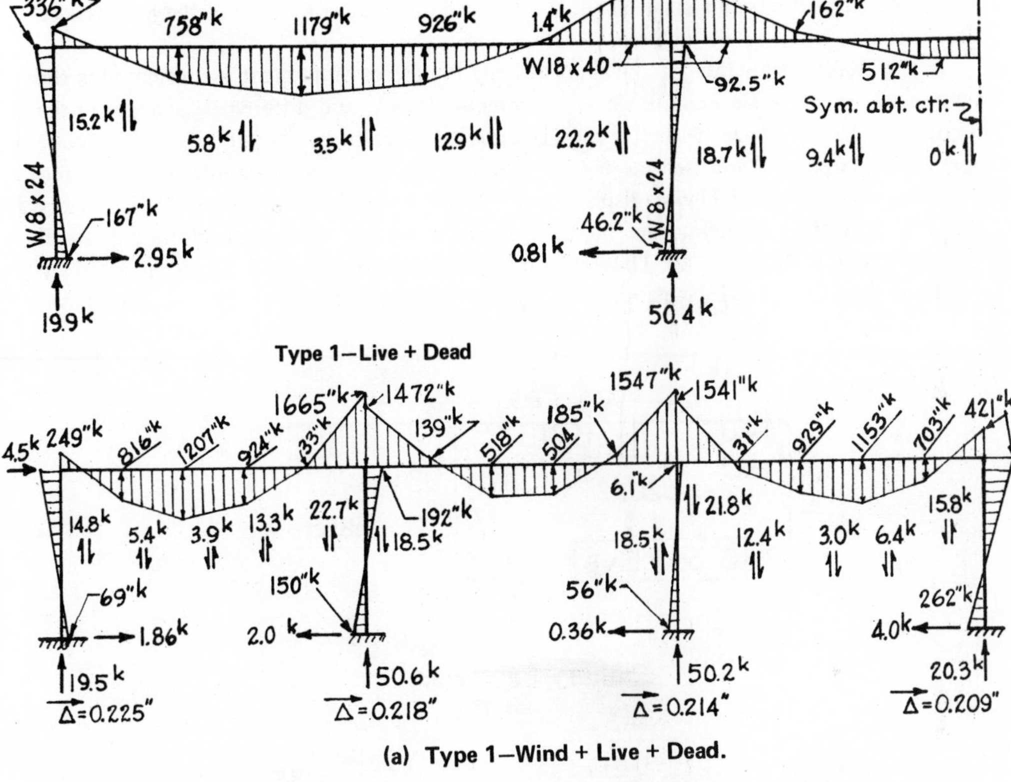

Type 1—Live + Dead

(a) Type 1—Wind + Live + Dead.

FIG. 2-5 Maximum Moments and Shears—Various Types of Construction—Design Example 1.

(f) Type 3 construction. Semi-rigid beam-column connections, including fixed column bases.

It will have been noted that the preliminary selections of sizes for columns and beams in the six analyses differ. In later chapters on design, the effect of varying yield point for Grades 36 and 50 will be compared. Due to space limitations, only the results of the analyses for critical load conditions are shown in Fig. 2-5. The computer-produced analyses shown in Fig. 2-5 include the effects of moment and axial shortening in computation of the lateral deflections, but the secondary effect ($P\Delta$ moment) was neglected. See the Appendix to Chapter 2 for the computer printouts showing separate effects and for the computer-produced analyses considering the ($P\Delta$) effect.

Type 1 Construction. Type 1 construction is the simplest expression of the concept of the completely elastic structure assumed for a classical elastic analysis. (See Fig. 2-5 (a).) Joints must be adequate to preserve the original angles of connected members (1.2). In application, this requirement usually demands that the most flexible member at a joint assemblage be fully developed. Elastic design for Type 1 construction is "unconditionally" permitted (1.2). There are some specific guides, if not conditions, imposed: (1) the use of center-to-center spans and heights for the analysis, and (2) the design for critical sections at the faces of the supports (joints). Thus, the stiffening effect of the joints is neglected in the frame analysis, but the reduced moment at the face of the joint is used for design.

In the traditional manual frame analysis, only the effect of moment is considered in the computation of displacements since the effects of shear and differential changes in the axial lengths are known to be of a lower order of magnitude. Traditionally, such computed displacements are used merely for establishing camber in horizontal members and limiting the computed lateral displacement ("drift") in columns. The secondary effect of this drift ($P\Delta$) is an increase in the column moments, causing an additional lateral displacement ("sidesway"). These secondary effects are provided for in an arbitrary fashion by the AISC Specifications by a moment magnifier term. This term,

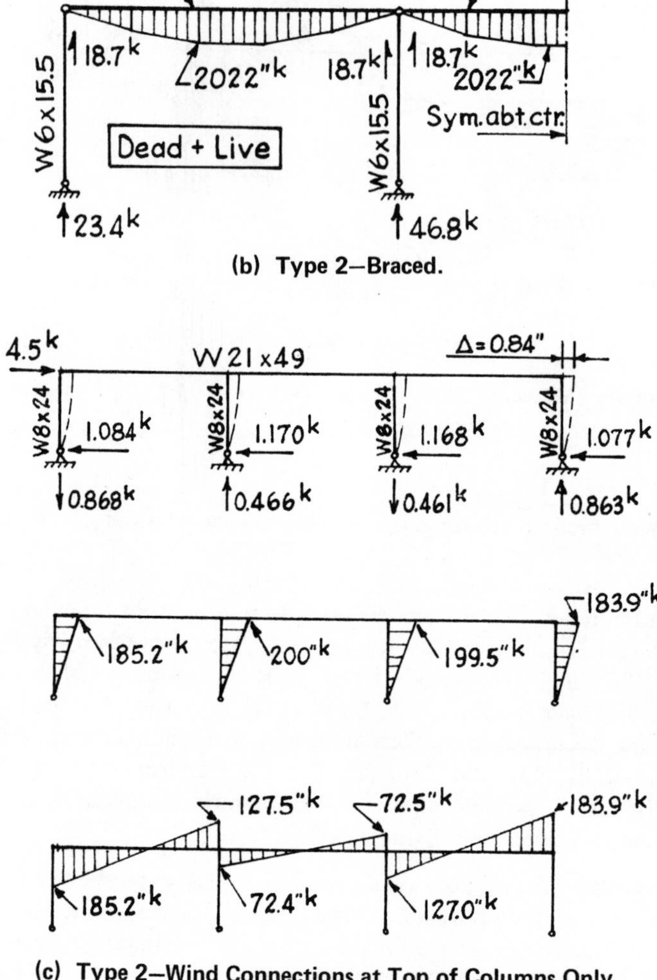

(b) Type 2—Braced.

(c) Type 2—Wind Connections at Top of Columns Only.

FIG. 2-5 (Continued)

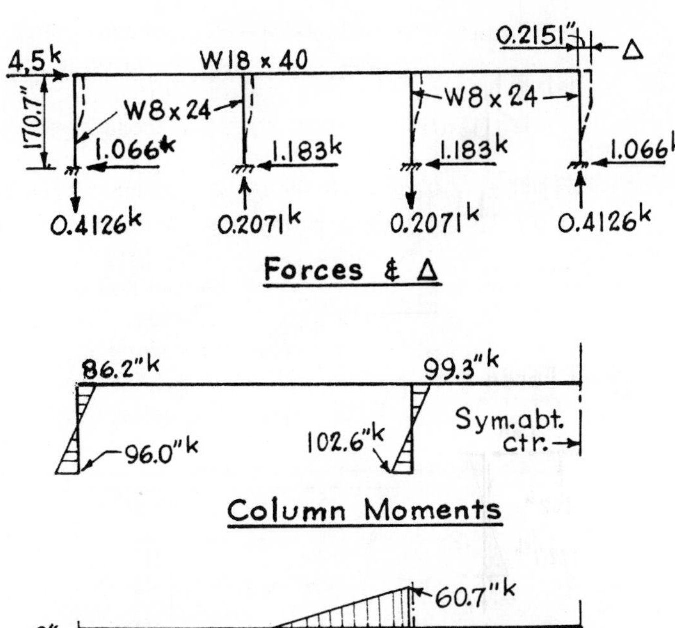

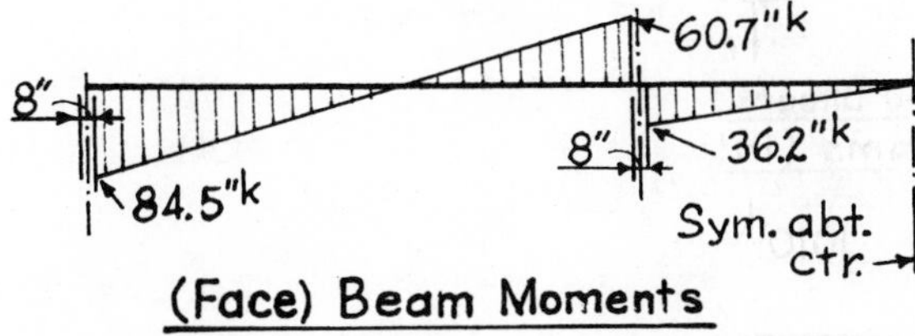

(d) Type 2—Wind Connections at Top and Bottom of Columns.

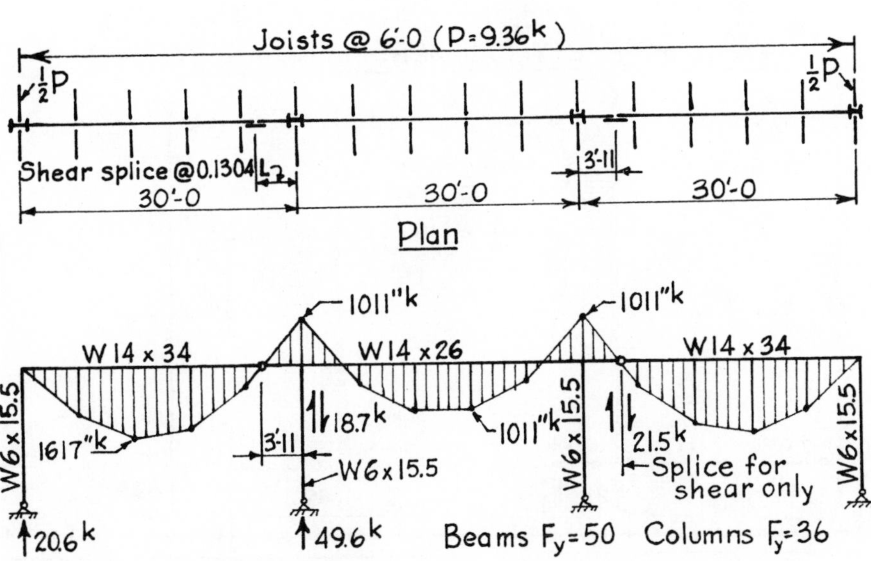

(e) Type 2—Cantilever—Suspended Span System, Braced (See Table 2-2).

FIG. 2-5 (Continued)

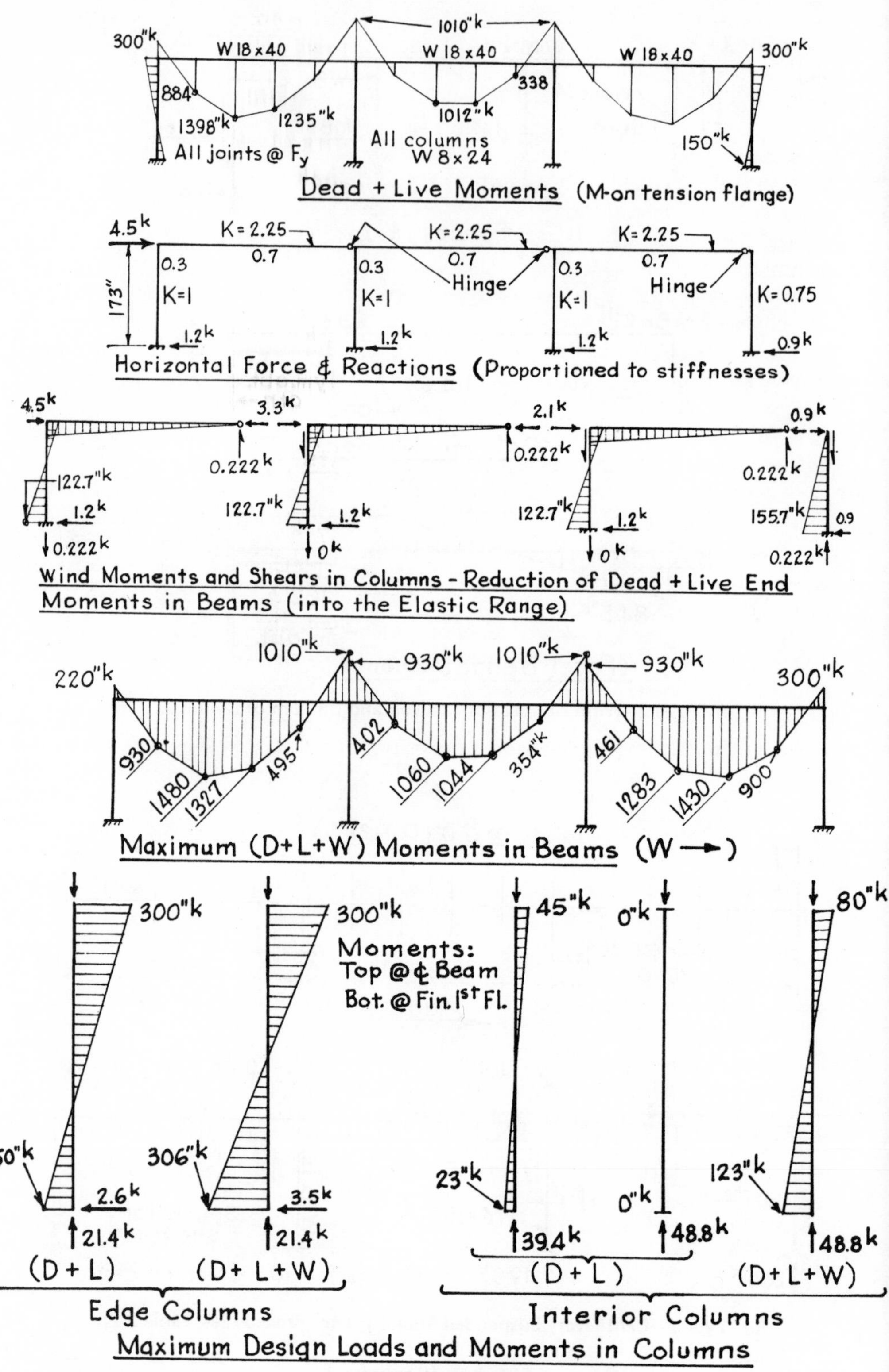

(f) Type 3—Semi-Rigid.

FIG. 2-5 (Continued)

$C_m/(1 - f_a/F'_e)$, with the denominator based upon the Euler buckling effect, is included in one of the prescribed column design formulas (1.6.1). The term C_m represents the effects of lateral bracing, if any, and end connection (moment) restraints (1.6.1). See Chapter 6, Columns, Figs. 6-2 and 6-6; and Table 6-1.

Recent developments of computer analysis programs, particularly for high rise building structures are more sophisticated. In one version commonly used, the effects of shear deformations and axial length changes are included and the drift is merely computed on this basis. In another, the "$P\Delta$" effects (secondary moments due to the drift; additional lateral displacement, sidesway, due to the secondary moments; third order increases in moments due to the additional second order sidesway; etc.) are included, but shear deformation and differential axial length changes are not included. Only for narrow high rise buildings are all of these effects important enough to justify the design time required. Results from both of these programs have been included among the computer-produced analyses for which the critical moments and shears are summarized in Fig. 2-5 for Design Example 1. Comparisons of the computer output, Appendix, Chapter 2, for Design Examples 1 and 2 (one story and 25 story buildings) show that these secondary effects begin to become significant for the 25 story building shown.

As noted, the analyses for Design Example 1 are summarized in Fig. 2-5. The results for Type 1 Construction with fixed column bases are taken from the computer analyses (Appendix-Chapter 2) for moment including the secondary effects of axial length changes in both the beams and the columns and shear deformations, but excluding the ($P\Delta$) effect. The ($P\Delta$) effect is excluded since the prescribed moment magnifier will be used in Chapter 6 for the design of the one-story columns of Design Example 1 (1.6.1). The ($P\Delta$) effects can be studied as the difference in lateral deformations between the two simple moment analyses both excluding shear and length change effects. (See Appendix, Chapter 2.)

Figure 2-5(d) for Type 2 Construction with wind moment connections at the top and bottom of the columns combines the "wind only" computer analysis moments from Type 1 Construction with the manual analysis dead plus live gravity load moments for Type 2 Construction as shown in Fig. 2-5(b). Manual analyses (moment only) were used for Fig. 2-5 (b), (c), (e), and (f).

Upon comparison of the analyses for the Type 1 frame with those from the other types of construction, the following observations are apparent:

1. Critical moments are less than for Type 2 Construction even when the latter is braced. Lighter members are to be expected with the Type 1 Construction.
2. Negative moment at interior supports is critical, and it is approximately 75 percent of the simple beam moment. (The ideal of balanced moments with the negative moment equal to positive moment, for design in Type 1 Construction, would occur only by chance perhaps with unusual framing, unequal spans, and/or unequal loads.)
3. The ten percent reduction in negative moment and increase in positive moment as a recognition of plastic behavior (1.5.1.4.1) is generally insufficient to achieve equal negative and positive moments for maximum reduction in tonnage.
4. The complete manual analysis including axial length effects, pattern loading, computation of sidesway, etc. is extremely time consuming even for the simplest one-story frame and economically not feasible for multi-story frames.
5. The straightforward Type 1 frame connections for this example (light wind loads; one story height) are unlikely to prove economical. Note that the increase in the critical beam moments due to wind load is generally far less than the one-third allowed (1.5.6).

Type 2 Construction. Type 2 Construction, traditionally the most popular because of its small demand upon design time, is perfectly suited to manual calculations for gravity load analyses. (See Fig. 2–5(b).) The provision permitting design of "wind connections" for wind only (1.2) extends the application of manual calculations to resist wind where bracing is architecturally undesirable without materially increasing the design time required. The computer solutions shown can be readily compared to manual solutions based upon simplifying assumptions that (1) reactions are proportional to the column stiffnesses, (2) axial changes in length are negligible, and (3) lateral deflections can be neglected. It will be concluded that the simplifying assumptions are eminently acceptable. (See Fig. 2-5 (c) and (d).) Theoretical analyses easily justify the allocation of the wind moment only to the connections and the assumption (implicit) of adequate rotational capacity to ensure safe behavior for designs based upon these assumptions.[1]

The following general observations from comparisons of the analyses for the various designs using Type 2 construction; Fig. 2-5 (b) braced; (c) wind connections at the top of the columns only; and (d) wind connections effective at both the top and the bottom of the columns; are pertinent:

1. The advantages of minimum size columns, minimum design time, feasibility of manual design, minimum cost shear connections, and minimum field labor required for the braced frame will often overcome the disadvantage of heavier total weight of steel required in the beams to achieve a minimum cost structure.

2. The slight added weight of the wind moment beam-column connections and/or the fixed bases will often be less than that required for separate (steel) wind bracing, but the added erection time and cost would be justified only by architectural requirements precluding the use of wind bracing.

Type 2 Construction (Modified). Laterally braced, cantilever-suspended span construction as shown in this example utilizes Type 2 construction shear connections throughout, where beams can be cantilevered into both spans over the interior supports. (See Fig. 2-5(e).) Where regular spans and uniform loads or concentrated loads regularly spaced occur, this modification of Type 2 construction adapts itself to a rapid manual analysis and results in low steel weight. See Table 2-2 (a) and (b). Simple shear connections and simple continuous-beam-over-column connections can be employed for roof construction. For intermediate floors in multi-story structures, the cantilever connection detail is more difficult, but can often be economical. See Chapter 3, Beams.

It will be noted from Fig. 2–5(e) that:

1. Critical positive and negative moments can be made equal in the interior spans.
2. Positive moment in the end span is critical; it is the largest in a set of equal spans.
3. Additional connections for wind only can be employed.
4. The only available moment capacity in the example shown which is not utilized is the negative moment capacity of edge columns. Since minimum size columns are used, this unused capacity is minor.
5. Low tonnage of steel can be achieved.
6. Fabrication and field erection time are both near minimum.

---

[1] "Wind Connections with Simple Framing," R. O. Disque, *AISC Engineering Journal*, July, 1964.

Type 3 Construction (Semi-rigid). Type 3 construction has never been a popular choice, perhaps because it is a compromise between elastic and plastic analyses. It permits the use of plastic theory for the design of the connections and analysis of their behavior and requires an elastic analysis and design otherwise. Most designers and developers of computer programs prefer the more formal approaches of Type 1, elastic, or Type 1, plastic for consistency. Type 1, plastic, however, until 1974 was limited in application to structures of two stories in height unless lateral bracing is provided (2.1). Under the 1974 Revision (Supplement No. 3) unbraced frames of more than two stories must be designed by rational analyses including *elastoplastic* ($P\Delta$) and axial deformation effects (2.3.2). These refinements require computer analysis for all but the simplest cases, and will no doubt stimulate development of suitable programs. No such limitation is placed upon Type 3 construction (1.2). For analysis, Type 3 construction gives the designer a flexible control to equalize the positive and negative moments to the positive and negative moment resisting capacity for maximum efficiency in reduced tonnage. It also avoids the need of deliberately designed hinges located at the points of low moment as in the cantilever suspended-span modification of Type 2 construction. The analysis can perhaps best be described as that for a simple structure, as with Type 2 construction, but with known added end moments. Instead of helplessly observing a computer print out fully rigid critical end moments much larger than the mid-span moments as for Type 1 construction, the designer resolves the analysis manually by preselecting beam end moments equal to approximately half the simple span moment for the interior spans. Approximately three-fourths of the "rigid connection" capacity at an end support of lesser stiffness than the beam, but not more than half the simple span beam moment at a more rigid end support, may be assigned to the connection at the edge column to reduce the positive moments in the end span.

The analysis here for pattern loading is simplified into a simple moment distribution at only one joint at a time, and this chore is required only when the "fixed-end" unloaded panel end moments (due to the dead loads) are less than the yield capacity used for the connection. Using the normal choice of an end connection capacity at $f_b = F_y$, equal to $(1/16)WL$, where W is the total (uniform) dead plus live loads, and L is the span, the pattern loading will not change the end moments unless $(1/12)W_dL$ is less, where W_d is the dead load moment. Ordinarily the difference $(0.0625\ WL - 0.0833\ W_dL)$ will require distribution into the column and to increase the positive moment in the unloaded panel.

The analysis for wind moments resolves into a simple distribution of the wind reaction forces proportional to the stiffnesses of a series of single column-beam frames and a single column. See Fig. 2-5(f). Where the wind moment reduces the gravity load end moment at a connection, the behavior is in the elastic range like any rigid joint; where the wind moment would tend to increase the gravity load moment at a connection, the connection can be considered as a hinge with an applied known end moment as well as shears. It will be noted from Fig. 2-5(e) versus (f) that Type 3 construction permits:

1. An end moment reducing critical midspan moments from 1617 to 1398 in the end spans.
2. Elimination of the hinge connections at the inflection points and the beam-on-column-top connections.
3. Beam-to-column-face end connections suitable for intermediate floors more easily than the Type 2 modified construction.
4. Equally feasible manual analysis.

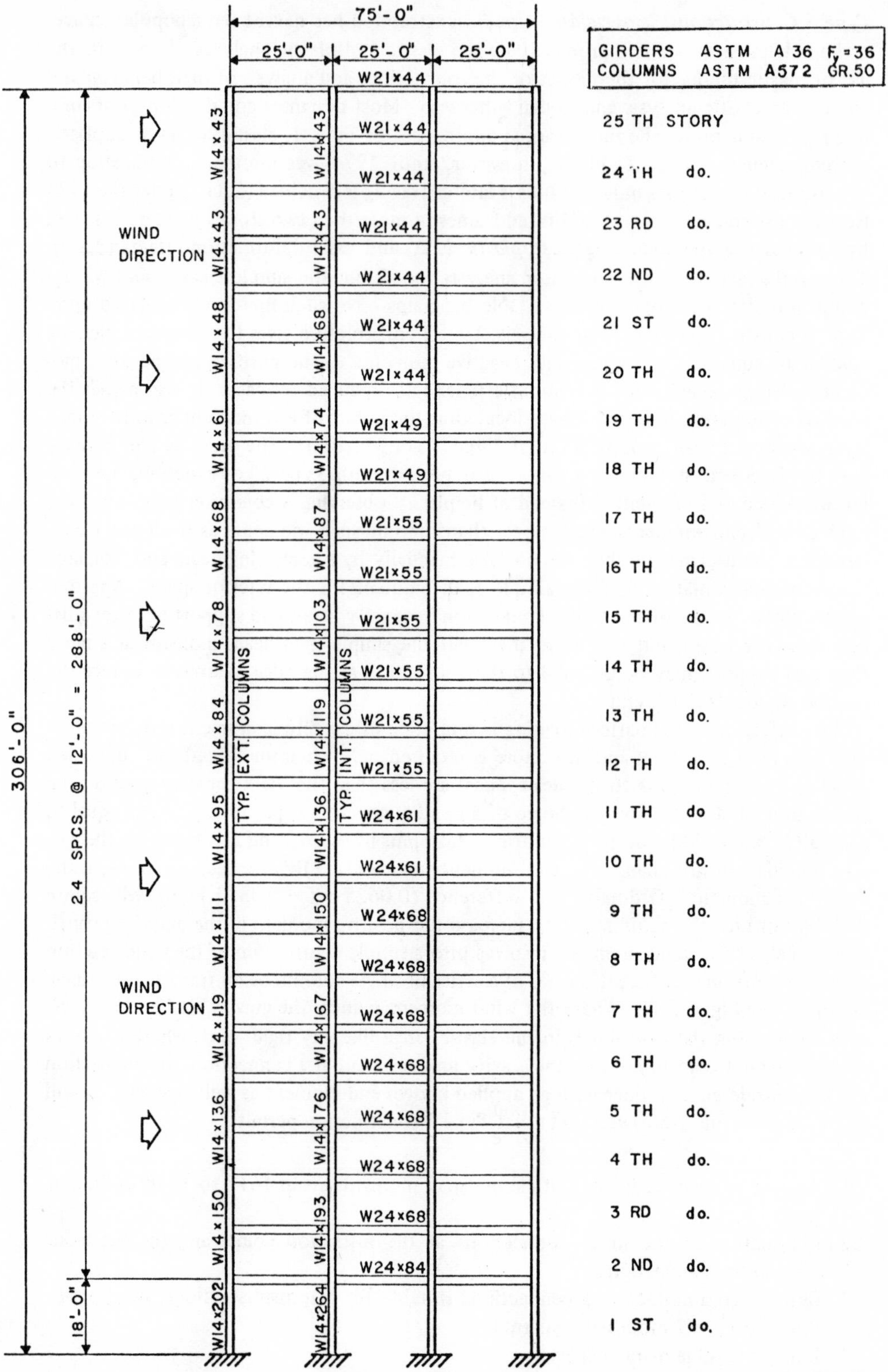

FIG. 2-6 Dimensions and Preliminary Member Sizes—Typical Moment-Resisting Rigid Frame; Type 1—25 Story Office Building—Design Example 2.

ELASTIC ANALYSES FOR MULTI-STORY BUILDING

Design Example 2

General. Design Example 2 represents a moderately high, high-rise (25 story) office building, three bays wide in the short direction and a large number in the long direction. See Fig. 2-6 for elevation and dimensions. Separate bracing against lateral displacement is assumed in the long direction. Resistance to lateral displacement in the short direction is to be provided only by rigid frames with moment-resistant connections, Type 1 Construction (1.2). For purposes of analysis, connections detailed as shown in Fig. 2-7 are assumed at all beam-column joints. The column bases are assumed to be fully

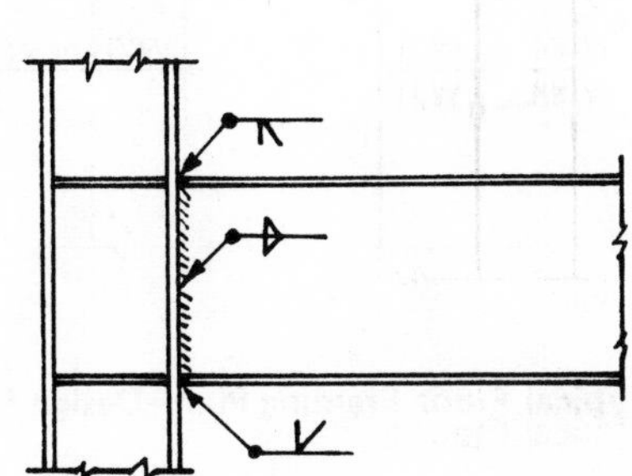

FIG. 2-7 Type 1 Connection Detail—Design Example 2.

fixed at the first floor (ground) level as shown in Fig. 2-6. In an actual structure, this condition would be achieved if the columns were carried through to the footings at a lower level, and moment-resistant first floor framing were provided to transmit all horizontal forces to the foundations. Connections for the floor framing in the long direction as shown in Fig. 2-8 are assumed to transmit vertical shears only to the rigid frames, Type 2 Construction (1.2). The panel size, 25 feet square, wind loads, live loads, and live load reductions were selected as those commonly used for typical office buildings. This building frame as described is called an "unbraced frame" in the short direction and a "braced frame" in the long direction.

The height (first floor 18′-0″ plus 24 typical stories at 12′-0″) was chosen deliberately for this example as somewhat more than the usual limit in practice at this time without separate lateral bracing. Separate analyses and designs were prepared for these three-bay frames (height/width ratio ≈ 4) braced frames, Type 2 Construction, and unbraced frames, Type 1 Construction to evaluate: (1) the tonnage premium replacing lateral bracing by a rigid frame, and (2) the total lateral displacement (drift plus sidesway) with the unbraced system. The reader is left to evaluate the *net* tonnage premium by deducting the weight required for his design of separate bracing.

If other materials were required for architectural or other reasons as fire walls, partitions, deep spandrels, etc. and could be utilized as stiffening elements for lateral bracing, the entire tonnage premium for wind resistance could, of course, be eliminated. On the other hand, almost any system of shear walls included solely for lateral bracing would, without question, cost more and require more field erection time than the rigidly connected Type 1 frame.

As for the total lateral displacement tolerable, again the reader is left to judge for himself. The Type 1 frames were designed to meet the usual rule of thumb limit, 0.0025

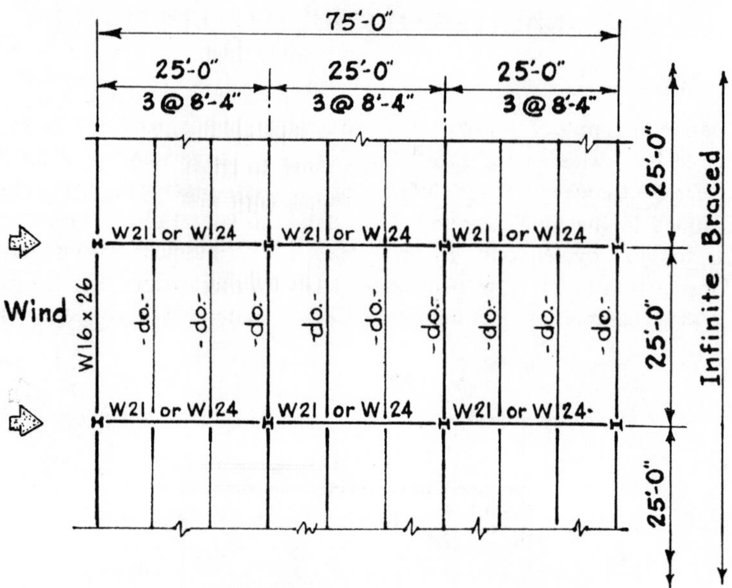

FIG. 2-8 Typical Floor Framing Plan—Design Example 2.

H, where H is the total height, and the displacement is calculated due to primary wind moments only (neglecting secondary effects of drift, sidesway, shear deformation, and axial length changes). Stiffening and damping elements such as walls, partitions, and deep spandrels parallel to the three-bay frames which would often be required in a particular building were not considered. Such elements could reduce the lateral displacements under service loads and the period of lateral displacements sufficiently so that the frame as designed might well be very satisfactory for one building and too flexible for another. (See Appendix-Chapter 2 for analyses; Fig. 2-12, 2-13, 2-14, and 2-15 for summaries of analyses; and Chapter 6 for design of columns.)

Loads. The loads for design were taken as follows. Wind load = 20 psf; live load = 50 psf for all floors above ground level; and live load reduction = $0.08\,A \leqslant 23(1 + D/L) \leqslant 60$ percent, where A is the area supported by an element, sq. ft. The dead loads were accumulated as:

$5\frac{3}{16}$ in. slab on metal deck	41 psf
ceiling	10 psf
partitions	20 psf
beams	4 psf
Total dead load on beams	**75 psf**
girders	2 psf
Total dead load on girders	**77 psf**

Columns. The columns were assumed continuous for two stories to reduce the number of field splices and to retain ease of fabrication, handling, and erection. Following this assumption, preliminary selections for analysis were made for critical conditions at the even stories. The results of the various computer-produced analyses with and without secondary effects are shown as printed out in the Appendix, Chapter 2.

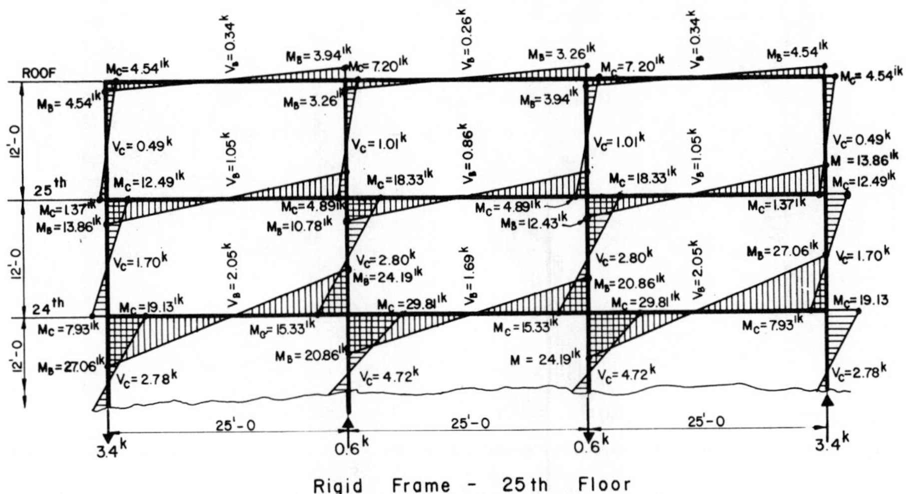

Rigid Frame - 25th Floor

Wind Only - Shears & Moments*

* Standard elastic analysis – primary moment effects only
Moments at joint centerlines in k-ft. Shears in kips

(a)

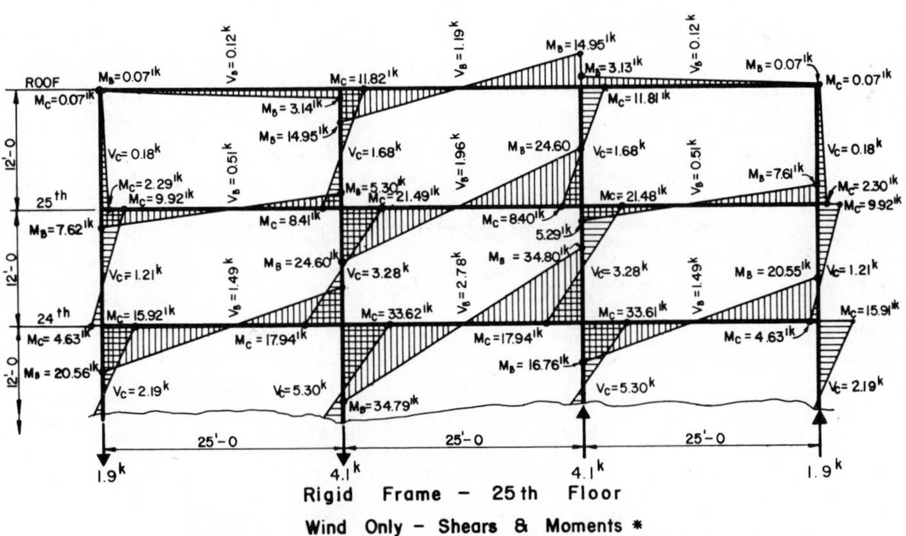

Rigid Frame - 25th Floor

Wind Only - Shears & Moments *

* Elastic analysis including axial shortening in columns and shear deformation ($P\Delta$) effect neglected. Moments at joint centerlines in k-ft. Shears in kips.

(b)

FIG. 2-9 Frame Analyses–25th Floor–Design Example 2.**

**All beams, W21 × 44, All columns W14 × 43.

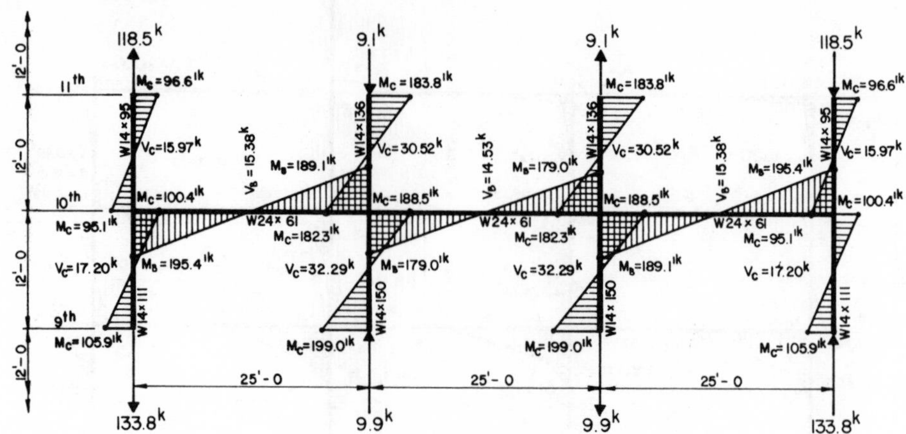

Rigid Frame – 10th Floor
Wind Only – Shears & Moments*

*Standard elastic analysis – primary moment effects only
Moments at joint centerlines in k-ft. Shears in kips.

(a)

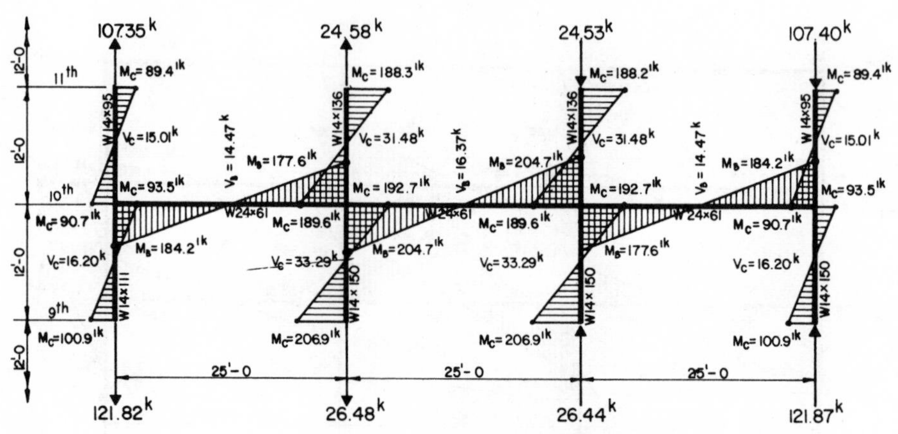

Rigid Frame – 10th Floor
Wind Only – Shears & Moments*

*Elastic analysis including axial shortening in columns and shear deformation. (PΔ) effect neglected. Moments at joint centerlines in k ft. Shears in kips.

(b)

FIG. 2-10 Frame Analyses—10th Floor—Design Example 2.

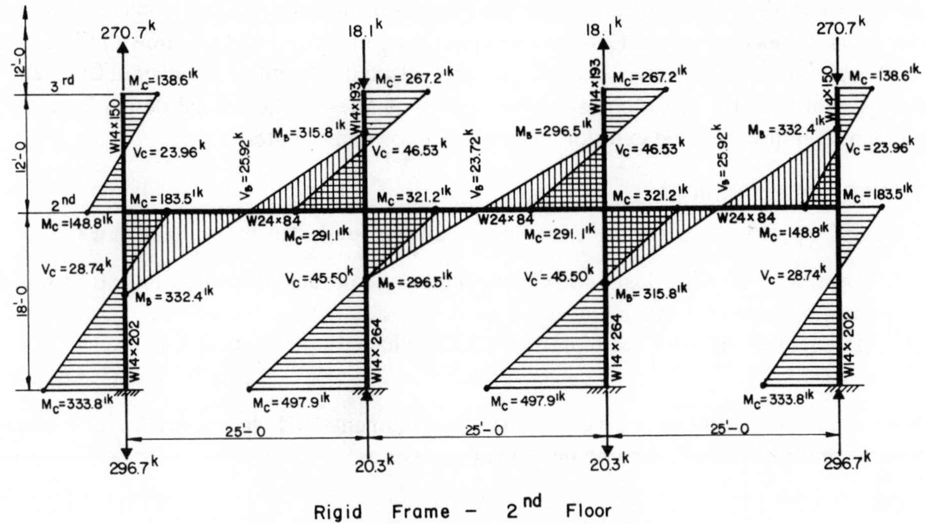

Rigid Frame – 2nd Floor
Wind Only – Shears & Moments*

* Standard elastic analysis – primary moment effects only
Moments at joint centerlines in k–ft. Shears in kips.

FIG. 2-11(a) Frame Analyses—2nd Floor—Design Example 2.

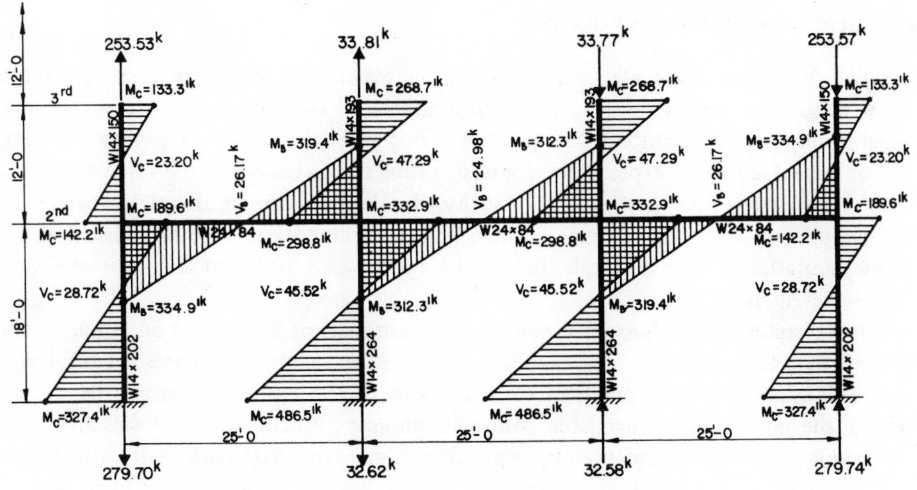

Rigid Frame – 2nd Floor
Wind Only – Shears & Moments*

* Elastic analysis including axial shortening in columns and shear deformation ($P\Delta$) effect neglected. Moments at joint centerlines in k–ft. Shears in kips.

Fig. 2-11(b) Frame Analyses—2nd Floor—Design Example 2.

Design Examples. The critical load conditions, moments, shears, and displacements for the design of selected stories (25th floor, 10th floor, and 2nd floor) are shown in Fig. 2-9, 2-10, and 2-11, respectively. Designs for these selected elements are shown in Chapter 3, Beams, and Chapter 6, Columns. Complete summaries of the critical design loads and moments for design of the exterior columns are shown in Chapter 6.

Design Data Required

The AISC Specifications prescribe loading conditions for design as two basic cases:

1. (Dead load plus live load) for which the sum of the actual-to-allowable stress ratios must not exceed 1.00, ... (1.6)
2. (Dead load plus live load plus wind) for which the sum must not exceed 1.33 . . . . (1.5.6; 1.6).

Ideally, the sum total of analysis data (critical moments and shears) desired can, therefore, be grouped under these two basic loading cases as:

Case 1—$(D + L)$
(a) $(D + L)$ full load on all spans
(b) (D) on all spans + all possible patterns of live loading
(c) $(P\Delta)$ effects for both (a) and (b)
(d) $(V + PL/AE)$ effects for both (a) and (b)

Case 2—$(D + L + W)$
(a), (b), (c), and (d) as above plus the effects of wind.

Practical Limitations on Analysis

General. In theory there are few limitations on elastic analysis except that credited to St. Venant. The authors regard "practical limitations" as limitations on design time and expense imposed by economics in practice. For example, the client paying for design time is not served by an expenditure for design time that saves less in the value of tonnage saved than design time cost spent. Equally, the designer cannot afford to spend more for design time than the fee paid. Thus, "practical limitations" are established as economic limitations. Like all ideals, the ideal total package of design data is beyond practical achievement.

In this chapter, the authors have presented for purposes of illustration only, much more analysis data than required for a practical design. Even for these purposes the total package of data is limited by practical considerations of the authors' time and page space. Within the limitations of available computer programs (including some simplifying assumptions therein) supplemented by some approximate manual solutions at critical points to fill in gaps created by these limitations, the analyses herein are presented in order to illustrate the relative importance of secondary effects and other refinements usually neglected in practice. The reader should thus be provided with some quantitative data upon which to exercise his own judgment for his own practice, but he will have to use caution in extrapolating this data to structures very dissimilar from the typical office building loads, spans, and story heights used in Design Example 2. The AISC Specifications require only that the maximum stresses due to "probable" combinations of load be satisfied (1.3.2).

Pattern Live Load. With three bays across the width and 25 stories in height, so many permutations of patterns for partial live loading are possible that full analyses of pattern live load are prohibitive even by computer. All analyses shown herein are for the same live load on all panels. These analyses contain some minor apparent inconsistencies of loading due to differing live load reductions permitted for the various structural elements. In addition to practical considerations, the procedure of neglecting partial load patterns here has theoretical justification recognized in the AISC Specifications for the very low statistical probability of a partial live load pattern occurring so as to develop overall critical conditions for the structure as a whole (1.3.2).

Live Load Reductions. Live load reductions used here are those commonly permitted in statutory building codes (1.3.2; 1.3.7).

$$R = 0.08\,A\ (\%) \leqslant 23(1 + D/L)\ (\%) \leqslant 60\ (\%)$$

where R = percent reduction in live load; A = the area supported, sq. ft.; D = dead load; and L = live load. The live load reduction is usually applicable only to uniform live loads less than 100 psf. In this example, live load reductions as applied to the different structural elements become:

slab	0%	interior beams	16.6%
spandrel beams	8.3%	girders	50.0%
exterior columns,		interior columns,	
24th story	26.0%	24th story	50.0%
23rd story	52.0%	23rd & below	60.0%
22nd & below	60.0%		

It will be noted that the effect of live load reductions exceeds that of partial live load patterns for equal spans and may create "partial live load patterns" with greatly unequal adjacent spans.

Manual versus Computer Analysis. Most of the designers' difficulties lie in the design of a structure, which is an art, rather than the analysis, which is essentially mathematics suitable for computerized calculation. Before the formal elastic analysis of a rigid-frame structure can even be programmed, the exercise of the designers' art is necessary to select preliminary member sizes. The usual procedure is to make this selection using a preliminary simplified analysis to guide the designer in applying his skill.

For the preliminary selection of interior column sections, the preliminary analysis may be simply an accumulation of $(D + L)$ axial loads on the tributary floor areas to each column similar to a Type 2 design. For the rigid-frame Type 1 structure, the preliminary sections for exterior columns can be selected for a lower axial stress, than that allowed for axial loads only, to leave a reserve capacity for $(D + L)$ rigid-frame moments. The skilled designer can compute the necessary capacity for moments manually at a few selected floors and estimate same for the intermediate floors.

With the preliminary sections selected, the frame can be subjected to a floor-by-floor manual or computer analysis for both load cases to complete the design; changing sizes from the preliminary selections where necessary. For such analyses, the traditional classic analysis is based upon assumptions that:

1. Moments are computed using member lengths from center-to-center of supports (joints).
2. All members at any joint meet at a point.
3. Stiffening effects in the depth of the joints or of any haunches are negligible.
4. Shear deformation is negligible.
5. $(P\Delta)$ effects can be neglected.
6. The effects of differential length changes in all members are negligible.

Displacements computed from this classical elastic analysis are traditionally used to establish required camber for horizontal members and to satisfy rules of thumb developed to limit deflection in horizontal members or lateral drift of the structure. The moment magnification factor $C_m/(1 - f_a/F'_e)$, allows for increases in the column moments due to secondary effects neglected in analysis, such as $(P\Delta)$. (1.6.1)

Within the limits on total height, spans, and height-to-width ratios in common use for structures without braced frames, these traditional methods of analysis and the associated design requirements have served well. As these limits are gradually increased and capability of computers permits more refinements in analysis, the quantitative evaluation of secondary effects becomes not only necessary but practically feasible.

Available Computer Programs. Recent developments of elastic analysis computer programs are becoming more sophisticated. Some versions commonly used include the effects of shear and axial length changes, either for all members or for columns only, but neglect the $(P\Delta)$ effect. These programs are "Stress," "Strudl," and "Fran." The $(P\Delta)$ effects are included, but axial length changes and shear deformation effects are neglected, in "Computer Program for Lateral Load Analysis of Multi-Story Frames with Shear Walls," A. J. Gouwens, Portland Cement Association, 1968.

Analyses Applied in Design Example 2. As previously noted, more analysis data than required for routine design was deliberately collected. Quantitative data on the sum of the effects of shear deformations and axial length changes is presented. To aid the reader in his decision when these effects become significant for tall buildings, the height-to-width ratio (4) was selected for this example because the authors believe these effects are significant for height-to-width ratios of 4 or more. Note that no effort was made to separate the effects of shear and axial length change since available programs include both.

The $(P\Delta)$ effect is important for all structures, being more dependent upon the story height and column stiffness than upon the overall height or height-to-width ratio. If the secondary moment $(P\Delta)$ is not included in column design, the design depends upon the specified moment magnification factor, usually conservative, to amplify the bending stress, f_b (1.6.1). If the sum of the $(P\Delta)$ moment and the primary column moment is used in design, it would seem logical to neglect the arbitrary moment magnification factor, $C_m/(1 - f_a/F'_e)$. Without question, where the ratio of the sum of $(P\Delta)$ plus primary moment to the primary moment exceeds the quantity, $C_m/(1 - f_a/F'_e)$, it would be prudent design to use the sum $(P\Delta) + M_p$.

A number of the simplifying assumptions used are, of necessity, commonly accepted for both manual and computer analyses. For the simple gravity load condition, $(D + L)$, the floor-by-floor solution, assuming far ends of columns fixed above and below, has been used in the analyses to be added to wind. See Fig. 2-9, 2-10, and 2-11. Axial loads in columns were accumulated for $(D + L)$ loads simply on the basis of supported area. Computer print-outs of the analyses described below are reproduced in Appendix-Chapter 2:

(a) $(D + L)$. Live loads used included a wall load on the exterior spandrel beams as well as full (reduced) live load on all beams.
(b) (Wind) including $(P\Delta)$. $P = \Sigma(D + L)$ on the entire floor applied to the sum of the four column stiffnesses.
(c) $(D + L)$ including $(V) + (PL/AE)$. The load, P, per column is based upon a uniform live plus dead load of 2.44 k/ft. upon all girders, accumulated floor-by-floor. The (PL/AE) effect is applied to columns only, assuming $P = 0$ in the girders.
(d) (Wind) including $(V) + (PL/AE)$. The load P for columns is due to wind; assumed $P = 0$ in girders.
(e) Combinations of above analyses.

The (PL/AE) effect is similar to that of differential settlement, and might assume more importance where differential settlement is possible and could create additive effects. Since the axial shortening of girders can occur due only to relatively small wind shear forces transmitted to columns, the differentials also are small and can usually well be neglected. See (manual) analysis, Chapter 6, for the second floor girders where this shear force is maximum. Such an approximate analysis is suggested in cases where it may be considered important. This analysis can then be used as an upper bound solution for all floors above since the stiffness requirements for deflection will ensure that the beam area is not successively reduced as rapidly as the linear reduction in wind shear for floor girders above.

Practical Combination of Analysis Solutions. The $(D + L)$ analysis for load Case 1 and the (W) analysis superimposed thereon for load Case 2, both based upon distributions caused by moment deformations only, constitute the basic practical package of design data. This data is usually considered sufficient for design. See Figs. 2-12 and 2-13.

Examination of the analyses for the effect of $(P\Delta)$ indicates that this effect becomes significant only in loading Case 2 $(D + L + W)$. Its principal value for the structure of Design Example 2 is to reassure the designer that the columns are sufficiently stiff to limit the increase in (drift + sidesway) to a small fraction, 10 to 15 percent, of the drift

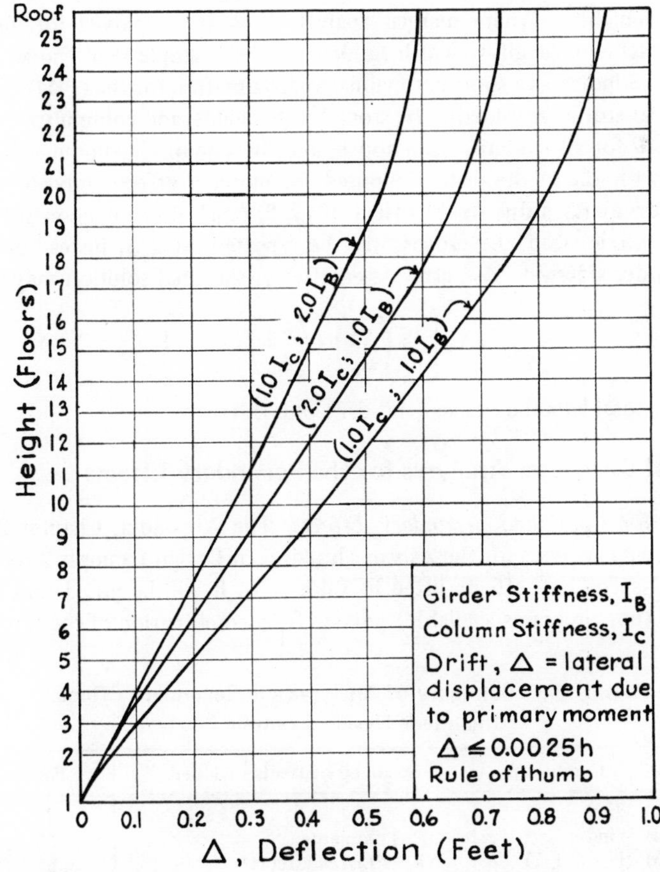

FIG. 2-12 Lateral Displacement Profiles—Girder versus Column Stiffness—Design Example 2.

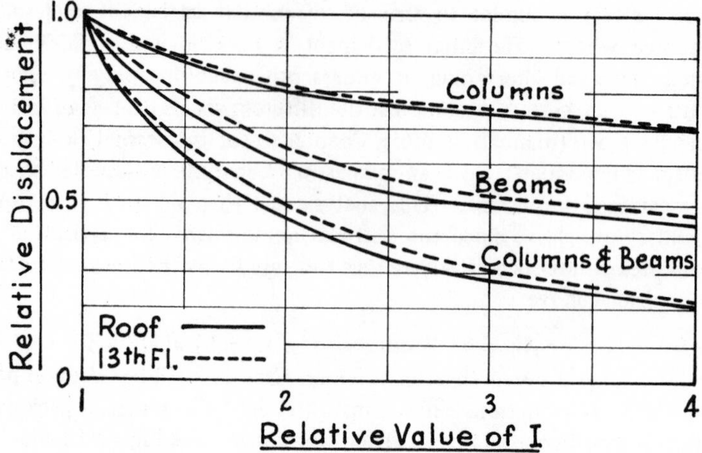

FIG. 2-13 Lateral Displacement at Top and Center of Height versus Ratios of Column-to-Girder Stiffnesses over the Full Range—Design Example 2.

computed without consideration of ($P\Delta$). The authors recommend use of a computer analysis program including the ($P\Delta$) effects for loading Case 2 where computer analysis facilities are available. Where manual analysis is to be employed for buildings with larger story heights or height-to-width ratios than the example rigid frame, an examination of Fig. 2-15 indicates a short-cut manual approximation for the ($P\Delta$) effect. For the probable critical stories considering H, story height; loads; and column size; use the total story load = ΣP for all columns from above, and the sum of all column stiffnesses in the story, $\Sigma I/H$ with Δ_0 as the drift computed for moment effects only, to compute the first order sidesway (Δ_1) due to the effect of (ΣP) (Δ_0) moment upon stiffness $\Sigma I/H$. As a closer approximation, the process may be repeated using Δ_1 instead of Δ_0, to compute second order sidesway, Δ_2, etc. Alternatively, the final solution may be taken as:

$$\Delta_s = \Delta_0 \left[1 + \left(\frac{\Delta_1}{\Delta_0} \right) + \left(\frac{\Delta_1}{\Delta_0} \right)^2 + \cdots \right]$$

where Δ_s = the total lateral displacement, drift plus sidesway.

Evaluation of Computer Analyses for the Secondary Effects

Superposition for Combined Secondary Effects. The Appendix, Chapter 2 shows the (computer output) analyses for the 25 story building of Design Example 2 for the loading conditions and secondary effects listed in Table 2-3. It will be noted that none of the analyses listed (nor programs available) provide for a combination of the three secondary

TABLE 2-3 Summary of Analyses and Secondary Effects Included—Design Example 2

Loading Condition	Secondary Effects Included	Page
(1) Wind	None	48
(2) Wind	($P\Delta$) moments	53
(3) ($W + D + L$)	($V + PL/AE$) effects	61
(4) ($D + L$)	None; column loads only	71
(5) Wind	($V + PL/AE$); P for wind only	74
(6) ($D + L$)	($V + PL/AE$)	79

effects. In order to evaluate the combined effect or to study the effects separately, a process of manual superposition is necessary.

Direct superposition of these results is valid only if the proper order is observed. Note that the gravity story load, P, was included in the analysis of case (2) only insofar as a term in the moment $(P\Delta)$. In case (5) vertical loads, P, in each column were those caused by wind only. In case (6) gravity loads were the sum of dead plus live loads applied uniformly and approximately equal to the story loads of case (2). One cannot add case (6) to case (2) since the Δ of case (6) will increase the $(P\Delta)$ effect of case (2). The additional drift computed due to the effects of shear deformation and differential axial shortening will increase the drift (Δ) used in the term $(P\Delta)$, but the effect of the $(P\Delta)$ sidesway will not directly increase the effects of $(V + PL/AE)$ likewise. Note that no effort has been made to separate the effects of V and PL/AE since the available computer programs include both.

Lateral Displacement as an Index to Significance of the Secondary Effects. As the simplest index to evaluate the importance of secondary effects, consider the lateral displacement. Let $\Delta_1, \Delta_2, \ldots, \Delta_6$ represent lateral displacements for each of the loading conditions analysed. See Table 2-3. Secondary effects for $(W + D + L)$ loading may be combined to compute the total lateral displacement resulting in either of the following procedures at the designer's option:

(A)
$$\Delta_2 - \Delta_1 = \Delta_{(P\Delta)}$$
$$\Delta_5 + \Delta_{(P\Delta)} \times \Delta_5/\Delta_1 = \Delta_{(\text{Total})}$$

This combination neglects the differential axial length changes due to gravity loads. It should be suitable where all columns have approximately equal axial stress, f_a, or where the designer wishes to evaluate the elastic effect of wind only, considering the long-time stress, f_a, to be self-adjusting.

(B)
$$\Delta_2 - \Delta_1 = \Delta_{(P\Delta)}$$
$$\Delta_3 + \Delta_{(P\Delta)} \times \Delta_3/\Delta_1 = \Delta_{(\text{Total})}$$

This combination considers axial length changes due to $(W + D + L)$ as though the entire loading were applied after the structure is complete, with no self-adjustment.

Computations for combination (A) have been performed. The results are tabulated in Table 2-4, and pictured in Fig. 2-14 (b). As the tabulation shows, Δ_3 and Δ_5 are so nearly identical that it is unnecessary in this example to combine displacements as in combination (B).

Practical Use of the Secondary Stress Analyses. Since the available computer programs do not include both $(P\Delta)$ and $(V + PL/AE)$ effects and practical applications of these programs are not now easily made compatible for superposition of the results, inclusion of all these secondary effects is not often practical. The authors consider that the $(P\Delta)$ effects are the more important and recommend analyses including same for all cases where computer facilities are to be used, and at least approximate calculations for same in manual analyses. The authors consider that the results herein show that the $(V + PL/AE)$ effects are not significant for structures within the conditions of Design Example 2 of story height, total height, and height-to-width ratio. Some other considerations mitigating neglect of these factors for general use are listed below:

1. Approximate manual solutions for the effect of differential changes in axial length of members considered critical in stories considered critical are not difficult.

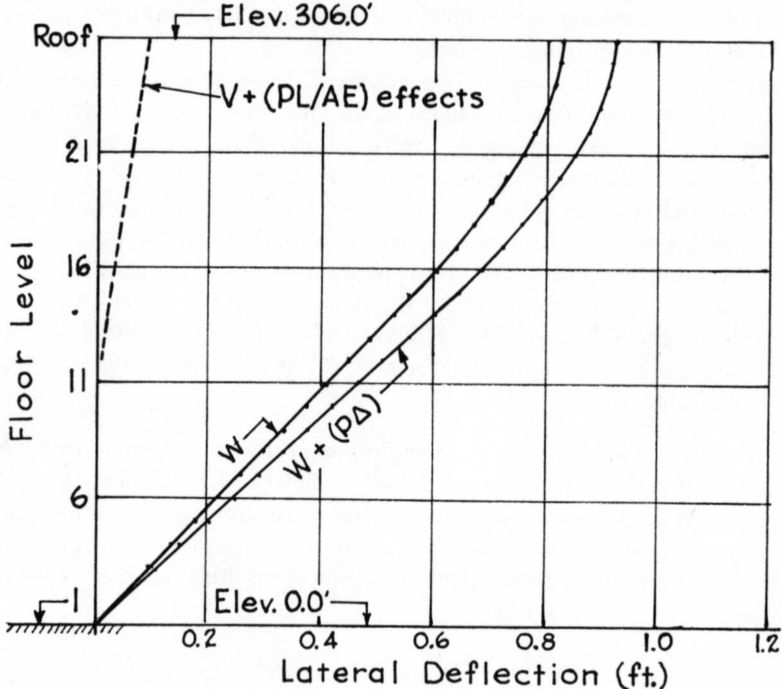

FIG. 2-14 (a) Secondary Stress Effects on Lateral Displacement Profiles (Separate)—Design Example 2.

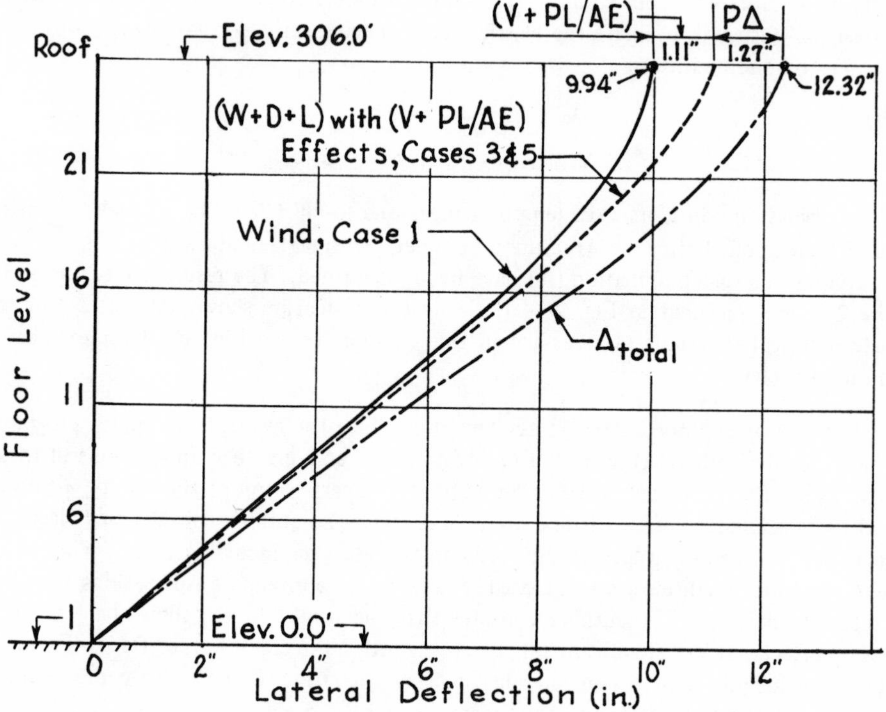

FIG. 2-14 (b) Secondary Stress Effects on Lateral Displacement Profiles (Superimposed)—Design Example 2.

TABLE 2-4 Secondary Effects Upon Lateral Deflection–Design Example 2.

Story (top)	Δ_1 (in.)	Δ_2 (in.)	$\Delta_2 - \Delta_1$ (in.)	Δ_5 (in.)	$\Delta(P\Delta) \times \frac{\Delta_5}{\Delta_1}$	Δ_{Total}	Δ_3 (in.)
1	0.61572	0.68400	0.06828	0.6248	0.06929	0.69409	0.62477
2	1.12548	1.2588	0.13332	1.1221	0.13292	1.25502	1.12209
3	1.6416	1.8456	0.2040	1.6321	0.20282	1.83492	1.63210
4	2.1564	2.4300	0.2736	2.1441	0.27204	2.41614	2.14409
5	2.6472	2.9844	0.3372	2.6412	0.33644	2.97764	2.64124
6	3.1272	3.5256	0.3984	3.1326	0.39909	3.53169	3.13258
7	3.5832	4.0368	0.4536	3.6076	0.45669	4.06429	3.60764
8	4.0332	4.5396	0.5064	4.0805	0.51234	4.59284	4.08054
9	4.4808	5.0400	0.5592	4.5566	0.56866	5.12526	4.55666
10	4.9476	5.5644	0.6168	5.0543	0.63010	5.68440	5.05427
11	5.4276	6.1056	0.6780	5.5838	0.69751	6.28131	5.58382
12	5.9376	6.6840	0.7464	6.1397	0.77181	6.91151	6.13970
13	6.4176	7.2252	0.8076	6.6720	0.83961	7.51161	6.67208
14	6.8772	7.7412	0.8640	7.1888	0.90315	8.09195	7.18828
15	7.2984	8.2104	0.9120	7.6719	0.95867	8.63057	7.67080
16	7.7100	8.6664	0.9564	8.1421	1.01000	9.15210	8.14106
17	8.0952	9.0912	0.9960	8.5911	1.05701	9.64811	8.59007
18	8.4720	9.5064	1.0344	9.0290	1.10241	10.13141	9.02833
19	8.8140	9.8784	1.0644	9.4380	1.13976	10.57776	9.43716
20	9.1296	10.2204	1.0908	9.8270	1.17413	11.00113	9.82612
21	9.3900	10.5000	1.1100	10.1660	1.20173	11.36773	10.16528
22	9.6276	10.7532	1.1256	10.4770	1.22491	11.70191	10.47626
23	9.7992	10.9332	1.1340	10.7280	1.24148	11.96948	10.72741
24	9.9024	11.0400	1.1376	10.9190	1.25439	12.17339	10.91798
25	9.9432	11.082	1.1388	11.0560	1.26625	12.32225	11.05501

2. The differential axial shortening effect in columns is uncertain since the gravity load portion will be applied gradually, differential temperature and settlement effects are involved, columns of tall buildings contain splices with uncertain local effects, and the elastic wind effects are somewhat dampened.

3. The differential axial shortening of girders is due mainly to wind shears and will be dampened or reduced by the diaphram effect in most floors.

4. Where considered possibly significant, the PL/AE effect due to differential gravity load axial stresses between exterior and interior columns can be minimized, if not eliminated, by the use of different grades of steel. In Design Example 2, if Grade 36 interior columns were used, the resulting increase in area required would nearly equalize the axial stresses to those in the Grade 50 exterior columns. This change would also add to the total stiffness, reducing Δ; see Fig. 2-14. The additional tonnage could be only partially offset by savings in the weight of girders if the designer did not regard the additional stiffness as needed. See Fig. 2-12 and 2-13 which show the lesser value per pound for stiffness contributed by columns versus that contributed by the girders.

Some Practical Observations From Analyses–Design Example 2

Grade of Steel. The relatively short unbraced lengths for both columns and beams in the moment resisting frames allow sufficient utilization of the higher stresses possible with A 572, Grade 50 steel to justify the higher cost of this grade by the reduction in tonnage. Although strength considerations could thereby be satisfied, the use of optimum shapes as Grade 50 beams for the least weight at the same depths would provide less lateral stiffness than the design required.

Lateral Stiffness. It is considered desirable to provide sufficient stiffness in a structure to limit the lateral displacement under design loads to approximately 0.0025 times the height. Preliminary analyses with both beams and columns in the main frames of Grade 50 proportioned for stress conditions only resulted in a calculated lateral displacement of approximately 0.003 times the height. It has been shown that increase in the girder stiffness is more efficient than increase in the column stiffness to control lateral displacement of frames with usual ranges of story height and spans.[1]

Comparative analyses for the structure of Design Example 2 are shown in summary form in Fig. 2-12 and 2-13.[2] Somewhat less efficient utilization of Grade 50 steel would result if the deeper sections used to provide the additional girder stiffness required were Grade 50. For this reason, design of the girders was based upon the lower cost A 36, Grade 36 steel. The lower allowable stresses then automatically resulted in stiffer members without sacrifice of efficiency as represented by the ratio of allowable stress- to-yield stress.

Secondary Effects upon Computed Lateral Displacement. Figure 2-14(a) shows separately the computed lateral deflections (a) by standard analysis, moment only, (b) standard analysis including the $(P\Delta)$ effects, and (c) standard analysis with the $(V + PL/AE)$ effects minus standard analysis with moment only as in (a). Figure 2-14(b) shows the computed lateral deflection due to wind by standard analysis, due to $(W + D + L)$ by standard analysis including the $(V + PL/AE)$ effects, and superimposed in this order the $(P\Delta)$ effects to give the combined total computed lateral displacement. Note that the rule of thumb limit, 0.0025 times height, is intended to apply to the first curve in each comparison, lateral displacement due to wind by standard analysis, neglecting all secondary effects. $0.0025 \times 308 \times 12 = 9.22$ in. compared to the computed 9.94 in.

$(P\Delta)$ Effects. The absolute value of the moments added due to the $(P\Delta)$ effects in the columns is plotted for the full 25 story height. Note that for the equal story heights (12′-0″) from the second floor to the roof, the curve is very close to a straight line, with minor divergencies due to use of two-story column lengths. (See Fig. 2-15.)

---

[1] "Drift in High Rise Steel Framing," J. B. Scalzi, Progressive Architecture, 1972.
[2] "Wind Effects on a 25 Story Rigid Frame," K. H. Hillmer, A. B. Itzkowitz, and C. A. Jacobson, Univ. of Ill., Chicago Circle Campus, June 1973.

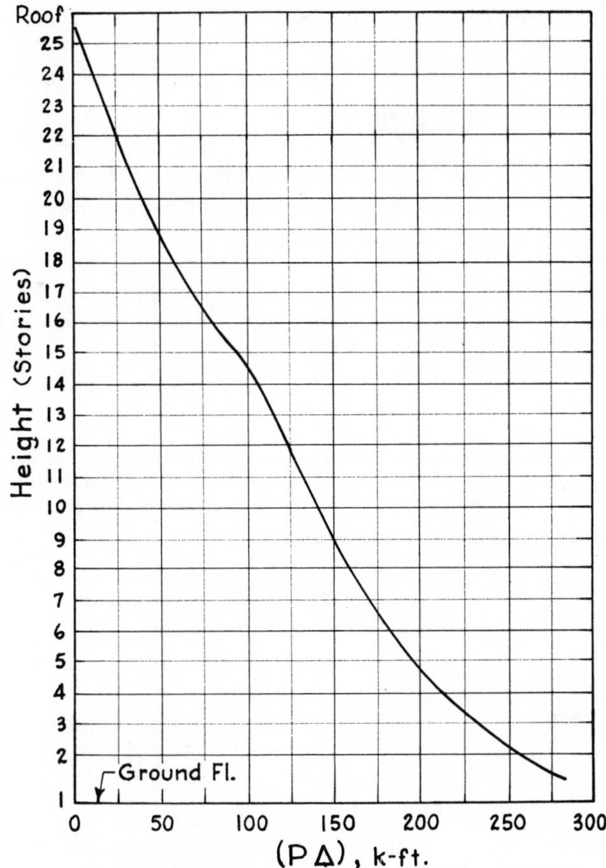

FIG. 2-15 Absolute Value of ($P\Delta$) Moment versus Height.

CHAPTER 2–APPENDIX: ELASTIC ANALYSES

EXAMPLE 1. ONE STORY FRAME, TYPE 1 CONSTRUCTION WITH FIXED COLUMN BASES (Computer Program I)

Elastic Analysis for Horizontal Load at Joint 5 (including the effects of V, PL/AE; neglecting effect of $P\Delta$) Units: forces in kips; moments in inch-kips; rotation in radians; displacement in inches. Signs: + right; clockwise; up.

Applied Joint Loads, Free Joints			
Joint	Force X	Force Y	Moment Z
5	4.499	0.000	−0.00
6	0.000	−0.000	0.00
7	−0.000	−0.000	0.00
8	0.000	−0.000	−0.00

Reactions, Applied Loads–Support Joints			
Joint	Force X	Force Y	Moment Z
1	−1.084	−0.414	97.92
2	−1.189	0.205	103.30
3	−1.175	−0.194	102.14
4	−1.050	0.403	94.85

Free Joint Displacements			
Joint	X- Disp.	Y- Disp.	Rotation
5	0.2206	0.0003	−0.0003
6	0.2170	−0.0001	−0.0001
7	0.2146	0.0001	−0.0001
8	0.2135	−0.0003	−0.0003

EXAMPLE 1 (Continued)

Member Forces				
Member	Joint	Axial Force	Shear Force	Moment
1	1	−0.414	1.084	97.92
1	5	0.414	−1.084	87.24
2	2	0.205	1.189	103.30
2	6	−0.205	−1.189	99.77
3	3	−0.194	1.175	102.14
3	7	0.194	−1.175	98.58
4	4	0.403	1.050	94.85
4	8	−0.403	−1.050	84.55
5	5	3.415	−0.414	−87.24
5	6	−3.415	0.414	−62.08
6	6	2.226	−0.209	−37.68
6	7	−2.226	0.209	−37.74
7	7	1.050	−0.403	−60.83
7	8	−1.050	0.403	−84.55

(Continued on page 43)

(Continued from page 42)

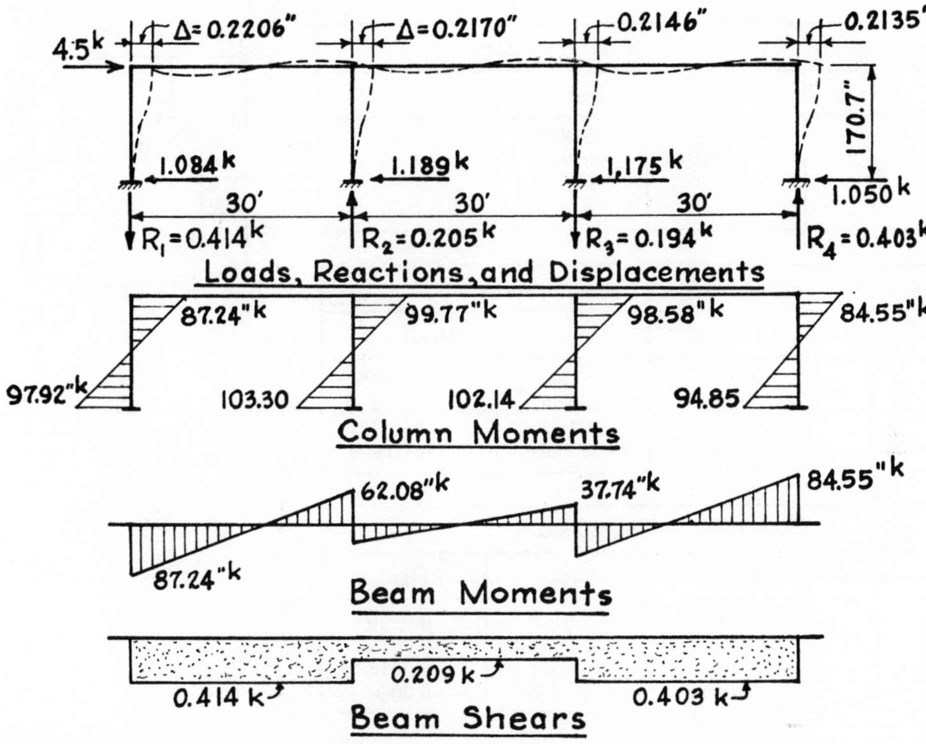

FIG. 2A-1 Moments, Shears, and Displacements for Wind Load plus Effects of Shear and Axial Length Changes—Design Example 1, One Story Building.

EXAMPLE 1. ONE STORY FRAME, TYPE 1 CONSTRUCTION WITH FIXED COLUMN BASES (Computer Program I)

Elastic Analysis for Live Plus Dead Load (including the effects of V, PL/AE; neglecting the effect of $P\Delta$)

Member Forces				
Member	Joint	Axial Force	Shear Force	Moment
1	1	15.204	-2.948	- 167.14
1	5	-15.204	2.948	- 336.36
2	2	40.955	0.814	46.52
2	6	-40.955	-0.814	92.49
3	3	40.955	-0.814	-46.52
3	7	-40.955	0.814	-92.49
4	4	15.204	2.948	167.14
4	8	-15.204	-2.948	336.36
5	5	2.948	15.204	336.36
5	6	-2.948	22.235	-1602.10
6	6	2.134	18.719	1509.60
6	7	-2.134	18.719	-1509.60
7	7	2.948	22.235	1602.10
7	8	-2.948	15.204	- 336.36

(Continued on page 44)

(Continued from page 43)

Applied Joint Loads, Free Joints			
Joint	Force X	Force Y	Moment Z
5	-0.000	0.000	0.00
6	0.000	0.000	-0.00
7	0.000	0.000	-0.00
8	0.000	0.000	0.00

Joints

Members

Reactions, Applied Loads–Support Joints			
Joint	Force X	Force Y	Moment Z
1	2.948	15.204	-167.14
2	-0.814	40.955	46.52
3	0.814	40.955	- 46.52
4	-2.948	15.204	167.14

Free Joint Displacements			
Joint	X-Disp.	Y-Disp.	Rotation
5	0.0042	-0.0126	-0.0060
6	0.0011	-0.0341	0.0016
7	-0.0011	-0.0341	-0.0016
8	-0.0042	-0.0126	0.0060

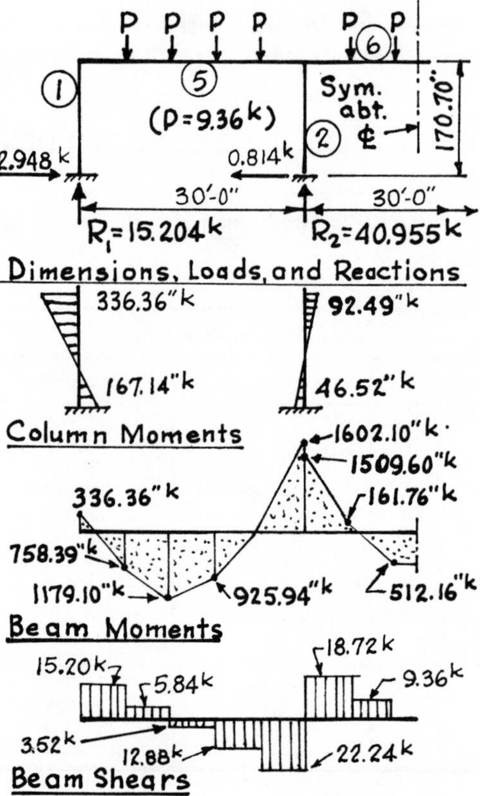

FIG. 2A-2 Moments, Shears, and Reactions for Live + Dead Loads plus Effects of Shear and Axial Length Changes—Design Example 1, One Story Building.

EXAMPLE 1. ONE STORY FRAME, TYPE 1 CONSTRUCTION WITH FIXED COLUMN BASES (Computer Program I)

Elastic Analysis for Wind plus Live plus Dead Loads (including the effects of $V, PL/AE$; neglecting the effect of $P\Delta$)

Member Forces				
Member	Joint	Axial Force	Shear Force	Moment
1	1	14.789	- 1.864	- 69.21
1	5	-14.789	1.864	- 249.12
2	2	41.161	2.003	149.83
2	6	-41.161	- 2.003	192.27
3	3	40.761	0.361	55.61
3	7	-40.761	- 0.361	6.08
4	4	15.607	3.999	261.99
4	8	-15.607	- 3.999	420.92
5	5	6.364	14.789	249.12
5	6	- 6.364	22.650	-1664.18
6	6	4.360	18.510	1471.91
6	7	- 4.360	18.929	-1547.34
7	7	3.999	21.832	1541.26
7	8	- 3.999	15.607	- 420.92

Applied Joint Loads, Free Joints			
Joint	Force X	Force Y	Moment Z
5	4.499	0.000	0.00
6	0.000	0.000	-0.00
7	-0.000	0.000	-0.00
8	0.000	0.000	0.00

Reactions, Applied Loads–Support Joints			
Joint	Force X	Force Y	Moment Z
1	1.864	14.789	- 69.21
2	-2.003	41.161	149.83
3	-0.361	40.761	55.61
4	-3.999	15.607	261.99

Free Joint Displacements			
Joint	X-Disp.	Y-Disp.	Rotation
5	0.2248	-0.0123	-0.0064
6	0.2181	-0.0343	0.0015
7	0.2135	-0.0339	-0.0017
8	0.2093	-0.0130	0.0056

EXAMPLE 1. ONE STORY FRAME, TYPE 1 CONSTRUCTION WITH FIXED COLUMN BASES

Elastic Analysis for Wind Load Only (neglecting the effects of $V, PL/AE$, and $P\Delta$). Computer program II.

Units: forces–kips; dimensions–in.; moment of inertia–(in.)^4; rotations–radians; moments–kip-ft.; joint rotation, + counterclockwise; end moment on member, + counterclockwise; lateral deflection, + to left; axial load, + tension.

(Continued on page 46)

Input:

```
     STORY
       1 DEPTH      0.8000E 01   0.8000E 01   0.8000E 01   0.8000E 01
       1 CL SPAN    0.3520E 03   0.3520E 03   0.3520E 03
       1 HEIGHT     0.1707E 03
       1 BEAM I     0.6120E 03   0.6120E 03   0.6120E 03
       1 VERT LD    0.0000E 00
       1 COLUM I    0.8250E 02   0.8250E 02   0.8250E 02   0.8250E 02
       1 HOR SHR    0.4499E 01
     FIXITY OF COLUMNS AT THE BASE
                    0.1000E 01   0.1000E 01   0.1000E 01   0.1000E 01
```

Output:

```
DEFLECTIONS AND ROTATIONS
STORY DEFLECTION   ROTATIONS
  1      -0.2151E 00 -0.3525E-03 -0.1168E-03 -0.1168E-03 -0.3525E-03
```

```
STORY
  MOMENTS
 1 COL TOP   0.8615E 02   0.9936E 02   0.9936E 02   0.8615E 02
 1 COL BOT   0.9603E 02   0.1026E 03   0.1026E 03   0.9603E 02
 1 BM L ND  -0.8450E 02  -0.3615E 02  -0.6073E 02
 1 BM R ND  -0.6073E 02  -0.3615E 02  -0.8450E 02
  SHEARS
 1 COLUMNS   0.1066E 01   0.1183E 01   0.1183E 01   0.1066E 01
 1 BEAMS    -0.4126E 00  -0.2054E 00  -0.4126E 00
 1 P*DELTA   0.0000E 00
```

```
AXIAL LOADS IN COLUMNS AND WALLS

  1 AXIAL LOAD  0.4126E 00 -0.2071E 00  0.2071E 00 -0.4126E 00
```

4.5 k
W 18 x 40
Cols. W8 x 24
8"
8"
30'-0"
30'-0"
30'-0"
170.7"
4.5 k
Δ = 0.2151"
1.066k
1.183k
1.183k
1.066k
0.4126k
0.2071 k
0.2071 k
0.4126 k
86.15"k
99.36"k
99.36"k
86.15"k
Bm. soffit

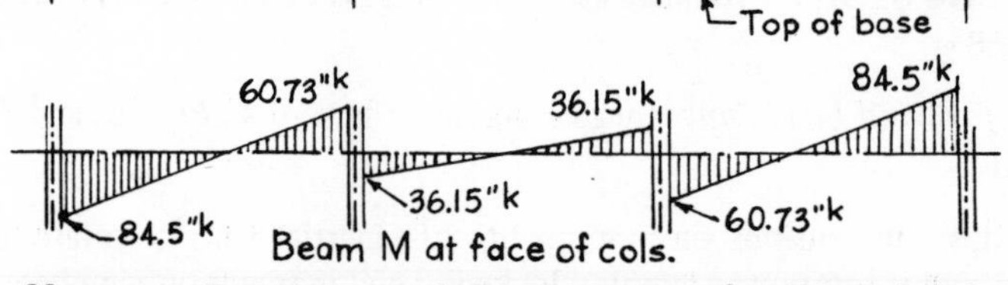

FIG. 2A-3 Moments, Reactions, and Displacements for Wind Load, Neglecting the Effects of *V*, *PL*/*AE*, and *P*Δ—Design Example 1, One Story Building.

EXAMPLE 1. ONE STORY FRAME, TYPE 1 CONSTRUCTION WITH FIXED COLUMN BASES (Computer program II)

Elastic Analysis for Wind Load (including the effects of $P\Delta$; neglecting the effects of $V, PL/AE$)

Input:

```
STORY
  1 DEPTH      0.8000E 01   0.8000E 01   0.8000E 01   0.8000E 01
  1 CL SPAN    0.3520E 03   0.3520E 03   0.3520E 03
  1 HEIGHT     0.1707E 03
  1 BEAM I     0.6120E 03   0.6120E 03   0.6120E 03
  1 VERT LD    0.1403E 03
  1 COLUM I    0.8250E 02   0.8250E 02   0.8250E 02   0.8250E 02
  1 HOR SHR    0.4499E 01
FIXITY OF COLUMNS AT THE BASE
               0.1000E 01   0.1000E 01   0.1000E 01   0.1000E 01
```

Output:

```
DEFLECTIONS AND ROTATIONS
STORY DEFLECTION  ROTATIONS
  1    -0.2239E 00 -0.3670E-03 -0.1216E-03 -0.1216E-03 -0.3670E-03

STORY
   MOMENTS
  1 COL TOP   0.8968E 02   0.1034E 03   0.1034E 03   0.8968E 02
  1 COL BOT   0.9996E 02   0.1068E 03   0.1068E 03   0.9996E 02
  1 BM L ND  -0.8796E 02  -0.3763E 02  -0.6322E 02
  1 BM R ND  -0.6322E 02  -0.3763E 02  -0.8796E 02
   SHEARS
  1 COLUMNS   0.1110E 01   0.1231E 01   0.1231E 01   0.1110E 01
  1 BEAMS    -0.4295E 00  -0.2138E 00  -0.4295E 00
  1 P*DELTA  -0.1841E 00

EQUILIBRIUM CHECK - UNBALANCED FORCES

STORY  SHEAR          MOMENTS
  1    -0.2444E-08    0.1788E-06 -0.5960E-07 -0.1192E-06   0.5960E-07

AXIAL LOADS IN COLUMNS AND WALLS

  1 AXIAL LOAD  0.4295E 00 -0.2156E 00   0.2156E 00 -0.4295E 00
```

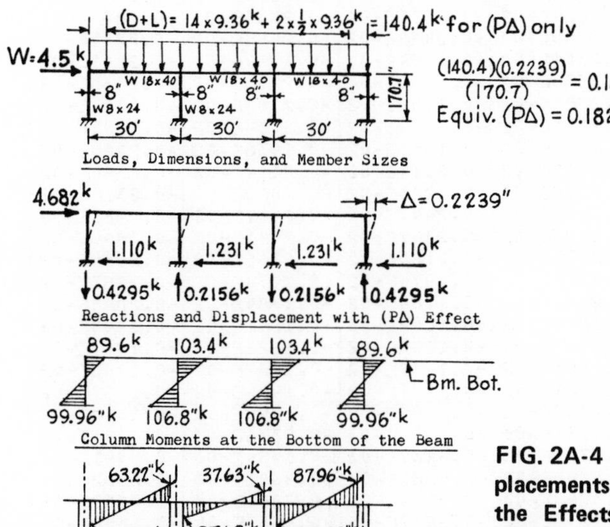

FIG. 2A-4 Moments, Reactions, and Displacements for Wind Load, Including the Effects of $P\Delta$, and Neglecting the Effects of V and PL/AE—Design Example 1, One Story Building.

EXAMPLE 2. ELASTIC ANALYSIS (Neglect V, PL/AE, AND $P\Delta$)

Loading Condition: WIND ONLY
(Program II)

Units: Moments in beams at column face (k-ft.)
Moments in columns at centerline of beams (k-ft.)
Shear and axial loads (kips); positive for tension
Deflections (ft.); positive toward the wind (left)
Rotations (radians); positive when joint rotates counterclockwise

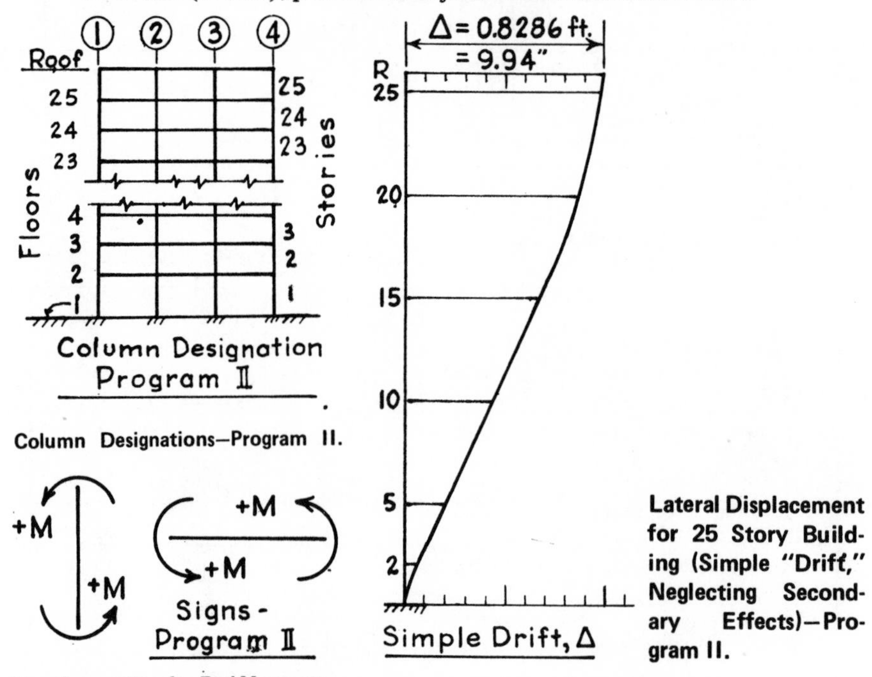

Column Designations—Program II.

Sign Convention for End Moments—Program II.

Lateral Displacement for 25 Story Building (Simple "Drift," Neglecting Secondary Effects)—Program II.

DEFLECTIONS AND ROTATIONS FOR LOAD CASE 1 – WIND ONLY

STORY	DEFLECTION	ROTATIONS			
1	-0.5131E-01	-0.2654E-02	-0.2239E-02	-0.2239E-02	-0.2654E-02
2	-0.9379E-01	-0.2824E-02	-0.2535E-02	-0.2535E-02	-0.2824E-02
3	-0.1368E 00	-0.2789E-02	-0.2464E-02	-0.2464E-02	-0.2789E-02
4	-0.1797E 00	-0.2681E-02	-0.2354E-02	-0.2354E-02	-0.2681E-02
5	-0.2206E 00	-0.2538E-02	-0.2255E-02	-0.2255E-02	-0.2538E-02
6	-0.2606E 00	-0.2384E-02	-0.2156E-02	-0.2156E-02	-0.2384E-02
7	-0.2986E 00	-0.2291E-02	-0.2034E-02	-0.2034E-02	-0.2291E-02
8	-0.3361E 00	-0.2229E-02	-0.1938E-02	-0.1938E-02	-0.2229E-02
9	-0.3734E 00	-0.2360E-02	-0.2114E-02	-0.2114E-02	-0.2360E-02
10	-0.4123E 00	-0.2316E-02	-0.2085E-02	-0.2085E-02	-0.2316E-02
11	-0.4523E 00	-0.2681E-02	-0.2447E-02	-0.2447E-02	-0.2681E-02
12	-0.4948E 00	-0.2596E-02	-0.2341E-02	-0.2341E-02	-0.2596E-02
13	-0.5348E 00	-0.2434E-02	-0.2165E-02	-0.2165E-02	-0.2434E-02
14	-0.5731E 00	-0.2253E-02	-0.1974E-02	-0.1974E-02	-0.2253E-02
15	-0.6082E 00	-0.2088E-02	-0.1784E-02	-0.1784E-02	-0.2088E-02
16	-0.6425E 00	-0.1944E-02	-0.1602E-02	-0.1602E-02	-0.1944E-02
17	-0.6746E 00	-0.1947E-02	-0.1622E-02	-0.1622E-02	-0.1947E-02
18	-0.7060E 00	-0.1785E-02	-0.1437E-02	-0.1437E-02	-0.1785E-02
19	-0.7345E 00	-0.1661E-02	-0.1404E-02	-0.1404E-02	-0.1661E-02
20	-0.7608E 00	-0.1360E-02	-0.1209E-02	-0.1209E-02	-0.1360E-02
21	-0.7825E 00	-0.1181E-02	-0.9336E-03	-0.9336E-03	-0.1181E-02
22	-0.8023E 00	-0.9741E-03	-0.6434E-03	-0.6434E-03	-0.9741E-03
23	-0.8166E 00	-0.6382E-03	-0.4381E-03	-0.4381E-03	-0.6382E-03
24	-0.8252E 00	-0.3260E-03	-0.2264E-03	-0.2264E-03	-0.3260E-03
25	-0.8286E 00	-0.1096E-03	-0.6837E-04	-0.6837E-04	-0.1096E-03
COL	ALL	①	②	③	④

EXAMPLE 2. (Continued)

LOAD CASE 1 - WIND ONLY

STORY				
MOMENTS AND SHEARS				
1 COL TOP	0.1835E 03	0.3212E 03	0.3212E 03	0.1835E 03
1 COL BOT	0.3338E 03	0.4979E 03	0.4979E 03	0.3338E 03
1 BM L ND	-0.3168E 03	-0.2823E 03	-0.3002E 03	
1 BM R ND	-0.3002E 03	-0.2823E 03	-0.3168E 03	
SHEARS				
1 COLUMNS	0.2874E 02	0.4550E 02	0.4550E 02	0.2874E 02
1 BEAMS	-0.2592E 02	-0.2372E 02	-0.2592E 02	
1 P*DELTA	0.0000E 00			
MOMENTS				
2 COL TOP	0.1386E 03	0.2672E 03	0.2672E 03	0.1386E 03
2 COL BOT	0.1488E 03	0.2911E 03	0.2911E 03	0.1488E 03
2 BM L ND	-0.2652E 03	-0.2467E 03	-0.2563E 03	
2 BM R ND	-0.2563E 03	-0.2467E 03	-0.2652E 03	
SHEARS				
2 COLUMNS	0.2396E 02	0.4653E 02	0.4653E 02	0.2396E 02
2 BEAMS	-0.2191E 02	-0.2073E 02	-0.2191E 02	
2 P*DELTA	0.0000E 00			
MOMENTS				
3 COL TOP	0.1418E 03	0.2670E 03	0.2670E 03	0.1418E 03
3 COL BOT	0.1397E 03	0.2613E 03	0.2613E 03	0.1397E 03
3 BM L ND	-0.2606E 03	-0.2398E 03	-0.2506E 03	
3 BM R ND	-0.2506E 03	-0.2398E 03	-0.2606E 03	
SHEARS				
3 COLUMNS	0.2346E 02	0.4403E 02	0.4403E 02	0.2346E 02
3 BEAMS	-0.2148E 02	-0.2015E 02	-0.2148E 02	
3 P*DELTA	0.0000E 00			
MOMENTS				
4 COL TOP	0.1374E 03	0.2564E 03	0.2564E 03	0.1374E 03
4 COL BOT	0.1316E 03	0.2484E 03	0.2484E 03	0.1316E 03
4 BM L ND	-0.2500E 03	-0.2291E 03	-0.2399E 03	
4 BM R ND	-0.2399E 03	-0.2291E 03	-0.2500E 03	
SHEARS				
4 COLUMNS	0.2242E 02	0.4207E 02	0.4207E 02	0.2242E 02
4 BEAMS	-0.2059E 02	-0.1925E 02	-0.2059E 02	
4 P*DELTA	0.0000E 00			
MOMENTS				
5 COL TOP	0.1326E 03	0.2437E 03	0.2437E 03	0.1326E 03
5 COL BOT	0.1249E 03	0.2365E 03	0.2365E 03	0.1249E 03
5 BM L ND	-0.2375E 03	-0.2194E 03	-0.2288E 03	
5 BM R ND	-0.2288E 03	-0.2194E 03	-0.2375E 03	
SHEARS				
5 COLUMNS	0.2147E 02	0.4002E 02	0.4002E 02	0.2147E 02
5 BEAMS	-0.1959E 02	-0.1844E 02	-0.1959E 02	
5 P*DELTA	0.0000E 00			
MOMENTS				
6 COL TOP	0.1237E 03	0.2341E 03	0.2341E 03	0.1237E 03
6 COL BOT	0.1166E 03	0.2274E 03	0.2274E 03	0.1166E 03
6 BM L ND	-0.2244E 03	-0.2098E 03	-0.2174E 03	
6 BM R ND	-0.2174E 03	-0.2098E 03	-0.2244E 03	
SHEARS				
6 COLUMNS	0.2003E 02	0.3846E 02	0.3846E 02	0.2003E 02
6 BEAMS	-0.1856E 02	-0.1763E 02	-0.1856E 02	
6 P*DELTA	0.0000E 00			
MOMENTS				
7 COL TOP	0.1161E 03	0.2231E 03	0.2231E 03	0.1161E 03
7 COL BOT	0.1118E 03	0.2148E 03	0.2148E 03	0.1118E 03
7 BM L ND	-0.2144E 03	-0.1979E 03	-0.2065E 03	
7 BM R ND	-0.2065E 03	-0.1979E 03	-0.2144E 03	
SHEARS				
7 COLUMNS	0.1900E 02	0.3649E 02	0.3649E 02	0.1900E 02
7 BEAMS	-0.1768E 02	-0.1663E 02	-0.1768E 02	
7 P*DELTA	0.0000E 00			
	COL ① BM ①-②	COL ② BM ②-③	COL ③ BM ③-④	COL ④ ~

(Continued on page 50)

EXAMPLE 2. (Continued)

MOMENTS				
8 COL TOP	0.1115E 03	0.2076E 03	0.2076E 03	0.1115E 03
8 COL BOT	0.1089E 03	0.2018E 03	0.2018E 03	0.1089E 03
8 BM L ND	-0.2073E 03	-0.1886E 03	-0.1983E 03	
8 BM R ND	-0.1983E 03	-0.1886E 03	-0.2073E 03	
SHEARS				
8 COLUMNS	0.1837E 02	0.3412E 02	0.3412E 02	0.1837E 02
8 BEAMS	-0.1704E 02	-0.1585E 02	-0.1704E 02	
8 P*DELTA	0.0000E 00			
MOMENTS				
9 COL TOP	0.1004E 03	0.1885E 03	0.1885E 03	0.1004E 03
9 COL BOT	0.1059E 03	0.1990E 03	0.1990E 03	0.1059E 03
9 BM L ND	-0.1862E 03	-0.1730E 03	-0.1799E 03	
9 BM R ND	-0.1799E 03	-0.1730E 03	-0.1862E 03	
SHEARS				
9 COLUMNS	0.1720E 02	0.3229E 02	0.3229E 02	0.1720E 02
9 BEAMS	-0.1538E 02	-0.1453E 02	-0.1538E 02	
9 P*DELTA	0.0000E 00			
MOMENTS				
10 COL TOP	0.9665E 02	0.1838E 03	0.1838E 03	0.9665E 02
10 COL BOT	0.9509E 02	0.1823E 03	0.1823E 03	0.9509E 02
10 BM L ND	-0.1831E 03	-0.1706E 03	-0.1771E 03	
10 BM R ND	-0.1771E 03	-0.1706E 03	-0.1831E 03	
SHEARS				
10 COLUMNS	0.1597E 02	0.3052E 02	0.3052E 02	0.1597E 02
10 BEAMS	-0.1513E 02	-0.1434E 02	-0.1513E 02	
10 P*DELTA	0.0000E 00			
MOMENTS				
11 COL TOP	0.8260E 02	0.1622E 03	0.1622E 03	0.8260E 02
11 COL BOT	0.9556E 02	0.1816E 03	0.1816E 03	0.9556E 02
11 BM L ND	-0.1582E 03	-0.1488E 03	-0.1537E 03	
11 BM R ND	-0.1537E 03	-0.1488E 03	-0.1582E 03	
SHEARS				
11 COLUMNS	0.1484E 02	0.2865E 02	0.2865E 02	0.1484E 02
11 BEAMS	-0.1310E 02	-0.1250E 02	-0.1310E 02	
11 P*DELTA	0.0000E 00			
MOMENTS				
12 COL TOP	0.8617E 02	0.1605E 03	0.1605E 03	0.8617E 02
12 COL BOT	0.8351E 02	0.1557E 03	0.1557E 03	0.8351E 02
12 BM L ND	-0.1526E 03	-0.1424E 03	-0.1477E 03	
12 BM R ND	-0.1477E 03	-0.1424E 03	-0.1526E 03	
SHEARS				
12 COLUMNS	0.1414E 02	0.2635E 02	0.2635E 02	0.1414E 02
12 BEAMS	-0.1262E 02	-0.1197E 02	-0.1262E 02	
12 P*DELTA	0.0000E 00			
MOMENTS				
13 COL TOP	0.7913E 02	0.1524E 03	0.1524E 03	0.7913E 02
13 COL BOT	0.7404E 02	0.1443E 03	0.1443E 03	0.7404E 02
13 BM L ND	-0.1424E 03	-0.1317E 03	-0.1372E 03	
13 BM R ND	-0.1372E 03	-0.1317E 03	-0.1424E 03	
SHEARS				
13 COLUMNS	0.1276E 02	0.2473E 02	0.2473E 02	0.1276E 02
13 BEAMS	-0.1175E 02	-0.1106E 02	-0.1175E 02	
13 P*DELTA	0.0000E 00			
MOMENTS				
14 COL TOP	0.7554E 02	0.1378E 03	0.1378E 03	0.7554E 02
14 COL BOT	0.7038E 02	0.1302E 03	0.1302E 03	0.7038E 02
14 BM L ND	-0.1312E 03	-0.1200E 03	-0.1258E 03	
14 BM R ND	-0.1258E 03	-0.1200E 03	-0.1312E 03	
SHEARS				
14 COLUMNS	0.1216E 02	0.2233E 02	0.2233E 02	0.1216E 02
14 BEAMS	-0.1080E 02	-0.1009E 02	-0.1080E 02	
14 P*DELTA	0.0000E 00			
MOMENTS				
15 COL TOP	0.6689E 02	0.1282E 03	0.1282E 03	0.6689E 02
15 COL BOT	0.6220E 02	0.1206E 03	0.1206E 03	0.6220E 02
15 BM L ND	-0.1207E 03	-0.1085E 03	-0.1148E 03	
15 BM R ND	-0.1148E 03	-0.1085E 03	-0.1207E 03	
SHEARS				
15 COLUMNS	0.1075E 02	0.2074E 02	0.2074E 02	0.1075E 02
15 BEAMS	-0.9899E 01	-0.9121E 01	-0.9899E 01	
15 P*DELTA	0.0000E 00			
	COL ① BM ①-②	COL ② BM ②-③	COL ③ BM ③-④	COL ④ –

(Continued on page 51)

EXAMPLE 2. (Continued)

MOMENTS				
16 COL TOP	0.6330E 02	0.1123E 03	0.1123E 03	0.6330E 02
16 COL BOT	0.5977E 02	0.1066E 03	0.1066E 03	0.5977E 02
16 BM L ND	-0.1111E 03	-0.9746E 02	-0.1045E 03	
16 BM R ND	-0.1045E 03	-0.9746E 02	-0.1111E 03	
SHEARS				
16 COLUMNS	0.1025E 02	0.1824E 02	0.1824E 02	0.1025E 02
16 BEAMS	-0.9063E 01	-0.8190E 01	-0.9063E 01	
16 P*DELTA	0.0000E 00			
MOMENTS				
17 COL TOP	0.5320E 02	0.9944E 02	0.9944E 02	0.5320E 02
17 COL BOT	0.5328E 02	0.1000E 03	0.1000E 03	0.5328E 02
17 BM L ND	-0.9544E 02	-0.8432E 02	-0.9008E 02	
17 BM R ND	-0.9008E 02	-0.8432E 02	-0.9544E 02	
SHEARS				
17 COLUMNS	0.8873E 01	0.1662E 02	0.1662E 02	0.8873E 01
17 BEAMS	-0.7795E 01	-0.7086E 01	-0.7795E 01	
17 P*DELTA	0.0000E 00			
MOMENTS				
18 COL TOP	0.5040E 02	0.8879E 02	0.8879E 02	0.5040E 02
18 COL BOT	0.4691E 02	0.8389E 02	0.8389E 02	0.4691E 02
18 BM L ND	-0.8662E 02	-0.7470E 02	-0.8087E 02	
18 BM R ND	-0.8087E 02	-0.7470E 02	-0.8662E 02	
SHEARS				
18 COLUMNS	0.8109E 01	0.1439E 02	0.1439E 02	0.8109E 01
18 BEAMS	-0.7037E 01	-0.6277E 01	-0.7037E 01	
18 P*DELTA	0.0000E 00			
MOMENTS				
19 COL TOP	0.4313E 02	0.7564E 02	0.7564E 02	0.4313E 02
19 COL BOT	0.4044E 02	0.7477E 02	0.7477E 02	0.4044E 02
19 BM L ND	-0.7134E 02	-0.6368E 02	-0.6765E 02	
19 BM R ND	-0.6765E 02	-0.6368E 02	-0.7134E 02	
SHEARS				
19 COLUMNS	0.6964E 01	0.1253E 02	0.1253E 02	0.6964E 01
19 BEAMS	-0.5839E 01	-0.5351E 01	-0.5839E 01	
19 P*DELTA	0.0000E 00			
MOMENTS				
20 COL TOP	0.3674E 02	0.6714E 02	0.6714E 02	0.3674E 02
20 COL BOT	0.3171E 02	0.6240E 02	0.6240E 02	0.3171E 02
20 BM L ND	-0.5934E 02	-0.5485E 02	-0.5718E 02	
20 BM R ND	-0.5718E 02	-0.5485E 02	-0.5934E 02	
SHEARS				
20 COLUMNS	0.5704E 01	0.1079E 02	0.1079E 02	0.5704E 01
20 BEAMS	-0.4896E 01	-0.4610E 01	-0.4896E 01	
20 P*DELTA	0.0000E 00			
MOMENTS				
21 COL TOP	0.2852E 02	0.5733E 02	0.5733E 02	0.2852E 02
21 COL BOT	0.2554E 02	0.5060E 02	0.5060E 02	0.2554E 02
21 BM L ND	-0.4974E 02	-0.4233E 02	-0.4617E 02	
21 BM R ND	-0.4617E 02	-0.4233E 02	-0.4974E 02	
SHEARS				
21 COLUMNS	0.4505E 01	0.8994E 01	0.8994E 01	0.4505E 01
21 BEAMS	-0.4029E 01	-0.3557E 01	-0.4029E 01	
21 P*DELTA	0.0000E 00			
MOMENTS				
22 COL TOP	0.2667E 02	0.3996E 02	0.3996E 02	0.2667E 02
22 COL BOT	0.2363E 02	0.3572E 02	0.3572E 02	0.2363E 02
22 BM L ND	-0.3904E 02	-0.2917E 02	-0.3429E 02	
22 BM R ND	-0.3429E 02	-0.2917E 02	-0.3904E 02	
SHEARS				
22 COLUMNS	0.4192E 01	0.6307E 01	0.6307E 01	0.4192E 01
22 BEAMS	-0.3081E 01	-0.2451E 01	-0.3081E 01	
22 P*DELTA	0.0000E 00			
MOMENTS				
23 COL TOP	0.1913E 02	0.2981E 02	0.2981E 02	0.1913E 02
23 COL BOT	0.1422E 02	0.2681E 02	0.2681E 02	0.1422E 02
23 BM L ND	-0.2583E 02	-0.1986E 02	-0.2296E 02	
23 BM R ND	-0.2296E 02	-0.1986E 02	-0.2583E 02	
SHEARS				
23 COLUMNS	0.2780E 01	0.4719E 01	0.4719E 01	0.2780E 01
23 BEAMS	-0.2050E 01	-0.1669E 01	-0.2050E 01	
23 P*DELTA	0.0000E 00			
	COL ①	COL ②	COL ③	COL ④
	BM ①-②	BM ②-③	BM ③-④	—

(Continued on page 52)

EXAMPLE 2. (Continued)

MOMENTS				
24 COL TOP	0.1249E 02	0.1833E 02	0.1833E 02	0.1249E 02
24 COL BOT	0.7933E 01	0.1523E 02	0.1523E 02	0.7933E 01
24 BM L ND	-0.1323E 02	-0.1026E 02	-0.1180E 02	
24 BM R ND	-0.1180E 02	-0.1026E 02	-0.1323E 02	
SHEARS				
24 COLUMNS	0.1702E 01	0.2797E 01	0.2797E 01	0.1702E 01
24 BEAMS	-0.1052E 01	-0.8626E 00	-0.1052E 01	
24 P*DELTA	0.0000E 00			
MOMENTS				
25 COL TOP	0.4536E 01	0.7199E 01	0.7199E 01	0.4536E 01
25 COL BOT	0.1374E 01	0.4889E 01	0.4889E 01	0.1374E 01
25 BM L ND	-0.4333E 01	-0.3100E 01	-0.3739E 01	
25 BM R ND	-0.3739E 01	-0.3100E 01	-0.4333E 01	
SHEARS				
25 COLUMNS	0.4926E 00	0.1007E 01	0.1007E 01	0.4926E 00
25 BEAMS	-0.3391E 00	-0.2605E 00	-0.3391E 00	
25 P*DELTA	0.0000E 00			
	COL ① BM ①-②	COL ② BM ②-③	COL ③ BM ③-④	COL ④

AXIAL LOADS IN COLUMNS -- LOAD CASE 1 - WIND ONLY

25 AXIAL LOAD	0.3391E 00	-0.7865E-01	0.7865E-01	-0.3391E 00
24 AXIAL LOAD	0.1391E 01	-0.2683E 00	0.2683E 00	-0.1391E 01
23 AXIAL LOAD	0.3441E 01	-0.6495E 00	0.6495E 00	-0.3441E 01
22 AXIAL LOAD	0.6523E 01	-0.1279E 01	0.1279E 01	-0.6523E 01
21 AXIAL LOAD	0.1055E 02	-0.1752E 01	0.1752E 01	-0.1055E 02
20 AXIAL LOAD	0.1544E 02	-0.2038E 01	0.2038E 01	-0.1544E 02
19 AXIAL LOAD	0.2128E 02	-0.2526E 01	0.2526E 01	-0.2128E 02
18 AXIAL LOAD	0.2832E 02	-0.3287E 01	0.3287E 01	-0.2832E 02
17 AXIAL LOAD	0.3612E 02	-0.3996E 01	0.3996E 01	-0.3612E 02
16 AXIAL LOAD	0.4518E 02	-0.4869E 01	0.4869E 01	-0.4518E 02
15 AXIAL LOAD	0.5508E 02	-0.5646E 01	0.5646E 01	-0.5508E 02
14 AXIAL LOAD	0.6588E 02	-0.6359E 01	0.6359E 01	-0.6588E 02
13 AXIAL LOAD	0.7764E 02	-0.7046E 01	0.7046E 01	-0.7764E 02
12 AXIAL LOAD	0.9026E 02	-0.7697E 01	0.7697E 01	-0.9026E 02
11 AXIAL LOAD	0.1033E 03	-0.8296E 01	0.8296E 01	-0.1033E 03
10 AXIAL LOAD	0.1185E 03	-0.9090E 01	0.9090E 01	-0.1185E 03
9 AXIAL LOAD	0.1338E 03	-0.9937E 01	0.9937E 01	-0.1338E 03
8 AXIAL LOAD	0.1509E 03	-0.1112E 02	0.1112E 02	-0.1509E 03
7 AXIAL LOAD	0.1686E 03	-0.1218E 02	0.1216E 02	-0.1686E 03
6 AXIAL LOAD	0.1871E 03	-0.1311E 02	0.1311E 02	-0.1871E 03
5 AXIAL LOAD	0.2067E 03	-0.1427E 02	0.1427E 02	-0.2067E 03
4 AXIAL LOAD	0.2273E 03	-0.1560E 02	0.1560E 02	-0.2273E 03
3 AXIAL LOAD	0.2488E 03	-0.1693E 02	0.1693E 02	-0.2488E 03
2 AXIAL LOAD	0.2707E 03	-0.1811E 02	0.1811E 02	-0.2707E 03
1 AXIAL LOAD	0.2967E 03	-0.2031E 02	0.2031E 02	-0.2967E 03
	COL ①	COL ②	COL ③	COL ④

EXAMPLE 2. ELASTIC ANALYSIS (Neglect *V* AND *PL/AE*)

Loading Condition: WIND + (PΔ) EFFECTS
(Program II)

Units: Moments
- –in beams, at column face (k-ft.)
- –in columns, at centerlines of beams (k-ft.)

Shear and axial loads (kips); positive for tension
Deflections (ft.); positive toward the wind (left)
Rotations (radians); positive when joint rotates counterclockwise
Dimensions (ft.); I (ft.^{4})

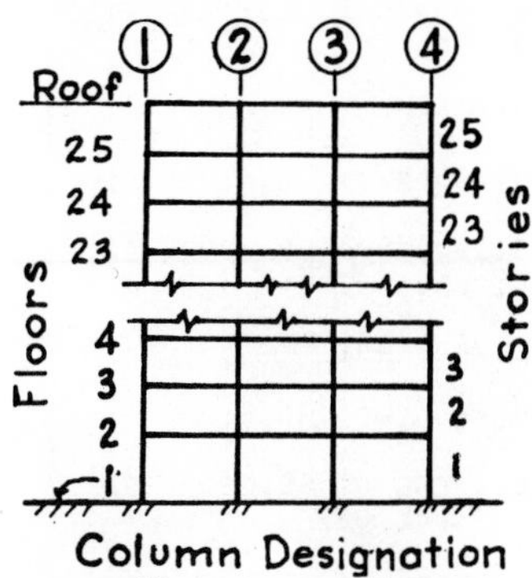

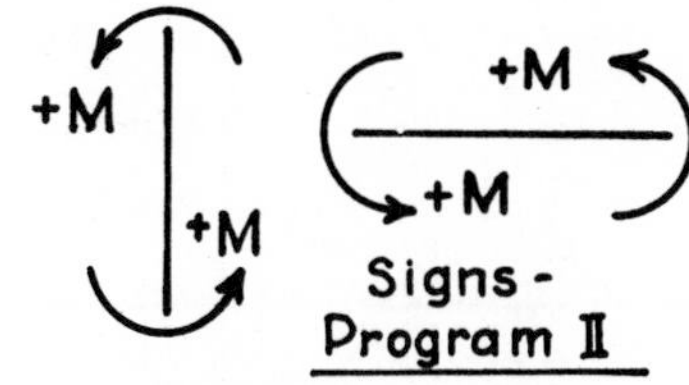

Column Designations—Program II.

Sign Convention for End Moments—Program II.

INPUT:

STORY	①&①-②	②&②-③	③&③-④	④
1 DEPTH	0.1200E 01	0.1200E 01	0.1200E 01	0.1200E 01
1 CL SPAN	0.2380E 02	0.2380E 02	0.2380E 02	
1 HEIGHT	0.1800E 02			
1 BEAM I	0.1140E 00	0.1140E 00	0.1140E 00	
1 VERT LD	0.4932E 04			
1 COLUM I	0.1220E 00	0.1700E 00	0.1700E 00	0.1220E 00
1 HOR SHR	0.1485E 03			
2 DEPTH	0.1200E 01	0.1200E 01	0.1200E 01	0.1200E 01
2 CL SPAN	0.2380E 02	0.2380E 02	0.2380E 02	
2 HEIGHT	0.1200E 02			
2 BEAM I	0.8800E-01	0.8800E-01	0.8800E-01	
2 VERT LD	0.4734E 04			
2 COLUM I	0.8600E-01	0.1160E 00	0.1160E 00	0.8600E-01
2 HOR SHR	0.1410E 03			
3 DEPTH	0.1200E 01	0.1200E 01	0.1200E 01	0.1200E 01
3 CL SPAN	0.2380E 02	0.2380E 02	0.2380E 02	
3 HEIGHT	0.1200E 02			
3 BEAM I	0.8800E-01	0.8800E-01	0.8800E-01	
3 VERT LD	0.4536E 04			
3 COLUM I	0.8600E-01	0.1160E 00	0.1160E 00	0.8600E-01
3 HOR SHR	0.1350E 03			
4 DEPTH	0.1200E 01	0.1200E 01	0.1200E 01	0.1200E 01
4 CL SPAN	0.2380E 02	0.2380E 02	0.2380E 02	
4 HEIGHT	0.1200E 02			
4 BEAM I	0.8800E-01	0.8800E-01	0.8800E-01	
4 VERT LD	0.4338E 04			
4 COLUM I	0.7700E-01	0.1040E 00	0.1040E 00	0.7700E-01
4 HOR SHR	0.1290E 03			
5 DEPTH	0.1200E 01	0.1200E 01	0.1200E 01	0.1200E 01
5 CL SPAN	0.2380E 02	0.2380E 02	0.2380E 02	
5 HEIGHT	0.1200E 02			
5 BEAM I	0.8800E-01	0.8800E-01	0.8800E-01	
5 VERT LD	0.4140E 04			
5 COLUM I	0.7700E-01	0.1040E 00	0.1040E 00	0.7700E-01
5 HOR SHR	0.1230E 03			
6 DEPTH	0.1200E 01	0.1200E 01	0.1200E 01	0.1200E 01
6 CL SPAN	0.2380E 02	0.2380E 02	0.2380E 02	
6 HEIGHT	0.1200E 02			
6 BEAM I	0.8800E-01	0.8800E-01	0.8800E-01	
6 VERT LD	0.3942E 04			
6 COLUM I	0.6600E-01	0.9800E-01	0.9800E-01	0.6600E-01
6 HOR SHR	0.1170E 03			

EXAMPLE 2. (Continued)

INPUT:

STORY	① ¢ ①-②	② ¢ ②-③	③ ¢ ③-④	④
7 DEPTH	0.1200E 01	0.1200E 01	0.1200E 01	0.1200E 01
7 CL SPAN	0.2380E 02	0.2380E 02	0.2380E 02	
7 HEIGHT	0.1200E 02			
7 BEAM I	0.8800E-01	0.8800E-01	0.8800E-01	
7 VERT LD	0.3744E 04			
7 COLUM I	0.6600E-01	0.9800E-01	0.9800E-01	0.6600E-01
7 HOR SHR	0.1109E 03			
8 DEPTH	0.1200E 01	0.1200E 01	0.1200E 01	0.1200E 01
8 CL SPAN	0.2380E 02	0.2380E 02	0.2380E 02	
8 HEIGHT	0.1200E 02			
8 BEAM I	0.8800E-01	0.8800E-01	0.8800E-01	
8 VERT LD	0.3545E 04			
8 COLUM I	0.6100E-01	0.8600E-01	0.8600E-01	0.6100E-01
8 HOR SHR	0.1050E 03			
9 DEPTH	0.1200E 01	0.1200E 01	0.1200E 01	0.1200E 01
9 CL SPAN	0.2380E 02	0.2380E 02	0.2380E 02	
9 HEIGHT	0.1200E 02			
9 BEAM I	0.7400E-01	0.7400E-01	0.7400E-01	
9 VERT LD	0.3348E 04			
9 COLUM I	0.6100E-01	0.8600E-01	0.8600E-01	0.6100E-01
9 HOR SHR	0.9900E 02			
10 DEPTH	0.1200E 01	0.1200E 01	0.1200E 01	0.1200E 01
10 CL SPAN	0.2380E 02	0.2380E 02	0.2380E 02	
10 HEIGHT	0.1200E 02			
10 BEAM I	0.7400E-01	0.7400E-01	0.7400E-01	
10 VERT LD	0.3150E 04			
10 COLUM I	0.5100E-01	0.7700E-01	0.7700E-01	0.5100E-01
10 HOR SHR	0.9300E 02			
11 DEPTH	0.1200E 01	0.1200E 01	0.1200E 01	0.1200E 01
11 CL SPAN	0.2380E 02	0.2380E 02	0.2380E 02	
11 HEIGHT	0.1200E 02			
11 BEAM I	0.5500E-01	0.5500E-01	0.5500E-01	
11 VERT LD	0.2951E 04			
11 COLUM I	0.5100E-01	0.7700E-01	0.7700E-01	0.5100E-01
11 HOR SHR	0.8700E 02			
12 DEPTH	0.1200E 01	0.1200E 01	0.1200E 01	0.1200E 01
12 CL SPAN	0.2380E 02	0.2380E 02	0.2380E 02	
12 HEIGHT	0.1200E 02			
12 BEAM I	0.5500E-01	0.5500E-01	0.5500E-01	
12 VERT LD	0.2754E 04			
12 COLUM I	0.4500E-01	0.6600E-01	0.6600E-01	0.4500E-01
12 HOR SHR	0.8100E 02			
13 DEPTH	0.1200E 01	0.1200E 01	0.1200E 01	0.1200E 01
13 CL SPAN	0.2380E 02	0.2380E 02	0.2380E 02	
13 HEIGHT	0.1200E 02			
13 BEAM I	0.5500E-01	0.5500E-01	0.5500E-01	
13 VERT LD	0.2556E 04			
13 COLUM I	0.4500E-01	0.6600E-01	0.6600E-01	0.4500E-01
13 HOR SHR	0.7500E 02			
14 DEPTH	0.1200E 01	0.1200E 01	0.1200E 01	0.1200E 01
14 CL SPAN	0.2380E 02	0.2380E 02	0.2380E 02	
14 HEIGHT	0.1200E 02			
14 BEAM I	0.5500E-01	0.5500E-01	0.5500E-01	
14 VERT LD	0.2358E 04			
14 COLUM I	0.4100E-01	0.5700E-01	0.5700E-01	0.4100E-01
14 HOR SHR	0.6899E 02			
15 DEPTH	0.1200E 01	0.1200E 01	0.1200E 01	0.1200E 01
15 CL SPAN	0.2380E 02	0.2380E 02	0.2380E 02	
15 HEIGHT	0.1200E 02			
15 BEAM I	0.5500E-01	0.5500E-01	0.5500E-01	
15 VERT LD	0.2160E 04			
15 COLUM I	0.4100E-01	0.5700E-01	0.5700E-01	0.4100E-01
15 HOR SHR	0.6299E 02			
16 DEPTH	0.1200E 01	0.1200E 01	0.1200E 01	0.1200E 01
16 CL SPAN	0.2380E 02	0.2380E 02	0.2380E 02	
16 HEIGHT	0.1200E 02			
16 BEAM I	0.5500E-01	0.5500E-01	0.5500E-01	

(Continued on page 55)

EXAMPLE 2. (Continued)

INPUT:

STORY	① ℄ ①-②	② ℄ ②-③	③ ℄ ③-④	④
16 VERT LD	0.1962E 04			
16 COLUM I	0.3500E-01	0.4500E-01	0.4500E-01	0.3500E-01
16 HOR SHR	0.5699E 02			
17 DEPTH	0.1200E 01	0.1200E 01	0.1200E 01	0.1200E 01
17 CL SPAN	0.2380E 02	0.2380E 02	0.2380E 02	
17 HEIGHT	0.1200E 02			
17 BEAM I	0.4700E-01	0.4700E-01	0.4700E-01	
17 VERT LD	0.1763E 04			
17 COLUM I	0.3500E-01	0.4500E-01	0.4500E-01	0.3500E-01
17 HOR SHR	0.5100E 02			
18 DEPTH	0.1200E 01	0.1200E 01	0.1200E 01	0.1200E 01
18 CL SPAN	0.2380E 02	0.2380E 02	0.2380E 02	
18 HEIGHT	0.1200E 02			
18 BEAM I	0.4700E-01	0.4700E-01	0.4700E-01	
18 VERT LD	0.1565E 04			
18 COLUM I	0.3100E-01	0.3800E-01	0.3800E-01	0.3100E-01
18 HOR SHR	0.4500E 02			
19 DEPTH	0.1200E 01	0.1200E 01	0.1200E 01	0.1200E 01
19 CL SPAN	0.2380E 02	0.2380E 02	0.2380E 02	
19 HEIGHT	0.1200E 02			
19 BEAM I	0.4100E-01	0.4100E-01	0.4100E-01	
19 VERT LD	0.1368E 04			
19 COLUM I	0.3100E-01	0.3800E-01	0.3800E-01	0.3100E-01
19 HOR SHR	0.3900E 02			
20 DEPTH	0.1200E 01	0.1200E 01	0.1200E 01	0.1200E 01
20 CL SPAN	0.2380E 02	0.2380E 02	0.2380E 02	
20 HEIGHT	0.1200E 02			
20 BEAM I	0.4100E-01	0.4100E-01	0.4100E-01	
20 VERT LD	0.1170E 04			
20 COLUM I	0.2400E-01	0.3500E-01	0.3500E-01	0.2400E-01
20 HOR SHR	0.3300E 02			
21 DEPTH	0.1200E 01	0.1200E 01	0.1200E 01	0.1200E 01
21 CL SPAN	0.2380E 02	0.2380E 02	0.2380E 02	
21 HEIGHT	0.1200E 02			
21 BEAM I	0.4100E-01	0.4100E-01	0.4100E-01	
21 VERT LD	0.9720E 03			
21 COLUM I	0.2400E-01	0.3500E-01	0.3500E-01	0.2400E-01
21 HOR SHR	0.2700E 02			
22 DEPTH	0.1200E 01	0.1200E 01	0.1200E 01	0.1200E 01
22 CL SPAN	0.2380E 02	0.2380E 02	0.2380E 02	
22 HEIGHT	0.1200E 02			
22 BEAM I	0.4100E-01	0.4100E-01	0.4100E-01	
22 VERT LD	0.7740E 03			
22 COLUM I	0.2100E-01	0.2100E-01	0.2100E-01	0.2100E-01
22 HOR SHR	0.2099E 02			
23 DEPTH	0.1200E 01	0.1200E 01	0.1200E 01	0.1200E 01
23 CL SPAN	0.2380E 02	0.2380E 02	0.2380E 02	
23 HEIGHT	0.1200E 02			
23 BEAM I	0.4100E-01	0.4100E-01	0.4100E-01	
23 VERT LD	0.5760E 03			
23 COLUM I	0.2100E-01	0.2100E-01	0.2100E-01	0.2100E-01
23 HOR SHR	0.1499E 02			
24 DEPTH	0.1200E 01	0.1200E 01	0.1200E 01	0.1200E 01
24 CL SPAN	0.2380E 02	0.2380E 02	0.2380E 02	
24 HEIGHT	0.1200E 02			
24 BEAM I	0.4100E-01	0.4100E-01	0.4100E-01	
24 VERT LD	0.3779E 03			
24 COLUM I	0.2100E-01	0.2100E-01	0.2100E-01	0.2100E-01
24 HOR SHR	0.9000E 01			
25 DEPTH	0.1200E 01	0.1200E 01	0.1200E 01	0.1200E 01
25 CL SPAN	0.2380E 02	0.2380E 02	0.2380E 02	
25 HEIGHT	0.1200E 02			
25 BEAM I	0.4100E-01	0.4100E-01	0.4100E-01	
25 VERT LD	0.1800E 03			
25 COLUM I	0.2100E-01	0.2100E-01	0.2100E-01	0.2100E-01
25 HOR SHR	0.3000E 01			
FIXITY OF COLUMNS AT THE BASE				
	0.1000E 01	0.1000E 01	0.1000E 01	0.1000E 01

EXAMPLE 2. (Continued)

Deflections and Rotations for Load Case 1 - WIND + (PΔ)

STORY	DEFLECTION	ROTATIONS---------------			
1	-0.5700E-01	-0.2970E-02	-0.2506E-02	-0.2506E-02	-0.2970E-02
2	-0.1049E 00	-0.3201E-02	-0.2872E-02	-0.2872E-02	-0.3201E-02
3	-0.1538E 00	-0.3167E-02	-0.2798E-02	-0.2798E-02	-0.3167E-02
4	-0.2025E 00	-0.3037E-02	-0.2667E-02	-0.2667E-02	-0.3037E-02
5	-0.2487E 00	-0.2863E-02	-0.2544E-02	-0.2544E-02	-0.2863E-02
6	-0.2938E 00	-0.2678E-02	-0.2422E-02	-0.2422E-02	-0.2678E-02
7	-0.3364E 00	-0.2565E-02	-0.2277E-02	-0.2277E-02	-0.2565E-02
8	-0.3783E 00	-0.2493E-02	-0.2167E-02	-0.2167E-02	-0.2493E-02
9	-0.4200E 00	-0.2645E-02	-0.2369E-02	-0.2369E-02	-0.2645E-02
10	-0.4637E 00	-0.2607E-02	-0.2347E-02	-0.2347E-02	-0.2607E-02
11	-0.5088E 00	-0.3033E-02	-0.2767E-02	-0.2767E-02	-0.3033E-02
12	-0.5570E 00	-0.2937E-02	-0.2649E-02	-0.2649E-02	-0.2937E-02
13	-0.6021E 00	-0.2739E-02	-0.2437E-02	-0.2437E-02	-0.2739E-02
14	-0.6451E 00	-0.2518E-02	-0.2206E-02	-0.2206E-02	-0.2518E-02
15	-0.6842E 00	-0.2320E-02	-0.1982E-02	-0.1982E-02	-0.2320E-02
16	-0.7222E 00	-0.2149E-02	-0.1772E-02	-0.1772E-02	-0.2149E-02
17	-0.7576E 00	-0.2144E-02	-0.1786E-02	-0.1786E-02	-0.2144E-02
18	-0.7922E 00	-0.1956E-02	-0.1574E-02	-0.1574E-02	-0.1956E-02
19	-0.8232E 00	-0.1807E-02	-0.1528E-02	-0.1528E-02	-0.1807E-02
20	-0.8517E 00	-0.1466E-02	-0.1304E-02	-0.1304E-02	-0.1466E-02
21	-0.8750E 00	-0.1262E-02	-0.9978E-03	-0.9978E-03	-0.1262E-02
22	-0.8961E 00	-0.1030E-02	-0.6807E-03	-0.6807E-03	-0.1030E-02
23	-0.9111E 00	-0.6654E-03	-0.4572E-03	-0.4572E-03	-0.6654E-03
24	-0.9200E 00	-0.3356E-03	-0.2333E-03	-0.2333E-03	-0.3356E-03
25	-0.9235E 00	-0.1119E-03	-0.6989E-04	-0.6989E-04	-0.1119E-03
JOINT	ALL	①	②	③	④

Moments and Shears for Load Case 1 - WIND + (PΔ)

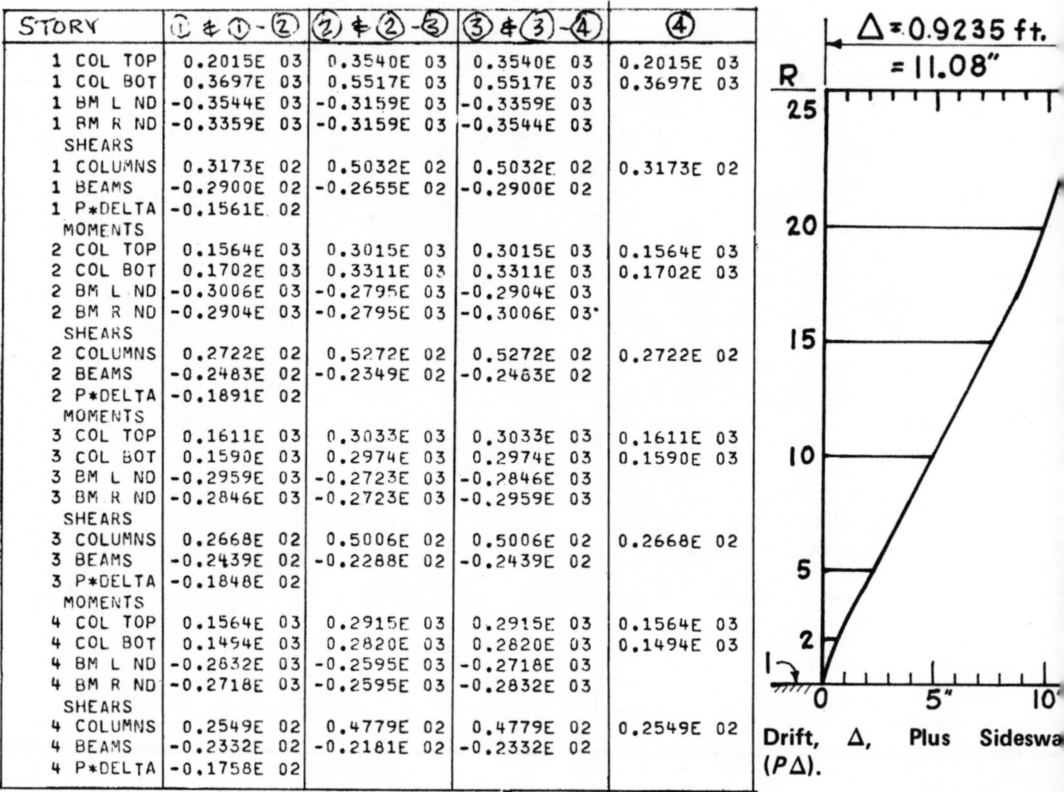

STORY	① ℄ ①-②	② ℄ ②-③	③ ℄ ③-④	④
1 COL TOP	0.2015E 03	0.3540E 03	0.3540E 03	0.2015E 03
1 COL BOT	0.3697E 03	0.5517E 03	0.5517E 03	0.3697E 03
1 BM L ND	-0.3544E 03	-0.3159E 03	-0.3359E 03	
1 BM R ND	-0.3359E 03	-0.3159E 03	-0.3544E 03	
SHEARS				
1 COLUMNS	0.3173E 02	0.5032E 02	0.5032E 02	0.3173E 02
1 BEAMS	-0.2900E 02	-0.2655E 02	-0.2900E 02	
1 P*DELTA	-0.1561E 02			
MOMENTS				
2 COL TOP	0.1564E 03	0.3015E 03	0.3015E 03	0.1564E 03
2 COL BOT	0.1702E 03	0.3311E 03	0.3311E 03	0.1702E 03
2 BM L ND	-0.3006E 03	-0.2795E 03	-0.2904E 03	
2 BM R ND	-0.2904E 03	-0.2795E 03	-0.3006E 03	
SHEARS				
2 COLUMNS	0.2722E 02	0.5272E 02	0.5272E 02	0.2722E 02
2 BEAMS	-0.2483E 02	-0.2349E 02	-0.2483E 02	
2 P*DELTA	-0.1891E 02			
MOMENTS				
3 COL TOP	0.1611E 03	0.3033E 03	0.3033E 03	0.1611E 03
3 COL BOT	0.1590E 03	0.2974E 03	0.2974E 03	0.1590E 03
3 BM L ND	-0.2959E 03	-0.2723E 03	-0.2846E 03	
3 BM R ND	-0.2846E 03	-0.2723E 03	-0.2959E 03	
SHEARS				
3 COLUMNS	0.2668E 02	0.5006E 02	0.5006E 02	0.2668E 02
3 BEAMS	-0.2439E 02	-0.2288E 02	-0.2439E 02	
3 P*DELTA	-0.1848E 02			
MOMENTS				
4 COL TOP	0.1564E 03	0.2915E 03	0.2915E 03	0.1564E 03
4 COL BOT	0.1494E 03	0.2820E 03	0.2820E 03	0.1494E 03
4 BM L ND	-0.2832E 03	-0.2595E 03	-0.2718E 03	
4 BM R ND	-0.2718E 03	-0.2595E 03	-0.2832E 03	
SHEARS				
4 COLUMNS	0.2549E 02	0.4779E 02	0.4779E 02	0.2549E 02
4 BEAMS	-0.2332E 02	-0.2181E 02	-0.2332E 02	
4 P*DELTA	-0.1758E 02			

Drift, Δ, Plus Sideswa
(PΔ).

(Continued on page 57)

EXAMPLE 2. (Continued)

Moments and Shears for Load Case 1 – WIND (PΔ)

STORY	① & ①-②	② & ②-③	③ & ③-④	④
MOMENTS				
5 COL TOP	0.1501E 03	0.2758E 03	0.2758E 03	0.1501E 03
5 COL BOT	0.1408E 03	0.2669E 03	0.2669E 03	0.1408E 03
5 BM L ND	-0.2680E 03	-0.2476E 03	-0.2582E 03	
5 BM R ND	-0.2582E 03	-0.2476E 03	-0.2680E 03	
SHEARS				
5 COLUMNS	0.2425E 02	0.4522E 02	0.4522E 02	0.2425E 02
5 BEAMS	-0.2211E 02	-0.2080E 02	-0.2211E 02	
5 P*DELTA	-0.1596E 02			
MOMENTS				
6 COL TOP	0.1396E 03	0.2641E 03	0.2641E 03	0.1396E 03
6 COL BOT	0.1311E 03	0.2558E 03	0.2558E 03	0.1311E 03
6 BM L ND	-0.2521E 03	-0.2357E 03	-0.2442E 03	
6 BM R ND	-0.2442E 03	-0.2357E 03	-0.2521E 03	
SHEARS				
6 COLUMNS	0.2257E 02	0.4332E 02	0.4332E 02	0.2257E 02
6 BEAMS	-0.2085E 02	-0.1981E 02	-0.2085E 02	
6 P*DELTA	-0.1479E 02			
MOMENTS				
7 COL TOP	0.1302E 03	0.2501E 03	0.2501E 03	0.1302E 03
7 COL BOT	0.1250E 03	0.2402E 03	0.2402E 03	0.1250E 03
7 BM L ND	-0.2400E 03	-0.2216E 03	-0.2311E 03	
7 BM R ND	-0.2311E 03	-0.2216E 03	-0.2400E 03	
SHEARS				
7 COLUMNS	0.2127E 02	0.4087E 02	0.4087E 02	0.2127E 02
7 BEAMS	-0.1980E 02	-0.1862E 02	-0.1980E 02	
7 P*DELTA	-0.1328E 02			
MOMENTS				
8 COL TOP	0.1247E 03	0.2322E 03	0.2322E 03	0.1247E 03
8 COL BOT	0.1217E 03	0.2256E 03	0.2256E 03	0.1217E 03
8 BM L ND	-0.2318E 03	-0.2109E 03	-0.2218E 03	
8 BM R ND	-0.2218E 03	-0.2109E 03	-0.2318E 03	
SHEARS				
8 COLUMNS	0.2054E 02	0.3815E 02	0.3815E 02	0.2054E 02
8 BEAMS	-0.1906E 02	-0.1772E 02	-0.1906E 02	
8 P*DELTA	-0.1240E 02			
MOMENTS				
9 COL TOP	0.1120E 03	0.2105E 03	0.2105E 03	0.1120E 03
9 COL BOT	0.1185E 03	0.2226E 03	0.2226E 03	0.1185E 03
9 BM L ND	-0.2087E 03	-0.1939E 03	-0.2016E 03	
9 BM R ND	-0.2016E 03	-0.1939E 03	-0.2087E 03	
SHEARS				
9 COLUMNS	0.1921E 02	0.3609E 02	0.3609E 02	0.1921E 02
9 BEAMS	-0.1724E 02	-0.1629E 02	-0.1724E 02	
9 P*DELTA	-0.1163E 02			
MOMENTS				
10 COL TOP	0.1083E 03	0.2062E 03	0.2062E 03	0.1083E 03
10 COL BOT	0.1070E 03	0.2050E 03	0.2050E 03	0.1070E 03
10 BM L ND	-0.2061E 03	-0.1921E 03	-0.1993E 03	
10 BM R ND	-0.1993E 03	-0.1921E 03	-0.2061E 03	
SHEARS				
10 COLUMNS	0.1794E 02	0.3427E 02	0.3427E 02	0.1794E 02
10 BEAMS	-0.1703E 02	-0.1614E 02	-0.1703E 02	
10 P*DELTA	-0.1145E 02			
MOMENTS				
11 COL TOP	0.9289E 02	0.1826E 03	0.1826E 03	0.9289E 02
11 COL BOT	0.1079E 03	0.2051E 03	0.2051E 03	0.1079E 03
11 BM L ND	-0.1789E 03	-0.1683E 03	-0.1738E 03	
11 BM R ND	-0.1738E 03	-0.1683E 03	-0.1789E 03	
SHEARS				
11 COLUMNS	0.1674E 02	0.3231E 02	0.3231E 02	0.1674E 02
11 BEAMS	-0.1482E 02	-0.1414E 02	-0.1482E 02	
11 P*DELTA	-0.1111E 02			

(Continued on page 58)

EXAMPLE 2. (Continued)

Moments and Shears for Load Case 1 – WIND and (PΔ)

STORY	① ℄ ①-②	② ℄ ②-③	③ ℄ ③-④	④
MOMENTS				
12 COL TOP	0.9797E 02	0.1823E 03	0.1823E 03	0.9797E 02
12 COL BOT	0.9498E 02	0.1769E 03	0.1769E 03	0.9498E 02
12 BM L ND	-0.1726E 03	-0.1611E 03	-0.1671E 03	
12 BM R ND	-0.1671E 03	-0.1611E 03	-0.1726E 03	
SHEARS				
12 COLUMNS	0.1607E 02	0.2994E 02	0.2994E 02	0.1607E 02
12 BEAMS	-0.1427E 02	-0.1354E 02	-0.1427E 02	
12 P*DELTA	-0.1104E 02			
MOMENTS				
13 COL TOP	0.8950E 02	0.1722E 03	0.1722E 03	0.8950E 02
13 COL BOT	0.8328E 02	0.1625E 03	0.1625E 03	0.8328E 02
13 BM L ND	-0.1603E 03	-0.1482E 03	-0.1545E 03	
13 BM R ND	-0.1545E 03	-0.1482E 03	-0.1603E 03	
SHEARS				
13 COLUMNS	0.1439E 02	0.2790E 02	0.2790E 02	0.1439E 02
13 BEAMS	-0.1322E 02	-0.1245E 02	-0.1322E 02	
13 P*DELTA	-0.9605E 01			
MOMENTS				
14 COL TOP	0.8507E 02	0.1550E 03	0.1550E 03	0.8507E 02
14 COL BOT	0.7877E 02	0.1458E 03	0.1458E 03	0.7877E 02
14 BM L ND	-0.1467E 03	-0.1342E 03	-0.1407E 03	
14 BM R ND	-0.1407E 03	-0.1342E 03	-0.1467E 03	
SHEARS				
14 COLUMNS	0.1365E 02	0.2507E 02	0.2507E 02	0.1365E 02
14 BEAMS	-0.1207E 02	-0.1127E 02	-0.1207E 02	
14 P*DELTA	-0.8456E 01			
MOMENTS				
15 COL TOP	0.7455E 02	0.1428E 03	0.1428E 03	0.7455E 02
15 COL BOT	0.6890E 02	0.1339E 03	0.1339E 03	0.6890E 02
15 BM L ND	-0.1341E 03	-0.1206E 03	-0.1276E 03	
15 BM R ND	-0.1276E 03	-0.1206E 03	-0.1341E 03	
SHEARS				
15 COLUMNS	0.1195E 02	0.2306E 02	0.2306E 02	0.1195E 02
15 BEAMS	-0.1099E 02	-0.1013E 02	-C.1099E 02	
15 P*DELTA	-0.7036E 01			
MOMENTS				
16 COL TOP	0.7034E 02	0.1246E 03	0.1246E 03	0.7034E 02
16 COL BOT	0.6618E 02	0.1180E 03	0.1180E 03	0.6618E 02
16 BM L ND	-0.1229E 03	-0.1077E 03	-0.1156E 03	
16 BM R ND	-0.1156E 03	-0.1077E 03	-0.1229E 03	
SHEARS				
16 COLUMNS	0.1137E 02	0.2023E 02	0.2023E 02	0.1137E 02
16 BEAMS	-0.1002E 02	-0.9057E 01	-0.1002E 02	
16 P*DELTA	-0.6218E 01			
MOMENTS				
17 COL TOP	0.5872E 02	0.1097E 03	0.1097E 03	0.5872E 02
17 COL BOT	0.5858E 02	0.1101E 03	0.1101E 03	0.5858E 02
17 BM L ND	-0.1050E 03	-0.9285E 02	-0.9920E 02	
17 BM R ND	-0.9920E 02	-0.9285E 02	-0.1050E 03	
SHEARS				
17 COLUMNS	0.9775E 01	0.1832E 02	0.1832E 02	0.9775E 01
17 BEAMS	-0.8584E 01	-0.7803E 01	-0.8584E 01	
17 P*DELTA	-0.5203E 01			
MOMENTS				
18 COL TOP	0.5558E 02	0.9776E 02	0.9776E 02	0.5558E 02
18 COL BOT	0.5152E 02	0.9216E 02	0.9216E 02	0.5152E 02
18 BM L ND	-0.9490E 02	-0.8184E 02	-0.8861E 02	
18 BM R ND	-0.8861E 02	-0.8184E 02	-0.9490E 02	
SHEARS				
18 COLUMNS	0.8925E 01	0.1582E 02	0.1582E 02	0.8925E 01
18 BEAMS	-0.7710E 01	-0.6877E 01	-0.7710E 01	
18 P*DELTA	-0.4506E 01			

(Continued on page 59)

EXAMPLE 2. (Continued)

Moments and Shears for Load Case 1 - WIND and (PΔ)

STORY	① ℄ ①-②	② ℄ ②-③	③ ℄ ③-④	④
MOMENTS				
19 COL TOP	0.4716E 02	0.8267E 02	0.8267E 02	0.4716E 02
19 COL BOT	0.4394E 02	0.8143E 02	0.8143E 02	0.4394E 02
19 BM L ND	-0.7761E 02	-0.6928E 02	-0.7360E 02	
19 BM R ND	-0.7360E 02	-0.6928E 02	-0.7761E 02	
SHEARS				
19 COLUMNS	0.7592E 01	0.1367E 02	0.1367E 02	0.7592E 01
19 BEAMS	-0.6353E 01	-0.5821E 01	-0.6353E 01	
19 P*DELTA	-0.3536E 01			
MOMENTS				
20 COL TOP	0.3995E 02	0.7295E 02	0.7295E 02	0.3995E 02
20 COL BOT	0.3426E 02	0.6751E 02	0.6751E 02	0.3426E 02
20 BM L ND	-0.6399E 02	-0.5915E 02	-0.6166E 02	
20 BM R ND	-0.6166E 02	-0.5915E 02	-0.6399E 02	
SHEARS				
20 COLUMNS	0.6184E 01	0.1170E 02	0.1170E 02	0.6184E 01
20 BEAMS	-0.5279E 01	-0.4971E 01	-0.5279E 01	
20 P*DELTA	-0.2781E 01			
MOMENTS				
21 COL TOP	0.3061E 02	0.6148E 02	0.6148E 02	0.3061E 02
21 COL BOT	0.2720E 02	0.5401E 02	0.5401E 02	0.2720E 02
21 BM L ND	-0.5314E 02	-0.4524E 02	-0.4933E 02	
21 BM R ND	-0.4933E 02	-0.4524E 02	-0.5314E 02	
SHEARS				
21 COLUMNS	0.4818E 01	0.9624E 01	0.9624E 01	0.4818E 01
21 BEAMS	-0.4306E 01	-0.3802E 01	-0.4306E 01	
21 P*DELTA	-0.1887E 01			
MOMENTS				
22 COL TOP	0.2849E 02	0.4259E 02	0.4259E 02	0.2849E 02
22 COL BOT	0.2511E 02	0.3796E 02	0.3796E 02	0.2511E 02
22 BM L ND	-0.4132E 02	-0.3086E 02	-0.3628E 02	
22 BM R ND	-0.3628E 02	-0.3086E 02	-0.4132E 02	
SHEARS				
22 COLUMNS	0.4467E 01	0.6713E 01	0.6713E 01	0.4467E 01
22 BEAMS	-0.3260E 01	-0.2593E 01	-0.3260E 01	
22 P*DELTA	-0.1360E 01			
MOMENTS				
23 COL TOP	0.2012E 02	0.3133E 02	0.3133E 02	0.2012E 02
23 COL BOT	0.1478E 02	0.2806E 02	0.2806E 02	0.1478E 02
23 BM L ND	-0.2694E 02	-0.2073E 02	-0.2395E 02	
23 BM R ND	-0.2395E 02	-0.2073E 02	-0.2694E 02	
SHEARS				
23 COLUMNS	0.2909E 01	0.4949E 01	0.4949E 01	0.2909E 01
23 BEAMS	-0.2138E 01	-0.1742E 01	-0.2138E 01	
23 P*DELTA	-0.7178E 00			
MOMENTS				
24 COL TOP	0.1292E 02	0.1896E 02	0.1896E 02	0.1292E 02
24 COL BOT	0.8106E 01	0.1568E 02	0.1568E 02	0.8106E 01
24 BM L ND	-0.1363E 02	-0.1057E 02	-0.1216E 02	
24 BM R ND	-0.1216E 02	-0.1057E 02	-0.1363E 02	
SHEARS				
24 COLUMNS	0.1752E 01	0.2887E 01	0.2887E 01	0.1752E 01
24 BEAMS	-0.1084E 01	-0.8890E 00	-0.1084E 01	
24 P*DELTA	-0.2798E 00			
MOMENTS				
25 COL TOP	0.4631E 01	0.7355E 01	0.7355E 01	0.4631E 01
25 COL BOT	0.1360E 01	0.4967E 01	0.4967E 01	0.1360E 01
25 BM L ND	-0.4423E 01	-0.3169E 01	-0.3819E 01	
25 BM R ND	-0.3819E 01	-0.3169E 01	-0.4423E 01	
SHEARS				
25 COLUMNS	0.4993E 00	0.1026E 01	0.1026E 01	0.4993E 00
25 BEAMS	-0.3463E 00	-0.2663E 00	-0.3463E 00	
25 P*DELTA	-0.5258E-01			

EXAMPLE 2. (Continued)

Axial Loads in Columns for Load Case 1 - WIND + (PΔ)

(The sum of the gravity loads at each story = P; P is *not* included in the axial column loads shown; the effects of $(P\Delta)$ are included in the axial loads.)

AXIAL LOADS IN COLUMNS		-- LOAD CASE 1		
25 AXIAL LOAD	0.3463E 00	-0.8004E-01	0.8004E-01	-0.3463E 00
24 AXIAL LOAD	0.1430E 01	-0.2750E 00	0.2750E 00	-0.1430E 01
23 AXIAL LOAD	0.3569E 01	-0.6715E 00	0.6715E 00	-0.3569E 01
22 AXIAL LOAD	0.6830E 01	-0.1338E 01	0.1338E 01	-0.6830E 01
21 AXIAL LOAD	0.1113E 02	-0.1842E 01	0.1842E 01	-0.1113E 02
20 AXIAL LOAD	0.1641E 02	-0.2151E 01	0.2151E 01	-0.1641E 02
19 AXIAL LOAD	0.2276E 02	-0.2683E 01	0.2683E 01	-0.2276E 02
18 AXIAL LOAD	0.3048E 02	-0.3516E 01	0.3516E 01	-0.3048E 02
17 AXIAL LOAD	0.3906E 02	-0.4297E 01	0.4297E 01	-0.3906E 02
16 AXIAL LOAD	0.4908E 02	-0.5262E 01	0.5262E 01	-0.4908E 02
15 AXIAL LOAD	0.6008E 02	-0.6125E 01	0.6125E 01	-0.6008E 02
14 AXIAL LOAD	0.7216E 02	-0.6922E 01	0.6922E 01	-0.7216E 02
13 AXIAL LOAD	0.8539E 02	-0.7694E 01	0.7694E 01	-0.8539E 02
12 AXIAL LOAD	0.9967E 02	-0.8432E 01	0.8432E 01	-0.9967E 02
11 AXIAL LOAD	0.1144E 03	-0.9110E 01	0.9110E 01	-0.1144E 03
10 AXIAL LOAD	0.1315E 03	-0.1000E 02	0.1000E 02	-0.1315E 03
9 AXIAL LOAD	0.1487E 03	-0.1095E 02	0.1095E 02	-0.1487E 03
8 AXIAL LOAD	0.1678E 03	-0.1228E 02	0.1228E 02	-0.1678E 03
7 AXIAL LOAD	0.1876E 03	-0.1346E 02	0.1346E 02	-0.1876E 03
6 AXIAL LOAD	0.2084E 03	-0.1451E 02	0.1451E 02	-0.2084E 03
5 AXIAL LOAD	0.2306E 03	-0.1581E 02	0.1581E 02	-0.2306E 03
4 AXIAL LOAD	0.2539E 03	-0.1733E 02	0.1733E 02	-0.2539E 03
3 AXIAL LOAD	0.2783E 03	-0.1883E 02	0.1883E 02	-0.2783E 03
2 AXIAL LOAD	0.3031E 03	-0.2018E 02	0.2018E 02	-0.3031E 03
1 AXIAL LOAD	0.3321E 03	-0.2263E 02	0.2263E 02	-0.3321E 03
COLUMN	(1)	(2)	(3)	(4)

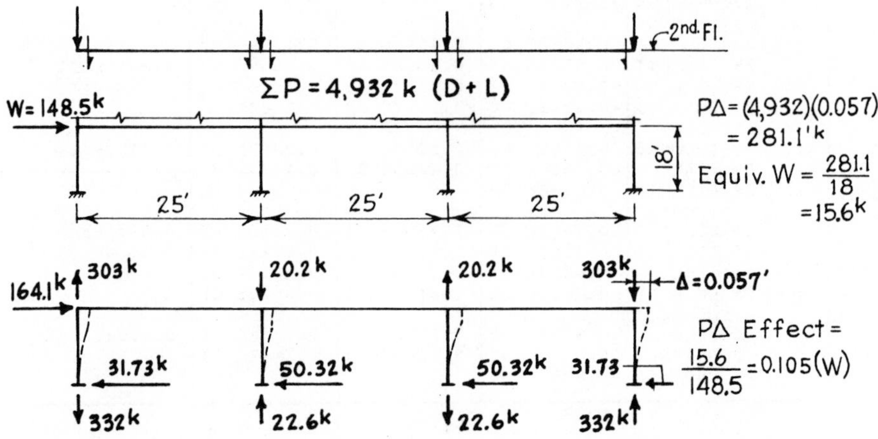

Lateral Displacements and Reactions Due to Wind Including Effects of (*PΔ*) Neglecting *V, PL/AE*—Design Example 2—Program II—25 Story Building.

EXAMPLE 2. ELASTIC ANALYSIS (Neglect $P\Delta$)

Loading Condition: Wind + Shear Effect + PL/AE Effect for W + D + L
(Note: *PL/AE* effects due to differential length changes in columns only are considered; changes in beam lengths are neglected.) Computer Program I.

INPUT—Typical Wind Bent | No. of Frames = 1 | No. of Story Levels = 25
No. of Loading Cases = 1 | Mod. of Elasticity (ksi) = 29000. | Total Height (ft) = 306.00

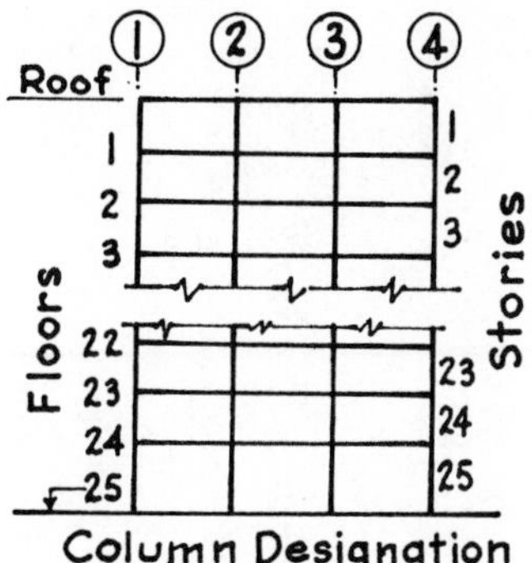

Column Designation Program I

+M +M +M +M

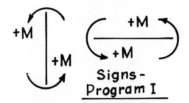

Signs - Program I

Member Designations—Program I —25 Story Building.

Lateral Displacements Due to Wind Including Effects of V and PL/AE and Neglecting ($P\Delta$)—Program I—25 Story Building.

FRAME DIMENSIONS AND MEMBER PROPERTIES - - - - NO. OF COL. LINES = 4

STORY LEVEL FROM TOP)	COL. LINE	STORY HEIGHT (FT)	BAY WIDTH (FT)	BEAM DEPTH (IN)	BEAM MOM. OF INERTIA (IN-4)	COLUMN WIDTH (IN)	COLUMN AREA (IN-2)	COL. MOM. OF INERTIA (IN-4)	EFF.COL. SHR AREA FACTOR
1	1	12.00	24.99	21.0	843.	14.0	12.	429.	0.33
	2		24.99	21.0	843.	14.0	12.	429.	0.33
	3		24.99	21.0	843.	14.0	12.	429.	0.33
	4					14.0	12.	429.	0.33
2	1	12.00	24.99	21.0	843.	14.0	12.	429.	0.33
	2		24.99	21.0	843.	14.0	12.	429.	0.33
	3		24.99	21.0	843.	14.0	12.	429.	0.33
	4					14.0	12.	429.	0.33
3	1	12.00	24.99	21.0	843.	14.0	12.	429.	0.33
	2		24.99	21.0	843.	14.0	12.	429.	0.33
	3		24.99	21.0	843.	14.0	12.	429.	0.33
	4					14.0	12.	429.	0.33
4	1	12.00	24.99	21.0	843.	14.0	12.	429.	0.33
	2		24.99	21.0	843.	14.0	12.	429.	0.33
	3		24.99	21.0	843.	14.0	12.	429.	0.33
	4					14.0	12.	429.	0.33
5	1	12.00	24.99	21.0	843.	14.0	14.	485.	0.33
	2		24.99	21.0	843.	14.0	20.	724.	0.23
	3		24.99	21.0	843.	14.C	20.	724.	0.23
	4					14.0	14.	485.	0.33
6	1	12.00	24.99	21.0	843.	14.0	14.	485.	0.33
	2		24.99	21.0	843.	14.0	20.	724.	0.23
	3		24.99	21.0	843.	14.0	20.	724.	0.23
	4					14.0	14.	485.	0.33
7	1	12.00	24.99	21.0	843.	14.0	17.	641.	0.29
	2		24.99	21.0	843.	14.0	21.	797.	0.29
	3		24.99	21.0	843.	14.0	21.	797.	0.29
	4					14.0	17.	641.	0.29
8	1	12.00	24.99	21.0	971.	14.0	17.	641.	0.29
	2		24.99	21.0	971.	14.0	21.	797.	0.29
	3		24.99	21.0	971.	14.0	21.	797.	0.29
	4					14.0	17.	641.	0.29
9	1	12.00	24.99	21.0	971.	14.0	20.	724.	0.23
	2		24.99	21.0	971.	14.0	25.	967.	0.23
	3		24.99	21.0	971.	14.0	25.	967.	0.23
	4					14.0	20.	724.	0.23
10	1	12.00	24.99	21.0	1140.	14.0	20.	724.	0.23
	2		24.99	21.0	1140.	14.0	25.	967.	0.23
	3		24.99	21.0	1140.	14.0	25.	967.	0.23
	4					14.0	20.	724.	0.23
11	1	12.00	24.99	21.0	1140.	14.0	22.	821.	0.26
	2		24.99	21.0	1140.	14.0	30.	1170.	0.23
	3		24.99	21.0	1140.	14.0	30.	1170.	0.23
	4					14.0	22.	851.	0.26

(Continued on page 62)

EXAMPLE 2. (Continued)

Input

STORY LEVEL (FROM TOP)	COL. LINE	STORY HEIGHT (FT)	BAY WIDTH (FT)	BEAM DEPTH (IN)	BEAM MOM. OF INERTIA (IN-4)	COLUMN WIDTH (IN)	COLUMN AREA (IN-2)	COL. MOM. OF INERTIA (IN-4)	EFF.COL SHR ARE FACTOR
12	1	12.00	24.99	21.0	1140.	14.0	22.	821.	0.26
	2		24.99	21.0	1140.	14.0	30.	1170.	0.23
	3		24.99	21.0	1140.	14.0	30.	1170.	0.23
	4					14.0	22.	851.	0.26
13	1	12.00	24.99	21.0	1140.	14.0	24.	928.	0.25
	2		24.99	21.0	1140.	14.0	35.	1370.	0.23
	3		24.99	21.0	1140.	14.0	35.	1370.	0.23
	4					14.0	24.	928.	0.25
14	1	12.00	24.99	21.0	1140.	14.0	24.	928.	0.25
	2		24.99	21.0	1140.	14.0	35.	1370.	0.23
	3		24.99	21.0	1140.	14.0	35.	1370.	0.23
	4					14.0	24.	928.	0.25
15	1	12.12	24.99	21.0	1140.	14.0	27.	1060.	0.23
	2		24.99	21.0	1140.	14.0	40.	1590.	0.24
	3		24.99	21.0	1140.	14.0	40.	1590.	0.24
	4					14.0	27.	1060.	0.23
16	1	12.00	24.99	24.0	1540.	14.0	27.	1060.	0.23
	2		24.99	24.0	1540.	14.0	40.	1590.	0.24
	3		24.99	24.0	1540.	14.0	40.	1590.	0.24
	4					14.0	27.	1060.	0.23
17	1	12.00	24.99	24.0	1540.	14.0	32.	1270.	0.23
	2		24.99	24.0	1540.	14.0	44.	1790.	0.23
	3		24.99	24.0	1540.	14.0	44.	1790.	0.23
	4					14.0	32.	1270.	0.23
18	1	12.00	24.99	24.0	1820.	14.0	32.	1270.	0.23
	2		24.99	24.0	1820.	14.0	44.	1790.	0.23
	3		24.99	24.0	1820.	14.0	44.	1790.	0.23
	4					14.0	32.	1270.	0.23
19	1	12.00	24.99	24.0	1820.	14.0	35.	1370.	0.23
	2		24.99	24.0	1820.	14.0	49.	2020.	0.24
	3		24.99	24.0	1820.	14.0	49.	2020.	0.24
	4					14.0	35.	1370.	0.23
20	1	12.00	24.99	24.0	1820.	14.0	35.	1370.	0.23
	2		24.99	24.0	1820.	14.0	49.	2020.	0.24
	3		24.99	24.0	1820.	14.0	49.	2020.	0.24
	4					14.0	35.	1370.	0.23
21	1	12.00	24.99	24.0	1820.	14.0	40.	1590.	0.24
	2		24.99	24.0	1820.	14.0	51.	2150.	0.24
	3		24.99	24.0	1820.	14.0	51.	2150.	0.24
	4					14.0	40.	1590.	0.24
22	1	12.00	24.99	24.0	1820.	14.0	40.	1590.	0.24
	2		24.99	24.0	1820.	14.0	51.	2150.	0.24
	3		24.99	24.0	1820.	14.0	51.	2150.	0.24
	4					14.0	40.	1590.	0.24
23	1	12.00	24.99	24.0	1820.	14.0	44.	1790.	0.23
	2		24.99	24.0	1820.	14.0	56.	2400.	0.24
	3		24.99	24.0	1820.	14.0	56.	2400.	0.24
	4					14.0	44.	1790.	0.23
24	1	12.00	24.99	24.0	1820.	14.0	44.	1790.	0.23
	2		24.99	24.0	1820.	14.0	56.	2400.	0.24
	3		24.99	24.0	1820.	14.0	56.	2400.	0.24
	4					14.0	44.	1790.	0.23
25	1	18.00	24.99	24.0	2370.	14.0	59.	2540.	0.24
	2		24.99	24.0	2370.	14.0	77.	3530.	0.25
	3		24.99	24.0	2370.	14.0	77.	3530.	0.25
	4					14.0	59.	2540.	0.24

---- STORY HEIGHTS AND BAY WIDTHS ARE CENTER-TO-CENTER DISTANCES - - - -

---- FINITE WIDTHS OF MEMBERS TO BE CONSIDERED - - - -

ALL COLUMN BASES AT LOWEST STORY LEVEL ARE FULLY FIXED - - - - -

EXAMPLE 2. (Continued)

Loads–Input

VERTICAL LOADS ON FRAME NO. (1)

*** NO CONC. VERTICAL LOADS ON BEAMS

STORY LEVEL	COL. LINE	UNIFORM BM LOAD (KIPS/FT)
1	1	2.44
	2	2.44
	3	2.44
2	1	2.44
	2	2.44
	3	2.44
3	1	2.44
	2	2.44
	3	2.44
4	1	2.44
	2	2.44
	3	2.44
5	1	2.44
	2	2.44
	3	2.44
6	1	2.44
	2	2.44
	3	2.44
7	1	2.44
	2	2.44
	3	2.44
8	1	2.44
	2	2.44
	3	2.44
9	1	2.44
	2	2.44
	3	2.44
10	1	2.44
	2	2.44
	3	2.44
11	1	2.44
	2	2.44
	3	2.44
12	1	2.44
	2	2.44
	3	2.44
13	1	2.44
	2	2.44
	3	2.44
14	1	2.44
	2	2.44
	3	2.44
15	1	2.44
	2	2.44
	3	2.44
16	1	2.44
	2	2.44
	3	2.44
17	1	2.44
	2	2.44
	3	2.44
18	1	2.44
	2	2.44
	3	2.44
19	1	2.44
	2	2.44
	3	2.44
20	1	2.44
	2	2.44
	3	2.44
21	1	2.44
	2	2.44
	3	2.44
22	1	2.44
	2	2.44
	3	2.44
23	1	2.44
	2	2.44
	3	2.44
24	1	2.44
	2	2.44
	3	2.44
25	1	2.44
	2	2.44
	3	2.44

CONCENTRATED LATERAL LOADS

(FOR ALL FRAMES)

STORY LEVEL (FROM TOP)	LATERAL LOAD (KIPS)
1	3.00
2	6.00
3	6.00
4	6.00
5	6.00
6	6.00
7	6.00
8	6.00
9	6.00
10	6.00
11	6.00
12	6.00
13	6.00
14	6.00
15	6.00
16	6.00
17	6.00
18	6.00
19	6.00
20	6.00
21	6.00
22	6.00
23	6.00
24	6.00
25	7.50

EXAMPLE 2. (Continued)

Fixed-End Moments and Shears in Beams and Resultant Forces on Joints (KIPS, In-KIPS)

STORY LEVEL	COL. LINE	FIXED-END MOMENTS LEFT	FIXED-END MOMENTS RIGHT	FIXED-END SHEARS LEFT	FIXED-END SHEARS RIGHT	RESULTANT FORCES ON JOINTS MOMENT	RESULTANT FORCES ON JOINTS VERT. FORCE
1	1	1385.60	-1385.60	29.07	-29.07	1589.10	-29.07
	2	1385.60	-1385.60	29.07	-29.07	0.00	-58.14
	3	1385.60	-1385.60	29.07	-29.07	0.00	-58.14
	4					-1589.10	-29.07
2	1	1385.60	-1385.60	29.07	-29.07	1589.10	-29.07
	2	1385.60	-1385.60	29.07	-29.07	0.00	-58.14
	3	1385.60	-1385.60	29.07	-29.07	0.00	-58.14
	4					-1589.10	-29.07
3	1	1385.60	-1385.60	29.07	-29.07	1589.10	-29.07
	2	1385.60	-1385.60	29.07	-29.07	0.00	-58.14
	3	1385.60	-1385.60	29.07	-29.07	0.00	-58.14
	4					-1589.10	-29.07
4	1	1385.60	-1385.60	29.07	-29.07	1589.10	-29.07
	2	1385.60	-1385.60	29.07	-29.07	0.00	-58.14
	3	1385.60	-1385.60	29.07	-29.07	0.00	-58.14
	4					-1589.10	-29.07
5	1	1385.60	-1385.60	29.07	-29.07	1589.10	-29.07
	2	1385.60	-1385.60	29.07	-29.07	0.00	-58.14
	3	1385.60	-1385.60	29.07	-29.07	0.00	-58.14
	4					-1589.10	-29.07
6	1	1385.60	-1385.60	29.07	-29.07	1589.10	-29.07
	2	1385.60	-1385.60	29.07	-29.07	0.00	-58.14
	3	1385.60	-1385.60	29.07	-29.07	0.00	-58.14
	4					-1589.10	-29.07
7	1	1385.60	-1385.60	29.07	-29.07	1589.10	-29.07
	2	1385.60	-1385.60	29.07	-29.07	0.00	-58.14
	3	1385.60	-1385.60	29.07	-29.07	0.00	-58.14
	4					-1589.10	-29.07
8	1	1385.60	-1385.60	29.07	-29.07	1589.10	-29.07
	2	1385.60	-1385.60	29.07	-29.07	0.00	-58.14
	3	1385.60	-1385.60	29.07	-29.07	0.00	-58.14
	4					-1589.10	-29.07
9	1	1385.60	-1385.60	29.07	-29.07	1589.10	-29.07
	2	1385.60	-1385.60	29.07	-29.07	0.00	-58.14
	3	1385.60	-1385.60	29.07	-29.07	0.00	-58.14
	4					-1589.10	-29.07
10	1	1385.60	-1385.60	29.07	-29.07	1589.10	-29.07
	2	1385.60	-1385.60	29.07	-29.07	0.00	-58.14
	3	1385.60	-1385.60	29.07	-29.07	0.00	-58.14
	4					-1589.10	-29.07
11	1	1385.60	-1385.60	29.07	-29.07	1589.10	-29.07
	2	1385.60	-1385.60	29.07	-29.07	0.00	-58.14
	3	1385.60	-1385.60	29.07	-29.07	0.00	-58.14
	4					-1589.10	-29.07
12	1	1385.60	-1385.60	29.07	-29.07	1589.10	-29.07
	2	1385.60	-1385.60	29.07	-29.07	0.00	-58.14
	3	1385.60	-1385.60	29.07	-29.07	0.00	-58.14
	4					-1589.10	-29.07
13	1	1385.60	-1385.60	29.07	-29.07	1589.10	-29.07
	2	1385.60	-1385.60	29.07	-29.07	0.00	-58.14
	3	1385.60	-1385.60	29.07	-29.07	0.00	-58.14
	4					-1589.10	-29.07
14	1	1385.60	-1385.60	29.07	-29.07	1589.10	-29.07
	2	1385.60	-1385.60	29.07	-29.07	0.00	-58.14
	3	1385.60	-1385.60	29.07	-29.07	0.00	-58.14
	4					-1589.10	-29.07
15	1	1385.60	-1385.60	29.07	-29.07	1589.10	-29.07
	2	1385.60	-1385.60	29.07	-29.07	0.00	-58.14
	3	1385.60	-1385.60	29.07	-29.07	0.00	-58.14
	4					-1589.10	-29.07

(Continued on page 65)

EXAMPLE 2. (Continued)

16	1	1385.60	-1385.60	29.07	-29.07	1589.10	-29.07
	2	1385.60	-1385.60	29.07	-29.07	0.00	-58.14
	3	1385.60	-1385.60	29.07	-29.07	0.00	-58.14
	4					-1589.10	-29.07
17	1	1385.60	-1385.60	29.07	-29.07	1589.10	-29.07
	2	1385.60	-1385.60	29.07	-29.07	0.00	-58.14
	3	1385.60	-1385.60	29.07	-29.07	0.00	-58.14
	4					-1589.10	-29.07
18	1	1385.60	-1385.60	29.07	-29.07	1589.10	-29.07
	2	1385.60	-1385.60	29.07	-29.07	0.00	-58.14
	3	1385.60	-1385.60	29.07	-29.07	0.00	-58.14
	4					-1589.10	-29.07
19	1	1385.60	-1385.60	29.07	-29.07	1589.10	-29.07
	2	1385.60	-1385.60	29.07	-29.07	0.00	-58.14
	3	1385.60	-1385.60	29.07	-29.07	0.00	-58.14
	4					-1589.10	-29.07
20	1	1385.60	-1385.60	29.07	-29.07	1589.10	-29.07
	2	1385.60	-1385.60	29.07	-29.07	0.00	-58.14
	3	1385.60	-1385.60	29.07	-29.07	0.00	-58.14
	4					-1589.10	-29.07
21	1	1385.60	-1385.60	29.07	-29.07	1589.10	-29.07
	2	1385.60	-1385.60	29.07	-29.07	0.00	-58.14
	3	1385.60	-1385.60	29.07	-29.07	0.00	-58.14
	4					-1589.10	-29.07
22	1	1385.60	-1385.60	29.07	-29.07	1589.10	-29.07
	2	1385.60	-1385.60	29.07	-29.07	0.00	-58.14
	3	1385.60	-1385.60	29.07	-29.07	0.00	-58.14
	4					-1589.10	-29.07
23	1	1385.60	-1385.60	29.07	-29.07	1589.10	-29.07
	2	1385.60	-1385.60	29.07	-29.07	0.00	-58.14
	3	1385.60	-1385.60	29.07	-29.07	0.00	-58.14
	4					-1589.10	-29.07
24	1	1385.60	-1385.60	29.07	-29.07	1589.10	-29.07
	2	1385.60	-1385.60	29.07	-29.07	0.00	-58.14
	3	1385.60	-1385.60	29.07	-29.07	0.00	-58.14
	4					-1589.10	-29.07
25	1	1385.60	-1385.60	29.07	-29.07	1589.10	-29.07
	2	1385.60	-1385.60	29.07	-29.07	0.00	-58.14
	3	1385.60	-1385.60	29.07	-29.07	0.00	-58.14
	4					-1589.10	-29.07

LATERAL DISPLACEMENT OR *DRIFT* (SAME FOR ALL FRAMES)

STORY LEVEL (FROM TOP)	RELATIVE DISPLMT. (INCHES)	TOTAL DISPLMT. (INCHES)
1	0.1370294E 00	0.1105501E 02
2	0.1905713E 00	0.1091798E 02
3	0.2511418E 00	0.1072741E 02
4	0.3109890E 00	0.1047626E 02
5	0.3391591E 00	0.1016528E 02
6	0.3889602E 00	0.9826120E 01
7	0.4088268E 00	0.9437160E 01
8	0.4382622E 00	0.9028333E 01
9	0.4490124E 00	0.8590071E 01
10	0.4702551E 00	0.8141059E 01
11	0.4825270E 00	0.7670803E 01
12	0.5161973E 00	0.7188276E 01
13	0.5323756E 00	0.6672079E 01
14	0.5558821E 00	0.6139703E 01
15	0.5295515E 00	0.5583821E 01
16	0.4976098E 00	0.5054270E 01
17	0.4761213E 00	0.4556660E 01
18	0.4729035E 00	0.4080538E 01
19	0.4750529E 00	0.3607635E 01
20	0.4913375E 00	0.3132582E 01
21	0.4971586E 00	0.2641244E 01
22	0.5119897E 00	0.2144086E 01
23	0.5100042E 00	0.1632096E 01
24	0.4973210E 00	0.1122092E 01
25	0.6247711E 00	0.6247711E 00

(DRIFT AT TOP)/(TOTAL HT.) = 0.00300081

EXAMPLE 2. (Continued)

VALUES OF DISPLACEMENT COMPONENTS OF JOINTS (INCHES, RADIANS)

SIGN CONVENTION - HOR. COMP. (U) IS POSITIVE TO THE RIGHT
- VERT. COMP. (V) IS POSITIVE UPWARD
- ROTATION (W) IS POSITIVE CLOCKWISE

STORY LEVEL (FROM TOP)	COL. LINE (FROM LEFT)	HOR. COMP. * U *	VERT. COMP. * V *	ROTATION * W *
1	1	0.110550E 02	-0.124216E 01	0.362139E-02
	2		-0.201303E 01	0.103330E-02
	3		-0.212044E 01	0.318894E-03
	4		-0.184206E 01	-0.182028E-02
2	1	0.109179E 02	-0.123207E 01	0.218476E-02
	2		-0.199416E 01	0.127874E-02
	3		-0.210069E 01	0.478140E-03
	4		-0.183205E 01	-0.102350E-03
3	1	0.107274E 02	-0.121116E 01	0.280803E-02
	2		-0.195742E 01	0.141664E-02
	3		-0.206207E 01	0.761151E-03
	4		-0.181088E 01	-0.982555E-04
4	1	0.104762E 02	-0.118000E 01	0.305377E-02
	2		-0.190251E 01	0.163676E-02
	3		-0.200443E 01	0.969629E-03
	4		-0.177844E 01	0.303865E-03
5	1	0.101652E 02	-0.113896E 01	0.317217E-02
	2		-0.182933E 01	0.183718E-02
	3		-0.192775E 01	0.134084E-02
	4		-0.173446E 01	0.621931E-03
6	1	0.982612E 01	-0.109378E 01	0.332519E-02
	2		-0.177174E 01	0.205509E-02
	3		-0.186741E 01	0.164508E-02
	4		-0.168457E 01	0.859134E-03
7	1	0.943716E 01	-0.104034E 01	0.342724E-02
	2		-0.170270E 01	0.225073E-02
	3		-0.179500E 01	0.185876E-02
	4		-0.162383E 01	0.127349E-02
8	1	0.902833E 01	-0.991920E 00	0.341912E-02
	2		-0.162886E 01	0.229692E-02
	3		-0.171754E 01	0.187572E-02
	4		-0.156718E 01	0.147485E-02
9	1	0.859007E 01	-0.937363E 00	0.351530E-02
	2		-0.154460E 01	0.242689E-02
	3		-0.162909E 01	0.208525E-02
	4		-0.150142E 01	0.166036E-02
10	1	0.814105E 01	-0.883239E 00	0.349963E-02
	2		-0.146396E 01	0.240336E-02
	3		-0.154437E 01	0.203871E-02
	4		-0.143428E 01	0.165717E-02
11	1	0.767080E 01	-0.824046E 00	0.353926E-02
	2		-0.137447E 01	0.252701E-02
	3		-0.145024E 01	0.222099E-02
	4		-0.135858E 01	0.189598E-02
12	1	0.718827E 01	-0.768094E 00	0.362547E-02
	2		-0.129138E 01	0.267017E-02
	3		-0.136274E 01	0.239762E-02
	4		-0.128486E 01	0.213699E-02
13	1	0.667207E 01	-0.708079E 00	0.373225E-02
	2		-0.120078E 01	0.282429E-02
	3		-0.126727E 01	0.259350E-02
	4		-0.120337E 01	0.231824E-02
14	1	0.613970E 01	-0.648876E 00	0.382768E-02
	2		-0.111584E 01	0.296292E-02
	3		-0.117768E 01	0.275605E-02
	4		-0.112050E 01	0.247864E-02
15	1	0.558382E 01	-0.586286E 00	0.381544E-02
	2		-0.102434E 01	0.301627E-02
	3		-0.108115E 01	0.284957E-02
	4		-0.103018E 01	0.256861E-02

(Continued on page 67)

EXAMPLE 2. (Continued)

STORY LEVEL (FROM TOP)	COL LINE (FROM LEFT)	HORIZ COMP U	VERT COMP V	ROTATION W
16	1	0.505427E 01	-0.527969E 00	0.342097E-02
	2		-0.938552E 00	0.266189E-02
	3		-0.990590E 00	0.245746E-02
	4		-0.943577E 00	0.222727E-02
17	1	0.455666E 01	-0.468467E 00	0.331816E-02
	2		-0.849341E 00	0.262203E-02
	3		-0.896303E 00	0.245595E-02
	4		-0.852221E 00	0.227095E-02
18	1	0.408053E 01	-0.415500E 00	0.313599E-02
	2		-0.763408E 00	0.243157E-02
	3		-0.805408E 00	0.225242E-02
	4		-0.768413E 00	0.216407E-02
19	1	0.360763E 01	-0.360524E 00	0.312200E-02
	2		-0.672460E 00	0.246953E-02
	3		-0.709157E 00	0.232937E-02
	4		-0.678529E 00	0.220808E-02
20	1	0.313258E 01	-0.307409E 00	0.315028E-02
	2		-0.586230E 00	0.254052E-02
	3		-0.617871E 00	0.242340E-02
	4		-0.588837E 00	0.227853E-02
21	1	0.264124E 01	-0.252698E 00	0.320311E-02
	2		-0.495415E 00	0.259334E-02
	3		-0.521733E 00	0.249144E-02
	4		-0.493357E 00	0.241950E-02
22	1	0.214408E 01	-0.203574E 00	0.324685E-02
	2		-0.404752E 00	0.264220E-02
	3		-0.425839E 00	0.256379E-02
	4		-0.404661E 00	0.255278E-02
23	1	0.163209E 01	-0.153354E 00	0.327816E-02
	2		-0.309603E 00	0.268988E-02
	3		-0.325353E 00	0.264072E-02
	4		-0.310748E 00	0.263256E-02
24	1	0.112209E 01	-0.106952E 00	0.320287E-02
	2		-0.218690E 00	0.271094E-02
	3		-0.229528E 00	0.267581E-02
	4		-0.220791E 00	0.266471E-02
25	1	0.624771E 00	-0.597933E-01	0.311010E-02
	2		-0.123576E 00	0.237592E-02
	3		-0.129485E 00	0.238701E-02
	4		-0.126048E 00	0.242536E-02

EXAMPLE 2. (Continued)

MEMBER-END FORCES (IN-KIPS, KIPS)

SIGN CONVENTION - MOMENTS - ARE POSITIVE WHEN COUNTERCLOCKWISE ON MEMBER ENDS
- SHEARS (A) IN COLUMNS ARE POSITIVE AT TOP WHEN DIRECTED TO RIGHT
AND AT BOTTOM WHEN DIRECTED TO LEFT
(B) IN BEAMS ARE POSITIVE AT LEFT END WHEN DIRECTED UPWARD
AND AT RIGHT WHEN DIRECTED DOWNWARD
- AXIAL FORCES (IN COLUMNS) ARE POSITIVE WHEN TENSILE

STORY	COL.	BEAM MOMENTS		BEAM SHEARS		COLUMN MOMENTS		COLUMN SHEARS		COL.AXIAL
LEVEL	LINE	LEFT	RIGHT	LEFT	RIGHT	TOP	BOTTOM	TOP	BOTTOM	FORCE
*****	****	********	********	********	********	********	********	********	********	*********
1	1	1294.84	-1033.84	29.98	-28.15	-1306.49	-1015.87	-18.88	-18.88	-29.98
	2	1153.41	-1495.63	27.87	-30.26	-96.81	-146.46	-1.97	-1.97	-56.03
	3	1107.29	-1298.14	28.40	-29.73	345.20	312.98	5.35	5.35	-58.67
	4					1311.99	964.46	18.50	18.50	-29.74
2	1	1743.38	-872.90	32.11	-26.02	-634.91	-760.99	-11.34	-11.34	-62.10
	2	1035.58	-1598.72	27.10	-31.04	-0.49	-28.39	-0.23	-0.23	-109.16
	3	753.00	-1918.94	24.99	-33.14	447.38	390.13	6.80	6.80	-114.71
	4					847.55	846.73	13.77	13.77	-62.88
3	1	1468.69	-1064.60	30.48	-27.65	-681.35	-731.06	-11.48	-11.48	-92.58
	2	931.40	-1727.72	26.28	-31.85	151.58	107.05	2.10	2.10	-163.11
	3	683.20	-1941.04	24.67	-33.47	543.90	501.73	8.50	8.50	-171.24
	4					1017.24	935.90	15.87	15.87	-96.36
4	1	1298.58	-1230.33	29.31	-28.83	-555.27	-579.22	-9.22	-9.22	-121.89
	2	810.19	-1846.93	25.44	-32.69	271.77	231.22	4.08	4.08	-217.39
	3	580.70	-2076.65	23.84	-34.30	635.19	560.10	9.71	9.71	-227.78
	4					1041.77	977.43	16.41	16.41	-130.66
5	1	1162.16	-1380.77	28.30	-29.83	-581.90	-616.90	-9.74	-9.74	-150.20
	2	664.73	-2021.59	24.32	-33.81	415.79	341.40	6.15	6.15	-271.56
	3	449.37	-2198.90	22.95	-35.19	849.89	746.02	12.97	12.97	-284.55
	4					1110.47	1056.22	17.61	17.61	-165.85
6	1	1045.66	-1508.37	27.45	-30.69	-441.27	-464.61	-7.36	-7.36	-177.66
	2	526.69	-2174.39	23.31	-34.83	540.68	473.89	8.24	8.24	-325.56
	3	316.72	-2320.08	22.06	-36.07	914.82	841.87	14.28	14.28	-341.45
	4					1144.18	1049.42	17.83	17.83	-201.93
7	1	945.59	-1624.44	26.69	-31.44	-504.60	-502.14	-8.18	-8.18	-204.36
	2	412.07	-2292.10	22.49	-35.64	611.57	594.21	9.80	9.80	-379.50
	3	185.85	-2485.27	21.03	-37.11	1039.74	1033.37	16.85	16.85	-398.13
	4					1292.85	1231.98	20.52	20.52	-239.04
8	1	819.83	-1730.36	25.88	-32.25	-350.62	-379.69	-5.93	-5.93	-230.24
	2	234.32	-2453.92	21.31	-36.83	751.39	702.55	11.82	11.82	-433.07
	3	-2.79	-2695.04	19.63	-38.50	1173.17	1094.42	18.43	18.43	-454.60
	4					1299.91	1243.84	20.68	20.68	-277.55
9	1	691.71	-1865.13	24.96	-33.17	-362.94	-357.59	-5.85	-5.85	-255.21
	2	128.29	-2575.61	20.51	-37.63	852.35	863.08	13.94	13.94	-486.76
	3	-80.71	-2768.23	19.10	-39.03	1277.97	1299.19	20.95	20.95	-511.34
	4					1349.91	1351.00	21.95	21.95	-316.59
10	1	519.95	-1997.76	23.90	-34.24	-228.08	-241.61	-3.81	-3.81	-279.12

EXAMPLE 2. (Continued)

STORY LEVEL	COL. LINE	BEAM MOMENTS		BEAM SHEARS		COLUMN MOMENTS		COLUMN SHEARS		COL. AXIAL FORCE
		LEFT	RIGHT	LEFT	RIGHT	TOP	BOTTOM	TOP	BOTTOM	
*****	****	********	********	********	********	********	********	********	********	*********
	2	-77.57	-2764.46	19.13	-39.01	1004.81	948.43	15.88	15.88	-540.13
	3	-270.16	-2953.14	17.80	-40.34	1427.31	1344.19	22.53	22.53	-568.15
	4					1418.72	1337.19	22.40	22.40	-356.93
11	1	396.77	-2140.37	22.97	-35.16	-233.22	-266.59	-4.06	-4.06	-302.09
	2	-193.36	-2893.80	18.27	-39.86	1147.61	1068.62	18.01	18.01	-593.58
	3	-372.10	-3068.15	17.04	-41.10	1582.96	1485.51	24.94	24.94	-625.07
	4					1530.39	1433.68	24.09	24.09	-398.03
12	1	254.08	-2296.23	21.93	-36.21	-80.98	-122.32	-1.65	-1.65	-324.02
	2	-316.52	-3024.70	17.38	-40.75	1276.10	1191.08	20.05	20.05	-647.18
	3	-483.19	-3194.12	16.21	-41.93	1658.34	1550.27	26.08	26.08	-682.04
	4					1543.60	1470.87	24.50	24.50	-439.96
13	1	90.46	-2470.79	20.74	-37.39	-78.97	-120.74	-1.62	-1.62	-344.77
	2	-450.86	-3168.70	16.41	-41.73	1430.28	1340.72	22.52	22.52	-700.99
	3	-588.16	-3295.71	15.49	-42.65	1815.38	1710.36	28.66	28.66	-739.26
	4					1599.06	1528.87	25.43	25.43	-482.62
14	1	-52.14	-2623.39	19.71	-38.42	43.96	49.32	0.75	0.75	-364.49
	2	-568.92	-3292.29	15.56	-42.57	1518.70	1484.24	24.41	24.41	-754.99
	3	-689.62	-3396.68	14.78	-43.36	1853.78	1793.36	29.65	29.65	-796.61
	4					1629.48	1590.11	26.17	26.17	-525.98
15	1	-129.63	-2716.04	19.12	-39.02	-66.90	130.25	0.51	0.51	-383.61
	2	-629.94	-3362.60	15.11	-43.03	1494.89	1760.58	26.46	26.46	-809.13
	3	-741.71	-3447.95	14.42	-43.72	1857.64	2151.63	32.59	32.59	-854.07
	4					1601.13	1771.74	27.42	27.42	-569.71
16	1	-377.04	-2911.14	17.57	-40.57	93.59	146.27	1.99	1.99	-401.18
	2	-991.79	-3699.13	12.66	-45.47	1680.83	1711.46	28.26	28.26	-862.37
	3	-1106.79	-3806.09	11.89	-46.25	2081.76	2082.92	34.70	34.70	-911.44
	4					1692.74	1670.37	28.02	28.02	-615.96
17	1	-394.44	-2948.20	17.38	-40.76	76.10	187.92	2.20	2.20	-418.57
	2	-982.10	-3701.42	12.69	-45.45	1716.33	1881.11	29.97	29.97	-915.82
	3	-1110.85	-3824.26	11.81	-46.33	2108.95	2285.03	36.61	36.61	-966.70
	4					1779.45	1845.05	30.20	30.20	-662.29
18	1	-630.85	-3142.02	15.87	-42.26	253.76	262.35	4.30	4.30	-434.45
	2	-1205.38	-3910.45	11.18	-46.96	1939.51	1906.67	32.05	32.05	-969.27
	3	-1339.18	-4077.76	10.12	-48.01	2324.74	2258.16	38.19	38.19	-1025.80
	4					1840.88	1813.87	30.45	30.45	-710.30
19	1	-774.49	-3304.84	14.80	-43.33	298.96	280.23	4.82	4.82	-449.25
	2	-1285.47	-4004.93	10.57	-47.57	2112.76	2043.46	34.63	34.63	-1023.18
	3	-1390.86	-4117.28	9.81	-48.33	2460.71	2368.90	40.24	40.24	-1083.18
	4					1900.78	1854.14	31.29	31.29	-758.64
20	1	-952.51	-3498.62	13.50	-44.63	436.08	401.10	6.97	6.97	-462.76
	2	-1396.64	-4124.61	9.76	-48.38	2237.87	2186.30	36.86	36.86	-1077.59
	3	-1484.57	-4202.29	9.18	-48.95	2531.93	2465.50	41.64	41.64	-1140.75
	4					1937.25	1843.91	31.50	31.50	-807.60
21	1	-1153.69	-3699.80	12.09	-46.04	489.61	455.99	7.88	7.88	-474.86
	2	-1484.64	-4218.22	9.12	-49.01	2349.34	2298.57	38.73	38.73	-1132.76
	3	-1589.95	-4334.60	8.35	-49.79	2612.56	2537.38	42.91	42.91	-1198.12
	4					2059.46	1957.03	33.47	33.47	-857.39
22	1	-1367.40	-3915.40	10.59	-47.54	620.88	596.82	10.14	10.14	-485.46
	2	-1570.96	-4313.22	8.49	-49.64	2501.09	2451.53	41.27	41.27	-1188.80
	3	-1670.26	-4437.40	7.71	-50.43	2693.88	2613.93	44.23	44.23	-1255.48
	4					2031.55	1970.25	33.34	33.34	-907.82

(Continued on page 70)

EXAMPLE 2. (Continued)

STORY LEVEL	COL. LINE	BEAM MOMENTS		BEAM SHEARS		COLUMN MOMENTS		COLUMN SHEARS		COL.AXIAL FORCE
		LEFT	RIGHT	LEFT	RIGHT	TOP	BOTTOM	TOP	BOTTOM	
*****	****	********	********	********	********	********	********	********	********	*********
23	1	-1584.25	-4138.30	9.06	-49.08	663.09	728.23	11.59	11.59	-494.52
	2	-1658.61	-4411.67	7.84	-50.30	2617.69	2593.27	43.42	43.42	-1245.73
	3	-1735.31	-4503.49	7.25	-50.88	2756.37	2715.67	45.60	45.60	-1313.04
	4					2076.72	2048.90	34.38	34.38	-958.71
24	1	-1707.35	-4296.96	8.07	-50.06	646.19	726.45	11.43	11.43	-502.60
	2	-1707.66	-4465.89	7.48	-50.66	2624.57	3013.20	46.98	46.98	-1303.29
	3	-1752.18	-4519.28	7.14	-51.00	2688.61	3023.61	47.60	47.60	-1370.84
	4					1995.11	2202.19	34.97	34.97	-1009.71
25	1	-2618.52	-5036.81	2.30	-55.84	1437.46	3683.44	25.10	25.10	-504.90
	2	-2184.36	-4960.89	4.08	-54.05	3459.25	5843.79	45.60	45.60	-1363.21
	3	-2262.33	-5051.97	3.49	-54.65	3437.06	5832.73	45.44	45.44	-1428.39
	4					2424.35	4175.85	32.35	32.35	-1064.37

EXAMPLE 2. ELASTIC ANALYSIS (Neglect shear and axial effects)

Loading Condition: Dead + Live.

Simplified analysis for gravity loads to columns proportional to the tributary areas supported; $K_y = 1.00$ (pin end, X-braced); K_x = values from nomograph for rigid end connections, Fig. 6-4.

Note: Column data are shown on lines between lines showing floor or roof levels supported; thus, the column under the story line "1" is the basement column supporting the first floor which is at Elevation 0.00′.

Interior Columns

Roof Input Data	Roof to Flr. Hgt. Feet	Slenderness Factors KX	KY	Exterior Trib. Area Sq. Ft.	Exterior Unit DL. Lbs. /S.F.	Loads Unit LL. Lbs./S.F.	Assume: All cols. W 14
	12.00	1.000	1.000	625.0	64.0	25.0	

Typical Floor Input Data

Flr. to Flr. Hgt. Feet	Slenderness Factors KX	KY	Add. Dead Kips	Conc. Lds. Live Kips	Floor Area Sq. Ft.	Area Unit DL. Lbs. /S.F.	No. 1 Unit LL. Lbs./S.F.
12.00	1.000	1.000	0.0	0.0	625.0	77.0	50.0

Non-Typical Floor Input Data	Flr. No.	Public Assembly Flag	Flr. to Flr. Hgt. Feet	Slenderness Factors KX	KY	Trib. Area Sq. Ft.	Flr. Unit Dead Load Lbs./S.F.	Flr. Unit Live Load Lbs./S.F.
	2.0	0	18.00	1.000	1.000	625.0	77.0	50.0
	1.0	0	12.00	1.000	1.000	625.0	115.0	100.0

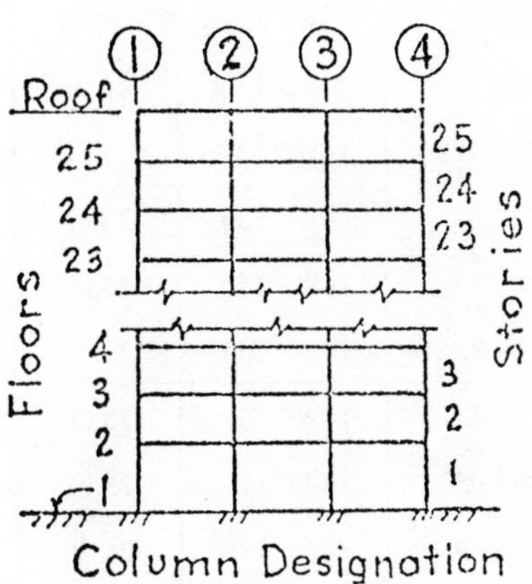

EXAMPLE 2. ELASTIC ANALYSIS (Continued)

Loading Condition: Dead + Live

Level	Actual D.L. Kips	Actual L.L. Kips	Reduced L.L. Kips	Act. D.L. & Red. L.L. Kips	Col. Size AISC Des. In.	Actual Stress Ksi
Roof	40.51	15.62	15.62	56.14	14WF 43.	4.43
25	89.15	46.87	28.12	117.28	14WF 43.	9.27
24	137.79	78.12	40.62	178.42	14WF 43.	14.10
23	186.43	109.37	53.12	239.56	14WF 43.	18.93
22	235.19	140.62	65.62	300.82	14WF 68.	15.04
21	284.14	171.87	78.12	362.26	14WF 68.	18.11
20	333.08	203.12	90.62	423.70	14WF 74.	19.44
19	382.09	234.37	103.12	485.21	14WF 74.	22.29
18	431.22	265.62	115.62	546.85	14WF 87.	21.36
17	480.36	296.87	128.12	608.48	14WF 87.	23.76
16	529.62	328.12	140.62	670.25	14WF 103.	22.12
15	578.98	359.37	153.12	732.11	14WF 103.	24.19
14	628.44	390.62	165.62	794.06	14WF 119.	22.69
13	677.99	421.87	178.12	856.12	14WF 119.	24.46
12	727.64	453.12	190.62	918.26	14WF 136.	22.96
11	777.40	484.37	203.12	980.52	14WF 136.	24.52
10	827.23	515.62	215.62	1042.85	14WF 150.	23.65
9	877.15	546.87	228.12	1105.27	14WF 150.	25.07
8	927.17	578.12	240.62	1167.80	14WF 167.	23.78
7	977.30	609.37	253.12	1230.42	14WF 167.	25.06
6	1027.43	640.62	265.62	1293.05	14WF 176.	25.01
5	1077.67	671.87	278.12	1355.79	14WF 176.	26.20
4	1128.00	703.12	290.62	1418.62	14WF 193.	25.02
3	1178.44	734.37	303.12	1481.56	14WF 193.	26.11
2	1230.67	765.62	315.62	1546.29	14WF 264.	19.93
1	1305.28	828.12	345.33	1650.61	14WF 264.	21.27

EXAMPLE 2. ELASTIC ANALYSIS

Exterior Columns

Roof Input Data	Roof to Flr. Hgt. Feet	Slenderness Factors KX	KY	Exterior Trib. Area Sq. Ft.	Exterior Unit DL. Lbs./S.F.	Loads Unit LL. Lbs./S.F.
	12.00	1.000	1.000	312.0	64.0	25.0

Typical Floor Input Data

Flr. to Flr. Hgt. Feet	Slenderness Factors KX	KY	Add. Dead Kips	Conc. Lds. Live Kips	Floor Area Sq. Ft.	Area Unit DL. Lbs./S.F.	No. 1 Unit LL. Lbs./S.F.
12.00	1.000	1.000	5.0	0.0	312.0	77.0	50.0

Additional dead load is for spandrel wall.

Non-Typical Floor Input Data	Flr. No.	Public Assembly Flag	Flr. to Flr. Hgt. Feet	Slenderness Factors KX	KY	Trib. Area Sq. Ft.	Flr. Unit Dead Load Lbs./S.F.	Flr. Unit Live Load Lbs./S.F.	Added Dead Load Kips
	2.0	0	18.00	1.000	1.000	312.0	77.0	50.0	5.0
	1.0	0	12.00	1.000	1.000	312.0	115.0	100.0	10.0

Level Roof	Actual D.L. Kips	Actual L.L. Kips	Reduced L.L. Kips	Act. D.L. & Red. L.L. Kips	Col. Size AISC Des. In.	Actual Stress Ksi
	25.48	7.80	7.80	33.28	14WF 43.	2.63
25	55.02	23.39	15.61	70.63	14WF 43.	5.58
24	84.56	38.99	20.27	104.84	14WF 43.	8.28
23	114.10	54.59	26.51	140.62	14WF 43.	11.11
22	143.64	70.19	32.75	176.40	14WF 43.	13.94
21	173.18	85.79	38.99	212.18	14WF 43.	16.77
20	202.72	101.39	45.23	247.96	14WF 53.	15.89
19	232.38	116.99	51.47	283.86	14WF 53.	18.20
18	262.13	132.59	57.71	319.85	14WF 61.	17.82
17	291.89	148.19	63.95	355.85	14WF 61.	19.83
16	321.65	163.79	70.19	391.85	14WF 68.	19.59
15	351.49	179.39	76.43	427.93	14WF 68.	21.39
14	381.40	194.99	82.67	464.08	14WF 74.	21.32
13	411.31	210.59	88.91	500.23	14WF 74.	22.98
12	441.34	226.19	95.15	536.50	14WF 84.	21.71
11	471.37	241.79	101.59	572.77	14WF 84.	23.18
10	501.41	257.39	107.63	609.05	14WF 95.	21.83
9	531.57	272.99	113.87	645.45	14WF 95.	23.10
8	561.73	288.59	120.11	681.85	14WF 95.	24.40
7	591.90	304.19	126.35	718.26	14WF 95.	25.70
6	622.16	319.79	132.59	754.76	14WF 111.	23.08
5	652.51	335.39	138.83	791.35	14WF 111.	24.23
4	682.87	350.99	145.07	827.95	14WF 119.	23.66
3	713.32	366.59	151.31	864.64	14WF 119.	24.71
2	744.79	382.19	157.55	902.35	14WF 136.	22.57
1	792.31	413.39	170.03	962.35	14WF 136.	24.07

EXAMPLE 2. ELASTIC ANALYSIS

Loading Condition: WIND (No gravity loads) · Computer program I.

Note: Moment, shear, and axial shortening effects included; ($P\Delta$) effects neglected; beam moments located at face of columns; column moments at face of beams.

INPUT. Dimensions, member sizes and properties same as for $W+D+L$ loading condition.

CONCENTRATED LATERAL LOADS

(FOR ALL FRAMES)

STORY LEVEL (FROM TOP)	LATERAL LOAD (KIPS)
1	3.00
2	6.00
3	6.00
4	6.00
5	6.00
6	6.00
7	6.00
8	6.00
9	6.00
10	6.00
11	6.00
12	6.00
13	6.00
14	6.00
15	6.00
16	6.00
17	6.00
18	6.00
19	6.00
20	6.00
21	6.00
22	6.00
23	6.00
24	6.00
25	7.50

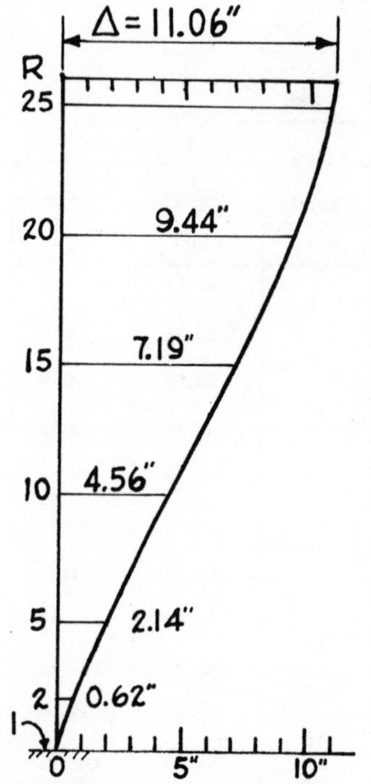

$\Delta = 11.06''$

R 25 · 20: 9.44″ · 15: 7.19″ · 10: 4.56″ · 5: 2.14″ · 2: 0.62″ · 1

0 5″ 10″

Drift, Δ, plus Side-sway due to V+PL/AE

LATERAL DISPLACEMENT OR *DRIFT* (SAME FOR ALL FRAMES)

STORY LEVEL (FROM TOP)	RELATIVE DISPLMT. (INCHES)	TOTAL DISPLMT. (INCHES)
1	0.1370240E 00	0.1105600E 02
2	0.1905664E 00	0.1091897E 02
3	0.2511367E 00	0.1072841E 02
4	0.3109839E 00	0.1047727E 02
5	0.3391546E 00	0.1016628E 02
6	0.3889555E 00	0.9827134E 01
7	0.4088219E 00	0.9438179E 01
8	0.4382556E 00	0.9029357E 01
9	0.4490062E 00	0.8591101E 01
10	0.4702088E 00	0.8142095E 01
11	0.4831088E 00	0.7671886E 01
12	0.5167445E 00	0.7188777E 01
13	0.5323518E 00	0.6672033E 01
14	0.5558779E 00	0.6139681E 01
15	0.5295475E 00	0.5583803E 01
16	0.4976070E 00	0.5054255E 01
17	0.4761187E 00	0.4556648E 01
18	0.4729013E 00	0.4080530E 01
19	0.4750511E 00	0.3607628E 01
20	0.4913361E 00	0.3132577E 01
21	0.4971574E 00	0.2641241E 01
22	0.5119888E 00	0.2144084E 01
23	0.5100036E 00	0.1632095E 01
24	0.4973206E 00	0.1122091E 01
25	0.6247708E 00	0.6247708E 00

(DRIFT AT TOP)/(TOTAL HT.) = 0.00300108

EXAMPLE 2. (Continued)

Loading Condition: WIND

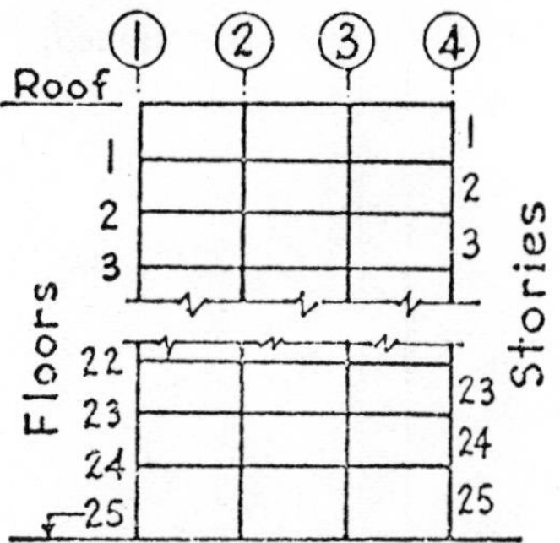

VALUES OF DISPLACEMENT COMPONENTS OF JOINTS (INCHES, RADIANS)

SIGN CONVENTION - HOR. COMP. (U) IS POSITIVE TO THE RIGHT
- VERT. COMP. (V) IS POSITIVE UPWARD
- ROTATION (W) IS POSITIVE CLOCKWISE

STORY LEVEL (FROM TOP)	COL. LINE (FROM LEFT)	HOR. COMP. * U *	VERT. COMP. * V *	ROTATION * W *
1	1	0.110560E 02	0.299906E 00	0.900363E-03
	2		0.537513E-01	0.676036E-03
	3		-0.537048E-01	0.676188E-03
	4		-0.299971E 00	0.900556E-03
2	1	0.109189E 02	0.299947E 00	0.104110E-02
	2		0.533075E-01	0.878385E-03
	3		-0.532608E-01	0.878476E-03
	4		-0.300013E 00	0.104119E-02
3	1	0.107284E 02	0.299814E 00	0.135476E-02
	2		0.523752E-01	0.108884E-02
	3		-0.523279E-01	0.108894E-02
	4		-0.299880E 00	0.135488E-02
4	1	0.104772E 02	0.299178E 00	0.167869E-02
	2		0.510086E-01	0.130313E-02
	3		-0.509603E-01	0.130324E-02
	4		-0.299245E 00	0.167880E-02
5	1	0.101662E 02	0.297702E 00	0.189693E-02
	2		0.492623E-01	0.158896E-02
	3		-0.492126E-01	0.158904E-02
	4		-0.297770E 00	0.189704E-02
6	1	0.982713E 01	0.295347E 00	0.209205E-02
	2		0.478859E-01	0.185004E-02
	3		-0.478353E-01	0.185011E-02
	4		-0.295417E 00	0.209216E-02
7	1	0.943817E 01	0.291695E 00	0.235024E-02
	2		0.462025E-01	0.205471E-02
	3		-0.461505E-01	0.205477E-02
	4		-0.291767E 00	0.235036E-02
8	1	0.902935E 01	0.287584E 00	0.244693E-02
	2		0.443924E-01	0.208624E-02
	3		-0.443390E-01	0.208634E-02
	4		-0.287657E 00	0.244696E-02
9	1	0.859110E 01	0.281977E 00	0.258740E-02
	2		0.423001E-01	0.225615E-02
	3		-0.422449E-01	0.225614E-02
	4		-0.282053E 00	0.258783E-02
10	1	0.814209E 01	0.275468E 00	0.257959E-02
	2		0.402659E-01	0.222041E-02
	3		-0.402091E-01	0.222074E-02
	4		-0.275546E 00	0.257825E-02
11	1	0.767188E 01	0.267215E 00	0.271047E-02
	2		0.379480E-01	0.237580E-02
	3		-0.378888E-01	0.237546E-02
	4		-0.267296E 00	0.271766E-02
12	1	0.718877E 01	0.258331E 00	0.286514E-02
	2		0.357447E-01	0.253845E-02
	3		-0.356855E-01	0.253714E-02
	4		-0.258411E 00	0.288267E-02

(Continued on page 76)

EXAMPLE 2. (Continued)

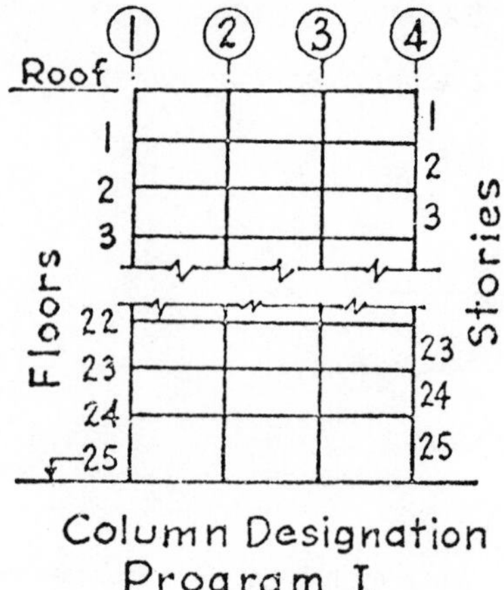

13	1	0.667203E 01	0.247599E 00	0.301916E-02
	2		0.332992E-01	0.271035E-02
	3		-0.332455E-01	0.271007E-02
	4		-0.247672E 00	0.302605E-02
14	1	0.613968E 01	0.235770E 00	0.315433E-02
	2		0.309736E-01	0.285900E-02
	3		-0.309265E-01	0.285925E-02
	4		-0.235834E 00	0.315292E-02
15	1	0.558380E 01	0.221910E 00	0.319167E-02
	2		0.284453E-01	0.293299E-02
	3		-0.284041E-01	0.293296E-02
	4		-0.221966E 00	0.319205E-02
16	1	0.505425E 01	0.207770E 00	0.282410E-02
	2		0.260570E-01	0.255962E-02
	3		-0.260209E-01	0.255969E-02
	4		-0.207819E 00	0.282411E-02
17	1	0.455664E 01	0.191848E 00	0.279448E-02
	2		0.235137E-01	0.253897E-02
	3		-0.234824E-01	0.253900E-02
	4		-0.191890E 00	0.279455E-02
18	1	0.408053E 01	0.176432E 00	0.264997E-02
	2		0.210282E-01	0.234197E-02
	3		-0.210012E-01	0.234201E-02
	4		-0.176468E 00	0.265002E-02
19	1	0.360762E 01	0.158982E 00	0.266499E-02
	2		0.183720E-01	0.239944E-02
	3		-0.183491E-01	0.239946E-02
	4		-0.159012E 00	0.266504E-02
20	1	0.313257E 01	0.140696E 00	0.271436E-02
	2		0.158401E-01	0.248194E-02
	3		-0.158208E-01	0.248197E-02
	4		-0.140722E 00	0.271440E-02
21	1	0.264124E 01	0.120315E 00	0.281128E-02
	2		0.131750E-01	0.254237E-02
	3		-0.131592E-01	0.254239E-02
	4		-0.120335E 00	0.281130E-02
22	1	0.214408E 01	0.100532E 00	0.289980E-02
	2		0.105562E-01	0.260298E-02
	3		-0.105436E-01	0.260300E-02
	4		-0.100548E 00	0.289981E-02
23	1	0.163209E 01	0.786888E-01	0.295534E-02
	2		0.788462E-02	0.266529E-02
	3		-0.787528E-02	0.266530E-02
	4		-0.787009E-01	0.295536E-02
24	1	0.112209E 01	0.569136E-01	0.293378E-02
	2		0.542576E-02	0.269337E-02
	3		-0.541932E-02	0.269338E-02
	4		-0.569220E-01	0.293379E-02
25	1	0.624770E 00	0.331245E-01	0.276772E-02
	2		0.295783E-02	0.238146E-02
	3		-0.295428E-02	0.238146E-02
	4		-0.331291E-01	0.276773E-02

EXAMPLE 2. (Continued)

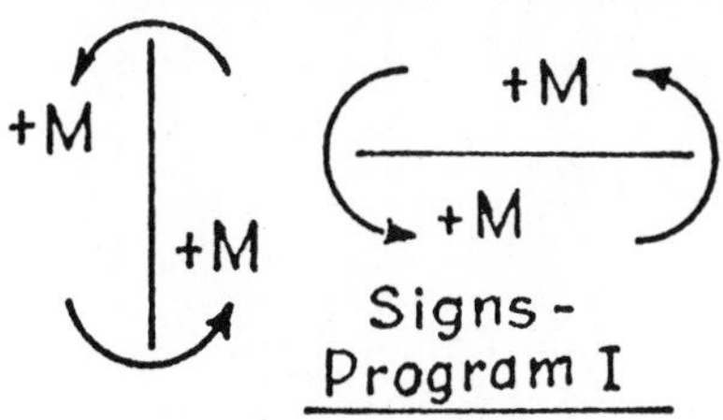

MEMBER-END FORCES (IN-KIPS, KIPS)

SIGN CONVENTION - MOMENTS - ARE POSITIVE WHEN COUNTERCLOCKWISE ON MEMBER ENDS
- SHEARS (A) IN COLUMNS ARE POSITIVE AT TOP WHEN DIRECTED TO RIGHT AND AT BOTTOM WHEN DIRECTED TO LEFT
(B) IN BEAMS ARE POSITIVE AT LEFT END WHEN DIRECTED UPWARD AND AT RIGHT WHEN DIRECTED DOWNWARD
- AXIAL FORCES (IN COLUMNS) ARE POSITIVE WHEN TENSILE

ORY EVEL	COL. LINE	BEAM MOMENTS		BEAM SHEARS		COLUMN MOMENTS		COLUMN SHEARS		COL. AXIAL FORCE
		LEFT	RIGHT	LEFT	RIGHT	TOP	BOTTOM	TOP	BOTTOM	
1	1	-1.73	36.62	0.12	0.12	2.82	-25.64	-0.18	-0.18	-0.12
	2	-171.02	-171.04	-1.19	-1.19	124.20	83.27	1.68	1.68	1.31
	3	36.73	-1.62	0.12	0.12	124.12	83.20	1.68	1.68	-1.31
	4					2.72	-25.72	-0.18	-0.18	0.12
2	1	-87.89	-60.07	-0.51	-0.51	106.36	42.91	1.21	1.21	0.39
	2	-281.48	-281.49	-1.96	-1.96	223.45	180.88	3.28	3.28	2.76
	3	-59.92	-87.74	-0.51	-0.51	223.39	180.82	3.28	3.28	-2.77
	4					106.30	42.85	1.21	1.21	-0.39
3	1	-236.28	-190.81	-1.49	-1.49	167.99	102.46	2.19	2.19	1.88
	2	-398.06	-398.08	-2.78	-2.78	347.75	304.40	5.30	5.30	4.06
	3	-190.67	-236.14	-1.49	-1.49	347.69	304.34	5.30	5.30	-4.06
	4					167.92	102.40	2.19	2.19	-1.88
4	1	-389.14	-324.93	-2.49	-2.49	243.30	199.15	3.59	3.59	4.38
	2	-518.27	-518.29	-3.62	-3.62	453.49	395.67	6.90	6.90	5.18
	3	-324.78	-389.00	-2.49	-2.49	453.43	395.62	6.90	6.90	-5.19
	4					243.23	199.08	3.59	3.59	-4.38
5	1	-518.49	-465.83	3.44	-3.44	264.34	219.71	3.93	3.93	7.82
	2	-678.32	-678.34	-4.74	-4.74	632.86	543.72	9.56	9.56	6.48
	3	-465.67	-518.33	-3.44	-3.44	632.78	543.65	9.56	9.56	-6.49
	4					264.26	219.64	3.93	3.93	-7.82
6	1	-637.34	-595.96	-4.31	-4.31	351.51	292.46	5.23	5.23	12.14
	2	-823.74	-823.75	-5.76	-5.76	727.76	657.89	11.26	11.26	7.93
	3	-595.79	-637.17	-4.31	-4.31	727.69	657.82	11.26	11.26	-7.94
	4					351.44	292.39	5.23	5.23	-12.13
7	1	-769.97	-719.43	-5.20	-5.20	394.18	364.95	6.17	6.17	17.34
	2	-939.90	-939.91	-6.57	-6.57	825.68	813.83	13.32	13.32	9.30
	3	-719.26	-769.80	-5.20	-5.20	825.59	813.73	13.32	13.32	-9.31
	4					394.10	364.90	6.17	6.17	-17.34
8	1	-937.78	-866.74	-6.31	-6.31	474.76	432.30	7.37	7.37	23.66
	2	-1109.65	-1109.67	-7.76	-7.76	962.27	898.41	15.12	15.12	10.75
	3	-866.53	-937.55	-6.30	-6.30	962.19	898.38	15.12	15.12	-10.76
	4					474.62	432.04	7.37	7.37	-23.65
9	1	-1038.32	-973.08	-7.03	-7.03	493.36	496.03	8.04	8.04	30.69
	2	-1223.58	-1223.57	-8.55	-8.55	1065.28	1081.58	17.45	17.45	12.27
	3	-972.90	-1038.22	-7.03	-7.03	1065.16	1081.30	17.45	17.45	-12.28
	4					493.48	496.75	8.05	8.05	-30.68
10	1	-1217.28	-1134.23	-8.22	-8.22	596.35	551.66	9.33	9.33	38.91
	2	-1420.49	-1420.57	-9.93	-9.93	1215.51	1144.65	19.18	19.18	13.98
	3	-1133.73	-1216.40	-8.21	-8.21	1215.36	1144.81	19.18	19.18	-14.00
	4					595.10	547.50	9.28	9.28	-38.90
11	1	-1332.99	-1255.61	-9.05	-9.05	642.16	582.29	9.95	9.95	47.96
	2	-1544.64	-1544.56	-10.80	-10.80	1367.32	1277.58	21.50	21.50	15.74
	3	-1256.89	-1336.01	-9.06	-9.06	1368.26	1279.06	21.52	21.52	-15.74
	4					649.24	583.03	10.01	10.01	-47.96
12	1	-1463.72	-1388.18	-9.97	-9.97	727.58	667.95	11.34	11.34	57.94
	2	-1673.44	-1673.14	-11.70	-11.70	1467.53	1372.69	23.09	23.09	17.47
	3	-1391.57	-1471.46	-10.01	-10.01	1468.99	1373.59	23.11	23.11	-17.43
	4					733.07	675.54	11.45	11.45	-57.98
13	1	-1600.33	-1528.93	-10.94	-10.94	764.38	705.23	11.94	11.94	68.88
	2	-1810.63	-1810.56	-12.66	-12.66	1621.07	1525.05	25.57	25.57	19.19
	3	-1530.19	-1603.25	-10.95	-10.95	1621.28	1524.90	25.57	25.57	-19.13
	4					759.27	703.76	11.89	11.89	-68.93

(Continued on page 78)

EXAMPLE 2. (Continued)

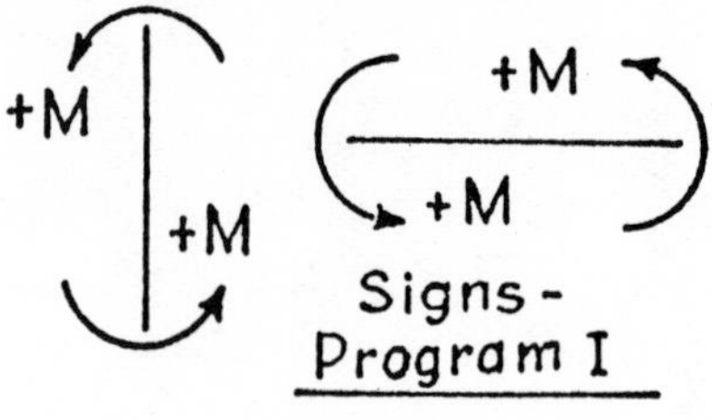

STORY LEVEL	COL. LINE	BEAM MOMENTS LEFT	BEAM MOMENTS RIGHT	BEAM SHEARS LEFT	BEAM SHEARS RIGHT	COLUMN MOMENTS TOP	COLUMN MOMENTS BOTTOM	COLUMN SHEARS TOP	COLUMN SHEARS BOTTOM	AXIAL FORCE
*****	****	*******	*******	*******	*******	*******	*******	*******	*******	********
14	1	-1725.08	-1656.79	-11.82	-11.82	835.86	819.52	13.45	13.45	80.71
	2	-1930.18	-1930.24	-13.49	-13.49	1686.74	1638.94	27.03	27.03	20.86
	3	-1656.29	-1724.20	-11.82	-11.82	1686.44	1638.83	27.03	27.03	-20.81
	4					836.87	819.75	13.46	13.46	-80.76
15	1	-1788.83	-1729.02	-12.30	-12.30	767.40	951.13	13.97	13.97	93.01
	2	-1996.20	-1996.20	-13.96	-13.96	1676.15	1956.09	29.53	29.53	22.52
	3	-1728.87	-1788.77	-12.30	-12.30	1676.14	1956.01	29.52	29.52	-22.47
	4					767.05	950.96	13.96	13.96	-93.06
16	1	-2091.76	-2009.15	-14.34	-14.34	893.19	908.37	15.01	15.01	107.35
	2	-2345.30	-2345.32	-16.40	-16.40	1881.34	1897.21	31.48	31.48	24.58
	3	-2008.92	-2091.51	-14.33	-14.33	1881.23	1897.13	31.48	31.48	-24.53
	4					893.16	908.30	15.01	15.01	-107.40
17	1	-2109.50	-2029.69	-14.47	-14.47	927.85	1016.56	16.20	16.20	121.82
	2	-2341.64	-2341.65	-16.37	-16.37	1912.65	2083.08	33.29	33.29	26.48
	3	-2029.49	-2109.31	-14.47	-14.47	1912.57	2083.00	33.29	33.29	-26.44
	4					927.75	1016.47	16.20	16.20	-121.87
18	1	-2354.46	-2240.76	-16.06	-16.06	1047.38	1038.16	17.37	17.37	137.89
	2	-2557.79	-2557.80	-17.88	-17.88	2132.14	2082.43	35.12	35.12	28.30
	3	-2240.56	-2354.26	-16.06	-16.06	2132.06	2082.36	35.12	35.12	-28.26
	4					1047.31	1038.09	17.37	17.37	-137.94
19	1	-2446.01	-2347.99	-16.76	-16.76	1099.92	1067.23	18.05	18.05	154.66
	2	-2645.09	-2645.10	-18.49	-18.49	2286.75	2206.19	37.44	37.44	30.04
	3	-2347.82	-2445.85	-16.76	-16.76	2286.68	2206.13	37.44	37.44	-30.00
	4					1099.85	1067.17	18.05	18.05	-154.70
20	1	-2577.50	-2491.71	-17.72	-17.72	1186.71	1122.54	19.24	19.24	172.39
	2	-2760.53	-2760.54	-19.30	-19.30	2384.91	2325.91	39.25	39.25	31.62
	3	-2491.57	-2577.37	-17.72	-17.72	2384.86	2325.86	39.25	39.25	-31.58
	4					1186.66	1122.49	19.24	19.24	-172.43
21	1	-2744.23	-2644.97	-18.84	-18.84	1274.57	1206.54	20.67	20.67	191.23
	2	-2851.36	-2851.37	-19.94	-19.94	2480.96	2417.98	40.82	40.82	32.71
	3	-2644.85	-2744.12	-18.84	-18.84	2480.92	2417.94	40.82	40.82	-32.67
	4					1274.52	1206.50	20.67	20.67	-191.27
22	1	-2902.47	-2792.91	-19.91	-19.91	1326.25	1283.56	21.74	21.74	211.15
	2	-2942.03	-2942.04	-20.57	-20.57	2597.49	2532.74	42.75	42.75	33.37
	3	-2792.81	-2902.38	-19.91	-19.91	2597.46	2532.71	42.75	42.75	-33.33
	4					1326.21	1283.53	21.74	21.74	-211.19
23	1	-3043.93	-2936.86	-20.91	-20.91	1369.93	1388.58	22.98	22.98	232.06
	2	-3035.10	-3035.10	-21.22	-21.22	2687.04	2654.47	44.51	44.51	33.69
	3	-2936.79	-3043.86	-20.91	-20.91	2687.01	2654.45	44.51	44.51	-33.65
	4					1369.90	1388.56	22.98	22.98	-232.10
24	1	-3113.35	-3024.61	-21.46	-21.46	1320.66	1464.33	23.20	23.20	253.53
	2	-3086.74	-3086.74	-21.58	-21.58	2656.60	3018.41	47.29	47.29	33.81
	3	-3024.56	-3113.31	-21.46	-21.46	2656.58	3018.39	47.29	47.29	-33.77
	4					1320.65	1464.32	23.20	23.20	-253.57
25	1	-3835.27	-3649.60	-26.17	-26.17	1930.91	3929.65	28.72	28.72	279.70
	2	-3572.60	-3572.61	-24.98	-24.98	3448.16	5838.26	45.52	45.52	32.62
	3	-3649.56	-3835.24	-26.17	-26.17	3448.15	5838.26	45.52	45.52	-32.58
	4					1930.90	3929.64	28.72	28.72	-279.74

EXAMPLE 2. ELASTIC ANALYSIS

Loading Condition: DEAD + LIVE · Computer program I.

Note: Moment, shear, and axial shortening effects included; ($P\Delta$) effects neglected; beam moments located at face of columns; column moments at face of beams.

INPUT. Dimensions, member sizes and properties same as for $W+D+L$ loading condition.

Story levels numbered 1 to 25 from the top.

GRAVITY LOADS ONLY

NO EXTERNAL FORCES APPLIED AT TOP OF FRAME

NO CONC. VERTICAL LOADS ON BEAMS

VERTICAL LOADS ON FRAME

STORY LEVEL	COL. LINE	UNIFORM BM LOAD (KIPS/FT)
*****	****	***************
1	1	2.44
	2	2.44
	3	2.44
2	1	2.44
	2	2.44
	3	2.44
3	1	2.44
	2	2.44
	3	2.44
4	1	2.44
	2	2.44
	3	2.44
5	1	2.44
	2	2.44
	3	2.44
6	1	2.44
	2	2.44
	3	2.44
7	1	2.44
	2	2.44
	3	2.44
8	1	2.44
	2	2.44
	3	2.44
9	1	2.44
	2	2.44
	3	2.44
10	1	2.44
	2	2.44
	3	2.44
11	1	2.44
	2	2.44
	3	2.44
12	1	2.44
	2	2.44
	3	2.44
13	1	2.44
	2	2.44
	3	2.44

VERTICAL LOADS ON FRAME

STORY LEVEL	COL. LINE	UNIFORM BM LOAD (KIPS/FT)
*****	****	***************
14	1	2.44
	2	2.44
	3	2.44
15	1	2.44
	2	2.44
	3	2.44
16	1	2.44
	2	2.44
	3	2.44
17	1	2.44
	2	2.44
	3	2.44
18	1	2.44
	2	2.44
	3	2.44
19	1	2.44
	2	2.44
	3	2.44
20	1	2.44
	2	2.44
	3	2.44
21	1	2.44
	2	2.44
	3	2.44
22	1	2.44
	2	2.44
	3	2.44
23	1	2.44
	2	2.44
	3	2.44
24	1	2.44
	2	2.44
	3	2.44
25	1	2.44
	2	2.44
	3	2.44

NO CONCENTRATED LATERAL LOADS

(Continued on page 80)

EXAMPLE 2. (Continued)

LATERAL DISPLACEMENT OR *DRIFT* (SAME FOR ALL FRAMES)

STORY LEVEL (FROM TOP)	RELATIVE DISPLMT. (INCHES)	TOTAL DISPLMT. (INCHES)
1	0.5311740E-05	-0.9892366E-03
2	0.4863214E-05	-0.9945484E-03
3	0.5022405E-05	-0.9994116E-03
4	0.5104138E-05	-0.1004434E-02
5	0.4522245E-05	-0.1009538E-02
6	0.4674570E-05	-0.1014060E-02
7	0.4966477E-05	-0.1018734E-02
8	0.6550323E-05	-0.1023701E-02
9	0.6274842E-05	-0.1030251E-02
10	0.4630558E-04	-0.1036526E-02
11	-0.5817566E-03	-0.1082632E-02
12	-0.5472368E-03	-0.5010755E-03
13	0.2376930E-04	0.4616124E-04
14	0.4173589E-05	0.2239193E-04
15	0.3951488E-05	0.1821834E-04
16	0.2896316E-05	0.1426685E-04
17	0.2587785E-05	0.1137054E-04
18	0.2180774E-05	0.8782757E-05
19	0.1776525E-05	0.6601982E-05
20	0.1445658E-05	0.4825456E-05
21	0.1182286E-05	0.3379798E-05
22	0.9031764E-06	0.2197511E-05
23	0.6454888E-06	0.1294335E-05
24	0.4132606E-06	0.6488466E-06
25	0.2355859E-06	0.2355859E-06

(DRIFT AT TOP)/(TOTAL HT.) = -0.00000026

VALUES OF DISPLACEMENT COMPONENTS OF JOINTS (INCHES, RADIANS)

---- SIGN CONVENTION
- HOR. COMP. (U) IS POSITIVE TO THE RIGHT
- VERT. COMP. (V) IS POSITIVE UPWARD
- ROTATION (W) IS POSITIVE CLOCKWISE

STORY LEVEL (FROM TOP)	COL. LINE (FROM LEFT)	HOR. COMP. * U *	VERT. COMP. * V *	ROTATION * W *
1	1	-0.989236E-03	-0.154207E 01	0.272102E-02
	2		-0.206678E 01	0.357271E-03
	3		-0.206673E	-0.357294E-03
	4		-0.154209E 01	-0.272084E-02
2	1	-0.994548E-03	-0.153202E 01	0.114365E-02
	2		-0.204747E 01	0.400356E-03
	3		-0.204743E 01	-0.400336E-03
	4		-0.153204E 01	-0.114354E-02
3	1	-0.999411E-03	-0.151098E 01	0.145326E-02
	2		-0.200979E 01	0.327799E-03
	3		-0.200975E 01	-0.327788E-03
	4		-0.151100E 01	-0.145313E-02
4	1	-0.100443E-02	-0.147918E 01	0.137507E-02
	2		-0.195352E 01	0.333625E-03
	3		-0.195347E 01	-0.333614E-03
	4		-0.147920E 01	-0.137494E-02
5	1	-0.100953E-02	-0.143667E 01	0.127524E-02
	2		-0.187859E 01	0.248221E-03
	3		-0.187854E 01	-0.248205E-03
	4		-0.143669E 01	-0.127511E-02

(Continued on page 81)

EXAMPLE 2. (Continued)

6	1	-0.101406E-02	-0.138913E 01	0.123314E-02
	2		-0.181963E 01	0.205052E-03
	3		-0.181957E 01	-0.205030E-03
	4		-0.138915E 01	-0.123302E-02
7	1	-0.101873E-02	-0.133203E 01	0.107700E-02
	2		-0.174890E 01	0.196026E-03
	3		-0.174885E 01	-0.196009E-03
	4		-0.133206E 01	-0.107686E-02
8	1	-0.102370E-02	-0.127950E 01	0.972187E-03
	2		-0.167325E 01	0.210686E-03
	3		-0.167320E 01	-0.210621E-03
	4		-0.127953E 01	-0.972114E-03
9	1	-0.103025E-02	-0.121934E 01	0.927899E-03
	2		-0.158690E 01	0.170743E-03
	3		-0.158684E 01	-0.170886E-03
	4		-0.121936E 01	-0.927468E-03
10	1	-0.103652E-02	-0.115870E 01	0.920037E-03
	2		-0.150422E 01	0.182950E-03
	3		-0.150416E 01	-0.182032E-03
	4		-0.115873E 01	-0.921087E-03
11	1	-0.106283E-02	-0.109126E 01	0.828789E-03
	2		-0.141242E 01	0.151212E-03
	3		-0.141235E 01	-0.154466E-03
	4		-0.109129E 01	-0.821677E-03
12	1	-0.501075E-03	-0.102642E 01	0.760332E-03
	2		-0.132712E 01	0.131722E-03
	3		-0.132705E 01	-0.139519E-03
	4		-0.102645E 01	-0.745674E-03
13	1	0.461612E-04	-0.955679E 00	0.713087E-03
	2		-0.123408E 01	0.113932E-03
	3		-0.123402E 01	-0.116573E-03
	4		-0.955706E 00	-0.707810E-03
14	1	0.223919E-04	-0.884646E 00	0.673349E-03
	2		-0.114681E 01	0.103924E-03
	3		-0.114676E 01	-0.103200E-03
	4		-0.884670E 00	-0.674276E-03
15	1	0.182183E-04	-0.808197E 00	0.623767E-03
	2		-0.105279E 01	0.832855E-04
	3		-0.105274E 01	-0.833910E-04
	4		-0.808218E 00	-0.623444E-03
16	1	0.142668E-04	-0.735740E 00	0.596872E-03
	2		-0.964609E 00	0.102275E-03
	3		-0.964569E 00	-0.102230E-03
	4		-0.735757E	-0.596832E-03
17	1	0.113705E-04	-0.660316E 00	0.523676E-03
	2		-0.872854E 00	0.830598E-04
	3		-0.872821E 00	-0.830487E-04
	4		-0.660331E 00	-0.523598E-03
18	1	0.878275E-05	-0.591932E 00	0.486015E-03
	2		-0.784436E 00	0.895989E-04
	3		-0.784407E 00	-0.895862E-04
	4		-0.591945E 00	-0.485953E-03
19	1	0.660198E-05	-0.519506E 00	0.457004E-03
	2		-0.690832E 00	0.700985E-04
	3		-0.690808E 00	-0.700879E-04
	4		-0.519517E 00	-0.456953E-03
20	1	0.482545E-05	-0.448106E 00	0.435911E-03
	2		-0.602070E 00	0.585719E-04
	3		-0.602050E 00	-0.585626E-04
	4		-0.448115E 00	-0.435869E-03
21	1	0.337979E-05	-0.373014E 00	0.391831E-03
	2		-0.508590E 00	0.509649E-04
	3		-0.508574E 00	-0.509572E-04
	4		-0.373021E 00	-0.391798E-03
22	1	0.219751E-05	-0.304107E 00	0.347055E-03
	2		-0.415308E 00	0.392157E-04
	3		-0.415295E 00	-0.392093E-04
	4		-0.304112E 00	-0.347030E-03
23	1	0.129433E-05	-0.232043E 00	0.322817E-03
	2		-0.317487E 00	0.245894E-04
	3		-0.317478E 00	-0.245845E-04
	4		-0.232047E 00	-0.322799E-03
24	1	0.648846E-06	-0.163866E 00	0.269088E-03
	2		-0.224115E 00	0.175720E-04
	3		-0.224109E 00	-0.175681E-04
	4		-0.163869E 00	-0.269076E-03
25	1	0.235585E-06	-0.929179E-01	0.342373E-03
	2		-0.126534E 00	-0.554431E-05
	3		-0.126530E 00	0.554538E-05
	4		-0.929195E-01	-0.342364E-03

EXAMPLE 2. (Continued)

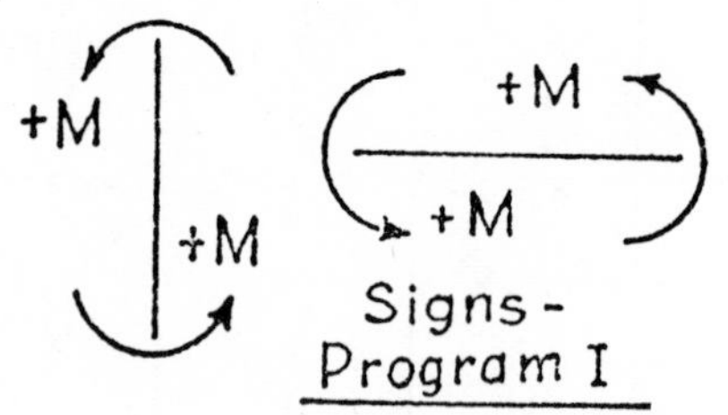

MEMBER-END FORCES (IN-KIPS, KIPS)

SIGN CONVENTION - MOMENTS - ARE POSITIVE WHEN COUNTERCLOCKWISE ON MEMBER ENDS
- SHEARS (A) IN COLUMNS ARE POSITIVE AT TOP WHEN DIRECTED TO RIGHT AND AT BOTTOM WHEN DIRECTED TO LEFT
(B) IN BEAMS ARE POSITIVE AT LEFT END WHEN DIRECTED UPWARD AND AT RIGHT WHEN DIRECTED DOWNWARD
- AXIAL FORCES (IN COLUMNS) ARE POSITIVE WHEN TENSILE

STORY LEVEL	COL. LINE	BEAM MOMENTS		BEAM SHEARS		COLUMN MOMENTS		COLUMN SHEARS		COL. AXIAL FORCE
		LEFT	RIGHT	LEFT	RIGHT	TOP	BOTTOM	TOP	BOTTOM	
*****	****	********	********	********	********	********	********	********	********	********
1	1	1296.57	-1070.46	29.86	-28.28	-1309.31	-990.22	-18.69	-18.69	-29.86
	2	1324.43	-1324.58	29.07	-29.07	-221.02	-229.73	-3.66	-3.66	-57.35
	3	1070.55	-1296.51	28.28	-29.86	221.07	229.77	3.66	3.66	-57.35
	4					1309.26	990.18	18.69	18.69	-29.86
2	1	1831.27	-812.83	32.63	-25.51	-741.28	-803.91	-12.56	-12.56	-62.49
	2	1317.06	-1317.23	29.07	-29.07	-223.95	-209.27	-3.52	-3.52	-111.93
	3	812.92	-1831.20	25.51	-32.63	223.98	209.30	3.52	3.52	-111.94
	4					741.25	803.87	12.56	12.56	-62.49
3	1	1704.97	-873.78	31.97	-26.16	-849.35	-833.53	-13.68	-13.68	-94.47
	2	1329.46	-1329.63	29.07	-29.07	-196.17	-197.35	-3.19	-3.19	-167.17
	3	873.88	-1704.90	26.16	-31.97	196.20	197.38	3.19	3.19	-167.18
	4					849.31	833.50	13.68	13.68	-94.47
4	1	1687.73	-905.39	31.80	-26.33	-798.57	-778.38	-12.82	-12.82	-126.28
	2	1328.46	-1328.64	29.07	-29.07	-181.72	-164.44	-2.81	-2.81	-222.58
	3	905.49	-1687.65	26.33	-31.80	181.75	164.48	2.81	2.81	-222.59
	4					798.54	778.34	12.82	12.82	-126.28
5	1	1680.65	-914.94	31.75	-26.39	-846.24	-836.62	-13.68	-13.68	-158.03
	2	1343.06	-1343.25	29.07	-29.07	-217.06	-202.32	-3.40	-3.40	-278.05
	3	915.04	-1680.57	26.39	-31.74	217.10	202.36	3.41	3.41	-278.05
	4					846.20	836.58	13.68	13.68	-158.03
6	1	1683.00	-912.41	31.76	-26.37	-792.78	-757.07	-12.60	-12.60	-189.80
	2	1350.44	-1350.64	29.07	-29.07	-187.08	-184.00	-3.01	-3.01	-333.50
	3	912.51	-1682.91	26.37	-31.76	187.12	184.04	3.01	3.01	-333.51
	4					792.74	757.03	12.59	12.59	-189.79
7	1	1715.56	-905.00	31.90	-26.23	-898.78	-867.10	-14.35	-14.35	-221.70
	2	1351.98	-1352.18	29.07	-29.07	-214.10	-219.61	-3.52	-3.52	-388.81
	3	905.11	-1715.47	26.23	-31.90	214.14	219.64	3.52	3.52	-388.82
	4					898.74	867.08	14.35	14.35	-221.70
8	1	1757.61	-863.61	32.19	-25.94	-825.38	-811.99	-13.31	-13.31	-253.90
	2	1343.97	-1344.24	29.07	-29.07	-210.87	-195.86	-3.30	-3.30	-443.82
	3	863.74	-1757.48	25.94	-32.19	210.97	196.04	3.30	3.30	-443.84
	4					825.29	811.80	13.30	13.30	-253.90
9	1	1730.03	-892.04	32.00	-26.14	-856.31	-853.63	-13.90	-13.90	-285.91
	2	1351.88	-1352.03	29.07	-29.07	-212.92	-218.49	-3.50	-3.50	-499.04
	3	892.19	-1730.00	26.14	-32.00	212.80	217.88	3.50	3.50	-499.06
	4					856.43	854.25	13.90	13.90	-285.90
10	1	1737.24	-863.52	32.12	-26.01	-824.43	-793.28	-13.15	-13.15	-318.03
	2	1342.91	-1343.89	29.06	-29.07	-210.69	-196.22	-3.30	-3.30	-554.12
	3	863.56	-1736.74	26.01	-32.12	211.95	199.38	3.34	3.34	-554.15
	4					823.62	789.68	13.11	13.11	-318.03

(Continued on page 83)

EXAMPLE 2. (Continued)

(For units and sign convention, see page 82.)

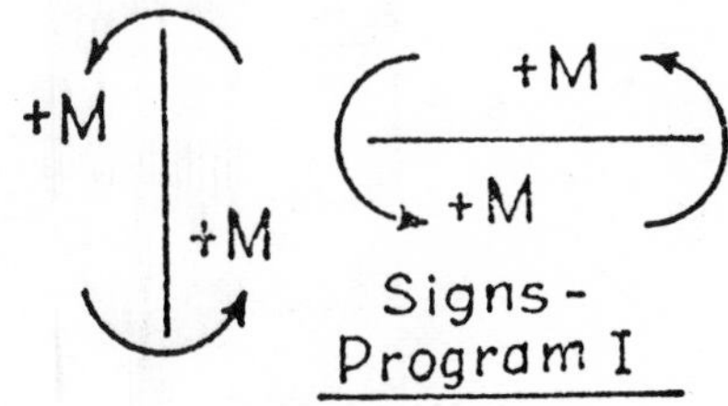

STORY LEVEL	COL. LINE	BEAM MOMENTS LEFT	BEAM MOMENTS RIGHT	BEAM SHEARS LEFT	BEAM SHEARS RIGHT	COLUMN MOMENTS TOP	COLUMN MOMENTS BOTTOM	COLUMN SHEARS TOP	COLUMN SHEARS BOTTOM	COL. AXIAL FORCE
*****	****	********	********	*******	********	********	********	*******	*******	********
11	1	1729.77	-884.75	32.02	-26.11	-875.38	-848.88	-14.01	-14.01	-350.06
	2	1351.28	-1349.23	29.07	-29.06	-219.70	-208.95	-3.48	-3.48	-609.32
	3	884.78	-1732.13	26.10	-32.03	214.69	206.44	3.42	3.42	-609.33
	4					881.14	850.65	14.07	14.07	-350.06
12	1	1717.80	-908.04	31.90	-26.24	-808.57	-790.28	-12.99	-12.99	-381.97
	2	1356.91	-1351.56	29.09	-29.05	-191.43	-181.61	-3.03	-3.03	-664.65
	3	908.38	-1722.66	26.22	-31.92	189.34	176.68	2.97	2.97	-664.60
	4					810.53	795.33	13.05	13.05	-381.98
13	1	1690.79	-941.86	31.69	-26.45	-843.36	-825.97	-13.57	-13.57	-413.66
	2	1359.76	-1358.13	29.07	-29.06	-190.79	-184.32	-3.04	-3.04	-720.19
	3	942.03	-1692.46	26.44	-31.69	194.10	185.46	3.08	3.08	-720.12
	4					839.78	825.11	13.53	13.53	-413.68
14	1	1672.93	-966.60	31.54	-26.60	-791.90	-770.20	-12.70	-12.70	-445.20
	2	1361.26	-1362.04	29.06	-29.07	-168.03	-154.70	-2.62	-2.62	-775.86
	3	966.66	-1672.48	26.60	-31.54	167.33	154.53	2.61	2.61	-775.80
	4					792.60	770.36	12.70	12.70	-445.22
15	1	1659.20	-987.02	31.42	-26.72	-834.31	-820.87	-13.45	-13.45	-476.62
	2	1366.25	-1366.40	29.07	-29.07	-181.26	-195.50	-3.06	-3.06	-831.65
	3	987.15	-1659.17	26.72	-31.42	181.49	195.62	3.06	3.06	-831.59
	4					834.07	820.77	13.45	13.45	-476.64
16	1	1714.72	-901.98	31.91	-26.23	-799.60	-762.09	-13.01	-13.01	-508.54
	2	1353.51	-1353.81	29.07	-29.07	-200.51	-185.74	-3.21	-3.21	-886.95
	3	902.13	-1714.57	26.23	-31.91	200.52	185.78	3.21	3.21	-886.90
	4					799.58	762.06	13.01	13.01	-508.56
17	1	1715.05	-918.51	31.85	-26.28	-851.75	-828.63	-14.00	-14.00	-540.40
	2	1359.54	-1359.77	29.07	-29.07	-196.31	-201.97	-3.31	-3.31	-942.31
	3	918.64	-1714.95	26.28	-31.85	196.37	202.02	3.31	3.31	-942.26
	4					851.69	828.58	14.00	14.00	-540.41
18	1	1723.60	-901.25	31.94	-26.19	-793.62	-775.81	-13.07	-13.07	-572.35
	2	1352.40	-1352.64	29.07	-29.07	-192.63	-175.76	-3.06	-3.06	-997.58
	3	901.38	-1723.49	26.19	-31.94	192.67	175.80	3.07	3.07	-997.53
	4					793.57	775.77	13.07	13.07	-572.36
19	1	1671.52	-956.85	31.57	-26.57	-800.96	-787.00	-13.23	-13.23	-603.92
	2	1359.62	-1359.82	29.07	-29.07	-173.98	-162.73	-2.80	-2.80	-1053.23
	3	956.96	-1671.43	26.57	-31.57	174.02	162.77	2.80	2.80	-1053.18
	4					800.92	786.96	13.23	13.23	-603.93
20	1	1624.99	-1006.90	31.23	-26.91	-750.63	-721.44	-12.26	-12.26	-635.15
	2	1363.89	-1364.06	29.07	-29.07	-147.04	-139.61	-2.38	-2.38	-1109.21
	3	1006.99	-1624.92	26.91	-31.23	147.07	139.64	2.38	2.38	-1109.16
	4					750.59	721.41	12.26	12.26	-635.17
21	1	1590.54	-1054.82	30.94	-27.19	-784.96	-750.55	-12.79	-12.79	-666.10
	2	1366.71	-1366.85	29.07	-29.07	-131.61	-119.41	-2.09	-2.09	-1165.48
	3	1054.90	-1590.47	27.19	-30.94	131.64	119.48	2.09	2.09	-1165.44
	4					784.93	750.53	12.79	12.79	-666.11
22	1	1535.07	-1122.49	30.51	-27.62	-705.36	-686.73	-11.60	-11.60	-696.61
	2	1371.07	-1371.17	29.07	-29.07	-96.40	-81.20	-1.48	-1.48	-1222.18
	3	1122.54	-1535.02	27.63	-30.51	96.42	81.22	1.48	1.48	-1222.14
	4					705.34	686.72	11.60	11.60	-696.63
23	1	1459.67	-1201.43	29.97	-28.16	-706.83	-660.35	-11.39	-11.39	-726.59
	2	1376.48	-1376.56	29.07	-29.07	-69.34	-61.20	-1.08	-1.08	-1279.43
	3	1201.48	-1459.63	28.16	-29.97	69.35	61.21	1.08	1.08	-1279.38
	4					706.81	660.33	11.39	11.39	-726.60
24	1	1406.00	-1272.35	29.53	-28.60	-674.47	-737.88	-11.76	-11.76	-756.13
	2	1379.08	-1379.14	29.07	-29.07	-32.02	-5.20	-0.31	-0.31	-1337.10
	3	1272.37	-1405.97	28.60	-29.53	32.03	5.22	0.31	0.31	-1337.06
	4					674.46	737.87	11.76	11.76	-756.14
25	1	1216.74	-1387.21	28.47	-29.66	-493.45	-246.20	-3.62	-3.62	-784.60
	2	1388.24	-1388.28	29.07	-29.07	11.09	5.52	0.08	0.08	-1395.84
	3	1387.23	-1216.72	29.66	-28.47	-11.08	-5.52	-0.08	-0.08	-1395.80
	4					493.44	246.20	3.62	3.62	-784.62

EXAMPLE 2. (Continued)

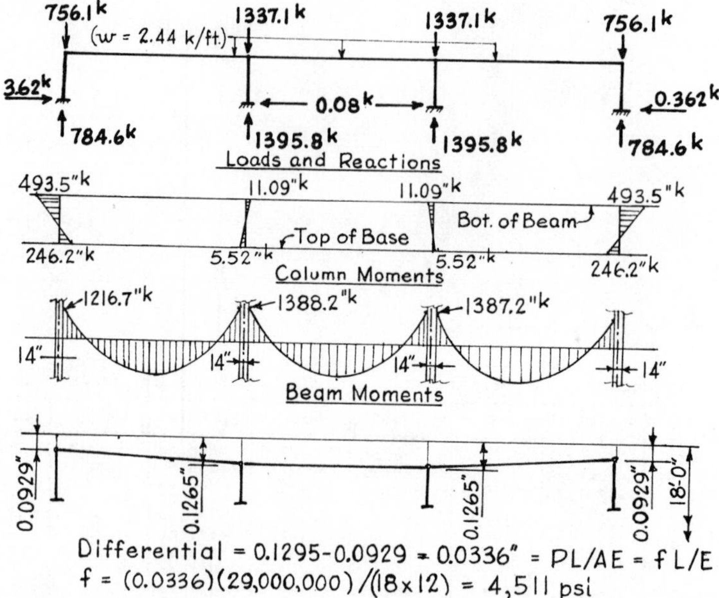

Displacements and Reactions Due to Dead and Live Loads, Including Effects of *V* and *PL/AE* and Neglecting Effects of *P*Δ—Program I—25 Story Building.

3

BEAMS AND GIRDERS AS FLEXURAL MEMBERS

GENERAL

The practical aspects of the design of structural steel beams and girders made from rolled shapes and built-up-members (except tapered girders) subjected only to shear and moment are developed here. For combined stresses from bending and axial force, see Chapter 6, Columns.

Beams and girders as flexural members are classified into three different types of construction, depending on the manner of their support and the rigidity of the connections to their supporting members (1.2).

Type 1 Construction. Rigid-frame (continuous frame) beams and girders with connections to columns sufficiently rigid to hold virtually unchanged the original angles between intersecting members are classified as Type 1 Construction (1.2).

Type 2 Construction. Simple-span beams and girders (assumed unrestrained or free ended) that are attached to their supporting members with flexible connections designed to transmit shear only and to allow the end of the beam or girder to rotate under gravity loads are classified as Type 2 Construction (1.2). Wind moment connections may be used with Type 2 Construction provided that the following conditions are satisfied:

1. The connections and connected members have the capacity to resist the wind moments.
2. The beams and girders are adequate to carry the full gravity load as simple span members.
3. The connections have adequate inelastic rotation capacity to avoid overstress of the fasteners or welds under combined gravity and wind loading.

Type 3 Construction. Beams and girders that are connected to their supports with "semi-rigid-connections" (partially restrained) that possess a dependable and known moment capacity intermediate between the rigidity of Type 1 Construction and flexibility of Type 2 Construction are classified as Type 3 Construction.

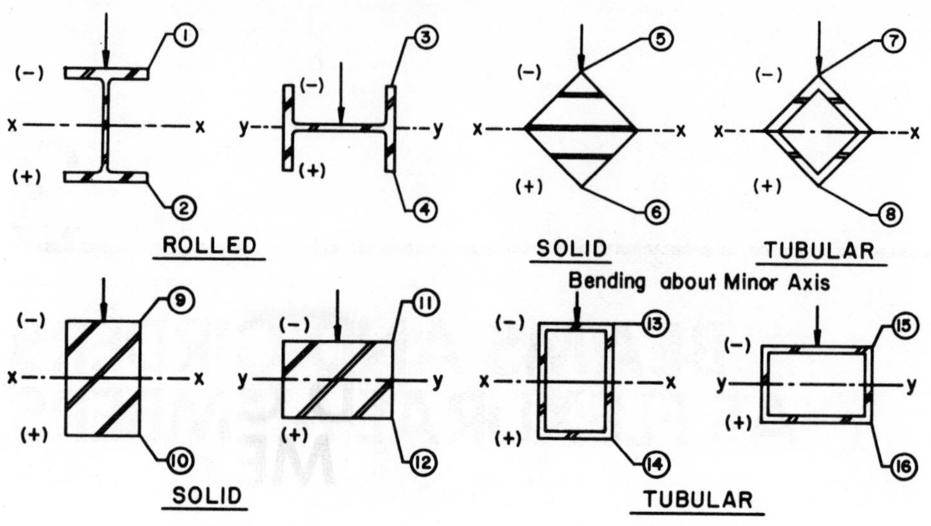

TABLE 3-A Maximum Allowable Stresses

Fig No.	COMPACT			SEMI-COMPACT			NON-COMPACT		
		Reference			Reference			Reference	
		AISC Section Reference	Rice-Hoffman Reference		AISC Section Reference	Rice-Hoffman Reference		AISC Section Reference	Rice-Hoffman Reference
1	$0.66\ F_y$	1.5.1.4.1	Table 3-1,-2 Col. (1)	1.5-5a	1.5.1.4.2	Table 3-1,-2 Col. (2)	1.5-6a 1.5-6b 1.5-7	1.5.1.4.6a	Table 3-1,-2 Cols.(4)(5)(5)
2	$0.66\ F_y$	1.5.1.4.1	Table 3-1,-2 Col. (1)	1.5-5a	1.5.1.4.2	Table 3-1,-2 Col. (2)	$0.60\ F_y$	1.5.1.4.5	-
3	$0.75\ F_y$	1.5.1.4.3	Table 3-1,-2 Col. (7)	1.5-5b	1.5.1.4.3	Table 3-1,-2 Col. (8)	$0.60\ F_y$	1.5.1.4.6b	Table 3-1,-2 Col. (9)
4	$0.75\ F_y$	1.5.1.4.3	Table 3-1,-2 Col. (7)	1.5-5b	1.5.1.4.3	Table 3-1,-2 Col. (8)	$0.60\ F_y$	1.5.1.4.5	-
5	-	-	-	-	-	-	$0.75\ F_y$	1.5.1.4.3	-
6	-	-	-	-	-	-	$0.75\ F_y$	1.5.1.4.3	-
7	*	-	*	**	-	*	$0.60\ F_y$	1.5.1.4.4	-
8	**	-	*	**	-	*	$0.60\ F_y$	1.5.1.4.4	-
9	$0.66\ F_y$	1.5.1.4.1	Fig. 3-6	1.5-5a	1.5.1.4.2	Fig. 3-6 Table 3-3	Fig. 3-6	1.5.1.4.4	Fig. 3-6
10	$0.66\ F_y$	1.5.1.4.1	-	1.5-5a	1.5.1.4.2	Table 3-3	$0.60\ F_y$	1.5.1.4.5	-
11	-	-	-	-	-	-	$0.75\ F_y$	1.5.1.4.3	-
12	-	-	-	-	-	-	$0.75\ F_y$	1.5.1.4.3	-
13	$0.66\ F_y$	1.5.1.4.1	Fig. 3-7	1.5-5a	1.5.1.4.2	Fig. 3-7 Table 3-3	Fig. 3-7	1.5.1.4.4	Fig. 3-7
14	$0.66\ F_y$	1.5.1.4.1	-	1.5-5a	1.5.1.4.2	Table 3-3	$0.60\ F_y$	1.5.1.4.5	-
15	-	-	-	-	-	-	$0.60\ F_y$	1.5.1.4.4	-
16	-	-	-	-	-	-	$0.60\ F_y$	1.5.1.4.4	-

(Continued on page 87)

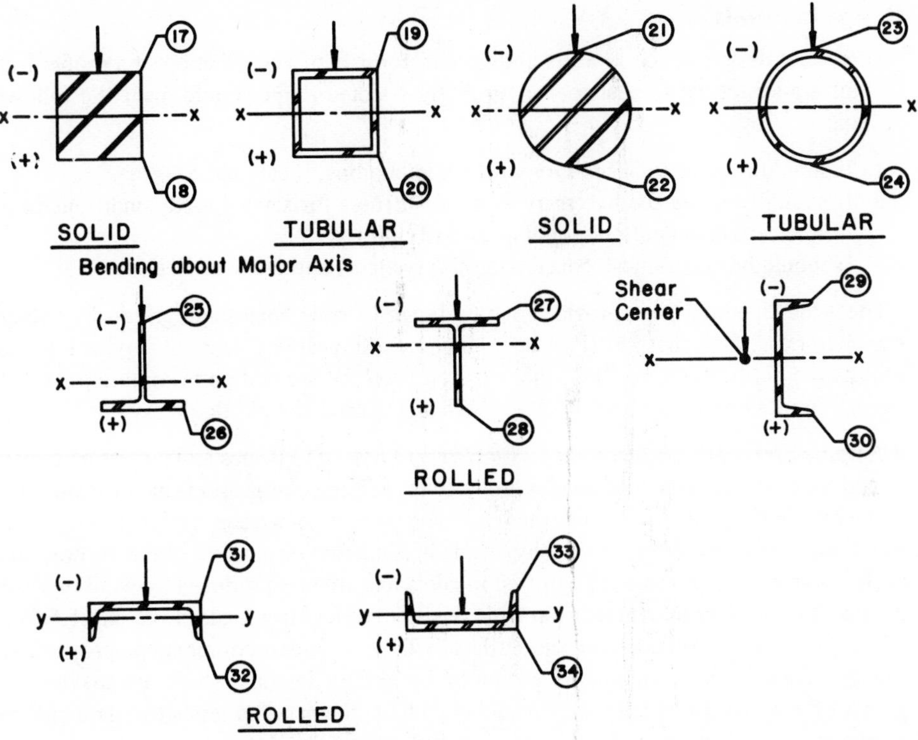

Fig No.	COMPACT			SEMI-COMPACT			NON-COMPACT		
		Reference			Reference			Reference	
		AISC Section Reference	Rice-Hoffman Reference		AISC Section Reference	Rice-Hoffman Reference		AISC Section Reference	Rice-Hoffman Reference
17	-	-	-	-	-	-	0.75 F_y	1.5.1.4.3	-
18	-	-	-	-	-	-	0.75 F_y	1.5.1.4.3	-
19	0.66 F_y	1.5.1.4.1	Fig. 3-7	1.5-5a	1.5.1.4.2	Fig. 3-7 Table 3-3	Fig. 3-7	1.5.1.4.4	-
20	0.66 F_y	1.5.1.4.1	-	1.5-5a	1.5.1.4.2	Table 3-3	0.60 F_y	1.5.1.4.4	-
21	-	-	-	-	-	-	0.75 F_y	1.5.1.4.3	-
22	-	-	-	-	-	-	0.75 F_y	1.5.1.4.3	-
23	-	-	-	-	-	-	0.60 F_y	1.5.1.4.6b	-
24	-	-	-	-	-	-	0.60 F_y	1.5.1.4.5	-
25	0.66 F_y	1.5.1.4.1	-	1.5-5a	1.5.1.4.2	Table 3-9	1.5-6a 1.5-6b 1.5-7	1.5.1.4.5	-
26	0.66 F_y	1.5.1.4.1	-	1.5-5a	1.5.1.4.2	-	0.60 F_y	1.5.1.4.5	-
27	0.66 F_y	1.5.1.4.1	-	1.5-5a	1.5.1.4.2	Table 3-8	1.5-6a 1.5-6b 1.5-7	1.5.1.4.5	-
28	0.66 F_y	1.5.1.4.1	-	1.5-5a	1.5.1.4.2	-	0.60 F_y	1.5.1.4.5	-
29	-	-	-	-	-	-	1.5-7	1.5.1.4.6a	-
30	-	-	-	-	-	-	0.60 F_y	1.5.1.4.5	-
31	-	-	-	-	-	-	0.60 F_y	1.5.1.4.6b	-
32	-	-	-	-	-	-	0.60 F_y	1.5.1.4.5	-
33	-	-	-	-	-	-	0.60 F_y	1.5.1.4.6b	-
34	-	-	-	-	-	-	0.60 F_y	1.5.1.4.5	-

* Reference equations cited for F_b give variable functions of F_y.

** As the hollow box walls become thicker, these allowable stresses approach 0.75 F_y as for solid sections.

Design Methods

Successful design of steel beams and girders consists of the selection of a proper rolled or built-up structural steel shape. A properly selected shape should meet the following criteria:

1. It should provide the necessary strength in bending, shear, and bearing.
2. It should provide the necessary flexural stiffness for service load conditions to prevent excessive deflection, ponding, and vibration.
3. It should be economical, considering size (weight) and grade of steel.

The AISC Specifications provide for the design of steel beams and girders by two general methods. One method (Part 1) is based on allowable stresses at service loads and utilizes the elastic theory. The other method (Part 2) is based on maximum predictable strength under minimum prescribed overload and utilizes the plastic theory.

Allowable Stress Design Method (Elastic Theory) (Part 1). Beams and girders of Types 1, 2, and 3 Construction within certain limitations (1.5) may be proportioned for shears and moments determined by elastic theory, using allowable stresses (1.5.1.2 and 1.5.1.4) when these stresses are produced by dead, live, impact, wind and crane runway loads (1.3). A one-third increase is permitted in allowable stresses produced by wind or seismic loads acting alone or in combination with service dead and live loads (1.3.5 and 1.5.6).

Except for cantilevers, beams and girders of Type 1 Construction may be proportioned for $\frac{9}{10}$ of the negative moments produced by gravity loading which are maximum at points of support, provided that, for such members, the maximum positive moment must be increased by $\frac{1}{10}$ of the average negative moments (1.5.1.4.1).

Strength Design Method (Plastic Theory) (Part 2). Beams and girders of Type 1 Construction that are within certain limitations may be proportioned on the basis of shears and moments determined by plastic theory for factored loads equal to 1.7 times the prescribed service dead, live, impact, and crane runway loads or 1.3 times prescribed service dead and live loads acting in conjunction with 1.3 times any specified wind or seismic loads (1.3 and 2.1). (See Chapter 7 for selected applications of plastic design.)

ALLOWABLE STRESS DESIGN

Allowable Stresses in Flexure

To approach the ideal of a uniform "safety factor" for maximum efficiency (minimum tonnage) the Specifications create elaborate adjustments of allowable stresses. A further complexity is due to semantics. The term "compact section" is not merely a misnomer, but an abominable perversion in specification language. When it is used to describe a rolled shape for the purpose of establishing an allowable stress, it is guaranteed to mislead a reader. In many cases it may mislead a reader, not once, but again and again. According to Webster, "compact" means "dense" or "much area enclosed in proportion to perimeter." To an Engineer, a "section" is a cross-section showing cross-sectional dimensions, when applied to a rolled shape. To an Engineer, the term "compact section" therefore must define a degree of compactness for a rolled section in terms of cross-sectional properties. The ultimate in compactness would be considered a round or square solid section, or a heavy H, depending upon whether area or section modulus is the index of "compact." Also, the term would normally be understood as a means of identifying shapes for once and for all, as compact or non-compact. Not so in the AISC Specifications! The AISC definition of compact, as an index to allowable stress, includes as essential criteria, not only the yield Grade and axis of bending, but the *unbraced length of the compression flange!*

A given rolled shape may qualify as a "compact section" or a "non-compact section." A slightly less compact shape may be classed as "semi-compact" qualifying for a transitional stress formula, or non-compact. Some shapes are forever non-compact. Other shapes are *too* compact to be called a "compact section," certain solid sections! The following definitions and limitations thereon deserve close study to establish allowable stresses for given conditions. The sketches and notes in Table 3-A will facilitate design by maximum allowable stresses at a glance for most conceivable sections. Tables 3-1 and 3-2 present semi-automatic solutions for moments of resistance of rolled shapes in Grades 36 and 50, respectively. See "Use and Derivations of Rolled Section Selection Tables," page 150, and "Design Examples," page 156.

Normal stresses in the extreme fibers of beams and girders due to bending may be calculated using the elastic section modulus (S_x) determined from gross section properties unless the area of bolt or rivet holes in either of the flanges exceeds fifteen per cent of the gross flange area (1.10.1). When the bolt or rivet hole area exceeds fifteen per cent net section properties must be used.

Compact Sections–Bending About the Major Axis.* The maximum allowable compression and tension stress in bending (F_b) on the extreme fibers of compact hot rolled shapes or built-up members that are symmetrical about, and loaded in the plane of, their minor axis is limited. (See Fig. 3-1.)

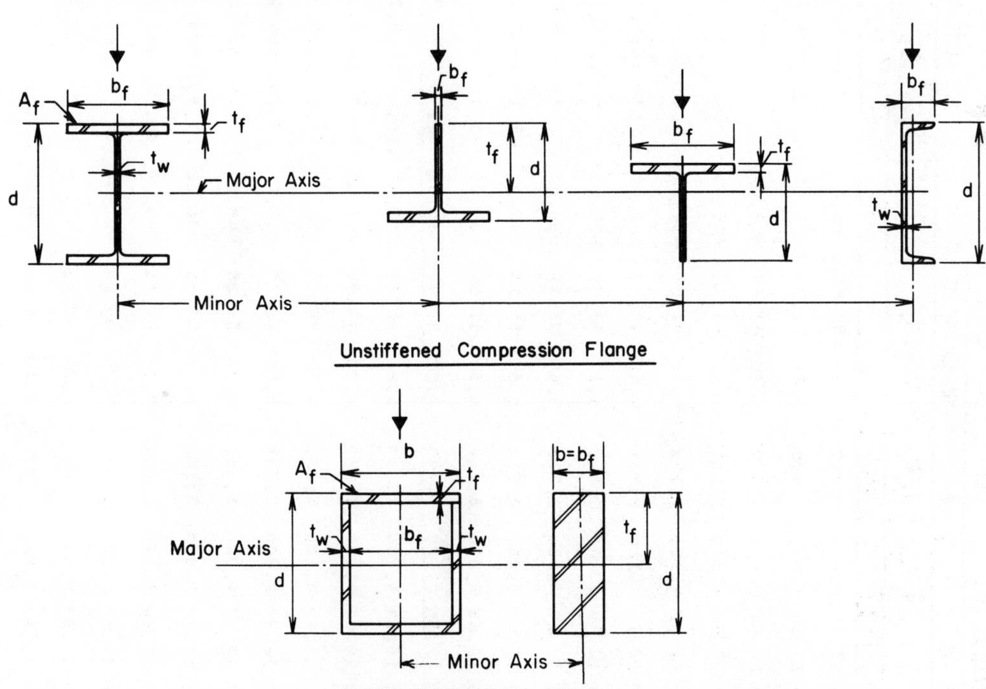

Fig. 3-1 Typical Cross Sections and Nomenclature for Bending.

*Exceptions are hybrid girders, tapered girders and members of A514 steel (1.5.1.4.1), and tension field action girders (1.10.7).

Continued on page 122

TABLE 3-1 Rolled Shape Resisting Moments, M_R (k-ft); $F_Y = 36$

F_y = 36 ksi	TABLE 3-1: RESISTING MOMENTS, M_R, (kip-ft.) - F_y = 36 ksi															
	SECTION PROPERTIES								MOMENT M_{RX}, k-ft.			M_{RX} FACTORS*			MOMENT, M_{RY}, k-ft.		
Section	Area	L_c	r_T	d/A_f	$b_f/2t_f$	d/t_w	S_{x-x}	S_{y-y}	(1) $\frac{0.66 F_y S_x}{12} = M_{RX}$	(2) Eq. 1.5-5 $= M_{RX}$	(3) $\frac{0.60 F_y S_x}{12} = M_{RX}$	(4) Eq. 1.5-6a =(MF)	(5) Eq. 1.5-7 =(MF)	(6) Eq. 1.5-6b =(MF)	(7) $\frac{0.75 F_y S_y}{12} = M_{RY}$	(8) Eq. 1.5-5b $= M_{RY}$	(9) $\frac{0.60 F_y S_y}{12} = M_{RY}$
	sq. in.	(ft.)	in.	in.$^{-1}$	–	–	in.^{3}	in.^{3}									
W 36 300.0	88.30	17.58	4.460	1.312	4.95	38.85	1110.00	156.00	2197.8	0.0	1997.9	39.389	845.81	312795.3	351.0	0.0	280.7
W 36 280.0	82.40	17.51	4.430	1.400	5.28	41.24	1030.00	144.00	2039.3	0.0	1853.9	37.047	735.22	286359.9	324.0	0.0	259.2
W 36 260.0	76.50	17.47	4.400	1.520	5.74	43.09	952.00	132.00	1884.9	0.0	1713.5	34.710	626.08	261101.8	297.0	0.0	237.5
W 36 245.0	72.10	17.42	4.380	1.617	6.11	44.96	894.00	123.00	1770.1	0.0	1609.1	32.894	552.64	242970.3	276.7	0.0	221.3
W 36 230.0	67.70	17.38	4.360	1.728	6.53	47.14	837.00	114.00	1657.2	0.0	1506.5	31.080	484.13	225406.3	256.5	0.0	205.1
W 36 194.0	57.20	12.79	3.120	2.389	4.80	47.37	665.00	61.90	1316.6	0.0	1197.0	48.221	278.31	91706.1	139.2	0.0	111.4
W 36 182.0	53.60	12.74	3.100	2.549	5.11	50.09	622.00	57.50	1231.5	0.0	1119.5	45.687	243.95	84680.1	129.3	0.0	103.4
W 36 170.0	50.00	12.69	3.080	2.733	5.46	53.17	580.00	53.20	1148.4	0.0	1044.0	43.157	212.20	77946.5	119.6	0.0	95.7
W 36 160.0	47.10	12.66	3.060	2.941	5.88	55.13	542.00	49.10	1073.1	0.0	975.5	40.859	184.28	71896.8	110.4	0.0	88.3
W 36 150.0	44.20	12.63	3.030	3.184	6.36	57.34	504.00	45.00	997.9	0.0	907.1	38.750	158.25	65551.6	101.2	0.0	81.0
W 36 135.0	39.80	12.35	2.970	3.748	7.52	59.44	440.00	37.90	871.2	0.0	791.9	35.210	117.38	54983.6	85.2	0.0	68.2
W 33 240.0	70.60	16.74	4.230	1.508	5.66	40.36	813.00	118.00	1609.7	0.0	1463.3	32.073	539.03	206081.4	265.5	0.0	212.3
W 33 220.0	64.80	16.68	4.200	1.649	6.20	42.90	742.00	106.00	1469.1	0.0	1335.5	29.691	449.83	185425.8	238.5	0.0	190.7
W 33 200.0	58.90	16.62	4.170	1.821	6.84	46.15	671.00	95.20	1328.5	0.0	1207.8	27.238	368.28	165295.9	214.2	0.0	171.3
W 33 152.0	44.80	12.20	2.980	2.745	5.48	52.75	487.00	47.20	964.2	0.0	876.5	38.710	177.37	61267.3	106.1	0.0	84.9
W 33 141.0	41.60	12.17	2.960	3.008	6.00	55.05	448.00	42.70	887.0	0.0	806.3	36.093	148.93	55606.9	96.0	0.0	76.8
W 33 130.0	38.30	12.14	2.920	3.363	6.73	57.06	406.00	37.90	803.8	0.0	730.7	33.611	120.70	49041.0	85.2	0.0	68.2
W 33 118.0	34.80	11.94	2.870	3.877	7.78	59.31	359.00	32.50	710.8	0.0	646.1	30.765	92.59	41891.5	73.1	0.0	58.4
W 30 210.0	61.90	15.94	4.050	1.529	5.74	39.20	651.00	100.00	1288.9	0.0	1171.8	28.015	425.63	151272.0	225.0	0.0	180.0
W 30 190.0	56.00	15.87	4.010	1.690	6.34	42.42	587.00	89.50	1162.2	0.0	1056.5	25.768	347.33	133719.4	201.3	0.0	161.0
W 30 172.0	50.70	15.81	3.980	1.872	7.03	45.61	530.00	79.80	1049.4	0.0	953.9	23.617	283.07	118934.9	179.5	0.0	143.6

W	30	132.0	38.90	11.13	2.720	2.871	5.27	49.26	380.00	37.20	752.4	0.0	684.0	36.255	132.32	39828.0	83.6	0.0	66.9
W	30	124.0	36.50	11.10	2.700	3.082	5.65	51.55	355.00	34.40	702.9	0.0	639.0	34.374	115.16	36662.6	77.3	0.0	61.9
W	30	116.0	34.20	11.08	2.680	3.361	6.17	53.19	329.00	31.30	651.4	0.0	592.1	32.333	97.87	33475.9	70.4	0.0	56.3
W	30	108.0	31.80	11.06	2.640	3.742	6.89	54.41	300.00	27.90	594.0	0.0	540.0	30.384	80.15	29620.8	62.7	0.0	50.2
W	30	99.0	29.10	10.94	2.610	4.230	7.80	56.78	270.00	24.50	534.5	0.0	485.9	27.977	63.82	26056.2	55.1	0.0	44.0
W	27	177.0	52.20	14.87	3.770	1.628	5.92	37.66	494.00	78.90	978.1	0.0	889.1	24.534	303.29	99466.6	177.5	0.0	142.0
W	27	160.0	47.10	14.80	3.750	1.796	6.52	41.15	446.00	70.60	883.0	0.0	802.7	22.387	248.27	88851.5	158.8	0.0	127.0
W	27	145.0	42.70	14.74	3.720	1.974	7.16	44.80	404.00	63.50	799.9	0.0	727.1	20.607	204.64	79201.7	142.8	0.0	114.2
W	27	114.0	33.60	10.62	2.620	2.906	5.40	47.85	300.00	31.60	594.0	0.0	540.0	30.849	103.21	29173.6	71.1	0.0	56.8
W	27	102.0	30.00	10.57	2.590	3.267	6.05	52.25	267.00	27.70	528.6	0.0	480.5	28.095	81.71	25373.3	62.3	0.0	49.8
W	27	94.0	27.70	10.54	2.560	3.606	6.68	54.91	243.00	24.90	481.1	0.0	437.3	26.173	67.38	22560.7	56.0	0.0	44.8
W	27	84.0	24.80	10.51	2.520	4.212	7.83	57.64	212.00	21.10	419.7	0.0	381.5	23.564	50.33	19072.3	47.4	0.0	37.9
W	24	160.0	47.10	14.87	3.820	1.545	6.20	37.68	414.00	75.20	819.7	0.0	745.1	20.026	267.84	85584.4	169.2	0.0	135.3
W	24	145.0	42.70	14.82	3.790	1.709	6.88	40.27	373.00	67.10	738.5	0.0	671.3	18.330	218.16	75902.2	150.9	0.0	120.7
W	24	130.0	38.30	14.77	3.760	1.924	7.77	42.92	332.00	58.90	657.3	0.0	597.5	16.576	172.50	66493.8	132.5	0.0	106.0
W	24	120.0	35.40	12.75	3.220	2.162	6.49	43.72	300.00	45.40	594.0	0.0	540.0	20.424	138.73	44065.6	102.1	0.0	81.7
W	24	110.0	32.50	12.71	3.200	2.346	7.04	47.37	276.00	41.40	546.4	0.0	496.7	19.025	117.61	40038.3	93.1	0.0	74.5
W	24	100.0	29.50	12.66	3.180	2.580	7.74	51.28	250.00	37.20	495.0	0.0	449.9	17.450	96.87	35814.7	83.6	0.0	66.9
W	24	94.0	27.70	9.56	2.370	3.074	5.19	47.07	221.00	23.90	437.5	0.0	397.7	27.773	71.88	17585.5	53.7	0.0	43.0
W	24	84.0	24.70	9.51	2.340	3.461	5.83	51.25	197.00	21.00	390.0	0.0	354.5	25.396	56.91	15281.4	47.2	0.0	37.7

(1) M_{rx}. Compact sections; $b_f/2t_f \leqslant 10.83$; beams, $f_a = 0$, $d/t_w \leqslant 106.7$, columns, $f_a > 0.16\ F_y$, $d/t_w \leqslant 36.3$; braced so that $L_b \leqslant L_c$.
(2) M_{rx}. Semi-compact sections; same as column (1) except $10.83 < b_f/2t_f < 15.83$.
(3) M_{rx}. No-compact sections for tension; all section not under (1), (2), (4), (6); for compression; upper limit for compression sections under (4), (5), (6).
*Cols (4), (5), (6): Non-compact sections; moment factors (*MF*) for compression.
*(4) For $53.2\sqrt{C_b} \leqslant l_b/r_T < 119\sqrt{C_b}$, except channels. Use *larger* M_{rx} from (4) or (5). . . . . (4) $M_{rx} = \text{col. (1)} - (MF \times 10^{-4})\ l_b^2/C_b \leqslant \text{Col. (3)}$.
*(5) For all values l_b/r_T. . . . . (5) $M_{rx} = 1000\ (MF)\ C_b/l_b \leqslant \text{Col. (3)}$.
*(6) For $l_b/r_T \geqslant 119\sqrt{C_b}$, except channels. Use larger M_{rx} from columns (5) or (6). . . . . (6) $M_{rx} = 1000\ (MF)\ C_b/l_b^2 \leqslant \text{Col. (3)}$.
(7) M_{ry}. For doubly symmetrical shapes; $b_f/2t_f \leqslant 10.83$.
(8) M_{ry}. For doubly symmetrical shapes; $10.83 < b_f/2t_f < 15.83$.
(9) M_{ry}. For non-compact sections; sections not under column (7) or (8).

F_y = 36 ksi	TABLE 3-1: RESISTING MOMENTS, M_R, (kip-ft.) - F_y = 36 ksi															
	SECTION PROPERTIES								MOMENT M_{RX}, k-ft.			M_{RX} FACTORS*			MOMENT, M_{RY}, k-ft.		
Section	Area	L_c	r_T	d/A_f	$b_f/2t_f$	d/t_w	S_{x-x}	S_{y-y}	(1) $\frac{0.66}{12}F_yS_x = M_{RX}$	(2) Eq. 1.5-5 $= M_{RX}$	(3) $\frac{0.60}{12}F_yS_x = M_{RX}$	(4) Eq. 1.5-6a =(MF)	(5) Eq. 1.5-7 =(MF)	(6) Eq. 1.5-6b =(MF)	(7) $\frac{0.75}{12}F_yS_y = M_{RY}$	(8) Eq. 1.5-5b $= M_{RY}$	(9) $\frac{0.60}{12}F_yS_y = M_{RY}$
	sq. in.	(ft.)	in.	in.$^{-1}$	–	–	in.^{3}	in.^{3}									
W 24 76.0	22.40	9.48	2.320	3.901	6.58	54.34	176.00	18.40	348.4	0.0	316.7	23.081	45.10	13420.1	41.4	0.0	33.1
W 24 68.0	20.00	9.45	2.280	4.546	7.69	56.99	153.00	15.60	302.9	0.0	275.3	20.775	33.65	11267.5	35.0	0.0	28.0
W 24 61.0	18.00	7.41	1.740	5.714	5.94	56.61	130.00	9.76	257.4	0.0	233.9	30.309	22.74	5575.8	21.9	0.0	17.5
W 24 55.0	16.20	6.92	1.700	6.688	6.95	59.46	114.00	8.25	225.7	0.0	205.1	27.844	17.04	4657.3	18.5	0.0	14.8
W 21 142.0	41.80	13.86	3.580	1.492	5.99	32.56	317.00	63.00	627.6	0.0	570.5	17.459	212.40	57556.3	141.7	0.0	113.3
W 21 127.0	37.40	13.78	3.550	1.650	6.62	36.12	284.00	56.10	562.3	0.0	511.1	15.907	172.01	50704.0	126.2	0.0	100.9
W 21 112.0	33.00	13.72	3.520	1.867	7.51	39.84	250.00	48.80	495.0	0.0	449.9	14.242	133.86	43882.6	109.8	0.0	87.8
W 21 96.0	28.30	9.54	2.390	2.501	4.83	36.76	198.00	25.50	392.0	0.0	356.3	24.468	79.14	16022.4	57.3	0.0	45.8
W 21 82.0	24.20	9.45	2.350	2.927	5.63	41.80	169.00	21.30	334.6	0.0	304.2	21.601	57.72	13221.7	47.9	0.0	38.3
W 21 73.0	21.50	8.75	2.160	3.460	5.60	46.68	151.00	17.00	298.9	0.0	271.7	22.845	43.63	9980.4	38.2	0.0	30.5
W 21 68.0	20.00	8.72	2.150	3.729	6.03	49.13	140.00	15.70	277.2	0.0	251.9	21.378	37.53	9167.9	35.3	0.0	28.2
W 21 62.0	18.30	8.69	2.130	4.142	6.69	52.47	127.00	13.90	251.4	0.0	228.5	19.759	30.66	8162.6	31.2	0.0	25.0
W 21 55.0	16.20	8.67	2.100	4.850	7.86	55.46	110.00	11.80	217.8	0.0	197.9	17.607	22.67	6872.2	26.5	0.0	21.2
W 21 49.0	14.40	6.88	1.630	6.002	6.12	56.57	93.30	7.57	184.7	0.0	167.9	24.787	15.54	3511.7	17.0	0.0	13.6
W 21 44.0	13.00	6.56	1.590	7.047	7.20	59.36	81.60	6.38	161.5	0.0	146.8	22.783	11.57	2922.4	14.3	0.0	11.4
W 18 114.0	33.50	12.49	3.230	1.575	5.97	31.05	220.00	46.30	435.6	0.0	395.9	14.885	139.60	32515.8	104.1	0.0	83.3
W 18 105.0	30.90	12.44	3.210	1.705	6.47	33.06	202.00	42.30	399.9	0.0	363.5	13.838	118.44	29486.8	95.1	0.0	76.1
W 18 96.0	28.20	12.40	3.190	1.859	7.06	35.46	185.00	38.30	366.2	0.0	333.0	12.832	99.47	26669.8	86.1	0.0	68.9
W 18 85.0	25.00	9.32	2.370	2.275	4.85	34.82	157.00	23.80	310.8	0.0	282.5	19.730	68.99	12492.9	53.5	0.0	42.8
W 18 77.0	22.70	9.27	2.360	2.486	5.28	38.23	142.00	21.40	281.1	0.0	255.5	17.996	57.09	11204.1	48.1	0.0	38.5
W 18 70.0	20.60	9.23	2.340	2.739	5.82	41.09	129.00	19.20	255.4	0.0	232.1	16.629	47.09	10006.6	43.1	0.0	34.5
W 18 64.0	18.90	9.19	2.320	2.989	6.35	44.34	118.00	17.40	233.6	0.0	212.3	15.475	39.47	8997.5	39.1	0.0	31.3

W	18	60.0	17.70	7.97	1.990	3.474	5.43	43.87	108.00	13.30	213.8	0.0	194.3	19.250	31.08	6058.9	29.9	0.0	23.9
W	18	55.0	16.20	7.95	1.980	3.818	5.97	46.46	98.40	11.90	194.8	0.0	177.1	17.717	25.76	5465.0	26.7	0.0	21.4
W	18	50.0	14.70	7.91	1.960	4.210	6.57	50.27	89.10	10.70	176.4	0.0	160.3	16.371	21.16	4849.0	24.0	0.0	19.2
W	18	45.0	13.20	7.89	1.940	4.786	7.49	53.31	79.00	9.32	156.4	0.0	142.1	14.816	16.50	4212.0	20.9	0.0	16.7
W	18	40.0	11.80	6.35	1.540	5.676	5.74	56.64	68.40	6.34	135.4	0.0	123.1	20.358	12.04	2298.0	14.2	0.0	11.4
W	18	35.0	10.30	6.33	1.510	6.880	6.99	59.42	57.90	5.16	114.6	0.0	104.2	17.924	8.41	1870.2	11.6	0.0	9.2
W	16	96.0	28.20	12.17	3.160	1.617	6.59	30.50	166.00	38.80	328.6	0.0	298.7	11.734	102.64	23482.8	87.3	0.0	69.8
W	16	88.0	25.90	12.14	3.140	1.767	7.23	32.06	151.00	35.10	298.9	0.0	271.7	10.810	85.44	21091.3	78.9	0.0	63.1
W	16	78.0	23.00	9.06	2.320	2.172	4.90	30.45	128.00	21.60	253.4	0.0	230.3	16.786	58.92	9760.0	48.5	0.0	38.8
W	16	71.0	20.90	9.01	2.300	2.379	5.37	33.25	116.00	19.40	229.6	0.0	208.7	15.478	48.75	8693.2	43.6	0.0	34.9
W	16	64.0	18.80	8.97	2.280	2.632	5.94	36.11	104.00	17.30	205.9	0.0	187.1	14.121	39.50	7658.9	38.9	0.0	31.1
W	16	58.0	17.10	8.93	2.260	2.905	6.56	38.96	94.40	15.40	186.9	0.0	169.9	13.046	32.49	6830.5	34.6	0.0	27.7
W	16	50.0	14.70	7.46	1.870	3.658	5.63	42.76	80.80	10.50	159.9	0.0	145.4	16.310	22.08	4002.7	23.6	0.0	18.8
W	16	45.0	13.30	7.43	1.850	4.067	6.25	46.58	72.50	9.32	143.5	0.0	130.5	14.952	17.82	3515.1	20.9	0.0	16.7
W	16	40.0	11.80	7.38	1.840	4.544	6.95	52.11	64.60	8.23	127.9	0.0	116.2	13.468	14.21	3098.3	18.5	0.0	14.8
W	16	36.0	10.60	7.38	1.810	5.296	8.16	53.01	56.50	6.99	111.8	0.0	101.6	12.173	10.66	2622.2	15.7	0.0	12.5
W	16	31.0	9.13	5.83	1.410	6.486	6.25	57.60	47.20	4.51	93.4	0.0	84.9	16.758	7.27	1329.3	10.1	0.0	8.1
W	16	26.0	7.67	5.61	1.380	8.247	7.97	62.60	38.30	3.49	75.8	0.0	68.9	14.196	4.64	1033.2	7.8	0.0	6.2
W	14	730.0	215.00	18.88	5.270	0.255	1.82	7.31	1280.00	527.00	2534.3	0.0	2304.0	32.532	5010.19	503615.1	1185.7	0.0	948.5
W	14	665.0	196.00	18.62	5.170	0.271	1.95	7.66	1150.00	472.00	2277.0	0.0	2070.0	30.370	4234.63	435458.1	1062.0	0.0	849.5

(1) M_{rx}. Compact sections; $b_f/2t_f \leqslant 10.83$; beams, $f_a = 0$, $d/t_w \leqslant 106.7$, columns, $f_a > 0.16\,F_y$, $d/t_w \leqslant 36.3$; braced so that $L_b \leqslant L_c$.
(2) M_{rx}. Semi-compact sections; same as column (1) except $10.83 < b_f/2t_f < 15.83$.
(3) M_{rx}. No-compact sections for tension; all section not under (1), (2), (4), (6); for compression; upper limit for compression sections under (4), (5), (6).
*Cols (4), (5), (6): Non-compact sections; moment factors (MF) for compression.
*(4) For $53.2\sqrt{C_b} \leqslant l_b/r_T < 119\sqrt{C_b}$, except channels. Use *larger* M_{rx} from (4) or (5). . . . . (4) M_{rx} = col. (1) − ($MF \times 10^{-4}$) $l_b^2/C_b \leqslant$ Col. (3).
*(5) For all values l_b/r_T. . . . . (5) $M_{rx} = 1000\,(MF)\,C_b/l_b \leqslant$ Col. (3).
*(6) For $l_b/r_T \geqslant 119\sqrt{C_b}$, except channels. Use larger M_{rx} from columns (5) or (6). . . . . (6) $M_{rx} = 1000\,(MF)\,C_b/l_b^2 \leqslant$ Col. (3).
(7) M_{ry}. For doubly symmetrical shapes; $b_f/2t_f \leqslant 10.83$.
(8) M_{ry}. For doubly symmetrical shapes; $10.83 < b_f/2t_f < 15.83$.
(9) M_{ry}. For non-compact sections; sections not under column (7) or (8).

TABLE 3-1: RESISTING MOMENTS, M_R, (kip-ft.) - F_y = 36 ksi

F_y = 36 ksi

Section	SECTION PROPERTIES								MOMENT M_{RX}, k-ft.			M_{RX} FACTORS*			MOMENT, M_{RY}, k-ft.		
	Area	L_c	r_T	d/A_f	$b_f/2t_f$	d/t_w	S_{x-x}	S_{y-y}	(1) $\frac{0.66 F_y S_x}{12} = M_{RX}$	(2) Eq. 1.5-5 $= M_{RX}$	(3) $\frac{0.60 F_y S_x}{12} = M_{RX}$	(4) Eq. 1.5-6a =(MF)	(5) Eq. 1.5-7 =(MF)	(6) Eq. 1.5-6b =(MF)	(7) $\frac{0.75 F_y S_y}{12} = M_{RY}$	(8) Eq. 1.5-5b $= M_{RY}$	(9) $\frac{0.60 F_y S_y}{12} = M_{RY}$
	sq. in.	(ft.)	in.	$in.^{-1}$	–	–	$in.^3$	$in.^3$									
W 14 605.0	178.00	18.38	5.080	0.289	2.09	8.06	1040.00	423.00	2059.2	0.0	1871.9	28.447	3596.12	380214.3	951.7	0.0	761.3
W 14 550.0	162.00	18.16	5.000	0.308	2.25	8.49	933.00	378.00	1847.3	0.0	1679.3	26.343	3025.22	330437.5	850.5	0.0	680.3
W 14 500.0	147.00	17.95	4.920	0.329	2.42	8.97	840.00	339.00	1663.1	0.0	1512.0	24.495	2548.02	288056.0	762.7	0.0	610.1
W 14 455.0	134.00	17.76	4.850	0.352	2.61	9.48	758.00	304.00	1500.8	0.0	1364.3	22.746	2151.38	252592.4	684.0	0.0	547.1
W 14 426.0	125.00	17.62	4.810	0.369	2.75	9.96	707.00	283.00	1399.8	0.0	1272.5	21.570	1915.44	231727.2	636.7	0.0	509.3
W 14 398.0	117.00	17.51	4.760	0.388	2.91	10.34	657.00	262.00	1300.8	0.0	1182.5	20.468	1692.38	210885.5	589.5	0.0	471.5
W 14 370.0	109.00	17.39	4.720	0.409	3.09	10.83	608.00	241.00	1203.8	0.0	1094.3	19.264	1484.09	191891.2	542.2	0.0	433.7
W 14 342.0	101.00	17.27	4.680	0.434	3.31	11.36	559.00	221.00	1106.8	0.0	1006.1	18.015	1285.72	173448.7	497.2	0.0	397.7
W 14 314.0	92.30	17.13	4.630	0.463	3.55	12.14	512.00	201.00	1013.7	0.0	921.5	16.859	1103.95	155488.9	452.2	0.0	361.7
W 14 287.0	84.40	17.02	4.580	0.497	3.85	12.83	465.00	182.00	920.7	0.0	836.9	15.647	933.87	138182.0	409.5	0.0	327.5
W 14 264.0	77.60	16.91	4.540	0.531	4.13	13.69	427.00	166.00	845.4	0.0	768.5	14.623	803.70	124683.0	373.5	0.0	298.7
W 14 246.0	72.30	16.83	4.510	0.562	4.39	14.44	397.00	154.00	786.0	0.0	714.5	13.777	706.25	114396.0	346.5	0.0	277.2
W 14 237.0	69.70	16.79	4.500	0.579	4.55	14.78	382.00	148.00	756.3	0.0	687.5	13.315	659.03	109586.2	333.0	0.0	266.3
W 14 228.0	67.10	16.74	4.480	0.597	4.69	15.31	368.00	142.00	728.6	0.0	662.3	12.942	615.94	104633.6	319.5	0.0	255.5
W 14 219.0	64.40	16.70	4.470	0.617	4.87	15.79	353.00	136.00	698.9	0.0	635.3	12.470	571.29	99921.1	306.0	0.0	244.7
W 14 211.0	62.10	16.67	4.460	0.637	5.05	16.07	339.00	130.00	671.2	0.0	610.1	12.029	531.53	95529.3	292.5	0.0	233.9
W 14 202.0	59.40	16.62	4.440	0.660	5.23	16.80	325.00	124.00	643.5	0.0	585.0	11.637	492.22	90764.6	279.0	0.0	223.1
W 14 193.0	56.70	16.58	4.420	0.686	5.46	17.41	310.00	118.00	613.8	0.0	558.0	11.200	451.81	85797.3	265.5	0.0	212.3
W 14 184.0	54.10	16.53	4.410	0.712	5.68	18.30	296.00	113.00	586.0	0.0	532.7	10.743	415.31	81552.3	254.2	0.0	203.3
W 14 176.0	51.70	16.50	4.390	0.742	5.95	18.59	282.00	107.00	558.3	0.0	507.5	10.328	379.73	76992.0	240.7	0.0	192.5
W 14 167.0	49.10	16.46	4.380	0.776	6.25	19.38	267.00	101.00	528.6	0.0	480.5	9.824	343.79	72564.9	227.2	0.0	181.7
W 14 158.0	46.50	16.41	4.360	0.811	6.54	20.54	253.00	95.80	500.9	0.0	455.3	9.394	311.58	68133.5	215.5	0.0	172.4

W	14	150.0	44.10	16.37	4.350	0.850	6.87	21.41	240.00	90.60	475.2	0.0	431.9	8.952	282.27	64336.4	203.8	0.0	163.0
W	14	142.0	41.80	16.36	4.330	0.895	7.29	21.69	227.00	85.20	449.4	0.0	408.5	8.546	253.57	60293.3	191.7	0.0	153.3
W	14	320.0	94.10	17.63	4.730	0.480	3.99	8.89	493.00	196.00	976.1	0.0	887.3	15.554	1025.71	156256.0	441.0	0.0	352.7
W	14	136.0	40.00	15.55	4.120	0.941	6.93	22.34	216.00	77.00	427.6	0.0	388.7	8.982	229.45	51941.6	173.2	0.0	138.6
W	14	127.0	37.30	15.50	4.100	0.997	7.35	23.96	202.00	71.80	399.9	0.0	363.5	8.482	202.56	48104.6	161.5	0.0	129.2
W	14	119.0	35.00	15.46	4.080	1.055	7.80	25.43	189.00	67.10	374.2	0.0	340.1	8.014	179.11	44570.7	150.9	0.0	120.7
W	14	111.0	32.70	15.43	4.070	1.125	8.37	26.61	176.00	62.20	348.4	0.0	316.7	7.499	156.32	41301.8	139.9	0.0	111.9
W	14	103.0	30.30	15.38	4.050	1.202	8.96	28.78	164.00	57.60	324.7	0.0	295.2	7.057	136.37	38108.4	129.6	0.0	103.6
W	14	95.0	27.90	15.35	4.040	1.297	9.72	30.36	151.00	52.80	298.9	0.0	271.7	6.530	116.34	34914.6	118.8	0.0	95.0
W	14	87.0	25.60	15.30	4.020	1.403	10.53	33.33	138.00	48.20	273.2	0.0	248.3	6.027	98.33	31593.5	108.4	0.0	86.7
W	14	84.0	24.70	12.69	3.320	1.515	7.72	31.44	131.00	37.50	259.3	0.0	235.7	8.389	86.41	20455.7	84.3	0.0	67.5
W	14	78.0	22.90	12.66	3.310	1.631	8.35	32.85	121.00	34.50	239.5	0.0	217.7	7.795	74.14	18780.5	77.6	0.0	62.0
W	14	74.0	21.80	10.63	2.760	1.799	6.43	31.53	112.00	26.50	221.7	0.0	201.5	10.378	62.24	12086.5	59.6	0.0	47.6
W	14	68.0	20.00	10.59	2.740	1.950	6.99	33.63	103.00	24.10	203.9	0.0	185.3	9.684	52.80	10954.8	54.2	0.0	43.3
W	14	61.0	17.90	10.55	2.730	2.163	7.77	36.79	92.20	21.50	182.5	0.0	165.9	8.732	42.62	9734.7	48.3	0.0	38.6
W	14	53.0	15.60	8.50	2.180	2.627	6.12	37.67	77.80	14.30	154.0	0.0	140.0	11.555	29.60	5237.9	32.1	0.0	25.7
W	14	48.0	14.10	8.47	2.160	2.899	6.77	40.73	70.20	12.80	138.9	0.0	126.3	10.620	24.20	4639.9	28.7	0.0	23.0
W	14	43.0	12.60	8.44	2.140	3.238	7.57	44.41	62.70	11.30	124.1	0.0	112.8	9.664	19.36	4067.8	25.4	0.0	20.3
W	14	38.0	11.20	7.15	1.800	4.062	6.60	45.11	54.70	7.86	108.3	0.0	98.4	11.917	13.46	2510.7	17.6	0.0	14.1
W	14	34.0	10.00	7.12	1.780	4.578	7.45	48.78	48.60	6.89	96.2	0.0	87.4	10.827	10.61	2181.4	15.5	0.0	12.4

(1) M_{rx}. Compact sections; $b_f/2t_f \leqslant 10.83$; beams, $f_a = 0$, $d/t_w \leqslant 106.7$, columns, $f_a > 0.16\ F_y$, $d/t_w \leqslant 36.3$; braced so that $L_b \leqslant L_c$.
(2) M_{rx}. Semi-compact sections; same as column (1) except $10.83 < b_f/2t_f < 15.83$.
(3) M_{rx}. No-compact sections for tension; all section not under (1), (2), (4), (6); for compression; upper limit for compression sections under (4), (5), (6).
*Cols (4), (5), (6): Non-compact sections; moment factors (MF) for compression.
*(4) For $53.2\sqrt{C_b} \leqslant l_b/r_T < 119\sqrt{C_b}$, except channels. Use *larger* M_{rx} from (4) or (5). . . . . (4) M_{rx} = col. (1) – $(MF \times 10^{-4})\ l_b^2/C_b \leqslant$ Col. (3).
*(5) For all values l_b/r_T. . . . . (5) $M_{rx} = 1000\ (MF)\ C_b/l_b \leqslant$ Col. (3).
*(6) For $l_b/r_T \geqslant 119\sqrt{C_b}$, except channels. Use larger M_{rx} from columns (5) or (6). . . . . (6) $M_{rx} = 1000\ (MF)\ C_b/l_b^2 \leqslant$ Col. (3).
(7) M_{ry}. For doubly symmetrical shapes; $b_f/2t_f \leqslant 10.83$.
(8) M_{ry}. For doubly symmetrical shapes; $10.83 < b_f/2t_f < 15.83$.
(9) M_{ry}. For non-compact sections; sections not under column (7) or (8).

F_y = 36 ksi	TABLE 3-1: RESISTING MOMENTS, M_R, (kip-ft.) - F_y = 36 ksi															
	SECTION PROPERTIES								MOMENT M_{RX}, k-ft.			M_{RX} FACTORS*			MOMENT, M_{RY}, k-ft.		
Section	Area	L_c	r_T	d/A_f	$b_f/2t_f$	d/t_w	S_{x-x}	S_{y-y}	(1) $\frac{0.66 F_y S_x}{12} = M_{RX}$	(2) Eq. 1.5-5 $= M_{RX}$	(3) $\frac{0.60 F_y S_x}{12} = M_{RX}$	(4) Eq. 1.5-6a =(MF)	(5) Eq. 1.5-7 =(MF)	(6) Eq. 1.5-6b =(MF)	(7) $\frac{0.75 F_y S_y}{12} = M_{RY}$	(8) Eq. 1.5-5b $= M_{RY}$	(9) $\frac{0.60 F_y S_y}{12} = M_{RY}$
	sq.in.	(ft.)	in.	$in.^{-1}$	–	–	$in.^3$	$in.^3$									
W 14 30.0	8.83	7.10	1.750	5.374	8.78	51.33	41.90	5.80	82.9	0.0	75.4	9.657	7.79	1817.8	13.0	0.0	10.4
W 14 26.0	7.67	5.30	1.290	6.612	6.01	54.47	35.10	3.53	69.4	0.0	63.1	14.888	5.30	827.4	7.9	0.0	6.3
W 14 22.0	6.49	5.27	1.260	8.191	7.46	59.65	28.90	2.80	57.2	0.0	52.0	12.849	3.52	649.9	6.3	0.0	5.0
W 12 190.0	55.90	13.37	3.590	0.653	3.64	13.56	263.00	93.10	520.7	0.0	473.3	14.404	402.27	48018.9	209.4	0.0	167.5
W 12 161.0	47.40	13.21	3.530	0.746	4.21	15.33	222.00	77.70	439.5	0.0	399.5	12.575	297.44	39189.5	174.8	0.0	139.8
W 12 133.0	39.10	13.05	3.470	0.875	5.00	17.72	183.00	63.10	362.3	0.0	329.3	10.728	209.02	31216.0	141.9	0.0	113.5
W 12 120.0	35.30	13.00	3.440	0.962	5.56	18.47	163.00	56.00	322.7	0.0	293.3	9.723	169.28	27325.7	126.0	0.0	100.7
W 12 106.0	31.20	12.90	3.410	1.068	6.20	20.77	145.00	49.20	287.1	0.0	261.0	8.802	135.75	23886.0	110.6	0.0	88.5
W 12 99.0	29.10	12.86	3.390	1.135	6.61	21.90	135.00	45.70	267.2	0.0	242.9	8.292	118.89	21978.6	102.8	0.0	82.2
W 12 92.0	27.10	12.83	3.380	1.212	7.09	23.15	125.00	42.20	247.5	0.0	224.9	7.723	103.05	20230.7	94.9	0.0	75.9
W 12 85.0	25.00	12.77	3.360	1.297	7.60	25.25	116.00	38.90	229.6	0.0	208.7	7.252	89.41	18552.5	87.5	0.0	70.0
W 12 79.0	23.20	12.75	3.340	1.392	8.20	26.34	107.00	35.80	211.8	0.0	192.5	6.770	76.84	16910.0	80.5	0.0	64.4
W 12 72.0	21.20	12.70	3.330	1.516	8.97	28.48	97.50	32.40	193.0	0.0	175.5	6.206	64.30	15316.5	72.8	0.0	58.3
W 12 65.0	19.10	12.66	3.310	1.666	9.90	31.07	88.00	29.10	174.2	0.0	158.3	5.669	52.80	13658.6	65.4	0.0	52.3
W 12 58.0	17.10	10.57	2.750	1.899	7.81	33.95	78.10	21.40	154.6	0.0	140.5	7.289	41.12	8367.2	48.1	0.0	38.5
W 12 53.0	15.60	10.55	2.740	2.093	8.68	34.95	70.70	19.20	139.9	0.0	127.2	6.647	33.76	7519.4	43.1	0.0	34.5
W 12 50.0	14.70	8.52	2.190	2.354	6.30	32.85	64.70	14.00	128.1	0.0	116.4	9.522	27.47	4396.0	31.5	0.0	25.1
W 12 45.0	13.20	8.48	2.180	2.603	6.98	35.89	58.20	12.40	115.2	0.0	104.7	8.644	22.35	3918.3	27.9	0.0	22.3
W 12 40.0	11.80	8.44	2.160	2.892	7.75	40.61	51.90	11.00	102.7	0.0	93.4	7.852	17.94	3430.3	24.7	0.0	19.7
W 12 36.0	10.60	6.92	1.770	3.452	6.07	40.13	46.00	7.77	91.0	0.0	82.7	10.364	13.32	2041.6	17.4	0.0	13.9
W 12 31.0	9.13	6.88	1.750	3.984	7.01	45.62	39.50	6.61	78.2	0.0	71.0	9.104	9.91	1713.7	14.8	0.0	11.8
W 12 27.0	7.95	6.85	1.740	4.602	8.12	50.46	34.20	5.63	67.7	0.0	61.5	7.973	7.43	1466.8	12.6	0.0	10.1

W	12	22.0	6.47	4.25	1.030	7.204	4.75	47.34	25.30	2.31	50.0	0.0	45.5	16.833	3.51	380.2	5.1	0.0	4.1
W	12	19.0	5.59	4.22	1.010	8.695	5.74	51.30	21.30	1.88	42.1	0.0	38.3	14.739	2.44	307.8	4.2	0.0	3.3
W	12	16.5	4.87	4.15	0.975	11.152	7.43	52.17	17.60	1.44	34.8	0.0	31.6	13.068	1.57	237.0	3.2	0.0	2.5
W	12	14.0	4.12	3.45	0.957	13.399	8.85	60.15	14.80	1.18	29.3	0.0	26.6	11.406	1.10	192.0	2.6	0.0	2.1
W	10	112.0	32.90	10.99	2.940	0.875	4.17	15.07	126.00	45.20	249.4	0.0	226.7	10.289	143.91	15428.8	101.6	0.0	81.3
W	10	100.0	29.40	10.91	2.910	0.961	4.62	16.23	112.00	39.90	221.7	0.0	201.5	9.336	116.48	13436.0	89.7	0.0	71.8
W	10	89.0	26.20	10.84	2.880	1.061	5.14	17.69	99.70	35.20	197.4	0.0	179.4	8.484	93.96	11715.1	79.1	0.0	63.3
W	10	77.0	22.70	10.76	2.850	1.200	5.87	19.85	86.10	30.10	170.4	0.0	154.9	7.482	71.74	9907.4	67.7	0.0	54.1
W	10	72.0	21.20	10.73	2.840	1.277	6.29	20.58	80.10	27.90	158.5	0.0	144.1	7.010	62.68	9152.4	62.7	0.0	50.2
W	10	66.0	19.40	10.67	2.820	1.371	6.76	22.71	73.70	25.50	145.9	0.0	132.6	6.541	53.73	8302.9	57.3	0.0	45.8
W	10	60.0	17.70	10.63	2.800	1.489	7.37	24.69	67.10	23.10	132.8	0.0	120.7	6.041	45.04	7452.5	51.9	0.0	41.5
W	10	54.0	15.90	10.58	2.780	1.632	8.11	27.50	60.40	20.70	119.5	0.0	108.7	5.516	36.98	6612.9	46.5	0.0	37.2
W	10	49.0	14.40	10.55	2.770	1.792	8.96	29.41	54.60	18.60	108.1	0.0	98.2	5.023	30.46	5934.9	41.8	0.0	33.4
W	10	45.0	13.20	8.46	2.210	2.041	6.49	28.91	49.10	13.30	97.2	0.0	88.3	7.096	24.05	3397.2	29.9	0.0	23.9
W	10	39.0	11.50	8.43	2.190	2.356	7.56	31.25	42.20	11.20	83.5	0.0	75.9	6.210	17.91	2867.2	25.2	0.0	20.1
W	10	33.0	9.71	8.40	2.160	2.827	9.19	33.39	35.00	9.16	69.3	0.0	62.9	5.295	12.37	2313.3	20.6	0.0	16.4
W	10	29.0	8.54	6.12	1.570	3.524	5.79	35.36	30.80	5.61	60.9	0.0	55.4	8.820	8.73	1075.5	12.6	0.0	10.0
W	10	25.0	7.36	6.08	1.560	4.068	6.70	40.00	26.50	4.76	52.4	0.0	47.6	7.686	6.51	913.6	10.7	0.0	8.5
W	10	21.0	6.20	6.06	1.530	5.063	8.45	41.25	21.50	3.75	42.5	0.0	38.6	6.483	4.24	712.9	8.4	0.0	6.7
W	10	19.0	5.61	4.24	1.050	6.471	5.10	41.00	18.80	2.13	37.2	0.0	33.8	12.036	2.90	293.6	4.7	0.0	3.8

(1) M_{rx}. Compact sections; $b_f/2t_f \leqslant 10.83$; beams, $f_a = 0$, $d/t_w \leqslant 106.7$, columns, $f_a > 0.16\ F_y$, $d/t_w \leqslant 36.3$; braced so that $L_b \leqslant L_c$.
(2) M_{rx}. Semi-compact sections; same as column (1) except $10.83 < b_f/2t_f < 15.83$.
(3) M_{rx}. No-compact sections for tension; all section not under (1), (2), (4), (6); for compression; upper limit for compression sections under (4), (5), (6).
*Cols (4), (5), (6): Non-compact sections; moment factors (*MF*) for compression.
*(4) For $53.2\sqrt{C_b} \leqslant l_b/r_T < 119\sqrt{C_b}$, except channels. Use *larger* M_{rx} from (4) or (5). . . . . (4) M_{rx} = col. (1) − $(MF \times 10^{-4})\ l_b^2/C_b \leqslant$ Col. (3).
*(5) For all values l_b/r_T. . . . . (5) $M_{rx} = 1000\ (MF)\ C_b/l_b \leqslant$ Col. (3).
*(6) For $l_b/r_T \geqslant 119\sqrt{C_b}$, except channels. Use larger M_{rx} from columns (5) or (6). . . . . (6) $M_{rx} = 1000\ (MF)\ C_b/l_b^2 \leqslant$ Col. (3).
(7) M_{ry}. For doubly symmetrical shapes; $b_f/2t_f \leqslant 10.83$.
(8) M_{ry}. For doubly symmetrical shapes; $10.83 < b_f/2t_f < 15.83$.
(9) M_{ry}. For non-compact sections; sections not under column (7) or (8).

F_y = 36 ksi	TABLE 3-1: RESISTING MOMENTS, M_R, (kip-ft.) - F_y = 36 ksi																
	SECTION PROPERTIES								MOMENT M_{RX}, k-ft.			M_{RX} FACTORS*			MOMENT, M_{RY}, k-ft.		
Section	Area	L_c	r_T	d/A_f	$b_f/2t_f$	d/t_w	S_{x-x}	S_{y-y}	(1) $\frac{0.66 F_y S_x}{12} = M_{RX}$	(2) Eq. 1.5-5 $= M_{RX}$	(3) $\frac{0.60 F_y S_x}{12} = M_{RX}$	(4) Eq. 1.5-6a =(MF)	(5) Eq. 1.5-7 =(MF)	(6) Eq. 1.5-6b =(MF)	(7) $\frac{0.75 F_y S_y}{12} = M_{RY}$	(8) Eq. 1.5-5b $= M_{RY}$	(9) $\frac{0.60 F_y S_y}{12} = M_{RY}$
	sq. in.	(ft.)	in.	$in.^{-1}$	–	–	$in.^3$	$in.^3$									
W 10 17.0	4.99	4.23	1.030	7.670	6.09	42.16	16.20	1.77	32.0	0.0	29.1	10.778	2.11	243.4	3.9	0.0	3.1
W 10 15.0	4.41	4.22	1.000	9.293	7.43	43.47	13.80	1.44	27.3	0.0	24.8	9.741	1.48	195.5	3.2	0.0	2.5
W 10 11.5	3.39	3.77	0.975	12.248	9.68	54.83	10.50	1.06	20.7	0.0	18.8	7.796	0.85	141.4	2.3	0.0	1.9
W 8 67.0	19.70	8.74	2.330	1.164	4.44	15.65	60.40	21.40	119.5	0.0	108.7	7.853	51.88	4645.3	48.1	0.0	38.5
W 8 58.0	17.10	8.67	2.310	1.317	5.08	17.15	52.00	18.20	102.9	0.0	93.5	6.878	39.48	3930.9	40.9	0.0	32.7
W 8 48.0	14.10	8.56	2.270	1.533	5.94	20.98	43.20	15.00	85.5	0.0	77.7	5.917	28.17	3153.5	33.7	0.0	26.9
W 8 40.0	11.80	8.52	2.240	1.830	7.23	22.60	35.50	12.10	70.2	0.0	63.8	4.994	19.39	2523.4	27.2	0.0	21.7
W 8 35.0	10.30	8.47	2.220	2.051	8.14	25.77	31.10	10.60	61.5	0.0	55.9	4.454	15.15	2171.3	23.8	0.0	19.0
W 8 31.0	9.12	8.44	2.210	2.309	9.23	27.77	27.40	9.24	54.2	0.0	49.3	3.960	11.86	1895.8	20.7	0.0	16.6
W 8 28.0	8.23	6.90	1.800	2.661	7.06	28.28	24.30	6.61	48.1	0.0	43.7	5.294	9.12	1115.3	14.8	0.0	11.8
W 8 24.0	7.06	6.86	1.780	3.065	8.16	32.36	20.80	5.61	41.1	0.0	37.4	4.633	6.78	933.6	12.6	0.0	10.0
W 8 20.0	5.89	5.56	1.420	4.087	6.96	32.82	17.00	3.50	33.6	0.0	30.5	5.951	4.15	485.6	7.8	0.0	6.2
W 8 17.0	5.01	5.54	1.400	4.947	8.52	34.78	14.10	2.83	27.9	0.0	25.3	5.078	2.84	391.5	6.3	0.0	5.0
W 8 15.0	4.43	4.23	1.040	6.440	6.39	33.14	11.80	1.69	23.3	0.0	21.2	7.701	1.83	180.8	3.8	0.0	3.0
W 8 13.0	3.83	4.22	1.020	7.874	7.87	34.78	9.90	1.36	19.6	0.0	17.8	6.716	1.25	145.9	3.0	0.0	2.4
W 8 10.0	2.96	4.15	1.000	9.828	9.65	46.47	7.80	1.06	15.4	0.0	14.0	5.505	0.79	110.5	2.3	0.0	1.9
W 6 25.0	7.35	6.41	1.690	2.297	6.66	19.90	16.70	5.62	33.0	0.0	30.0	4.127	7.26	675.7	12.6	0.0	10.1
W 6 20.0	5.88	6.35	1.660	2.807	8.19	24.03	13.40	4.43	26.5	0.0	24.1	3.432	4.77	523.1	9.9	0.0	7.9
W 6 15.5	4.56	6.32	1.630	3.720	11.14	25.53	10.00	3.23	19.7	19.6	18.0	2.656	2.68	376.3	0.0	7.1	5.8
W 6 16.0	4.72	4.25	1.100	3.838	4.98	24.03	10.20	2.19	20.1	0.0	18.3	5.950	2.65	174.8	4.9	0.0	3.9
W 6 12.0	3.54	4.22	1.070	5.376	7.16	26.08	7.25	1.49	14.3	0.0	13.0	4.469	1.34	117.5	3.3	0.0	2.6
W 6 8.5	2.51	4.15	1.040	7.627	10.15	34.29	5.08	1.01	10.0	0.0	9.1	3.315	0.66	77.8	2.2	0.0	1.8

W	5	18.5	5.43	5.30	1.400	2.425	5.98	19.32	9.94	3.54	19.6	0.0	17.8	3.579	4.09	276.0	7.9	0.0	6.3
W	5	16.0	4.70	5.27	1.390	2.777	6.94	20.83	8.53	3.00	16.8	0.0	15.3	3.116	3.07	233.4	6.7	0.0	5.3
W	4	13.0	3.82	4.28	1.110	2.969	5.88	14.85	5.45	1.85	10.7	0.0	9.8	3.122	1.83	95.1	4.1	0.0	3.3
M	14	17.2	5.05	3.59	0.925	12.867	7.35	66.66	21.10	1.33	41.7	0.0	37.9	17.407	1.63	255.7	2.9	0.0	2.3
M	12	11.8	3.47	2.66	0.690	17.400	6.81	67.79	12.00	0.63	23.7	0.0	21.5	17.791	0.68	80.9	1.4	0.0	1.1
M	10	29.1	8.56	6.26	1.400	4.277	7.63	23.13	26.60	3.76	52.6	0.0	47.8	9.579	6.21	738.5	8.4	0.0	6.7
M	10	22.9	6.73	6.07	1.400	4.415	7.39	40.82	23.60	3.48	46.7	0.0	42.4	8.499	5.34	655.2	7.8	0.0	6.2
M	10	9.0	2.65	2.56	0.616	18.045	6.52	63.69	7.76	0.45	15.3	0.0	13.9	14.435	0.43	41.7	1.0	0.0	0.8
M	8	34.3	10.10	8.44	2.080	2.177	8.71	21.16	29.10	8.73	57.6	0.0	52.3	4.747	13.36	1783.5	19.6	0.0	15.7
M	8	32.6	9.58	8.38	2.080	2.195	8.64	25.39	28.40	8.58	56.2	0.0	51.1	4.633	12.93	1740.6	19.3	0.0	15.4
M	8	22.5	6.60	5.69	1.280	4.200	7.64	21.33	17.10	2.77	33.8	0.0	30.7	7.367	4.07	396.9	6.2	0.0	4.9
M	8	18.5	5.44	5.54	1.280	4.316	7.43	34.78	15.50	2.60	30.6	0.0	27.8	6.677	3.59	359.7	5.8	0.0	4.6
M	8	6.5	1.92	2.40	0.535	18.556	6.03	59.25	4.62	0.30	9.1	0.0	8.3	11.393	0.24	18.7	0.6	0.0	0.5
M	7	5.5	1.62	2.19	0.493	18.696	5.77	54.68	3.44	0.23	6.8	0.0	6.1	9.990	0.18	11.8	0.5	0.0	0.4
M	6	22.5	6.62	6.39	1.550	2.612	7.99	16.12	13.70	4.08	27.1	0.0	24.6	4.025	5.24	466.2	9.1	0.0	7.3
M	6	20.0	5.89	6.26	1.540	2.666	7.83	24.00	13.00	3.90	25.7	0.0	23.3	3.869	4.87	436.7	8.7	0.0	7.0
M	6	4.4	1.29	1.94	0.444	19.028	5.39	52.63	2.40	0.17	4.7	0.0	4.3	8.593	0.12	6.7	0.4	0.0	0.3
M	5	18.9	5.55	5.28	1.320	2.402	6.01	15.82	9.63	3.14	19.0	0.0	17.3	3.901	4.00	237.7	7.0	0.0	5.6
M	4	13.8	4.06	4.22	1.050	2.695	5.39	12.77	5.42	1.79	10.7	0.0	9.7	3.470	2.01	84.6	4.0	0.0	3.2
M	4	13.0	3.81	4.15	1.040	2.736	5.30	15.74	5.24	1.71	10.3	0.0	9.4	3.419	1.91	80.2	3.8	0.0	3.0

(1) M_{rx}. Compact sections; $b_f/2t_f \leqslant 10.83$; beams, $f_a = 0$, $d/t_w \leqslant 106.7$, columns, $f_a > 0.16\,F_y$, $d/t_w \leqslant 36.3$; braced so that $L_b \leqslant L_c$.
(2) M_{rx}. Semi-compact sections; same as column (1) except $10.83 < b_f/2t_f < 15.83$.
(3) M_{rx}. No-compact sections for tension; all section not under (1), (2), (4), (6); for compression; upper limit for compression sections under (4), (5), (6).
*Cols (4), (5), (6): Non-compact sections; moment factors (*MF*) for compression.
*(4) For $53.2\sqrt{C_b} \leqslant l_b/r_T < 119\sqrt{C_b}$, except channels. Use *larger* M_{rx} from (4) or (5). . . . . (4) M_{rx} = col. (1) − $(MF \times 10^{-4})\, l_b^2/C_b \leqslant$ Col. (3).
*(5) For all values l_b/r_T. . . . . (5) $M_{rx} = 1000\,(MF)\,C_b/l_b \leqslant$ Col. (3).
*(6) For $l_b/r_T \geqslant 119\sqrt{C_b}$, except channels. Use larger M_{rx} from columns (5) or (6). . . . . (6) $M_{rx} = 1000\,(MF)\,C_b/l_b^2 \leqslant$ Col. (3).
(7) M_{ry}. For doubly symmetrical shapes; $b_f/2t_f \leqslant 10.83$.
(8) M_{ry}. For doubly symmetrical shapes; $10.83 < b_f/2t_f < 15.83$.
(9) M_{ry}. For non-compact sections; sections not under column (7) or (8).

TABLE 3-1: RESISTING MOMENTS, M_R, (kip-ft.) - F_y = 36 ksi

F_y = 36 ksi

Section	SECTION PROPERTIES								MOMENT M_{RX}, k-ft.			M_{RX} FACTORS*			MOMENT, M_{RY}, k-ft.		
	Area	L_c	r_T	d/A_f	$b_f/2t_f$	d/t_w	S_{x-x}	S_{y-y}	(1) $\frac{0.66 F_y S_x}{12} = M_{RX}$	(2) Eq. 1.5-5 $= M_{RX}$	(3) $\frac{0.60 F_y S_x}{12} = M_{RX}$	(4) Eq. 1.5-6a =(MF)	(5) Eq. 1.5-7 =(MF)	(6) Eq. 1.5-6b =(MF)	(7) $\frac{0.75 F_y S_y}{12} = M_{RY}$	(8) Eq. 1.5-5b $= M_{RY}$	(9) $\frac{0.60 F_y S_y}{12} = M_{RY}$
	sq. in.	(ft.)	in.	in.$^{-1}$	–	–	in.^{3}	in.^{3}									
S 24 120.0	35.30	8.49	1.930	2.706	3.65	30.07	252.00	20.90	498.9	0.0	453.5	47.754	93.12	13297.8	47.0	0.0	37.6
S 24 105.9	31.10	8.31	1.930	2.765	3.57	38.40	236.00	19.80	467.2	0.0	424.7	44.722	85.33	12453.5	44.5	0.0	35.6
S 24 100.0	29.40	7.64	1.650	3.802	4.16	32.12	199.00	13.20	394.0	0.0	358.1	51.596	52.33	7675.1	29.7	0.0	23.7
S 24 90.0	26.50	7.51	1.650	3.867	4.08	38.46	187.00	12.60	370.2	0.0	336.5	48.484	48.34	7212.3	28.3	0.0	22.6
S 24 79.9	23.50	7.38	1.660	3.935	4.01	47.90	175.00	12.10	346.5	0.0	315.0	44.828	44.46	6831.5	27.2	0.0	21.7
S 20 95.0	27.90	7.59	1.700	3.032	3.93	25.00	161.00	13.80	318.7	0.0	289.7	39.324	53.09	6591.6	31.0	0.0	24.8
S 20 85.0	25.00	7.44	1.690	3.095	3.84	30.62	152.00	13.10	300.9	0.0	273.5	37.566	49.10	6150.1	29.4	0.0	23.5
S 20 75.0	22.10	6.74	1.480	3.966	4.05	31.20	128.00	9.28	253.4	0.0	230.3	41.249	32.27	3971.9	20.8	0.0	16.7
S 20 65.4	19.20	6.59	1.480	4.055	3.96	40.00	118.00	8.77	233.6	0.0	212.3	38.026	29.09	3661.6	19.7	0.0	15.7
S 18 70.0	20.60	6.59	1.410	4.167	4.52	25.31	103.00	7.72	203.9	0.0	185.3	36.570	24.71	2900.9	17.3	0.0	13.8
S 18 54.7	16.10	6.33	1.410	4.340	4.34	39.04	89.40	6.94	177.0	0.0	160.9	31.741	20.59	2517.9	15.6	0.0	12.4
S 15 50.0	14.70	5.95	1.310	4.275	4.53	27.27	64.80	5.57	128.3	0.0	116.6	26.654	15.15	1575.3	12.5	0.0	10.0
S 15 42.9	12.60	5.80	1.310	4.383	4.42	36.49	59.60	5.23	118.0	0.0	107.2	24.515	13.59	1448.9	11.7	0.0	9.4
S 12 50.0	14.70	5.78	1.310	3.324	4.15	17.46	50.80	5.74	100.5	0.0	91.4	20.895	15.27	1235.0	12.9	0.0	10.3
S 12 40.8	12.00	5.54	1.280	3.467	3.98	25.97	45.40	5.16	89.8	0.0	81.7	19.559	13.09	1053.7	11.6	0.0	9.2
S 12 35.0	10.30	5.36	1.200	4.344	4.66	28.03	38.20	3.89	75.6	0.0	68.7	18.725	8.79	779.2	8.7	0.0	7.0
S 12 31.8	9.35	5.27	1.200	4.411	4.59	34.28	36.40	3.74	72.0	0.0	65.5	17.843	8.25	742.5	8.4	0.0	6.7
S 10 35.0	10.30	5.21	1.150	4.119	5.03	16.83	29.40	3.38	58.2	0.0	52.9	15.692	7.13	550.8	7.6	0.0	6.0
S 10 25.4	7.46	4.91	1.130	4.369	4.74	32.15	24.70	2.91	48.9	0.0	44.4	13.654	5.65	446.8	6.5	0.0	5.2
S 8 23.0	6.77	4.40	0.987	4.512	4.90	18.14	16.20	2.07	32.0	0.0	29.1	11.738	3.58	223.5	4.6	0.0	3.7
S 8 18.4	5.41	4.22	0.973	4.704	4.70	29.52	14.40	1.86	28.5	0.0	25.9	10.736	3.06	193.1	4.1	0.0	3.3
S 7 20.0	5.88	4.07	0.914	4.626	4.92	15.55	12.10	1.64	23.9	0.0	21.7	10.224	2.61	143.2	3.6	0.0	2.9

S	7	15.3	4.50	3.86	0.894	4.876	4.67	27.77	10.50	1.44	20.7	0.0	18.8	9.273	2.15	118.8	3.2	0.0	2.5
S	6	17.2	5.07	3.76	0.845	4.688	4.96	12.90	8.77	1.30	17.3	0.0	15.7	8.669	1.87	88.7	2.9	0.0	2.3
S	6	12.5	3.67	3.51	0.817	5.015	4.64	25.86	7.37	1.09	14.5	0.0	13.2	7.793	1.46	69.6	2.4	0.0	1.9
S	5	14.75	4.34	3.46	0.781	4.670	5.03	10.12	6.09	1.01	12.0	0.0	10.9	7.047	1.30	52.6	2.2	0.0	1.8
S	5	10.0	2.94	3.17	0.741	5.105	4.60	23.36	4.92	0.80	9.7	0.0	8.8	6.325	0.96	38.2	1.8	0.0	1.4
S	4	9.5	2.79	2.95	0.684	4.882	4.77	12.26	3.39	0.64	6.7	0.0	6.1	5.114	0.69	22.4	1.4	0.0	1.1
S	4	7.7	2.26	2.81	0.662	5.126	4.54	20.72	3.04	0.57	6.0	0.0	5.4	4.896	0.59	18.8	1.2	0.0	1.0
S	3	7.5	2.21	2.64	0.621	4.598	4.82	8.59	1.95	0.46	3.8	0.0	3.5	3.569	0.42	10.6	1.0	0.0	0.8
S	3	5.7	1.67	2.45	0.585	4.952	4.48	17.64	1.68	0.39	3.3	0.0	3.0	3.465	0.33	8.1	0.8	0.0	0.7
HP	14	117.0	34.40	15.71	4.060	1.187	9.24	17.67	173.00	59.50	342.5	0.0	311.3	7.408	145.67	40398.5	133.8	0.0	107.0
HP	14	102.0	30.00	15.60	4.020	1.348	10.50	19.92	150.00	51.30	297.0	0.0	270.0	6.551	111.27	34340.8	115.4	0.0	92.3
HP	14	89.0	26.20	15.51	3.990	1.531	11.92	22.50	131.00	44.40	259.3	254.2	235.7	5.808	85.56	29545.0	0.0	95.5	79.9
HP	14	73.0	21.50	15.39	3.940	1.848	14.41	26.95	108.00	35.90	213.8	199.9	194.3	4.910	58.43	23751.1	0.0	69.2	64.6
HP	12	74.0	21.80	12.89	3.310	1.634	10.06	19.96	93.40	30.20	184.9	0.0	168.1	6.017	57.14	14496.7	67.9	0.0	54.3
HP	12	53.0	15.60	12.71	3.230	2.242	13.81	27.01	66.90	21.10	132.4	125.2	120.4	4.526	29.82	9887.7	0.0	41.8	37.9
HP	10	57.0	16.80	10.79	2.780	1.735	9.06	17.74	58.80	19.70	116.4	0.0	105.8	5.370	33.87	6437.7	44.3	0.0	35.4
HP	10	42.0	12.40	10.63	2.720	2.307	12.05	23.25	43.40	14.20	85.9	84.0	78.1	4.140	18.80	4548.7	0.0	30.3	25.5
HP	8	36.0	10.60	8.61	2.220	2.206	9.14	18.00	29.90	9.91	59.2	0.0	53.8	4.282	13.54	2087.5	22.2	0.0	17.8
C	15	50.0	14.70	0.00	0.000	6.210	0.00	0.00	53.80	3.78	0.0	0.0	96.8	0.000	8.66	0.0	0.0	0.0	6.8
C	15	40.0	11.80	0.00	0.000	6.555	0.00	0.00	46.50	3.36	0.0	0.0	83.6	0.000	7.09	0.0	0.0	0.0	6.0

(1) M_{rx}. Compact sections; $b_f/2t_f \leqslant 10.83$; beams, $f_a = 0$, $d/t_w \leqslant 106.7$, columns, $f_a > 0.16\ F_y$, $d/t_w \leqslant 36.3$; braced so that $L_b \leqslant L_c$.
(2) M_{rx}. Semi-compact sections; same as column (1) except $10.83 < b_f/2t_f < 15.83$.
(3) M_{rx}. No-compact sections for tension; all section not under (1), (2), (4), (6); for compression; upper limit for compression sections under (4), (5), (6).
*Cols (4), (5), (6): Non-compact sections; moment factors (*MF*) for compression.
*(4) For $53.2\sqrt{C_b} \leqslant l_b/r_T < 119\sqrt{C_b}$, except channels. Use *larger* M_{rx} from (4) or (5). . . . . (4) M_{rx} = col. (1) – $(MF \times 10^{-4})\ l_b^2/C_b \leqslant$ Col. (3).
*(5) For all values l_b/r_T. . . . . (5) $M_{rx} = 1000\ (MF)\ C_b/l_b \leqslant$ Col. (3).
*(6) For $l_b/r_T \geqslant 119\sqrt{C_b}$, except channels. Use larger M_{rx} from columns (5) or (6). . . . . (6) $M_{rx} = 1000\ (MF)\ C_b/l_b^2 \leqslant$ Col. (3).
(7) M_{ry}. For doubly symmetrical shapes; $b_f/2t_f \leqslant 10.83$.
(8) M_{ry}. For doubly symmetrical shapes; $10.83 < b_f/2t_f < 15.83$.
(9) M_{ry}. For non-compact sections; sections not under column (7) or (8).

F_y = 36 ksi	TABLE 3-1: RESISTING MOMENTS, M_R, (kip-ft.) - F_y = 36 ksi																
	SECTION PROPERTIES								MOMENT M_{RX}, k-ft.			M_{RX} FACTORS*			MOMENT, M_{RY}, k-ft.		
Section	Area	L_c	r_T	d/A_f	$b_f/2t_f$	d/t_w	S_{x-x}	S_{y-y}	(1) $\frac{0.66 F_y S_x}{12} = M_{RX}$	(2) Eq. 1.5-5 $= M_{RX}$	(3) $\frac{0.60 F_y S_x}{12} = M_{RX}$	(4) Eq. 1.5-6a =(MF)	(5) Eq. 1.5-7 =(MF)	(6) Eq. 1.5-6 b =(MF)	(7) $\frac{0.75 F_y S_y}{12} = M_{RY}$	(8) Eq. 1.5-5b $= M_{RY}$	(9) $\frac{0.60 F_y S_y}{12} = M_{RY}$
	sq. in.	(ft.)	in.	$in.^{-1}$	–	–	$in.^3$	$in.^3$									
C 15 33.9	9.96	0.00	0.000	6.787	0.00	0.00	42.00	3.11	0.0	0.0	75.5	0.000	6.18	0.0	0.0	0.0	5.5
C 12 30.0	8.82	0.00	0.000	7.555	0.00	0.00	27.00	2.06	0.0	0.0	48.5	0.000	3.57	0.0	0.0	0.0	3.7
C 12 25.0	7.35	0.00	0.000	7.860	0.00	0.00	24.10	1.88	0.0	0.0	43.3	0.000	3.06	0.0	0.0	0.0	3.3
C 12 20.7	6.09	0.00	0.000	8.141	0.00	0.00	21.50	1.73	0.0	0.0	38.6	0.000	2.64	0.0	0.0	0.0	3.1
C 10 30.0	8.82	0.00	0.000	7.562	0.00	0.00	20.70	1.65	0.0	0.0	37.2	0.000	2.73	0.0	0.0	0.0	2.9
C 10 25.0	7.35	0.00	0.000	7.947	0.00	0.00	18.20	1.48	0.0	0.0	32.7	0.000	2.29	0.0	0.0	0.0	2.6
C 10 20.0	5.88	0.00	0.000	8.373	0.00	0.00	15.80	1.32	0.0	0.0	28.4	0.000	1.88	0.0	0.0	0.0	2.3
C 10 15.3	4.49	0.00	0.000	8.821	0.00	0.00	13.50	1.16	0.0	0.0	24.2	0.000	1.53	0.0	0.0	0.0	2.0
C 9 20.0	5.88	0.00	0.000	8.229	0.00	0.00	13.50	1.17	0.0	0.0	24.2	0.000	1.64	0.0	0.0	0.0	2.1
C 9 15.0	4.41	0.00	0.000	8.769	0.00	0.00	11.30	1.01	0.0	0.0	20.3	0.000	1.28	0.0	0.0	0.0	1.8
C 9 13.4	3.94	0.00	0.000	8.956	0.00	0.00	10.60	0.96	0.0	0.0	19.0	0.000	1.18	0.0	0.0	0.0	1.7
C 8 18.7	5.51	0.00	0.000	8.117	0.00	0.00	11.00	1.01	0.0	0.0	19.7	0.000	1.35	0.0	0.0	0.0	1.8
C 8 13.7	4.04	0.00	0.000	8.754	0.00	0.00	9.03	0.85	0.0	0.0	16.2	0.000	1.03	0.0	0.0	0.0	1.5
C 8 11.5	3.38	0.00	0.000	9.076	0.00	0.00	8.14	0.78	0.0	0.0	14.6	0.000	0.89	0.0	0.0	0.0	1.4
C 7 14.7	4.33	0.00	0.000	8.319	0.00	0.00	7.78	0.77	0.0	0.0	14.0	0.000	0.93	0.0	0.0	0.0	1.4
C 7 12.2	3.60	0.00	0.000	8.717	0.00	0.00	6.93	0.70	0.0	0.0	12.4	0.000	0.79	0.0	0.0	0.0	1.2
C 7 9.8	2.87	0.00	0.000	9.151	0.00	0.00	6.08	0.62	0.0	0.0	10.9	0.000	0.66	0.0	0.0	0.0	1.1
C 6 13.0	3.83	0.00	0.000	8.109	0.00	0.00	5.80	0.64	0.0	0.0	10.4	0.000	0.71	0.0	0.0	0.0	1.1
C 6 10.5	3.09	0.00	0.000	8.600	0.00	0.00	5.06	0.56	0.0	0.0	9.1	0.000	0.58	0.0	0.0	0.0	1.0
C 6 8.2	2.40	0.00	0.000	9.110	0.00	0.00	4.38	0.49	0.0	0.0	7.8	0.000	0.48	0.0	0.0	0.0	0.8
C 5 9.0	2.64	0.00	0.000	8.289	0.00	0.00	3.56	0.44	0.0	0.0	6.4	0.000	0.42	0.0	0.0	0.0	0.8
C 5 6.7	1.97	0.00	0.000	8.928	0.00	0.00	3.00	0.37	0.0	0.0	5.3	0.000	0.33	0.0	0.0	0.0	0.6

C	4	7.2	2.13	0.00	0.000	7.852	0.00	0.00	2.29	0.34	0.0	0.0	4.1	0.000	0.29	0.0	0.0	0.0	0.6
C	4	5.4	1.59	0.00	0.000	8.531	0.00	0.00	1.93	0.28	0.0	0.0	3.4	0.000	0.22	0.0	0.0	0.0	0.5
C	3	6.0	1.76	0.00	0.000	6.885	0.00	0.00	1.38	0.26	0.0	0.0	2.4	0.000	0.20	0.0	0.0	0.0	0.4
C	3	5.0	1.47	0.00	0.000	7.335	0.00	0.00	1.24	0.23	0.0	0.0	2.2	0.000	0.16	0.0	0.0	0.0	0.4
C	3	4.1	1.21	0.00	0.000	7.793	0.00	0.00	1.10	0.20	0.0	0.0	1.9	0.000	0.14	0.0	0.0	0.0	0.3
MC	18	58.0	17.10	0.00	0.000	6.857	0.00	0.00	75.10	5.32	0.0	0.0	135.1	0.000	10.95	0.0	0.0	0.0	9.5
MC	18	51.9	15.30	0.00	0.000	7.024	0.00	0.00	69.70	5.07	0.0	0.0	125.4	0.000	9.92	0.0	0.0	0.0	9.1
MC	18	45.8	13.50	0.00	0.000	7.200	0.00	0.00	64.30	4.82	0.0	0.0	115.7	0.000	8.93	0.0	0.0	0.0	8.6
MC	18	42.7	12.60	0.00	0.000	7.291	0.00	0.00	61.60	4.69	0.0	0.0	110.8	0.000	8.44	0.0	0.0	0.0	8.4
MC	13	50.0	14.70	0.00	0.000	4.830	0.00	0.00	48.40	4.79	0.0	0.0	87.1	0.000	10.01	0.0	0.0	0.0	8.6
MC	13	40.0	11.80	0.00	0.000	5.092	0.00	0.00	42.00	4.26	0.0	0.0	75.5	0.000	8.24	0.0	0.0	0.0	7.6
MC	13	35.0	10.30	0.00	0.000	5.233	0.00	0.00	30.80	3.99	0.0	0.0	69.8	0.000	7.41	0.0	0.0	0.0	7.1
MC	13	31.8	9.35	0.00	0.000	5.327	0.00	0.00	36.80	3.81	0.0	0.0	66.2	0.000	6.90	0.0	0.0	0.0	6.8
MC	12	50.0	14.70	0.00	0.000	4.145	0.00	0.00	44.90	5.65	0.0	0.0	80.8	0.000	10.83	0.0	0.0	0.0	10.1
MC	12	45.0	13.20	0.00	0.000	4.272	0.00	0.00	42.00	5.33	0.0	0.0	75.5	0.000	9.82	0.0	0.0	0.0	9.5
MC	12	40.0	11.80	0.00	0.000	4.406	0.00	0.00	39.00	5.00	0.0	0.0	70.1	0.000	8.84	0.0	0.0	0.0	9.0
MC	12	35.0	10.30	0.00	0.000	4.550	0.00	0.00	36.10	4.67	0.0	0.0	64.9	0.000	7.93	0.0	0.0	0.0	8.4
MC	12	37.0	10.90	0.00	0.000	5.555	0.00	0.00	34.20	3.59	0.0	0.0	61.5	0.000	6.15	0.0	0.0	0.0	6.4
MC	12	32.9	9.67	0.00	0.000	5.714	0.00	0.00	31.80	3.39	0.0	0.0	57.2	0.000	5.56	0.0	0.0	0.0	6.1
MC	12	30.9	9.07	0.00	0.000	5.797	0.00	0.00	30.60	3.28	0.0	0.0	55.0	0.000	5.27	0.0	0.0	0.0	5.9

(1) M_{rx}. Compact sections; $b_f/2t_f \leqslant 10.83$; beams, $f_a = 0$, $d/t_w \leqslant 106.7$, columns, $f_a > 0.16\,F_y$, $d/t_w \leqslant 36.3$; braced so that $L_b \leqslant L_c$.
(2) M_{rx}. Semi-compact sections; same as column (1) except $10.83 < b_f/2t_f < 15.83$.
(3) M_{rx}. No-compact sections for tension; all section not under (1), (2), (4), (6); for compression; upper limit for compression sections under (4), (5), (6).
*Cols (4), (5), (6): Non-compact sections; moment factors (*MF*) for compression.
*(4) For $53.2\sqrt{C_b} \leqslant l_b/r_T < 119\sqrt{C_b}$, except channels. Use *larger* M_{rx} from (4) or (5). . . . . (4) M_{rx} = col. (1) − $(MF \times 10^{-4})\, l_b^2/C_b \leqslant$ Col. (3).
*(5) For all values l_b/r_T. . . . . (5) $M_{rx} = 1000\,(MF)\, C_b/l_b \leqslant$ Col. (3).
*(6) For $l_b/r_T \geqslant 119\sqrt{C_b}$, except channels. Use larger M_{rx} from columns (5) or (6). . . . . (6) $M_{rx} = 1000\,(MF)\, C_b/l_b^2 \leqslant$ Col. (3).
(7) M_{ry}. For doubly symmetrical shapes; $b_f/2t_f \leqslant 10.83$.
(8) M_{ry}. For doubly symmetrical shapes; $10.83 < b_f/2t_f < 15.83$.
(9) M_{ry}. For non-compact sections; sections not under column (7) or (8).

TABLE 3-1: RESISTING MOMENTS, M_R, (kip-ft.) - F_y = 36 ksi

F_y = 36 ksi

Section	Area	L_c	r_T	d/A_f	$b_f/2t_f$	d/t_w	S_{x-x}	S_{y-y}	(1) $\frac{0.66 F_y S_x}{12} = M_{RX}$	(2) Eq. 1.5-5 $= M_{RX}$	(3) $\frac{0.60 F_y S_x}{12} = M_{RX}$	(4) Eq. 1.5-6a =(MF)	(5) Eq. 1.5-7 =(MF)	(6) Eq. 1.5-6b =(MF)	(7) $\frac{0.75 F_y S_y}{12} = M_{RY}$	(8) Eq. 1.5-5b $= M_{RY}$	(9) $\frac{0.60 F_y S_y}{12} = M_{RY}$
	SECTION PROPERTIES								MOMENT M_{RX}, k-ft.			M_{RX} FACTORS*			MOMENT, M_{RY}, k-ft.		
	sq. in.	(ft.)	in.	in.$^{-1}$	–	–	in.^{3}	in.^{3}									
MC 12 10.6	3.10	0.00	0.000	25.889	0.00	0.00	9.23	0.31	0.0	0.0	16.6	0.000	0.35	0.0	0.0	0.0	0.5
MC 10 41.1	12.10	0.00	0.000	4.024	0.00	0.00	31.50	4.88	0.0	0.0	56.6	0.000	7.82	0.0	0.0	0.0	8.7
MC 10 33.6	9.87	0.00	0.000	4.241	0.00	0.00	27.80	4.38	0.0	0.0	50.0	0.000	6.55	0.0	0.0	0.0	7.8
MC 10 28.5	8.37	0.00	0.000	4.402	0.00	0.00	25.30	4.02	0.0	0.0	45.5	0.000	5.74	0.0	0.0	0.0	7.2
MC 10 28.3	8.32	0.00	0.000	4.966	0.00	0.00	23.60	3.20	0.0	0.0	42.4	0.000	4.75	0.0	0.0	0.0	5.7
MC 10 25.3	7.43	0.00	0.000	5.633	0.00	0.00	21.40	2.89	0.0	0.0	38.5	0.000	3.79	0.0	0.0	0.0	5.2
MC 10 24.9	7.32	0.00	0.000	5.112	0.00	0.00	22.00	2.99	0.0	0.0	39.5	0.000	4.30	0.0	0.0	0.0	5.3
MC 10 21.9	6.43	0.00	0.000	5.797	0.00	0.00	19.70	2.70	0.0	0.0	35.4	0.000	3.39	0.0	0.0	0.0	4.8
MC 10 8.4	2.46	0.00	0.000	23.809	0.00	0.00	6.40	0.27	0.0	0.0	11.5	0.000	0.26	0.0	0.0	0.0	0.4
MC 10 6.5	1.91	0.00	0.000	43.926	0.00	0.00	4.42	0.11	0.0	0.0	7.9	0.000	0.10	0.0	0.0	0.0	0.2
MC 9 25.4	7.47	0.00	0.000	4.675	0.00	0.00	19.60	3.02	0.0	0.0	35.2	0.000	4.19	0.0	0.0	0.0	5.4
MC 9 23.9	7.02	0.00	0.000	4.743	0.00	0.00	18.90	2.93	0.0	0.0	34.0	0.000	3.98	0.0	0.0	0.0	5.2
MC 8 22.8	6.70	0.00	0.000	4.351	0.00	0.00	16.00	2.84	0.0	0.0	28.7	0.000	3.67	0.0	0.0	0.0	5.1
MC 8 21.4	6.28	0.00	0.000	4.416	0.00	0.00	15.40	2.74	0.0	0.0	27.7	0.000	3.48	0.0	0.0	0.0	4.9
MC 8 20.0	5.88	0.00	0.000	5.289	0.00	0.00	13.60	2.05	0.0	0.0	24.4	0.000	2.57	0.0	0.0	0.0	3.6
MC 8 18.7	5.50	0.00	0.000	5.372	0.00	0.00	13.10	1.97	0.0	0.0	23.5	0.000	2.43	0.0	0.0	0.0	3.5
MC 8 8.5	2.50	0.00	0.000	13.726	0.00	0.00	5.83	0.43	0.0	0.0	10.4	0.000	0.42	0.0	0.0	0.0	0.7
MC 7 22.7	6.67	0.00	0.000	3.885	0.00	0.00	13.60	2.85	0.0	0.0	24.4	0.000	3.50	0.0	0.0	0.0	5.1
MC 7 19.1	5.61	0.00	0.000	4.055	0.00	0.00	12.30	2.57	0.0	0.0	22.1	0.000	3.03	0.0	0.0	0.0	4.6
MC 7 17.6	5.17	0.00	0.000	4.912	0.00	0.00	10.80	1.89	0.0	0.0	19.4	0.000	2.19	0.0	0.0	0.0	3.4
MC 6 18.0	5.29	0.00	0.000	3.604	0.00	0.00	9.91	2.48	0.0	0.0	17.8	0.000	2.74	0.0	0.0	0.0	4.4
MC 6 16.3	4.79	0.00	0.000	4.210	0.00	0.00	8.68	1.84	0.0	0.0	15.6	0.000	2.06	0.0	0.0	0.0	3.3

MC	6	15.3	4.50	0.00	0.000	4.452	0.00	0.00	8.47	2.03	0.0	0.0	15.2	0.000	1.90	0.0	0.0	0.0	3.6
MC	6	15.1	4.44	0.00	0.000	4.294	0.00	0.00	8.52	1.75	0.0	0.0	14.9	0.000	1.93	0.0	0.0	0.0	3.1
MC	6	12.0	3.53	0.00	0.000	6.407	0.00	0.00	6.24	1.04	0.0	0.0	11.2	0.000	0.97	0.0	0.0	0.0	1.8
MC	3	9.0	2.65	0.00	0.000	4.027	0.00	0.00	2.10	0.67	0.0	0.0	3.7	0.000	0.52	0.0	0.0	0.0	1.2
MC	3	7.1	2.09	0.00	0.000	4.410	0.00	0.00	1.82	0.56	0.0	0.0	3.2	0.000	0.41	0.0	0.0	0.0	1.0

(1) M_{rx}. Compact sections; $b_f/2t_f \leqslant 10.83$; beams, $f_a = 0$, $d/t_w \leqslant 106.7$, columns, $f_a > 0.16\,F_y$, $d/t_w \leqslant 36.3$; braced so that $L_b \leqslant L_c$.
(2) M_{rx}. Semi-compact sections; same as column (1) except $10.83 < b_f/2t_f < 15.83$.
(3) M_{rx}. No-compact sections for tension; all section not under (1), (2), (4), (6); for compression; upper limit for compression sections under (4), (5), (6).
*Cols (4), (5), (6): Non-compact sections; moment factors (MF) for compression.
*(4) For $53.2\sqrt{C_b} \leqslant l_b/r_T < 119\sqrt{C_b}$, except channels. Use *larger* M_{rx} from (4) or (5). . . . . (4) M_{rx} = col. (1) − $(MF \times 10^{-4})\, l_b^2/Cb \leqslant$ Col. (3)
*(5) For all values l_b/r_T. . . . . (5) $M_{rx} = 1000\,(MF)\,C_b/l_b \leqslant$ Col. (3).
*(6) For $l_b/r_T \geqslant 119\sqrt{C_b}$, except channels. Use larger M_{rx} from columns (5) or (6). . . . . (6) $M_{rx} = 1000\,(MF)\,C_b/l_b^2 \leqslant$ Col. (3).
(7) M_{ry}. For doubly symmetrical shapes; $b_f/2t_f \leqslant 10.83$.
(8) M_{ry}. For doubly symmetrical shapes; $10.83 < b_f/2t_f < 15.83$.
(9) M_{ry}. For non-compact sections; sections not under column (7) or (8).

TABLE 3-2 Rolled Shape Resisting Moments, M_R (k-ft); $F_Y = 50$

F_y = 50 ksi	TABLE 3-2: RESISTING MOMENTS, M_R, (kip-ft.) - F_y = 50 ksi															
	SECTION PROPERTIES								MOMENT M_{RX}, k-ft.			M_{RX} FACTORS*			MOMENT, M_{RY}, k-ft.		
Section	Area	L_c	r_T	d/A_f	$b_f/2t_f$	d/t_w	S_{x-x}	S_{y-y}	(1) $\frac{0.66}{12}F_yS_x = M_{RX}$	(2) Eq. 1.5-5 $= M_{RX}$	(3) $\frac{0.60}{12}F_yS_x = M_{RX}$	(4) Eq. 1.5-6a =(MF)	(5) Eq. 1.5-7 =(MF)	(6) Eq. 1.5-6b =(MF)	(7) $\frac{0.75}{12}F_yS_y = M_{RY}$	(8) Eq. 1.5-5b $= M_{RY}$	(9) $\frac{0.60}{12}F_yS_y = M_{RY}$
	sq. in.	(ft.)	in.	in.$^{-1}$	–	–	in.^{3}	in.^{3}									
W 36 300.0	88.30	14.91	4.460	1.312	4.95	38.85	1110.00	156.00	3052.5	0.0	2775.0	75.983	845.81	312795.3	487.5	0.0	389.9
W 36 280.0	82.40	14.86	4.430	1.400	5.28	41.24	1030.00	144.00	2832.5	0.0	2575.0	71.465	735.22	286359.9	450.0	0.0	360.0
W 36 260.0	76.50	14.82	4.400	1.520	5.74	43.09	952.00	132.00	2618.0	0.0	2380.0	66.957	626.08	261101.8	412.5	0.0	330.0
W 36 245.0	72.10	14.78	4.380	1.617	6.11	44.96	894.00	123.00	2458.5	0.0	2235.0	63.453	552.64	242970.3	384.3	0.0	307.5
W 36 230.0	67.70	14.75	4.360	1.728	6.53	47.14	837.00	114.00	2301.7	0.0	2092.5	59.954	484.13	225406.3	356.2	0.0	285.0
W 36 194.0	57.20	10.85	3.120	2.389	4.80	47.37	665.00	61.90	1828.7	0.0	1662.4	93.020	278.31	91706.1	193.4	0.0	154.7
W 36 182.0	53.60	10.81	3.100	2.549	5.11	50.09	622.00	57.50	1710.5	0.0	1554.9	88.132	243.95	84680.1	179.6	0.0	143.7
W 36 170.0	50.00	10.77	3.080	2.733	5.46	53.17	580.00	53.20	1595.0	0.0	1450.0	83.251	212.20	77946.5	166.2	0.0	133.0
W 36 160.0	47.10	10.74	3.060	2.941	5.88	55.13	542.00	49.10	1490.5	0.0	1355.0	78.817	184.28	71896.8	153.4	0.0	122.7
W 36 150.0	44.20	10.46	3.030	3.184	6.36	57.34	504.00	45.00	1386.0	0.0	1260.0	74.750	158.25	65551.6	140.6	0.0	112.4
W 36 135.0	39.80	8.89	2.970	3.748	7.52	59.44	440.00	37.90	1210.0	0.0	1100.0	67.921	117.38	54983.6	118.4	0.0	94.7
W 33 240.0	70.60	14.20	4.230	1.508	5.66	40.36	813.00	118.00	2235.7	0.0	2032.4	61.869	539.03	206081.4	368.7	0.0	295.0
W 33 220.0	64.80	14.16	4.200	1.649	6.20	42.90	742.00	106.00	2040.5	0.0	1854.9	57.276	449.83	185425.8	331.2	0.0	265.0
W 33 200.0	58.90	14.10	4.170	1.821	6.84	46.15	671.00	95.20	1845.2	0.0	1677.4	52.543	368.28	165295.9	297.5	0.0	237.9
W 33 152.0	44.80	10.35	2.980	2.745	5.48	52.75	487.00	47.20	1339.2	0.0	1217.5	74.673	177.37	61267.3	147.5	0.0	117.9
W 33 141.0	41.60	10.33	2.960	3.008	6.00	55.05	448.00	42.70	1232.0	0.0	1120.0	69.624	148.93	55606.9	133.4	0.0	106.7
W 33 130.0	38.30	9.91	2.920	3.363	6.73	57.06	406.00	37.90	1116.5	0.0	1014.9	64.837	120.70	49041.0	118.4	0.0	94.7
W 33 118.0	34.80	8.59	2.870	3.877	7.78	59.31	359.00	32.50	987.2	0.0	897.4	59.346	92.59	41891.5	101.5	0.0	81.2
W 30 210.0	61.90	13.52	4.050	1.529	5.74	39.20	651.00	100.00	1790.2	0.0	1627.4	54.042	425.63	151272.0	312.5	0.0	249.9
W 30 190.0	56.00	13.47	4.010	1.690	6.34	42.42	587.00	89.50	1614.2	0.0	1467.5	49.706	347.33	133719.4	279.6	0.0	223.7
W 30 172.0	50.70	13.42	3.980	1.872	7.03	45.61	530.00	79.80	1457.5	0.0	1325.0	45.559	283.07	118934.9	249.3	0.0	199.4

W	30	132.0	38.90	9.45	2.720	2.871	5.27	49.26	380.00	37.20	1045.0	0.0	949.9	69.937	132.32	39828.0	116.2	0.0	92.9
W	30	124.0	36.50	9.42	2.700	3.082	5.65	51.55	355.00	34.40	976.2	0.0	887.4	66.308	115.16	36662.6	107.4	0.0	85.9
W	30	116.0	34.20	9.40	2.680	3.361	6.17	53.19	329.00	31.30	904.7	0.0	822.4	62.372	97.87	33475.9	97.8	0.0	78.2
W	30	108.0	31.80	8.90	2.640	3.742	6.89	54.41	300.00	27.90	825.0	0.0	750.0	58.611	80.15	29620.8	87.1	0.0	69.7
W	30	99.0	29.10	7.87	2.610	4.230	7.80	56.78	270.00	24.50	742.5	0.0	675.0	53.969	63.82	26056.2	76.5	0.0	61.2
W	27	177.0	52.20	12.61	3.770	1.628	5.92	37.66	494.00	78.90	1358.5	0.0	1235.0	47.327	303.29	99466.6	246.5	0.0	197.2
W	27	160.0	47.10	12.55	3.750	1.796	6.52	41.15	446.00	70.60	1226.5	0.0	1115.0	43.185	248.27	88851.5	220.6	0.0	176.4
W	27	145.0	42.70	12.50	3.720	1.974	7.16	44.80	404.00	63.50	1111.0	0.0	1009.9	39.752	204.64	79201.7	198.4	0.0	158.7
W	27	114.0	33.60	9.01	2.620	2.906	5.40	47.85	300.00	31.60	825.0	0.0	750.0	59.509	103.21	29173.6	98.7	0.0	78.9
W	27	102.0	30.00	8.97	2.590	3.267	6.05	52.25	267.00	27.70	734.2	0.0	667.5	54.197	81.71	25373.3	86.5	0.0	69.2
W	27	94.0	27.70	8.94	2.560	3.606	6.68	54.91	243.00	24.90	668.2	0.0	607.5	50.488	67.38	22560.7	77.8	0.0	62.2
W	27	84.0	24.80	7.91	2.520	4.212	7.83	57.64	212.00	21.10	583.0	0.0	530.0	45.457	50.33	19072.3	65.9	0.0	52.7
W	24	160.0	47.10	12.62	3.820	1.545	6.20	37.68	414.00	75.20	1138.5	0.0	1035.0	38.631	267.84	85584.4	235.0	0.0	187.9
W	24	145.0	42.70	12.57	3.790	1.709	6.88	40.27	373.00	67.10	1025.7	0.0	932.4	35.358	218.16	75902.2	209.6	0.0	167.7
W	24	130.0	38.30	12.53	3.760	1.924	7.77	42.92	332.00	58.90	913.0	0.0	829.9	31.976	172.50	66493.8	184.0	0.0	147.2
W	24	120.0	35.40	10.82	3.220	2.162	6.49	43.72	300.00	45.40	825.0	0.0	750.0	39.398	138.73	44065.6	141.8	0.0	113.4
W	24	110.0	32.50	10.78	3.200	2.346	7.04	47.37	276.00	41.40	759.0	0.0	690.0	36.700	117.61	40038.3	129.3	0.0	103.4
W	24	100.0	29.50	10.74	3.180	2.580	7.74	51.28	250.00	37.20	687.5	0.0	625.0	33.663	96.87	35814.7	116.2	0.0	92.9
W	24	94.0	27.70	8.11	2.370	3.074	5.19	47.07	221.00	23.90	607.7	0.0	552.5	53.575	71.88	17585.5	74.6	0.0	59.7
W	24	84.0	24.70	8.07	2.340	3.461	5.83	51.25	197.00	21.00	541.7	0.0	492.4	48.989	56.91	15281.4	65.6	0.0	52.4

(1) M_{rx}. Compact sections; $b_f/2t_f \leqslant 9.19$; beams, $f_a = 0$, $d/t_w \leqslant 90.5$; columns, $f_a > 0.16\ F_y$, $d/t_w \leqslant 36.3$; braced so that $L_b \leqslant L_c$.
(2) M_{rx}. Semi-compact sections; same as column (1) except $9.19 < b_f/2t_f < 13.44$.
(3) M_{rx}. No-compact sections for tension; all sections not under (1), (2), (4), (6); for compression; upper limit for compression sections under (4), (5), (6).
*Cols (4), (5), (6): Non-compact sections; moment factors (*MF*) for compression.
*(4) For $45.2\sqrt{C_b} \leqslant l_b/r_T < 101\sqrt{C_b}$, except channels. Use *larger* M_{rx} from (4) or (5). . . . . (4) $M_{rx} = \text{Col. (1)} - (MF \times 10^{-4})\ l_b^2/C_b \leqslant \text{Col. (3)}$.
*(5) For all values l_b/r_T. . . . . (5) $M_{rx} = 1000\ (MF)\ C_b/l_b \leqslant \text{Col. (3)}$.
*(6) For $l_b/r_T \geqslant 101\sqrt{C_b}$, except channels. Use larger M_{rx} from columns (5) or (6). . . . . (6) $M_{rx} = 1000\ (MF)\ C_b/l_b^2 \leqslant \text{Col. (3)}$.
(7) M_{ry}. For doubly symmetrical shapes; $b_f/2t_f \leqslant 9.19$.
(8) M_{ry}. For doubly symmetrical shapes; $9.19 < b_f/2t_f < 13.44$.
(9) M_{ry}. For non-compact sections; sections not under column (7) or (8).

F_y = 50 ksi	TABLE 3-2: RESISTING MOMENTS, M_R, (kip-ft.) - F_y = 50 ksi																
	SECTION PROPERTIES								MOMENT M_{RX}, k-ft.			M_{RX} FACTORS*			MOMENT, M_{RY}, k-ft.		
Section	Area	L_c	r_T	d/A_f	$b_f/2t_f$	d/t_w	S_{x-x}	S_{y-y}	(1) $\frac{0.66}{12} F_y S_x = M_{RX}$	(2) Eq. 1.5-5 $= M_{RX}$	(3) $\frac{0.60}{12} F_y S_x = M_{RX}$	(4) Eq. 1.5-6a =(MF)	(5) Eq. 1.5-7 =(MF)	(6) Eq. 1.5-6b =(MF)	(7) $\frac{0.75}{12} F_y S_y = M_{RY}$	(8) Eq 1.5-5b $= M_{RY}$	(9) $\frac{0.60}{12} F_y S_y = M_{RY}$
	sq.in.	(ft.)	in.	$in.^{-1}$	–	–	$in.^3$	$in.^3$									
W 24 76.0	22.40	8.04	2.320	3.901	6.58	54.34	176.00	18.40	484.0	0.0	439.9	44.525	45.10	13420.1	57.5	0.0	45.9
W 24 68.0	20.00	7.33	2.280	4.546	7.69	56.99	153.00	15.60	420.7	0.0	382.5	40.076	33.65	11267.5	48.7	0.0	38.9
W 24 61.0	18.00	5.83	1.740	5.714	5.94	56.61	130.00	9.76	357.5	0.0	325.0	58.467	22.74	5575.8	30.4	0.0	24.3
W 24 55.0	16.20	4.98	1.700	6.688	6.95	59.46	114.00	8.25	313.5	0.0	285.0	53.712	17.04	4667.3	25.7	0.0	20.6
W 21 142.0	41.80	11.76	3.580	1.492	5.99	32.56	317.00	63.00	871.7	0.0	792.4	33.679	212.40	57556.3	196.8	0.0	157.5
W 21 127.0	37.40	11.69	3.550	1.650	6.62	36.12	284.00	56.10	781.0	0.0	710.0	30.685	172.01	50704.0	175.3	0.0	140.2
W 21 112.0	33.00	11.64	3.520	1.867	7.51	39.84	250.00	48.80	687.5	0.0	625.0	27.474	133.86	43882.6	152.5	0.0	121.9
W 21 96.0	28.30	8.09	2.390	2.501	4.83	36.76	198.00	25.50	544.5	0.0	494.9	47.199	79.14	16022.4	79.6	0.0	63.7
W 21 82.0	24.20	8.02	2.350	2.927	5.63	41.80	169.00	21.30	464.7	0.0	422.4	41.669	57.72	13221.7	66.5	0.0	53.2
W 21 73.0	21.50	7.42	2.160	3.460	5.60	46.68	151.00	17.00	415.2	0.0	377.5	44.069	43.63	9980.4	53.1	0.0	42.5
W 21 68.0	20.00	7.40	2.150	3.729	6.03	49.13	140.00	15.70	385.0	0.0	350.0	41.239	37.53	9167.9	49.0	0.0	39.2
W 21 62.0	18.30	7.38	2.130	4.142	6.69	52.47	127.00	13.90	349.2	0.0	317.5	38.116	30.66	8162.6	43.4	0.0	34.7
W 21 55.0	16.20	6.87	2.100	4.850	7.86	55.46	110.00	11.80	302.5	0.0	275.0	33.964	22.67	6872.2	36.8	0.0	29.4
W 21 49.0	14.40	5.55	1.630	6.002	6.12	56.57	93.30	7.57	256.5	0.0	233.2	47.816	15.54	3511.7	23.6	0.0	18.9
W 21 44.0	13.00	4.72	1.590	7.047	7.20	59.36	81.60	6.38	224.3	0.0	203.9	43.950	11.57	2922.4	19.9	0.0	15.9
W 18 114.0	33.50	10.59	3.230	1.575	5.97	31.05	220.00	46.30	605.0	0.0	550.0	28.713	139.60	32515.8	144.6	0.0	115.7
W 18 105.0	30.90	10.56	3.210	1.705	6.47	33.06	202.00	42.30	555.5	0.0	504.9	26.693	118.44	29486.8	132.1	0.0	105.7
W 18 96.0	28.20	10.52	3.190	1.859	7.06	35.46	185.00	38.30	508.7	0.0	462.4	24.754	99.47	26669.8	119.6	0.0	95.7
W 18 85.0	25.00	7.91	2.370	2.275	4.85	34.82	157.00	23.80	431.7	0.0	392.4	38.060	68.99	12492.9	74.3	0.0	59.4
W 18 77.0	22.70	7.87	2.360	2.486	5.28	38.23	142.00	21.40	390.5	0.0	355.0	34.716	57.09	11204.1	66.8	0.0	53.4
W 18 70.0	20.60	7.83	2.340	2.739	5.82	41.09	129.00	19.20	354.7	0.0	322.5	32.079	47.09	10006.6	59.9	0.0	47.9
W 18 64.0	18.90	7.80	2.320	2.989	6.35	44.34	118.00	17.40	324.5	0.0	295.0	29.851	39.47	8997.5	54.3	0.0	43.4

W	18	60.0	17.70	6.76	1.990	3.474	5.43	43.87	108.00	13.30	297.0	0.0	270.0	37.135	31.08	6058.9	41.5	0.0	33.2
W	18	55.0	16.20	6.74	1.980	3.818	5.97	46.46	98.40	11.90	270.5	0.0	245.9	34.176	25.76	5465.0	37.1	0.0	29.7
W	18	50.0	14.70	6.71	1.960	4.210	6.57	50.27	89.10	10.70	245.0	0.0	222.7	31.581	21.16	4849.0	33.4	0.0	26.7
W	18	45.0	13.20	6.69	1.940	4.786	7.49	53.31	79.00	9.32	217.2	0.0	197.4	28.581	16.50	4212.0	29.1	0.0	23.2
W	18	40.0	11.80	5.39	1.540	5.676	5.74	56.64	68.40	6.34	188.0	0.0	170.9	39.271	12.04	2298.0	19.8	0.0	15.8
W	18	35.0	10.30	4.84	1.510	6.880	6.99	59.42	57.90	5.16	159.2	0.0	144.7	34.577	8.41	1870.2	16.1	0.0	12.8
W	16	96.0	28.20	10.32	3.160	1.617	6.59	30.50	166.00	38.80	456.5	0.0	414.9	22.636	102.64	23482.8	121.2	0.0	96.9
W	16	88.0	25.90	10.30	3.140	1.767	7.23	32.06	151.00	35.10	415.2	0.0	377.5	20.853	85.44	21091.3	109.6	0.0	87.7
W	16	78.0	23.00	7.69	2.320	2.172	4.90	30.85	128.00	21.60	352.0	0.0	320.0	32.381	58.92	9760.0	67.5	0.0	53.9
W	16	71.0	20.90	7.65	2.300	2.379	5.37	33.25	116.00	19.40	319.0	0.0	290.0	29.858	48.75	8693.2	60.6	0.0	48.4
W	16	64.0	18.80	7.61	2.280	2.632	5.94	36.11	104.00	17.30	286.0	0.0	260.0	27.241	39.50	7658.9	54.0	0.0	43.2
W	16	58.0	17.10	7.58	2.260	2.905	6.56	38.96	94.40	15.40	259.5	0.0	235.9	25.166	32.49	6830.5	48.1	0.0	38.5
W	16	50.0	14.70	6.33	1.870	3.658	5.63	42.76	80.80	10.50	222.1	0.0	201.9	31.462	22.08	4002.7	32.8	0.0	26.2
W	16	45.0	13.30	6.30	1.850	4.067	6.25	46.58	72.50	9.32	199.3	0.0	181.2	28.844	17.82	3515.1	29.1	0.0	23.2
W	16	40.0	11.80	6.26	1.840	4.544	6.95	52.11	64.60	8.23	177.6	0.0	161.4	25.981	14.21	3098.3	25.7	0.0	20.5
W	16	36.0	10.60	6.26	1.810	5.296	8.16	53.01	56.50	6.99	155.3	0.0	141.2	23.483	10.66	2622.2	21.8	0.0	17.4
W	16	31.0	9.13	4.94	1.410	6.486	6.25	57.60	47.20	4.51	129.7	0.0	117.9	32.327	7.27	1329.3	14.0	0.0	11.2
W	16	26.0	7.67	4.04	1.380	8.247	7.97	62.60	38.30	3.49	105.3	0.0	95.7	27.384	4.64	1033.2	10.9	0.0	8.7
W	14	730.0	215.00	16.02	5.270	0.255	1.82	7.31	1280.00	527.00	3520.0	0.0	3199.9	62.756	5010.19	503615.1	1646.8	0.0	1317.5
W	14	665.0	196.00	15.80	5.170	0.271	1.95	7.66	1150.00	472.00	3162.5	0.0	2875.0	58.584	4234.63	435458.1	1475.0	0.0	1180.0

(1) M_{rx}. Compact sections; $b_f/2t_f \leqslant 9.19$; beams, $f_a = 0$, $d/t_w \leqslant 90.5$; columns, $f_a > 0.16\, F_y$, $d/t_w \leqslant 36.3$; braced so that $L_b \leqslant L_c$.
(2) M_{rx}. Semi-compact sections; same as column (1) except $9.19 < b_f/2t_f < 13.44$.
(3) M_{rx}. No-compact sections for tension; all sections not under (1), (2), (4), (6); for compression; upper limit for compression sections under (4), (5), (6).
*Cols (4), (5), (6): Non-compact sections; moment factors (*MF*) for compression.
*(4) For $45.2\sqrt{C_b} \leqslant l_b/r_T < 101\sqrt{C_b}$, except channels. Use *larger* M_{rx} from (4) or (5). . . . . (4) $M_{rx} = \text{Col. (1)} - (MF \times 10^{-4})\, l_b^2/C_b \leqslant \text{Col. (3)}$.
*(5) For all values l_b/r_T. . . . . (5) $M_{rx} = 1000\,(MF)\, C_b/l_b \leqslant \text{Col. (3)}$.
*(6) For $l_b/r_T \geqslant 101\sqrt{C_b}$, except channels. Use larger M_{rx} from columns (5) or (6). . . . . (6) $M_{rx} = 1000\,(MF)\, C_b/l_b^2 \leqslant \text{Col. (3)}$.
(7) M_{ry}. For doubly symmetrical shapes; $b_f/2t_f \leqslant 9.19$.
(8) M_{ry}. For doubly symmetrical shapes; $9.19 < b_f/2t_f < 13.44$.
(9) M_{ry}. For non-compact sections; sections not under column (7) or (8).

TABLE 3-2: RESISTING MOMENTS, M_R, (kip-ft) - F_y = 50 ksi

F_y = 50 ksi

Section	SECTION PROPERTIES								MOMENT M_{RX}, k-ft.			M_{RX} FACTORS*			MOMENT, M_{RY}, k-ft.		
	Area	L_c	r_T	d/A_f	$b_f/2t_f$	d/t_w	S_{x-x}	S_{y-y}	(1) $\frac{0.66 F_y S_x}{12} = M_{RX}$	(2) Eq. 1.5-5 $= M_{RX}$	(3) $\frac{0.60 F_y S_x}{12} = M_{RX}$	(4) Eq. 1.5-6a =(MF)	(5) Eq. 1.5-7 =(MF)	(6) Eq. 1.5-6b =(MF)	(7) $\frac{0.75 F_y S_y}{12} = M_{RY}$	(8) Eq. 1.5-5b $= M_{RY}$	(9) $\frac{0.60 F_y S_y}{12} = M_{RY}$
	sq. in.	(ft.)	in.	in.$^{-1}$	–	–	in.^{3}	in.^{3}									
W 14 605.0	178.00	15.60	5.080	0.289	2.09	8.06	1040.00	423.00	2860.0	0.0	2600.0	54.874	3596.12	380214.3	1321.8	0.0	1057.5
W 14 550.0	162.00	15.41	5.000	0.308	2.25	8.49	933.00	378.00	2565.7	0.0	2332.5	50.816	3025.22	330437.5	1181.2	0.0	944.9
W 14 500.0	147.00	15.23	4.920	0.329	2.42	8.97	840.00	339.00	2310.0	0.0	2100.0	47.251	2548.02	288056.0	1059.3	0.0	847.4
W 14 455.0	134.00	15.07	4.850	0.352	2.61	9.48	758.00	304.00	2084.5	0.0	1894.9	43.878	2151.38	252592.4	950.0	0.0	760.0
W 14 426.0	125.00	14.95	4.810	0.369	2.75	9.96	707.00	283.00	1944.2	0.0	1767.4	41.609	1915.44	231727.2	884.3	0.0	707.5
W 14 398.0	117.00	14.85	4.760	0.388	2.91	10.34	657.00	262.00	1806.7	0.0	1642.4	39.483	1692.38	210885.5	818.7	0.0	655.0
W 14 370.0	109.00	14.75	4.720	0.409	3.09	10.83	608.00	241.00	1672.0	0.0	1520.0	37.160	1484.09	191891.2	753.1	0.0	602.5
W 14 342.0	101.00	14.65	4.680	0.434	3.31	11.36	559.00	221.00	1537.2	0.0	1397.5	34.752	1285.72	173448.7	690.6	0.0	552.5
W 14 314.0	92.30	14.54	4.630	0.463	3.55	12.14	512.00	201.00	1408.0	0.0	1280.0	32.521	1103.95	155488.9	628.1	0.0	502.4
W 14 287.0	84.40	14.44	4.580	0.497	3.85	12.83	465.00	182.00	1278.7	0.0	1162.5	30.184	933.87	138182.0	568.7	0.0	454.9
W 14 264.0	77.60	14.35	4.540	0.531	4.13	13.69	427.00	166.00	1174.2	0.0	1067.5	28.208	803.70	124683.0	518.7	0.0	414.9
W 14 246.0	72.30	14.28	4.510	0.562	4.39	14.44	397.00	154.00	1091.7	0.0	992.4	26.576	706.25	114396.0	481.2	0.0	384.9
W 14 237.0	69.70	14.25	4.500	0.579	4.55	14.78	382.00	148.00	1050.5	0.0	954.9	25.686	659.03	109586.2	462.5	0.0	370.0
W 14 228.0	67.10	14.20	4.480	0.597	4.69	15.31	368.00	142.00	1012.0	0.0	919.9	24.966	615.94	104633.6	443.7	0.0	355.0
W 14 219.0	64.40	14.17	4.470	0.617	4.87	15.79	353.00	136.00	970.7	0.0	882.4	24.056	571.29	99921.1	425.0	0.0	340.0
W 14 211.0	62.10	14.15	4.460	0.637	5.05	16.07	339.00	130.00	932.2	0.0	847.4	23.205	531.53	95529.3	406.2	0.0	325.0
W 14 202.0	59.40	14.10	4.440	0.660	5.23	16.80	325.00	124.00	893.7	0.0	812.4	22.448	492.22	90764.6	387.5	0.0	310.0
W 14 193.0	56.70	14.07	4.420	0.686	5.46	17.41	310.00	118.00	852.5	0.0	774.9	21.606	451.81	85797.3	368.7	0.0	295.0
W 14 184.0	54.10	14.02	4.410	0.712	5.68	18.30	296.00	113.00	814.0	0.0	740.0	20.724	415.31	81552.3	353.1	0.0	282.5
W 14 176.0	51.70	14.00	4.390	0.742	5.95	18.59	282.00	107.00	775.5	0.0	705.0	19.924	379.73	76992.0	334.3	0.0	267.5
W 14 167.0	49.10	13.97	4.380	0.776	6.25	19.38	267.00	101.00	734.2	0.0	667.5	18.950	343.79	72564.9	315.6	0.0	252.4
W 14 158.0	46.50	13.92	4.360	0.811	6.54	20.54	253.00	95.80	695.7	0.0	632.5	18.122	311.58	68133.5	299.3	0.0	239.4

W	14	150.0	44.10	13.89	4.350	0.850	6.87	21.41	240.00	90.60	660.0	0.0	600.0	17.270	282.27	64336.4	283.1	0.0	226.4
W	14	142.0	41.80	13.88	4.330	0.895	7.29	21.69	227.00	85.20	624.2	0.0	567.5	16.486	253.57	60293.3	266.2	0.0	212.9
W	14	320.0	94.10	14.96	4.730	0.480	3.99	8.89	493.00	196.00	1355.7	0.0	1232.5	30.004	1025.71	156256.0	612.5	0.0	489.9
W	14	136.0	40.00	13.20	4.120	0.941	6.93	22.34	216.00	77.00	594.0	0.0	540.0	17.327	229.45	51941.6	240.6	0.0	192.4
W	14	127.0	37.30	13.15	4.100	0.997	7.35	23.96	202.00	71.80	555.5	0.0	504.9	16.362	202.56	48104.6	224.3	0.0	179.4
W	14	119.0	35.00	13.12	4.080	1.055	7.80	25.43	189.00	67.10	519.7	0.0	472.4	15.459	179.11	44570.7	209.6	0.0	167.7
W	14	111.0	32.70	13.09	4.070	1.125	8.37	26.61	176.00	62.20	484.0	0.0	439.9	14.467	156.32	41301.8	194.3	0.0	155.4
W	14	103.0	30.30	13.05	4.050	1.202	8.96	28.78	164.00	57.60	451.0	0.0	409.9	13.614	136.37	38108.4	180.0	0.0	144.0
W	14	95.0	27.90	13.02	4.040	1.297	9.72	30.36	151.00	52.80	415.2	410.5	377.5	12.597	116.34	34914.6	0.0	160.8	131.9
W	14	87.0	25.60	12.98	4.020	1.403	10.53	33.33	138.00	48.20	379.5	368.5	345.0	11.627	98.33	31593.5	0.0	141.0	120.4
W	14	84.0	24.70	10.76	3.320	1.515	7.72	31.44	131.00	37.50	360.2	0.0	327.5	16.183	86.41	20455.7	117.1	0.0	93.7
W	14	78.0	22.90	10.74	3.310	1.631	8.35	32.85	121.00	34.50	332.7	0.0	302.5	15.038	74.14	18780.5	107.8	0.0	86.2
W	14	74.0	21.80	9.02	2.760	1.799	6.43	31.53	112.00	26.50	308.0	0.0	280.0	20.020	62.24	12086.5	82.8	0.0	66.2
W	14	68.0	20.00	8.99	2.740	1.950	6.99	33.63	103.00	24.10	283.2	0.0	257.5	18.681	52.80	10954.8	75.3	0.0	60.2
W	14	61.0	17.90	8.95	2.730	2.163	7.77	36.79	92.20	21.50	253.5	0.0	230.4	16.845	42.62	9734.7	67.1	0.0	53.7
W	14	53.0	15.60	7.22	2.180	2.627	6.12	37.67	77.80	14.30	213.9	0.0	194.4	22.291	29.60	5237.9	44.6	0.0	35.7
W	14	48.0	14.10	7.19	2.160	2.899	6.77	40.73	70.20	12.80	193.0	0.0	175.4	20.487	24.20	4639.9	40.0	0.0	31.9
W	14	43.0	12.60	7.16	2.140	3.238	7.57	44.41	62.70	11.30	172.4	0.0	156.7	18.642	19.36	4067.8	35.3	0.0	28.2
W	14	38.0	11.20	6.06	1.800	4.062	6.60	45.11	54.70	7.96	150.4	0.0	136.7	22.988	13.46	2510.7	24.5	0.0	19.6
W	14	34.0	10.00	6.04	1.780	4.578	7.45	48.78	48.60	6.89	133.6	0.0	121.4	20.886	10.61	2181.4	21.5	0.0	17.2

(1) M_{rx}. Compact sections; $b_f/2t_f \leqslant 9.19$; beams, $f_a = 0$, $d/t_w \leqslant 90.5$; columns, $f_a > 0.16\ F_y$, $d/t_w \leqslant 36.3$; braced so that $L_b \leqslant L_c$.

(2) M_{rx}. Semi-compact sections; same as column (1) except $9.19 < b_f/2t_f < 13.44$.

(3) M_{rx}. No-compact sections for tension; all sections not under (1), (2), (4), (6); for compression; upper limit for compression sections under (4), (5), (6).

*Cols (4), (5), (6): Non-compact sections; moment factors (*MF*) for compression.

*(4) For $45.2\sqrt{C_b} \leqslant l_b/r_T < 101\sqrt{C_b}$, except channels. Use *larger* M_{rx} from (4) or (5). . . . . (4) M_{rx} = Col. (1) − $(MF \times 10^{-4})\ l_b^2/C_b \leqslant$ Col. (3).

*(5) For all values l_b/r_T. . . . . (5) $M_{rx} = 1000\ (MF)\ C_b/l_b \leqslant$ Col. (3).

*(6) For $l_b/r_T \geqslant 101\sqrt{C_b}$, except channels. Use larger M_{rx} from columns (5) or (6). . . . . (6) $M_{rx} = 1000\ (MF)\ C_b/l_b^2 \leqslant$ Col. (3).

(7) M_{ry}. For doubly symmetrical shapes; $b_f/2t_f \leqslant 9.19$.

(8) M_{ry}. For doubly symmetrical shapes; $9.19 < b_f/2t_f < 13.44$.

(9) M_{ry}. For non-compact sections; sections not under column (7) or (8).

TABLE 3-2: RESISTING MOMENTS, M_R, (kip-ft.) - F_y = 50 ksi

F_y = 50 ksi

Section	Area	L_c	r_T	d/A_f	$b_f/2t_f$	d/t_w	S_{x-x}	S_{y-y}	(1) $\frac{0.66}{12}F_yS_x = M_{RX}$	(2) Eq. 1.5-5 $= M_{RX}$	(3) $\frac{0.60}{12}F_yS_x = M_{RX}$	(4) Eq. 1.5-6a =(MF)	(5) Eq. 1.5-7 =(MF)	(6) Eq. 1.5-6b =(MF)	(7) $\frac{0.75}{12}F_yS_y = M_{RY}$	(8) Eq. 1.5-5b $= M_{RY}$	(9) $\frac{0.60}{12}F_yS_y = M_{RY}$
	SECTION PROPERTIES								MOMENT M_{RX}, k-ft.			M_{RX} FACTORS*			MOMENT, M_{RY}, k-ft.		
	sq. in.	(ft.)	in.	in.$^{-1}$	–	–	in.^{3}	in.^{3}									
W 14 30.0	8.83	6.03	1.750	5.374	8.78	51.33	41.90	5.80	115.2	0.0	104.7	18.629	7.79	1817.8	18.1	0.0	14.4
W 14 26.0	7.67	4.50	1.290	6.612	6.01	54.47	35.10	3.53	96.5	0.0	87.7	28.720	5.30	827.4	11.0	0.0	8.8
W 14 22.0	6.49	4.06	1.260	8.191	7.46	59.65	28.90	2.80	79.4	0.0	72.2	24.786	3.52	649.9	8.7	0.0	6.9
W 12 190.0	55.90	11.34	3.590	0.653	3.64	13.56	263.00	93.10	723.2	0.0	657.5	27.786	402.27	48018.9	290.9	0.0	232.7
W 12 161.0	47.40	11.20	3.530	0.746	4.21	15.33	222.00	77.70	610.5	0.0	555.0	24.258	297.44	39189.5	242.8	0.0	194.2
W 12 133.0	39.10	11.07	3.470	0.875	5.00	17.72	183.00	63.10	503.2	0.0	457.4	20.694	209.02	31216.0	197.1	0.0	157.7
W 12 120.0	35.30	11.03	3.440	0.962	5.56	18.47	163.00	56.00	448.2	0.0	407.4	18.755	169.28	27325.7	175.0	0.0	140.0
W 12 106.0	31.20	10.95	3.410	1.068	6.20	20.77	145.00	49.20	398.7	0.0	362.5	16.979	135.75	23886.0	153.7	0.0	122.9
W 12 99.0	29.10	10.91	3.390	1.135	6.61	21.90	135.00	45.70	371.2	0.0	337.5	15.995	118.89	21978.6	142.8	0.0	114.2
W 12 92.0	27.10	10.88	3.380	1.212	7.09	23.15	125.00	42.20	343.7	0.0	312.5	14.898	103.05	20230.7	131.8	0.0	105.4
W 12 85.0	25.00	10.84	3.360	1.297	7.60	25.25	116.00	38.90	319.0	0.0	290.0	13.990	89.41	18552.5	121.5	0.0	97.2
W 12 79.0	23.20	10.81	3.340	1.392	8.20	26.34	107.00	35.80	294.2	0.0	267.5	13.060	76.84	16910.0	111.8	0.0	89.4
W 12 72.0	21.20	10.78	3.330	1.516	8.97	28.48	97.50	32.40	268.1	0.0	243.7	11.972	64.30	15316.5	101.2	0.0	80.9
W 12 65.0	19.10	10.74	3.310	1.666	9.90	31.07	88.00	29.10	242.0	238.3	219.9	10.936	52.80	13658.6	0.0	87.8	72.7
W 12 58.0	17.10	8.96	2.750	1.899	7.81	33.95	78.10	21.40	214.7	0.0	195.2	14.062	41.12	8367.2	66.8	0.0	53.4
W 12 53.0	15.60	8.95	2.740	2.093	8.68	34.95	70.70	19.20	194.4	0.0	176.7	12.822	33.76	7519.4	59.9	0.0	47.9
W 12 50.0	14.70	7.23	2.190	2.354	6.30	32.85	64.70	14.00	177.9	0.0	161.7	18.368	27.47	4396.0	43.7	0.0	35.0
W 12 45.0	13.20	7.20	2.180	2.603	6.98	35.89	58.20	12.40	160.0	0.0	145.4	16.675	22.35	3918.3	38.7	0.0	30.9
W 12 40.0	11.80	7.16	2.160	2.892	7.75	40.61	51.90	11.00	142.7	0.0	129.7	15.147	17.94	3430.3	34.3	0.0	27.4
W 12 36.0	10.60	5.88	1.770	3.452	6.07	40.13	46.00	7.77	126.5	0.0	114.9	19.993	13.32	2041.6	24.2	0.0	19.4
W 12 31.0	9.13	5.84	1.750	3.984	7.01	45.62	39.50	6.61	108.6	0.0	98.7	17.562	9.91	1713.7	20.6	0.0	16.5
W 12 27.0	7.95	5.81	1.740	4.602	8.12	50.46	34.20	5.63	94.0	0.0	85.4	15.381	7.43	1466.8	17.5	0.0	14.0

W	12	22.0	6.47	3.60	1.030	7.204	4.75	47.34	25.30	2.31	69.5	0.0	63.2	32.472	3.51	380.2	7.2	0.0	5.7
W	12	19.0	5.59	3.58	1.010	8.695	5.74	51.30	21.30	1.88	58.5	0.0	53.2	28.431	2.44	307.8	5.8	0.0	4.6
W	12	16.5	4.87	2.98	0.975	11.152	7.43	52.17	17.60	1.44	48.3	0.0	43.9	25.209	1.57	237.0	4.5	0.0	3.5
W	12	14.0	4.12	2.48	0.957	13.399	8.85	60.15	14.80	1.18	40.6	0.0	36.9	22.004	1.10	192.0	3.6	0.0	2.9
W	10	112.0	32.90	9.32	2.940	0.875	4.17	15.07	126.00	45.20	346.5	0.0	315.0	19.849	143.91	15428.8	141.2	0.0	112.9
W	10	100.0	29.40	9.26	2.910	0.961	4.62	16.23	112.00	39.90	308.0	0.0	280.0	18.009	116.48	13436.0	124.6	0.0	99.7
W	10	89.0	26.20	9.20	2.880	1.061	5.14	17.69	99.70	35.20	274.1	0.0	249.2	16.367	93.96	11715.1	110.0	0.0	87.9
W	10	77.0	22.70	9.13	2.850	1.200	5.87	19.85	86.10	30.10	236.7	0.0	215.2	14.433	71.74	9907.4	94.0	0.0	75.2
W	10	72.0	21.20	9.10	2.840	1.277	6.29	20.58	80.10	27.90	220.2	0.0	200.2	13.522	62.68	9152.4	87.1	0.0	69.7
W	10	66.0	19.40	9.06	2.820	1.371	6.76	22.71	73.70	25.50	202.6	0.0	184.2	12.619	53.73	8302.9	79.6	0.0	63.7
W	10	60.0	17.70	9.02	2.800	1.489	7.37	24.69	67.10	23.10	184.5	0.0	167.7	11.653	45.04	7452.5	72.1	0.0	57.7
W	10	54.0	15.90	8.98	2.780	1.632	8.11	27.50	60.40	20.70	166.0	0.0	150.9	10.641	36.98	6612.9	64.6	0.0	51.7
W	10	49.0	14.40	8.95	2.770	1.792	8.96	29.41	54.60	18.60	150.1	0.0	136.5	9.689	30.46	5934.9	58.1	0.0	46.4
W	10	45.0	13.20	7.18	2.210	2.041	6.49	28.91	49.10	13.30	135.0	0.0	122.7	13.688	24.05	3397.2	41.5	0.0	33.2
W	10	39.0	11.50	7.15	2.190	2.356	7.56	31.25	42.20	11.20	116.0	0.0	105.4	11.980	17.91	2867.2	35.0	0.0	27.9
W	10	33.0	9.71	7.13	2.160	2.827	9.19	33.39	35.00	9.16	96.2	96.2	87.5	10.214	12.37	2313.3	0.0	28.6	22.8
W	10	29.0	8.54	5.19	1.570	3.524	5.79	35.36	30.80	5.61	84.6	0.0	77.0	17.014	8.73	1075.5	17.5	0.0	14.0
W	10	25.0	7.36	5.16	1.560	4.068	6.70	40.00	26.50	4.76	72.8	0.0	66.2	14.827	6.51	913.6	14.8	0.0	11.8
W	10	21.0	6.20	5.15	1.530	5.063	8.45	41.25	21.50	3.75	59.1	0.0	53.7	12.506	4.24	712.9	11.7	0.0	9.3
W	10	19.0	5.61	3.60	1.050	6.471	5.10	41.00	18.80	2.13	51.6	0.0	46.9	23.219	2.90	293.6	6.6	0.0	5.3

(1) M_{rx}. Compact sections; $b_f/2t_f \leqslant 9.19$; beams, $f_a = 0$, $d/t_w \leqslant 90.5$; columns, $f_a > 0.16\ F_y$, $d/t_w \leqslant 36.3$; braced so that $L_b \leqslant L_c$.
(2) M_{rx}. Semi-compact sections; same as column (1) except $9.19 < b_f/2t_f < 13.44$.
(3) M_{rx}. No-compact sections for tension; all sections not under (1), (2), (4), (6); for compression; upper limit for compression sections under (4), (5), (6).
*Cols (4), (5), (6): Non-compact sections; moment factors (MF) for compression.
*(4) For $45.2\sqrt{C_b} \leqslant l_b/r_T < 101\sqrt{C_b}$, except channels. Use *larger* M_{rx} from (4) or (5). . . . . (4) $M_{rx} = \text{Col. (1)} - (MF \times 10^{-4})\ l_b^2/C_b \leqslant \text{Col. (3)}$.
*(5) For all values l_b/r_T. . . . . (5) $M_{rx} = 1000\ (MF)\ C_b/l_b \leqslant \text{Col. (3)}$.
*(6) For $l_b/r_T \geqslant 101\sqrt{C_b}$, except channels. Use larger M_{rx} from columns (5) or (6). . . . . (6) $M_{rx} = 1000\ (MF)\ C_b/l_b^2 \leqslant \text{Col. (3)}$.
(7) M_{ry}. For doubly symmetrical shapes; $b_f/2t_f \leqslant 9.19$.
(8) M_{ry}. For doubly symmetrical shapes; $9.19 < b_f/2t_f < 13.44$.
(9) M_{ry}. For non-compact sections; sections not under column (7) or (8).

TABLE 3-2: RESISTING MOMENTS, M_R, (kip-ft.) - F_y = 50 ksi

F_y = 50 ksi

Section	Area	L_c	r_T	d/A_f	$b_f/2t_f$	d/t_w	S_{x-x}	S_{y-y}	(1) $\frac{0.66 F_y S_x}{12} = M_{RX}$	(2) Eq. 1.5-5 $= M_{RX}$	(3) $\frac{0.60 F_y S_x}{12} = M_{RX}$	(4) Eq. 1.5-6a =(MF)	(5) Eq. 1.5-7 =(MF)	(6) Eq. 1.5-6b =(MF)	(7) $\frac{0.75 F_y S_y}{12} = M_{RY}$	(8) Eq. 1.5-5b $= M_{RY}$	(9) $\frac{0.60 F_y S_y}{12} = M_{RY}$
	SECTION PROPERTIES								MOMENT M_{RX}, k-ft.			M_{RX} FACTORS*			MOMENT, M_{RY}, k-ft.		
	sq. in.	(ft.)	in.	$in.^{-1}$	–	–	$in.^3$	$in.^3$									
W 10 17.0	4.99	3.59	1.030	7.670	6.09	42.16	16.20	1.77	44.5	0.0	40.4	20.792	2.11	243.4	5.5	0.0	4.4
W 10 15.0	4.41	3.58	1.000	9.293	7.43	43.47	13.80	1.44	37.9	0.0	34.5	18.790	1.48	195.5	4.5	0.0	3.5
W 10 11.5	3.39	2.72	0.975	12.248	9.68	54.83	10.50	1.06	28.8	28.5	26.2	15.039	0.85	141.4	0.0	3.2	2.6
W 8 67.0	19.70	7.42	2.330	1.164	4.44	15.65	60.40	21.40	166.0	0.0	150.9	15.149	51.88	4645.3	66.8	0.0	53.4
W 8 58.0	17.10	7.36	2.310	1.317	5.08	17.15	52.00	18.20	143.0	0.0	130.0	13.269	39.48	3930.9	56.8	0.0	45.4
W 8 48.0	14.10	7.27	2.270	1.533	5.94	20.98	43.20	15.00	118.7	0.0	107.9	11.415	28.17	3153.5	46.8	0.0	37.5
W 8 40.0	11.80	7.23	2.240	1.830	7.23	22.60	35.50	12.10	97.6	0.0	88.7	9.633	19.39	2523.4	37.8	0.0	30.2
W 8 35.0	10.30	7.18	2.220	2.051	8.14	25.77	31.10	10.60	85.5	0.0	77.7	8.592	15.15	2171.3	33.1	0.0	26.4
W 8 31.0	9.12	7.16	2.210	2.309	9.23	27.77	27.40	9.24	75.3	75.2	68.4	7.638	11.86	1895.8	0.0	28.8	23.0
W 8 28.0	8.23	5.85	1.800	2.661	7.06	28.28	24.30	6.61	66.8	0.0	60.7	10.212	9.12	1115.3	20.6	0.0	16.5
W 8 24.0	7.06	5.82	1.780	3.065	8.16	32.36	20.80	5.61	57.1	0.0	51.9	8.939	6.78	933.6	17.5	0.0	14.0
W 8 20.0	5.89	4.71	1.420	4.087	6.96	32.82	17.00	3.50	46.7	0.0	42.5	11.479	4.15	485.6	10.9	0.0	8.7
W 8 17.0	5.01	4.70	1.400	4.947	8.52	34.78	14.10	2.83	38.7	0.0	35.2	9.795	2.84	391.5	8.8	0.0	7.0
W 8 15.0	4.43	3.59	1.040	6.440	6.39	33.14	11.80	1.69	32.4	0.0	29.4	14.855	1.83	180.8	5.2	0.0	4.2
W 8 13.0	3.83	3.58	1.020	7.874	7.87	34.78	9.90	1.36	27.2	0.0	24.7	12.956	1.25	145.9	4.2	0.0	3.3
W 8 10.0	2.96	3.39	1.000	9.828	9.65	46.47	7.80	1.06	21.4	21.2	19.4	10.620	0.79	110.5	0.0	3.2	2.6
W 6 25.0	7.35	5.44	1.690	2.297	6.66	19.90	16.70	5.62	45.9	0.0	41.7	7.961	7.26	675.7	17.5	0.0	14.0
W 6 20.0	5.88	5.39	1.660	2.807	8.19	24.03	13.40	4.43	36.8	0.0	33.5	6.621	4.77	523.1	13.8	0.0	11.0
W 6 15.5	4.56	5.36	1.630	3.720	11.14	25.53	10.00	3.23	27.5	26.3	24.9	5.124	2.68	376.3	0.0	9.1	8.0
W 6 16.0	4.72	3.60	1.100	3.838	4.98	24.03	10.20	2.19	28.0	0.0	25.4	11.478	2.65	174.8	6.8	0.0	5.4
W 6 12.0	3.54	3.58	1.070	5.376	7.16	26.08	7.25	1.49	19.9	0.0	18.1	8.622	1.34	117.5	4.6	0.0	3.7
W 6 8.5	2.51	3.52	1.040	7.627	10.15	34.29	5.08	1.01	13.9	13.6	12.6	6.395	0.66	77.8	0.0	3.0	2.5

W	5	18.5	5.43	4.50	1.400	2.425	5.98	19.32	9.94	3.54	27.3	0.0	24.8	6.905	4.09	276.0	11.0	0.0	8.8
W	5	16.0	4.70	4.47	1.390	2.777	6.94	20.83	8.53	3.00	23.4	0.0	21.3	6.011	3.07	233.4	9.3	0.0	7.4
W	4	13.0	3.82	3.63	1.110	2.969	5.88	14.85	5.45	1.85	14.9	0.0	13.6	6.023	1.83	95.1	5.7	0.0	4.6
M	14	17.2	5.05	2.59	0.925	12.867	7.35	66.66	21.10	1.33	58.0	0.0	52.7	33.578	1.63	255.7	4.1	0.0	3.3
M	12	11.8	3.47	1.91	0.690	17.400	6.81	67.79	12.00	0.63	33.0	0.0	29.9	34.320	0.68	80.9	1.9	0.0	1.5
M	10	29.1	8.56	5.31	1.400	4.277	7.63	23.13	26.60	3.76	73.1	0.0	66.5	18.479	6.21	738.5	11.7	0.0	9.3
M	10	22.9	6.73	5.15	1.400	4.415	7.39	40.82	23.60	3.48	64.8	0.0	58.9	16.395	5.34	655.2	10.8	0.0	8.6
M	10	9.0	2.65	1.84	0.616	18.045	6.52	63.69	7.76	0.45	21.3	0.0	19.3	27.846	0.43	41.7	1.4	0.0	1.1
M	8	34.3	10.10	7.16	2.080	2.177	8.71	21.16	29.10	8.73	80.0	0.0	72.7	9.158	13.36	1783.5	27.2	0.0	21.8
M	8	32.6	9.58	7.11	2.080	2.195	8.64	25.39	28.40	8.58	78.0	0.0	70.9	8.938	12.93	1740.6	26.8	0.0	21.4
M	8	22.5	6.60	4.83	1.280	4.200	7.64	21.33	17.10	2.77	47.0	0.0	42.7	14.211	4.07	396.9	8.6	0.0	6.9
M	8	18.5	5.44	4.70	1.280	4.316	7.43	34.78	15.50	2.60	42.6	0.0	38.7	12.881	3.59	359.7	8.1	0.0	6.4
M	8	6.5	1.92	1.79	0.535	18.556	6.03	59.25	4.62	0.30	12.7	0.0	11.5	21.978	0.24	18.7	0.9	0.0	0.7
M	7	5.5	1.62	1.78	0.493	18.696	5.77	54.68	3.44	0.23	9.4	0.0	8.5	19.272	0.18	11.8	0.7	0.0	0.5
M	6	22.5	6.62	5.42	1.550	2.612	7.99	16.12	13.70	4.08	37.6	0.0	34.2	7.764	5.24	466.2	12.7	0.0	10.1
M	6	20.0	5.89	5.31	1.540	2.666	7.83	24.00	13.00	3.90	35.7	0.0	32.5	7.463	4.87	436.7	12.1	0.0	9.7
M	6	4.4	1.29	1.65	0.444	19.028	5.39	52.63	2.40	0.17	6.5	0.0	5.9	16.577	0.12	6.7	0.5	0.0	0.4
M	5	18.9	5.55	4.48	1.320	2.402	6.01	15.82	9.63	3.14	26.4	0.0	24.0	7.525	4.00	237.7	9.8	0.0	7.8
M	4	13.8	4.06	3.58	1.050	2.695	5.39	12.77	5.42	1.79	14.9	0.0	13.5	6.694	2.01	84.6	5.5	0.0	4.4
M	4	13.0	3.81	3.52	1.040	2.736	5.30	15.74	5.24	1.71	14.4	0.0	13.0	6.596	1.91	80.2	5.3	0.0	4.2

(1) M_{rx}. Compact sections; $b_f/2t_f \leqslant 9.19$; beams, $f_a = 0$, $d/t_w \leqslant 90.5$; columns, $f_a > 0.16\,F_y$, $d/t_w \leqslant 36.3$; braced so that $L_b \leqslant L_c$.
(2) M_{rx}. Semi-compact sections; same as column (1) except $9.19 < b_f/2t_f < 13.44$.
(3) M_{rx}. No-compact sections for tension; all sections not under (1), (2), (4), (6); for compression; upper limit for compression sections under (4), (5), (6).
*Cols (4), (5), (6): Non-compact sections; moment factors (*MF*) for compression.
*(4) For $45.2\sqrt{C_b} \leqslant l_b/r_T < 101\sqrt{C_b}$, except channels. Use *larger* M_{rx} from (4) or (5). . . . . (4) $M_{rx} = \text{Col. (1)} - (MF \times 10^{-4})\, l_b^2/C_b \leqslant \text{Col. (3)}$.
*(5) For all values l_b/r_T. . . . . (5) $M_{rx} = 1000\,(MF)\, C_b/l_b \leqslant \text{Col. (3)}$.
*(6) For $l_b/r_T \geqslant 101\sqrt{C_b}$, except channels. Use larger M_{rx} from columns (5) or (6). . . . . (6) $M_{rx} = 1000\,(MF)\, C_b/l_b^2 \leqslant \text{Col. (3)}$.
(7) M_{ry}. For doubly symmetrical shapes; $b_f/2t_f \leqslant 9.19$.
(8) M_{ry}. For doubly symmetrical shapes; $9.19 < b_f/2t_f < 13.44$.
(9) M_{ry}. For non-compact sections; sections not under column (7) or (8).

F_y = 50 ksi	TABLE 3-2: RESISTING MOMENTS, M_R, (kip-ft.) - F_y = 50 ksi																
	SECTION PROPERTIES								MOMENT M_{RX}, k-ft.			M_{RX} FACTORS*			MOMENT, M_{RY}, k-ft.		
Section	Area	L_c	r_T	d/A_f	$b_f/2t_f$	d/t_w	S_{x-x}	S_{y-y}	(1) $\frac{0.66 F_y S_x}{12} = M_{RX}$	(2) Eq. 1.5-5 $= M_{RX}$	(3) $\frac{0.60 F_y S_x}{12} = M_{RX}$	(4) Eq. 1.5-6a =(MF)	(5) Eq. 1.5-7 =(MF)	(6) Eq. 1.5-6b =(MF)	(7) $\frac{0.75 F_y S_y}{12} = M_{RY}$	(8) Eq. 1.5-5b $= M_{RY}$	(9) $\frac{0.60 F_y S_y}{12} = M_{RY}$
	sq. in.	(ft.)	in.	$in.^{-1}$	–	–	$in.^3$	$in.^3$									
S 24 120.0	35.30	7.20	1.930	2.706	3.65	30.07	252.00	20.90	693.0	0.0	630.0	92.119	93.12	13297.8	65.3	0.0	52.2
S 24 105.9	31.10	7.05	1.930	2.765	3.57	38.40	236.00	19.80	649.0	0.0	590.0	86.270	85.33	12453.5	61.8	0.0	49.4
S 24 100.0	29.40	6.49	1.650	3.802	4.16	32.12	199.00	13.20	547.2	0.0	497.4	99.529	52.33	7675.1	41.2	0.0	32.9
S 24 90.0	26.50	6.38	1.650	3.867	4.08	38.46	187.00	12.60	514.2	0.0	467.4	93.527	48.34	7212.3	39.3	0.0	31.4
S 24 79.9	23.50	6.27	1.660	3.935	4.01	47.90	175.00	12.10	481.2	0.0	437.4	86.474	44.46	6831.5	37.8	0.0	30.2
S 20 95.0	27.90	6.44	1.700	3.032	3.93	25.00	161.00	13.80	442.7	0.0	402.4	75.856	53.09	6591.6	43.1	0.0	34.5
S 20 85.0	25.00	6.31	1.690	3.095	3.84	30.62	152.00	13.10	418.0	0.0	380.0	72.466	49.10	6150.1	40.9	0.0	32.7
S 20 75.0	22.10	5.72	1.480	3.966	4.05	31.20	128.00	9.28	352.0	0.0	320.0	79.570	32.27	3971.9	29.0	0.0	23.1
S 20 65.4	19.20	5.59	1.480	4.055	3.96	40.00	118.00	8.77	324.5	0.0	295.0	73.354	29.09	3661.6	27.4	0.0	21.9
S 18 70.0	20.60	5.59	1.410	4.167	4.52	25.31	103.00	7.72	283.2	0.0	257.5	70.545	24.71	2900.9	24.1	0.0	19.2
S 18 54.7	16.10	5.37	1.410	4.340	4.34	39.04	89.40	6.94	245.8	0.0	223.4	61.230	20.59	2517.9	21.6	0.0	17.3
S 15 50.0	14.70	5.05	1.310	4.275	4.53	27.27	64.80	5.57	178.1	0.0	161.9	51.416	15.15	1575.3	17.4	0.0	13.9
S 15 42.9	12.60	4.92	1.310	4.383	4.42	36.49	59.60	5.23	163.8	0.0	149.0	47.290	13.59	1448.9	16.3	0.0	13.0
S 12 50.0	14.70	4.90	1.310	3.324	4.15	17.46	50.80	5.74	139.6	0.0	126.9	40.307	15.27	1235.0	17.9	0.0	14.3
S 12 40.8	12.00	4.70	1.280	3.467	3.98	25.97	45.40	5.16	124.8	0.0	113.4	37.731	13.09	1053.7	16.1	0.0	12.8
S 12 35.0	10.30	4.54	1.200	4.344	4.66	28.03	38.20	3.89	105.0	0.0	95.4	36.121	8.79	779.2	12.1	0.0	9.7
S 12 31.8	9.35	4.47	1.200	4.411	4.59	34.28	36.40	3.74	100.0	0.0	90.9	34.419	8.25	742.5	11.6	0.0	9.3
S 10 35.0	10.30	4.42	1.150	4.119	5.03	16.83	29.40	3.38	80.8	0.0	73.4	30.270	7.13	550.8	10.5	0.0	8.4
S 10 25.4	7.46	4.17	1.130	4.369	4.74	32.15	24.70	2.91	67.9	0.0	61.7	26.339	5.65	446.8	9.0	0.0	7.2
S 8 23.0	6.77	3.73	0.987	4.512	4.90	18.14	16.20	2.07	44.5	0.0	40.4	22.643	3.58	223.5	6.4	0.0	5.1
S 8 18.4	5.41	3.58	0.973	4.704	4.70	29.52	14.40	1.86	39.5	0.0	36.0	20.711	3.06	193.1	5.8	0.0	4.6
S 7 20.0	5.88	3.45	0.914	4.626	4.92	15.55	12.10	1.64	33.2	0.0	30.2	19.722	2.61	143.2	5.1	0.0	4.0

S	7	15.3	4.50	3.27	0.894	4.876	4.67	27.77	10.50	1.44	28.8	0.0	26.2	17.888	2.15	118.8	4.5	0.0	3.5
S	6	17.2	5.07	3.19	0.845	4.688	4.96	12.90	8.77	1.30	24.1	0.0	21.9	16.724	1.87	88.7	4.0	0.0	3.2
S	6	12.5	3.67	2.98	0.817	5.015	4.64	25.86	7.37	1.09	20.2	0.0	18.4	15.034	1.46	69.6	3.4	0.0	2.7
S	5	14.75	4.34	2.94	0.781	4.670	5.03	10.12	6.09	1.01	16.7	0.0	15.2	13.595	1.30	52.6	3.1	0.0	2.5
S	5	10.0	2.94	2.69	0.741	5.105	4.60	23.36	4.92	0.80	13.5	0.0	12.2	12.201	0.96	38.2	2.5	0.0	2.0
S	4	9.5	2.79	2.50	0.684	4.882	4.77	12.26	3.39	0.64	9.3	0.0	8.4	9.866	0.69	22.4	2.0	0.0	1.6
S	4	7.7	2.26	2.38	0.662	5.126	4.54	20.72	3.04	0.57	8.3	0.0	7.5	9.445	0.59	18.8	1.7	0.0	1.4
S	3	7.5	2.21	2.24	0.621	4.598	4.82	8.59	1.95	0.46	5.3	0.0	4.8	6.885	0.42	10.6	1.4	0.0	1.1
S	3	5.7	1.67	2.08	0.585	4.952	4.48	17.64	1.68	0.39	4.6	0.0	4.1	6.684	0.33	8.1	1.2	0.0	0.9
HP	14	117.0	34.40	13.33	4.060	1.187	9.24	17.67	173.00	59.50	475.7	475.2	432.4	14.290	145.67	40398.5	0.0	185.4	148.7
HP	14	102.0	30.00	13.24	4.020	1.348	10.50	19.92	150.00	51.30	412.5	400.9	375.0	12.638	111.27	34340.8	0.0	150.4	128.2
HP	14	89.0	26.20	13.16	3.990	1.531	11.92	22.50	131.00	44.40	360.2	339.1	327.5	11.204	85.56	29545.0	0.0	120.8	110.9
HP	14	73.0	21.50	0.00	3.940	1.848	14.41	26.95	108.00	35.90	297.0	0.0	270.0	9.473	58.43	23751.1	0.0	0.0	89.7
HP	12	74.0	21.80	10.94	3.310	1.634	10.06	19.96	93.40	30.20	256.8	252.0	233.4	11.608	57.14	14496.7	0.0	90.4	75.4
HP	12	53.0	15.60	0.00	3.230	2.242	13.81	27.01	66.90	21.10	183.9	0.0	167.2	8.731	29.82	9887.7	0.0	0.0	52.7
HP	10	57.0	16.80	9.15	2.780	1.735	9.06	17.74	58.80	19.70	161.6	0.0	146.9	10.359	33.87	6437.7	61.5	0.0	49.2
HP	10	42.0	12.40	9.02	2.720	2.307	12.05	23.25	43.40	14.20	119.3	112.0	108.4	7.987	18.80	4548.7	0.0	38.3	35.4
HP	8	36.0	10.60	7.30	2.220	2.206	9.14	18.00	29.90	9.91	82.2	0.0	74.7	8.260	13.54	2087.5	30.9	0.0	24.7
C	15	50.0	14.70	0.00	0.000	6.210	0.00	0.00	53.80	3.78	0.0	0.0	134.4	0.000	8.66	0.0	0.0	0.0	9.4
C	15	40.0	11.80	0.00	0.000	6.555	0.00	0.00	46.50	3.36	0.0	0.0	116.2	0.000	7.09	0.0	0.0	0.0	8.3

(1) M_{rx}. Compact sections; $b_f/2t_f \leqslant 9.19$; beams, $f_a = 0$, $d/t_w \leqslant 90.5$; columns, $f_a > 0.16\ F_y$, $d/t_w \leqslant 36.3$; braced so that $L_b \leqslant L_c$.
(2) M_{rx}. Semi-compact sections; same as column (1) except $9.19 < b_f/2t_f < 13.44$.
(3) M_{rx}. No-compact sections for tension; all sections not under (1), (2), (4), (6); for compression; upper limit for compression sections under (4), (5), (6).
*Cols (4), (5), (6): Non-compact sections; moment factors (*MF*) for compression.
*(4) For $45.2\sqrt{C_b} \leqslant l_b/r_T < 101\sqrt{C_b}$, except channels. Use *larger* M_{rx} from (4) or (5). . . . . (4) M_{rx} = Col. (1) − ($MF \times 10^{-4}$) $l_b^2/C_b \leqslant$ Col. (3).
*(5) For all values l_b/r_T. . . . . (5) M_{rx} = 1000 (MF) $C_b/l_b \leqslant$ Col. (3).
*(6) For $l_b/r_T \geqslant 101\sqrt{C_b}$, except channels. Use larger M_{rx} from columns (5) or (6). . . . . (6) M_{rx} = 1000 (MF) $C_b/l_b^2 \leqslant$ Col. (3).
(7) M_{ry}. For doubly symmetrical shapes; $b_f/2t_f \leqslant 9.19$.
(8) M_{ry}. For doubly symmetrical shapes; $9.19 < b_f/2t_f < 13.44$.
(9) M_{ry}. For non-compact sections; sections not under column (7) or (8).

TABLE 3-2: RESISTING MOMENTS, M_R, (kip-ft.) - F_y = 50 ksi

F_y = 50 ksi	SECTION PROPERTIES								MOMENT M_{RX}, k-ft.			M_{RX} FACTORS*			MOMENT, M_{RY}, k-ft.		
Section	Area	L_c	r_T	d/A_f	$b_f/2t_f$	d/t_w	S_{x-x}	S_{y-y}	(1) $\frac{0.66 F_y S_x}{12} = M_{RX}$	(2) Eq. 1.5-5 $= M_{RX}$	(3) $\frac{0.60 F_y S_x}{12} = M_{RX}$	(4) Eq. 1.5-6a =(MF)	(5) Eq. 1.5-7 =(MF)	(6) Eq. 1.5-6b =(MF)	(7) $\frac{0.75 F_y S_y}{12} = M_{RY}$	(8) Eq. 1.5-5b $= M_{RY}$	(9) $\frac{0.60 F_y S_y}{12} = M_{RY}$
	sq. in.	(ft.)	in.	in.$^{-1}$	–	–	in.^{3}	in.^{3}									
C 15 33.9	9.96	0.00	0.000	6.787	0.00	0.00	42.00	3.11	0.0	0.0	104.9	0.000	6.18	0.0	0.0	0.0	7.7
C 12 30.0	8.82	0.00	0.000	7.555	0.00	0.00	27.00	2.06	0.0	0.0	67.5	0.000	3.57	0.0	0.0	0.0	5.1
C 12 25.0	7.35	0.00	0.000	7.860	0.00	0.00	24.10	1.88	0.0	0.0	60.2	0.000	3.06	0.0	0.0	0.0	4.6
C 12 20.7	6.09	0.00	0.000	8.141	0.00	0.00	21.50	1.73	0.0	0.0	53.7	0.000	2.64	0.0	0.0	0.0	4.3
C 10 30.0	8.82	0.00	0.000	7.562	0.00	0.00	20.70	1.65	0.0	0.0	51.7	0.000	2.73	0.0	0.0	0.0	4.1
C 10 25.0	7.35	0.00	0.000	7.947	0.00	0.00	18.20	1.48	0.0	0.0	45.4	0.000	2.29	0.0	0.0	0.0	3.6
C 10 20.0	5.88	0.00	0.000	8.373	0.00	0.00	15.80	1.32	0.0	0.0	39.4	0.000	1.88	0.0	0.0	0.0	3.2
C 10 15.3	4.49	0.00	0.000	8.821	0.00	0.00	13.50	1.16	0.0	0.0	33.7	0.000	1.53	0.0	0.0	0.0	2.8
C 9 20.0	5.88	0.00	0.000	8.229	0.00	0.00	13.50	1.17	0.0	0.0	33.7	0.000	1.64	0.0	0.0	0.0	2.9
C 9 15.0	4.41	0.00	0.000	8.769	0.00	0.00	11.30	1.01	0.0	0.0	28.2	0.000	1.28	0.0	0.0	0.0	2.5
C 9 13.4	3.94	0.00	0.000	8.956	0.00	0.00	10.60	0.96	0.0	0.0	26.4	0.000	1.18	0.0	0.0	0.0	2.4
C 8 18.7	5.51	0.00	0.000	8.117	0.00	0.00	11.00	1.01	0.0	0.0	27.4	0.000	1.35	0.0	0.0	0.0	2.5
C 8 13.7	4.04	0.00	0.000	8.754	0.00	0.00	9.03	0.85	0.0	0.0	22.5	0.000	1.03	0.0	0.0	0.0	2.1
C 8 11.5	3.38	0.00	0.000	9.076	0.00	0.00	8.14	0.78	0.0	0.0	20.3	0.000	0.89	0.0	0.0	0.0	1.9
C 7 14.7	4.33	0.00	0.000	8.319	0.00	0.00	7.78	0.77	0.0	0.0	19.4	0.000	0.93	0.0	0.0	0.0	1.9
C 7 12.2	3.60	0.00	0.000	8.717	0.00	0.00	6.93	0.70	0.0	0.0	17.3	0.000	0.79	0.0	0.0	0.0	1.7
C 7 9.8	2.87	0.00	0.000	9.151	0.00	0.00	6.08	0.62	0.0	0.0	15.1	0.000	0.66	0.0	0.0	0.0	1.5
C 6 13.0	3.83	0.00	0.000	8.109	0.00	0.00	5.80	0.64	0.0	0.0	14.4	0.000	0.71	0.0	0.0	0.0	1.6
C 6 10.5	3.09	0.00	0.000	8.600	0.00	0.00	5.06	0.56	0.0	0.0	12.6	0.000	0.58	0.0	0.0	0.0	1.4
C 6 8.2	2.40	0.00	0.000	9.110	0.00	0.00	4.38	0.49	0.0	0.0	10.9	0.000	0.48	0.0	0.0	0.0	1.2
C 5 9.0	2.64	0.00	0.000	8.289	0.00	0.00	3.56	0.44	0.0	0.0	8.8	0.000	0.42	0.0	0.0	0.0	1.1
C 5 6.7	1.97	0.00	0.000	8.928	0.00	0.00	3.00	0.37	0.0	0.0	7.4	0.000	0.33	0.0	0.0	0.0	0.9

C	4	7.2	2.13	0.00	0.000	7.852	0.00	0.00	2.29	0.34	0.0	0.0	5.7	0.000	0.29	0.0	0.0	0.0	0.8
C	4	5.4	1.59	0.00	0.000	8.531	0.00	0.00	1.93	0.28	0.0	0.0	4.8	0.000	0.22	0.0	0.0	0.0	0.7
C	3	6.0	1.76	0.00	0.000	6.885	0.00	0.00	1.38	0.26	0.0	0.0	3.4	0.000	0.20	0.0	0.0	0.0	0.6
C	3	5.0	1.47	0.00	0.000	7.335	0.00	0.00	1.24	0.23	0.0	0.0	3.0	0.000	0.16	0.0	0.0	0.0	0.5
C	3	4.1	1.21	0.00	0.000	7.793	0.00	0.00	1.10	0.20	0.0	0.0	2.7	0.000	0.14	0.0	0.0	0.0	0.5
MC	18	58.0	17.10	0.00	0.000	6.857	0.00	0.00	75.10	5.32	0.0	0.0	187.7	0.000	10.95	0.0	0.0	0.0	13.2
MC	18	51.9	15.30	0.00	0.000	7.024	0.00	0.00	69.70	5.07	0.0	0.0	174.2	0.000	9.92	0.0	0.0	0.0	12.6
MC	18	45.8	13.50	0.00	0.000	7.200	0.00	0.00	64.30	4.82	0.0	0.0	160.7	0.000	8.93	0.0	0.0	0.0	12.0
MC	18	42.7	12.60	0.00	0.000	7.291	0.00	0.00	61.60	4.69	0.0	0.0	154.0	0.000	8.44	0.0	0.0	0.0	11.7
MC	13	50.0	14.70	0.00	0.000	4.830	0.00	0.00	48.40	4.79	0.0	0.0	120.9	0.000	10.01	0.0	0.0	0.0	11.9
MC	13	40.0	11.80	0.00	0.000	5.092	0.00	0.00	42.00	4.26	0.0	0.0	104.9	0.000	8.24	0.0	0.0	0.0	10.6
MC	13	35.0	10.30	0.00	0.000	5.233	0.00	0.00	38.80	3.99	0.0	0.0	96.9	0.000	7.41	0.0	0.0	0.0	9.9
MC	13	31.8	9.35	0.00	0.000	5.327	0.00	0.00	36.80	3.81	0.0	0.0	91.9	0.000	6.90	0.0	0.0	0.0	9.5
MC	12	50.0	14.70	0.00	0.000	4.145	0.00	0.00	44.90	5.65	0.0	0.0	112.2	0.000	10.83	0.0	0.0	0.0	14.1
MC	12	45.0	13.20	0.00	0.000	4.272	0.00	0.00	42.00	5.33	0.0	0.0	104.9	0.000	9.82	0.0	0.0	0.0	13.3
MC	12	40.0	11.80	0.00	0.000	4.406	0.00	0.00	39.00	5.00	0.0	0.0	97.4	0.000	8.84	0.0	0.0	0.0	12.4
MC	12	35.0	10.30	0.00	0.000	4.550	0.00	0.00	36.10	4.67	0.0	0.0	90.2	0.000	7.93	0.0	0.0	0.0	11.6
MC	12	37.0	10.90	0.00	0.000	5.555	0.00	0.00	34.20	3.59	0.0	0.0	85.4	0.000	6.15	0.0	0.0	0.0	8.9
MC	12	32.9	9.67	0.00	0.000	5.714	0.00	0.00	31.80	3.39	0.0	0.0	79.5	0.000	5.56	0.0	0.0	0.0	8.4
MC	12	30.9	9.07	0.00	0.000	5.797	0.00	0.00	30.60	3.28	0.0	0.0	76.4	0.000	5.27	0.0	0.0	0.0	8.1

(1) M_{rx}. Compact sections; $b_f/2t_f \leqslant 9.19$; beams, $f_a = 0$, $d/t_w \leqslant 90.5$; columns, $f_a > 0.16\,F_y$, $d/t_w \leqslant 36.3$; braced so that $L_b \leqslant L_c$.
(2) M_{rx}. Semi-compact sections; same as column (1) except $9.19 < b_f/2t_f < 13.44$.
(3) M_{rx}. No-compact sections for tension; all sections not under (1), (2), (4), (6); for compression; upper limit for compression sections under (4), (5), (6).
*Cols (4), (5), (6): Non-compact sections; moment factors (*MF*) for compression.
*(4) For $45.2\sqrt{C_b} \leqslant l_b/r_T < 101\sqrt{C_b}$, except channels. Use *larger* M_{rx} from (4) or (5). . . . . (4) $M_{rx} = \text{Col. (1)} - (MF \times 10^{-4})\, l_b^2/C_b \leqslant \text{Col. (3)}$.
*(5) For all values l_b/r_T. . . . . (5) $M_{rx} = 1000\,(MF)\,C_b/l_b \leqslant \text{Col. (3)}$.
*(6) For $l_b/r_T \geqslant 101\sqrt{C_b}$, except channels. Use larger M_{rx} from columns (5) or (6). . . . . (6) $M_{rx} = 1000\,(MF)\,C_b/l_b^2 \leqslant \text{Col. (3)}$.
(7) M_{ry}. For doubly symmetrical shapes; $b_f/2t_f \leqslant 9.19$.
(8) M_{ry}. For doubly symmetrical shapes; $9.19 < b_f/2t_f < 13.44$.
(9) M_{ry}. For non-compact sections; sections not under column (7) or (8).

F_y = 50 ksi	TABLE 3-2: RESISTING MOMENTS, M_R, (kip-ft.) - F_y = 50 ksi																
	SECTION PROPERTIES								MOMENT M_{RX}, k-ft.			M_{RX} FACTORS*			MOMENT, M_{RY}, k-ft.		
Section	Area	L_c	r_T	d/A_f	$b_f/2t_f$	d/t_w	S_{x-x}	S_{y-y}	(1) $\frac{0.66 F_y S_x}{12} = M_{RX}$	(2) Eq. 1.5-5 $= M_{RX}$	(3) $\frac{0.60 F_y S_x}{12} = M_{RX}$	(4) Eq. 1.5-6a =(MF)	(5) Eq. 1.5-7 =(MF)	(6) Eq. 1.5-6b =(MF)	(7) $\frac{0.75 F_y S_y}{12} = M_{RY}$	(8) Eq. 1.5-5b $= M_{RY}$	(9) $\frac{0.60 F_y S_y}{12} = M_{RY}$
	sq.in.	(ft.)	in.	in.$^{-1}$	–	–	in.^{3}	in.^{3}									
MC 12 10.6	3.10	0.00	0.000	25.889	0.00	0.00	9.23	0.31	0.0	0.0	23.0	0.000	0.35	0.0	0.0	0.0	0.7
MC 10 41.1	12.10	0.00	0.000	4.024	0.00	0.00	31.50	4.88	0.0	0.0	78.7	0.000	7.82	0.0	0.0	0.0	12.1
MC 10 33.6	9.87	0.00	0.000	4.241	0.00	0.00	27.80	4.38	0.0	0.0	69.5	0.000	6.55	0.0	0.0	0.0	10.9
MC 10 28.5	8.37	0.00	0.000	4.402	0.00	0.00	25.30	4.02	0.0	0.0	63.2	0.000	5.74	0.0	0.0	0.0	10.0
MC 10 28.3	8.32	0.00	0.000	4.966	0.00	0.00	23.60	3.20	0.0	0.0	58.9	0.000	4.75	0.0	0.0	0.0	7.9
MC 10 25.3	7.43	0.00	0.000	5.633	0.00	0.00	21.40	2.89	0.0	0.0	53.4	0.000	3.79	0.0	0.0	0.0	7.2
MC 10 24.9	7.32	0.00	0.000	5.112	0.00	0.00	22.00	2.99	0.0	0.0	54.9	0.000	4.30	0.0	0.0	0.0	7.4
MC 10 21.9	6.43	0.00	0.000	5.797	0.00	0.00	19.70	2.70	0.0	0.0	49.2	0.000	3.39	0.0	0.0	0.0	6.7
MC 10 8.4	2.46	0.00	0.000	23.809	0.00	0.00	6.40	0.27	0.0	0.0	15.9	0.000	0.26	0.0	0.0	0.0	0.6
MC 10 6.5	1.91	0.00	0.000	43.926	0.00	0.00	4.42	0.11	0.0	0.0	11.0	0.000	0.10	0.0	0.0	0.0	0.2
MC 9 25.4	7.47	0.00	0.000	4.675	0.00	0.00	19.60	3.02	0.0	0.0	48.9	0.000	4.19	0.0	0.0	0.0	7.5
MC 9 23.9	7.02	0.00	0.000	4.743	0.00	0.00	18.90	2.93	0.0	0.0	47.2	0.000	3.98	0.0	0.0	0.0	7.3
MC 8 22.8	6.70	0.00	0.000	4.351	0.00	0.00	16.00	2.84	0.0	0.0	40.0	0.000	3.67	0.0	0.0	0.0	7.0
MC 8 21.4	6.28	0.00	0.000	4.416	0.00	0.00	15.40	2.74	0.0	0.0	38.5	0.000	3.48	0.0	0.0	0.0	6.8
MC 8 20.0	5.88	0.00	0.000	5.289	0.00	0.00	13.60	2.05	0.0	0.0	33.9	0.000	2.57	0.0	0.0	0.0	5.1
MC 8 18.7	5.50	0.00	0.000	5.372	0.00	0.00	13.10	1.97	0.0	0.0	32.7	0.000	2.43	0.0	0.0	0.0	4.9
MC 8 8.5	2.50	0.00	0.000	13.726	0.00	0.00	5.83	0.43	0.0	0.0	14.5	0.000	0.42	0.0	0.0	0.0	1.0
MC 7 22.7	6.67	0.00	0.000	3.885	0.00	0.00	13.60	2.85	0.0	0.0	33.9	0.000	3.50	0.0	0.0	0.0	7.1
MC 7 19.1	5.61	0.00	0.000	4.055	0.00	0.00	12.30	2.57	0.0	0.0	30.7	0.000	3.03	0.0	0.0	0.0	6.4
MC 7 17.6	5.17	0.00	0.000	4.912	0.00	0.00	10.80	1.89	0.0	0.0	26.9	0.000	2.19	0.0	0.0	0.0	4.7
MC 6 18.0	5.29	0.00	0.000	3.604	0.00	0.00	9.91	2.48	0.0	0.0	24.7	0.000	2.74	0.0	0.0	0.0	6.1
MC 6 16.3	4.79	0.00	0.000	4.210	0.00	0.00	8.68	1.84	0.0	0.0	21.6	0.000	2.06	0.0	0.0	0.0	4.5

MC	6	15.3	4.50	0.00	0.000	4.452	0.00	0.00	8.47	2.03	0.0	0.0	21.1	0.000	1.90	0.0	0.0	0.0	5.0
MC	6	15.1	4.44	0.00	0.000	4.294	0.00	0.00	8.32	1.75	0.0	0.0	20.7	0.000	1.93	0.0	0.0	0.0	4.3
MC	6	12.0	3.53	0.00	0.000	6.407	0.00	0.00	6.24	1.04	0.0	0.0	15.5	0.000	0.97	0.0	0.0	0.0	2.5
MC	3	9.0	2.65	0.00	0.000	4.027	0.00	0.00	2.10	0.67	0.0	0.0	5.2	0.000	0.52	0.0	0.0	0.0	1.6
MC	3	7.1	2.09	0.00	0.000	4.410	0.00	0.00	1.82	0.56	0.0	0.0	4.5	0.000	0.41	0.0	0.0	0.0	1.4

(1) M_{rx}. Compact sections; $b_f/2t_f \leqslant 9.19$; beams, $f_a = 0$, $d/t_w \leqslant 90.5$; columns, $f_a > 0.16\ F_y$, $d/t_w \leqslant 36.3$; braced so that $L_b \leqslant L_c$.
(2) M_{rx}. Semi-compact sections; same as column (1) except $9.19 < b_f/2t_f < 13.44$.
(3) M_{rx}. No-compact sections for tension; all sections not under (1), (2), (4), (6); for compression; upper limit for compression sections under (4), (5), (6).
*Cols (4), (5), (6): Non-compact sections; moment factors (*MF*) for compression.
*(4) For $45.2\sqrt{C_b} \leqslant l_b/r_T < 101\sqrt{C_b}$, except channels. Use *larger* M_{rx} from (4) or (5). . . . . (4) $M_{rx} = \text{Col. (1)} - (MF \times 10^{-4})\ l_b^2/C_b \leqslant \text{Col. (3)}$.
*(5) For all values l_b/r_T. . . . . (5) $M_{rx} = 1000\ (MF)\ C_b/l_b \leqslant \text{Col. (3)}$.
*(6) For $l_b/r_T \geqslant 101\sqrt{C_b}$, except channels. Use larger M_{rx} from columns (5) or (6). . . . . (6) $M_{rx} = 1000\ (MF)\ C_b/l_b^2 \leqslant \text{Col. (3)}$.
(7) M_{ry}. For doubly symmetrical shapes; $b_f/2t_f \leqslant 9.19$.
(8) M_{ry}. For doubly symmetrical shapes; $9.19 < b_f/2t_f < 13.44$.
(9) M_{ry}. For non-compact sections; sections not under column (7) or (8).

$Fb \leqslant 0.66\, F_y$ . . . . . . (1.5.1.4)

Grade 36, $F_B \leqslant 24$ ksi

Grade 50, $F_b \leqslant 33$ ksi

In order to qualify as a compact section, a member must meet the following criteria: (See Fig. 3-1.)

1. The flange must be continuously connected to the web. For built-up members, this eliminates the use of intermittent fillet welds to connect the flanges to the web (1.5.1.4.1a).
2. The width-thickness ratio (b_f/t_f) of an unstiffened compression flange is limited to:

   $b_f/t_f \leqslant 65/\sqrt{F_y}$ . . . . . . (1.5.1.4.1b)

   Grade 36, $b_f/t_f \leqslant 10.83$ — See Fig. 3-1.

   Grade 50, $b_f/t_f \leqslant 9.19$

3. The width thickness ratio ($b_f/2t_f$) of a stiffened compression flange (1.9.2.1) is limited to:

   $b_f/2t_f \leqslant 190/\sqrt{F_y}$ . . . . . . (1.5.1.4.1c)

   Grade 36, $b_f/2t_f \leqslant 31.7$ — See Fig. 3-1.

   Grade 50, $b_f/2t_f \leqslant 26.9$

4. The depth thickness ratio (d/t_w) of the web or webs of flexural members without axial load is limited to:

   $d/t_w \leqslant 640/\sqrt{F_y}$ . . . . . . (1.5.1.4.1d)

   Grade 36, $d/t_w \leqslant 106.7$ — See Fig. 3-1.

   Grade 50, $d/t_w \leqslant 90.5$

5. Unbraced length, l_b, limitations:
   (a) For rectangular box-type members with $d \leqslant 6b$ (See Fig. 3-1), the laterally unsupported length of the compression flange, where $t_f \leqslant 2t_w$, is limited to:

   $(1950 + 1200\, M_1/M_2)\,(b/F_y) \geqslant 1200\,(b/F_y)$ . . . . . . (1.5.1.4.1-f),

   where:

   M_1 = smaller moment at one end of unbraced length, taken about strong axis of member.
   M_2 = larger moment at the other end of unbraced length about the same axis.
   M_1/M_2 is positive for reverse curvature; negative for single curvature.*
   M_1/M_2 = 1 when bending moment at any location within the unbraced length is larger than at both ends.* See Fig. 3-2.

   (b) For members other than box-type members, the compression flange must be laterally supported at intervals, l_b, not greater than $76\, b_f/\sqrt{F_y}$ nor $20{,}000/[(d/A_f)\, F_y]$, where A_f is the area of the compression flange (1.5.1.4.1e). (See Fig. 3-1.)

---

*For simple beams, $M_1 = M_2 = 0$, $(M_1/M_2) = -1$.

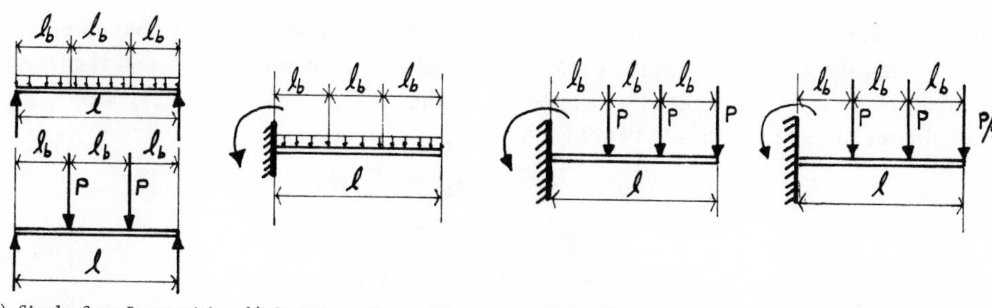

a) Simple Span Beams with Uniform Load, w, and/or equal Concentrated Loads at Bracing Points.

l_b	C_b
l	1.000
$l/2$	1.750
$l/3$	1.000
$l/4$	1.131
$l/5$	1.000
$l/6$	1.054
$l/7$	1.000
$l/8$	1.029

b) Cantilever Beam with Uniform Load

l_b	C_b
l	1.000
$l/2$	1.506
$l/3$	1.343
$l/4$	1.254
$l/5$	1.201
$l/6$	1.166
$l/7$	1.141
$l/8$	1.122

c) Cantilever Beam with Equal Concentrated Loads

l_b	C_b
l	1.750
$l/2$	1.433
$l/3$	1.300
$l/4$	1.228
$l/5$	1.183
$l/6$	1.153
$l/7$	1.131
$l/8$	1.115

d) Cantilever Beam with Unequal concentrated Loads

l_b	C_b
l	1.750
$l/2$	1.506
$l/3$	1.343
$l/4$	1.254
$l/5$	1.201
$l/6$	1.166
$l/7$	1.141
$l/8$	1.122

Fig. 3-2 Bending Coefficient C_b, for Simple and Cantilever Spans.

$$\text{Grade 36, } l_b \leqslant 12.67\, b_f$$

$$l_b \leqslant 556/(d/A_f)$$

$$\text{Grade 50, } l_b \leqslant 10.74\, b_f$$

$$l_b \leqslant 400/(d/A_f)$$

$b_f/2t_f$ ratios for rolled shapes vary from a minimum of 1.82 for a W 14 X 730 section to a maximum of 14.4 for a HP 14 X 73 section and d/t_w ratios vary from a minimum of 7.31 for a W 14 X 730 section to a maximum of 67.8 for a M 12 X 11.8 section.

Compact section parameters, $b_f/2t_f$, d/t_w and d/A_f, for rolled shapes are readily available in the "Properties for Designing" section of the AISC Manual and in the authors' Tables 3-1 and 3-2 of rolled-shape resisting moments.

Tables for "Properties for Designing" and the "Allowable Stress Design Selection" by the AISC list the theoretical maximum yield stress for the $b_f/2t_f$ ratio, F_y', and the d/t_w ratio, F_y'', that conform to compact section criteria.

$$F_y' = [65/(b_f/2t_f)]^2$$

$$F_y'' = [640/(d/t_w)]^2$$

The absence of a value for F_y' or F_y'' in these AISC tables denotes that the theoretical maximum yield stress that can be effectively utilized is greater than 65 ksi for flange or web.

Tables 3-1 and 3-2 herein and the AISC "Allowable Stress Design Selection Table" list the maximum unbraced length of the compression flange of rolled shapes, L_c, beyond which the section is no longer compact.

*Compact Sections–Bending About the Minor Axis.** The maximum allowable compression and tension stress (F_b) on the extreme fibers of doubly-symmetrical I- and H-shaped members with webs continuously connected to the flanges and with $b_f/2t_f \leq 65/\sqrt{F_y}$ and solid rectangular sections in bending about the minor axis is:

$$F_b = 0.75\,F_y \;\ldots\ldots (1.5.1.4.3).$$

$$\text{Grade 36}, F_b = 27.0 \text{ ksi}$$

$$\text{Grade 50}, F_b = 37.5 \text{ ksi}$$

These stresses also apply to solid round and square bars. The increase in allowable stress for doubly-symmetrical I- and H-shaped members bending about their minor axis over that about the major axis is due to the favorable shape factor. The ratio $Z/S = 1.50$, (ratio of plastic section modulus to elastic section modulus) for a rectangular section versus 1.12 for a wide flange member bending about its major axis. The limitation on unbraced length does not apply to compact doubly-symmetrical shaped sections bending about their minor axis because these sections are not subject to lateral torsional buckling. The limitation on depth/thickness ratio of the web also does not apply because for doubly-symmetrical sections the web is on the neutral axis (1.5.1.5.1d).

*Semi-Compact Sections–Bending About the Major Axis.** The term "semi-compact" can be applied to members, bending about their major axis, that meet all compact section criteria, except that

$$65/\sqrt{F_y} < b_f/2t_f < 95/F_y \;\ldots\ldots (1.5.1.4.2).$$

$$\text{Grade 36}, 10.83 < b_f/2t_f < 15.83$$

$$\text{Grade 50}, 9.19 < b_f/2t_f < 13.44$$

The allowable compression and tension stress in bending about the major axis for semi-compact sections can be obtained from Table 3-3 or calculated from the expression:

$$F_b = F_y\,[0.79 - 0.002\,(b_f/2t_f)\,\sqrt{F_y}] \;\ldots\ldots (1.5.1.4.2).$$

where $65/\sqrt{F_y} < b_f/2t_f < 95/\sqrt{F_y}$.

TABLE 3-3 Allowable Bending Stress, F_b, for Semi-Compact Sections Bending About the Major Axis

$b_f/2t_f$	F_b–Grade 36*	F_b–Grade 50*
9.19	–	33.00
10	–	32.43
10.83	23.76	–
11	23.69	31.72
12	23.26	31.01
13	22.82	30.31
13.44	–	30.00
14	22.39	–
15	21.96	–
15.83	21.60	–

*When $65/\sqrt{F_y} \leq b_f/2t_f \leq 95/\sqrt{F_y}$
$F_b = F_y\;[0.79 - 0.002\,(b_f 2t_f)\sqrt{F_y}] \ldots\ldots (1.5.1.4.2)$

*Except for hybrid girders, tapered girders, and rolled shapes and girders of A514 steel (1.5.1.4.1), and tension field action girders (1.10.7).

TABLE 3-4 Allowable Bending Stress, F_b, for Semi-Compact Sections Bending About the Minor Axis

$b_f/2t_f$	F_b–Grade 36*	F_b–Grade 50*
9.19	–	37.50
10.00	–	36.07
10.83	27.00	–
11	26.82	34.30
12	25.74	32.54
13	24.66	30.77
13.44	–	30.00
13	23.58	–
15	22.50	–
15.83	21.60	–

*When $65/\sqrt{F_y} < b_f/2t_f < 95/\sqrt{F_y}$

$F_b = F_y\,[1.075 - 0.005\,(b_f/2t_f)\sqrt{F_y}]$ ...... (1.5.1.4.3)

Semi-Compact Sections–Bending About the Minor Axis.* The term "semi-compact" can also be applied to doubly-symmetrical I- and H-shaped members bending about their minor axis if the flanges are continuously connected to their webs and $65/\sqrt{F_y} < b_f/2t_f < 95/\sqrt{F_y}$. The allowable compression and tension stress in bending about the minor axis for semi-compact sections can be obtained from Table 3-4 or calculated from the expression:

$$F_b = F_y\,[1.075 - 0.005\,(b_f/2t_f)\sqrt{F_y}] \quad \ldots\ldots (1.5.1.4.3)$$

Non-compact Box-type, Non-slender Flexural Members. Non-compact, non-slender box-type flexural members are those whose width-thickness ratios of compression flange and web, $(b_f/2t_f)$ and (d/t_w), do not meet compact section criteria (1.5.1.4.1), but do con-

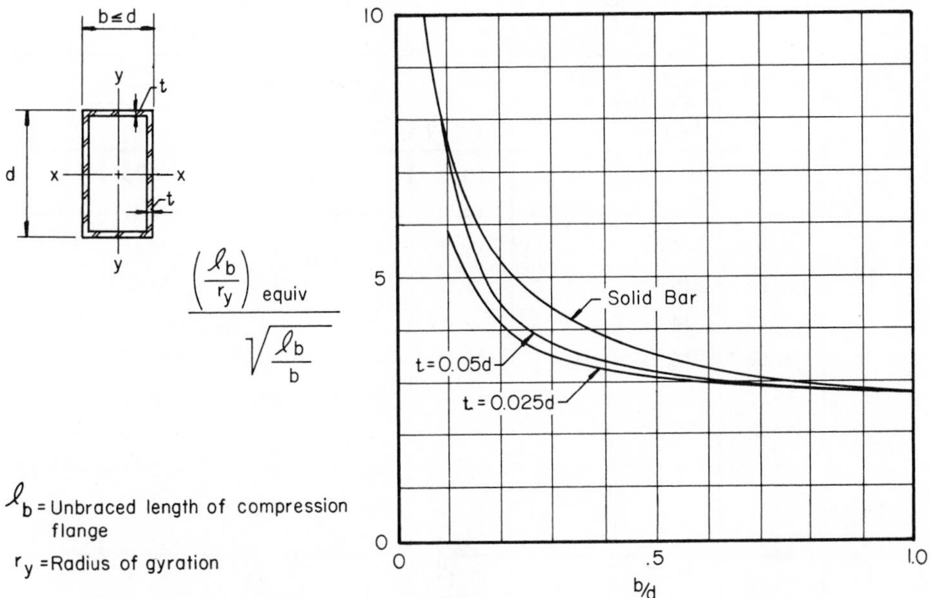

Fig. 3-3 (l_b/r) Equivalent for Stiffened Box and Rectangular Sections.

*Exceptions are hybrid girders, tapered girders, and members of A514 steel (1.5.1.4.1), and tension field action girders (1.10.7).

form to AISC (1.9). Lateral torsional buckling need not be investigated for such members if their depth is less than 6 times their width, $d < 6b$ (1.5.1.4.4). (See Table 3-5.)

The maximum allowable stress for tension and compression in bending is

$$F_b = 0.60\,F_y \;\ldots\ldots\; (1.5.1.4.4)$$

Lateral support requirements for non-compact box-type non-slender flexural members with $d > 6b$ must be determined by special analysis. (See Figs. 3-3, 3-4 and 3-5.)

TABLE 3-5 Maximum Flange Widths, b, and depths, d, for Compact Box-Sections with $F_b = 0.66\,F_Y$ and Non-Compact Box Sections with $F_b = 0.60\,F_Y$

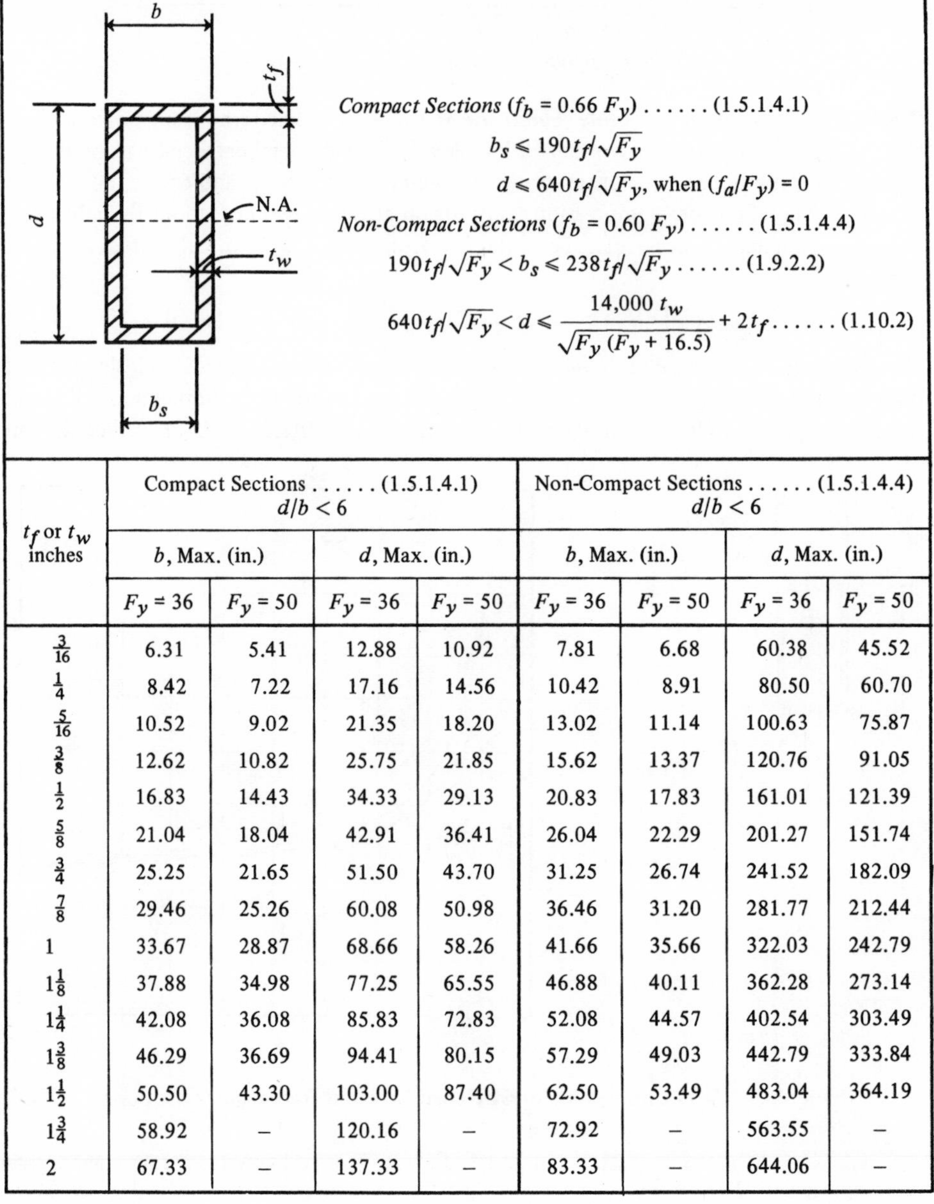

t_f or t_w inches	Compact Sections . . . . . . (1.5.1.4.1) $d/b < 6$				Non-Compact Sections . . . . . . (1.5.1.4.4) $d/b < 6$			
	b, Max. (in.)		d, Max. (in.)		b, Max. (in.)		d, Max. (in.)	
	$F_y = 36$	$F_y = 50$	$F_y = 36$	$F_y = 50$	$F_y = 36$	$F_y = 50$	$F_y = 36$	$F_y = 50$
$\frac{3}{16}$	6.31	5.41	12.88	10.92	7.81	6.68	60.38	45.52
$\frac{1}{4}$	8.42	7.22	17.16	14.56	10.42	8.91	80.50	60.70
$\frac{5}{16}$	10.52	9.02	21.35	18.20	13.02	11.14	100.63	75.87
$\frac{3}{8}$	12.62	10.82	25.75	21.85	15.62	13.37	120.76	91.05
$\frac{1}{2}$	16.83	14.43	34.33	29.13	20.83	17.83	161.01	121.39
$\frac{5}{8}$	21.04	18.04	42.91	36.41	26.04	22.29	201.27	151.74
$\frac{3}{4}$	25.25	21.65	51.50	43.70	31.25	26.74	241.52	182.09
$\frac{7}{8}$	29.46	25.26	60.08	50.98	36.46	31.20	281.77	212.44
1	33.67	28.87	68.66	58.26	41.66	35.66	322.03	242.79
$1\frac{1}{8}$	37.88	34.98	77.25	65.55	46.88	40.11	362.28	273.14
$1\frac{1}{4}$	42.08	36.08	85.83	72.83	52.08	44.57	402.54	303.49
$1\frac{3}{8}$	46.29	36.69	94.41	80.15	57.29	49.03	442.79	333.84
$1\frac{1}{2}$	50.50	43.30	103.00	87.40	62.50	53.49	483.04	364.19
$1\frac{3}{4}$	58.92	–	120.16	–	72.92	–	563.55	–
2	67.33	–	137.33	–	83.33	–	644.06	–

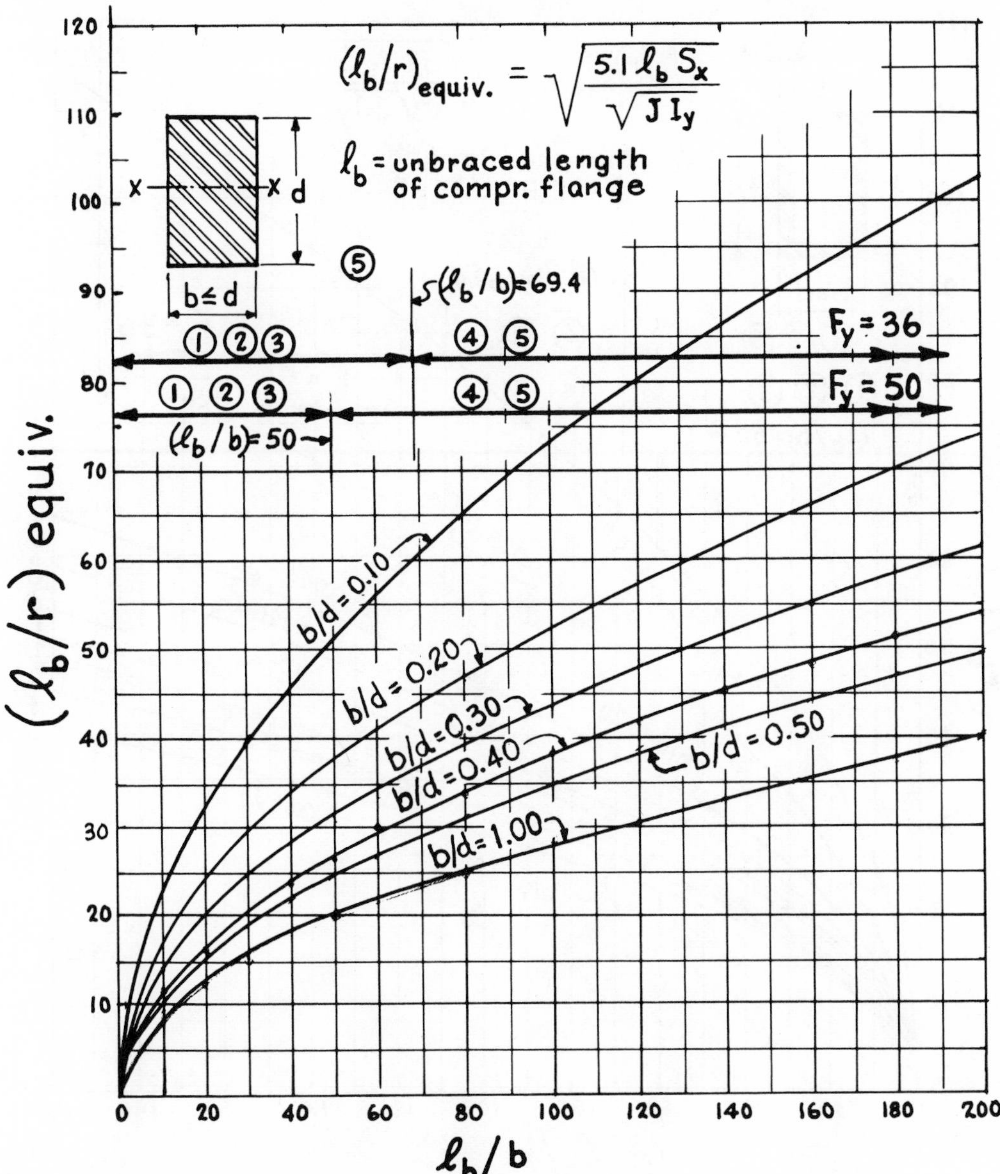

Fig. 3-4 **(l_b/r) Equivalent for Solid Rectangular Shapes for Allowable Bending Stress, F_b, (Eq. 1.5-1).**

① For *compact* sections bending about the major axis, use $F_b = 0.66\,F_y$ when $d \leqslant 6b$ and $l_b/b \leqslant (1950 + 1200\,M_1/M_2)/F_y \geqslant 1200\,b/F_y$. Use $F_b = 0.75\,F_y$ for bending about the minor axis.

② For *non-compact* sections bending about the major axis, use $F_b = 0.60\,F_y$ when $d \leqslant 6b$.

③ For *non-compact* sections bending about the major axis, use $F_b = 0.60\,F_y$ when $6b < d \leqslant 10b$ and $l_b/b \leqslant 2500/F_y$.

④ For *non-compact* sections bending about the major axis, determine F_b from (l_b/r) equiv. using Fig. 6-5 (AISC Eq. 1.5-1) when $d > 10b$ and $l_b/b > 2500/F_y$. See Fig. 3-3.

⑤ For all solid round, square and rectangular sections bending about the minor axis, use $F_b = 0.75\,F_y$.

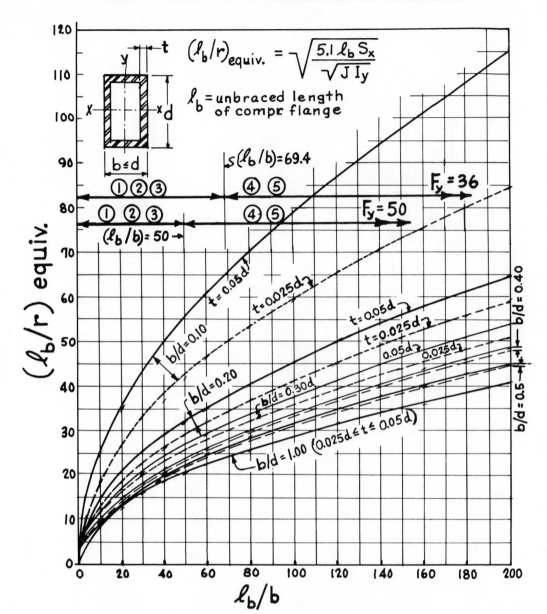

Fig. 3-5 (l_b/r) Equivalent for Stiffened Box Sections of Uniform Thickness for Allowable Bending Stress, F_b, (Eq. 1.5-1).

① For *compact* sections bending about the major axis, use $F_b = 0.66\ F_y$ when $d \leqslant 6b$ and $l_b/b \leqslant (1950 + 1200\ M_1/M_2)/F_y \geqslant 1200\ b/F_y$. Use $F_b = 0.75\ F_y$ for bending about the minor axis.

② For *non-compact* sections bending about the major axis, use $F_b = 0.60\ F_y$ when $d \leqslant 6b$.

③ For *non-compact* sections bending about the major axis, use $F_b = 0.60\ F_y$ when $6b < d \leqslant 10b$ and $l_b/b \leqslant 2500/F_y$.

④ For *non-compact* sections bending about the major axis, determine F_b from (l_b/r) equiv. using Fig. 6-5 (AISC Eq. 1.5-1) when $d > 10b$ and $l_b/b > 2500/F_y$. See Fig. 3-3.

⑤ For all solid round, square and rectangular sections bending about the minor axis, use $F_b = 0.75\ F_y$.

Compression–Channel Shaped Sections Bending About the Major Axis. The maximum allowable compression stress in bending on the extreme fiber,

$F_b = 12{,}000\ C_b/(l_b d/A_f) \leqslant 0.60 F_y$ . . . . . . (1.5.1.4.6a, Eq. 1.5-7).

(See Table 3-6 for solutions of F_b.)

Tension–Non-Compact and Non-Semi-Compact Sections. The maximum allowable tension stress in bending on the extreme fiber of all non-compact and non-semi-compact sections,

$$F_b = 0.60\ F_y \ldots\ldots (1.5.1.4.4).$$

Compression–Non-Compact and Non-Semi-Compact Sections Bending About Their Major Axis and With an Axis of Symmetry in the Plane of Their Web. The allowable compression stress in bending on the extreme fiber of sections loaded in the plane of the web with width-thickness ratios (1.9.1.1) of compression flanges of beams, $b_f/2t_f \leqslant 95/F_y$, and width-thickness ratios of stems of tees, $d/t_w \leqslant 127/F_y$, must not be greater than the larger of the following formulas, and not more than 0.60 F_y unless a more precise analysis* is made (1.9.1.2; 1.5.1.4.6a).

a. When $\sqrt{102{,}000\ C_b/F_y} \leqslant l_b \leqslant \sqrt{510{,}000\ C_b/F_y}$:
   for Grade 36, $53.2\sqrt{C_b} \leqslant l_b \leqslant 119.0\sqrt{C_b}$;
   for Grade 50, $45.2\sqrt{C_b} \leqslant l_b \leqslant 101.0\sqrt{C_b}$;
   $F_b = [(2/3) - F_y(l_b/r_T)^2/(1{,}530{,}000\ C_b)]\ F_y \leqslant 0.60\ F_y$ . . . . . . (Eq. 1.5-6a).

b. When $l_b/r_T > \sqrt{510{,}000\ C_b/F_y}$:
   for Grade 36, $l_b/r_T \geqslant 119.0\sqrt{C_b}$;
   for Grade 50, $l_b/r_T \geqslant 101.0\sqrt{C_b}$;
   $F_b = 170{,}000\ C_b/(l_b/r_T)^2 \leqslant 0.60\ F_y$ . . . . . . (Eq. 1.5-6b).
   (See Tables 3-7(a) and (b) for Grades 36 and 50, respectively.)
c. For a compression flange that is solid and approximately rectangular in cross section, and with an area that is not less than that of the tension flange. (See Table 3-6.)

$$F_b = 12{,}000\ C_b/(l_b d/A_f) \leqslant 0.60\ F_y \ldots\ldots (\text{Eq. 1.5-7}).$$

Special nomenclature for the above three formulas is:

C_b = Bending coefficient, which can conservatively be taken as unity, that is dependent upon moment gradient and equal to: $1.75 + 1.05\ M_1/M_2 + 0.3\ (M_1/M_2)^2 \leqslant 2.3$. For solutions, see Fig. 6-2.
M_1 = smaller moment at one end of unbraced length, taken about strong axis of member.
M_2 = larger moment at the other end of unbraced length about the same axis.
M_1/M_2 is positive for reverse curvature; negative for single curvature.
M_1/M_2 = 1 when bending moment at any location within the unbraced length is larger than at both ends.

Values of C_b can be computed or obtained from Table 6-2 for different ratios of M_1/M_2. For simple span and cantilever beams with uniform and concentrated loads, values of C_b can also be obtained from the tabulations in Fig. 3-2.

---

*Commentary (1.5.1.4.5; and last two paragraphs of 1.5.1.4.6).

TABLE 3-6 Allowable Compressive Stress in Bending, F_b vs. $(l_b d/A_f)$ . . . . . (Eq. 1.5-7); Grades 36 and 50

$$F_b = 12{,}000\, C_b / (\ell_b d/A_f) \leq 0.60\, F_y \quad \text{(Eq. 1.5-7)}$$

$$C_b = 1.75 + 1.05(M_1/M_2) + 0.3(M_1/M_2)^2$$

$(\ell_b d/A_f)$	$\geq$1.0	1.2	1.4	1.6	1.8	2.0	2.2	$\leq$2.3
400.	29.99							
450.	26.66	29.99						
500.	24.00	28.79						
550.	21.81	26.18	29.99					
600.	20.00	24.00	28.00	29.99				
650.	18.46	22.15	25.84	29.53				
700.	17.14	20.57	24.00	27.42	29.99			
750.	16.00	19.20	22.39	25.59	28.79			
800.	15.00	18.00	21.00	24.00	27.00	29.99		
850.	14.11	16.94	19.76	22.58	25.41	28.23	29.99	
900.	13.33	15.99	18.66	21.33	24.00	26.66	29.33	29.99
950.	12.63	15.15	17.68	20.21	22.73	25.26	27.78	29.05
1000.	12.00	14.39	16.79	19.20	21.60	24.00	26.39	27.59
1050.	11.42	13.71	16.00	18.28	20.57	22.85	25.14	26.28
1100.	10.90	13.09	15.27	17.45	19.63	21.81	24.00	25.09
1150.	10.43	12.52	14.60	16.69	18.78	20.86	22.95	24.00
1200.	10.00	12.00	14.00	16.00	18.00	20.00	22.00	23.00
1250.	9.60	11.51	13.43	15.36	17.28	19.20	21.11	22.07
1300.	9.23	11.07	12.92	14.76	16.61	18.46	20.30	21.23
1350.	8.88	10.66	12.44	14.22	16.00	17.77	19.55	20.44
1400.	8.57	10.28	12.00	13.71	15.42	17.14	18.85	19.71
1450.	8.27	9.93	11.58	13.24	14.89	16.55	18.20	19.03
1500.	8.00	9.60	11.19	12.79	14.39	15.99	17.59	18.39
1550.	7.74	9.29	10.83	12.38	13.93	15.48	17.03	17.80
1600.	7.50	9.00	10.50	12.00	13.50	15.00	16.50	17.25
1650.	7.27	8.72	10.18	11.63	13.09	14.54	16.00	16.72
1700.	7.05	8.47	9.88	11.29	12.70	14.11	15.52	16.23
1750.	6.85	8.22	9.60	10.97	12.34	13.71	15.08	15.77
1800.	6.66	7.99	9.33	10.66	12.00	13.33	14.66	15.33
1850.	6.48	7.78	9.08	10.37	11.67	12.97	14.27	14.91
1900.	6.31	7.57	8.84	10.10	11.36	12.63	13.89	14.52
1950.	6.15	7.38	8.61	9.84	11.07	12.30	13.53	14.15
2000.	6.00	7.19	8.39	9.60	10.80	12.00	13.19	13.79

F_b = 30 ksi ...Grade 50

F_b = 21.6 ksi .... Grade 36

All Grades

TABLE 3-7 Allowable Compressive Stress in Bending, F_b, vs. (l_b/r_T) . . . . . (Eqs. 1.5-6a and 1.5-6b); Grades 36 and 50

(a) Grade 36

(ℓ_b/r_T)	$C_b = 1.75 + 1.05(M_1/M_2) + 0.3(M_1/M_2)^2$							
	(Min.) 1.0	1.2	1.4	1.6	1.8	2.0	2.2	(Max.) 2.3
50.	21.59	21.59	21.59	21.59	21.59	21.59	21.59	21.59
60.	20.95	21.45	21.59	21.59	21.59	21.59	21.59	21.59
70.	19.84	20.54	21.03	21.40	21.59	21.59	21.59	21.59
80.	18.57	19.48	20.12	20.61	20.98	21.28	21.53	21.59
90.	17.13	18.28	19.09	19.71	20.18	20.56	20.88	21.01
100.	15.52	16.94	17.94	18.70	19.29	19.76	20.14	20.31
110.	13.75	15.45	16.67	17.59	18.30	18.87	19.34	19.54
120.	11.80	13.83	15.28	16.37	17.22	17.90	18.45	18.69
130.	10.05	12.07	13.77	15.05	16.04	16.84	17.49	17.77
140.	8.67	10.40	12.14	13.62	14.77	15.69	16.45	16.78
150.	7.55	9.06	10.57	12.08	13.41	14.47	15.33	15.71
160.	6.64	7.96	9.29	10.62	11.95	13.15	14.14	14.57
170.	5.88	7.05	8.23	9.41	10.58	11.76	12.87	13.35
180.	5.24	6.29	7.34	8.39	9.44	10.49	11.54	12.06
190.	4.70	5.65	6.59	7.53	8.47	9.41	10.36	10.83
200.	4.25	5.09	5.94	6.79	7.64	8.50	9.34	9.77

When $53\sqrt{C_b} \leq (\ell_b/r_T) \leq 119\sqrt{C_b}$, $F_b = 24 - (8.47)(\ell_b/r_T)^2/10{,}000\ C_b$

When $(\ell_b/r_T) \geq 119\sqrt{C_b}$, $F_b = 170{,}000\ C_b/(\ell_b/r_T)^2$

(b) Grade 50

(ℓ_b/r_T)	(Min.) 1.0	1.2	1.4	1.6	1.8	2.0	2.2	(Max.) 2.3
40.	29.99	29.99	29.99	29.99	29.99	29.99	29.99	29.99
50.	29.24	29.92	29.99	29.99	29.99	29.99	29.99	29.99
60.	27.45	28.43	29.13	29.65	29.99	29.99	29.99	29.99
70.	25.32	26.66	27.61	28.32	28.88	29.33	29.69	29.85
80.	22.87	24.61	25.86	26.79	27.52	28.10	28.57	28.78
90.	20.09	22.30	23.87	25.06	25.98	26.71	27.31	27.57
100.	16.99	19.71	21.66	23.12	24.25	25.16	25.90	26.22
110.	14.04	16.85	19.21	20.97	22.34	23.44	24.34	24.73
120.	11.80	14.16	16.52	18.62	20.26	21.56	22.63	23.10
130.	10.05	12.07	14.08	16.09	17.99	19.52	20.78	21.32
140.	8.67	10.40	12.14	13.87	15.61	17.32	18.77	19.40
150.	7.55	9.06	10.57	12.08	13.59	15.11	16.62	17.34
160.	6.64	7.96	9.29	10.62	11.95	13.28	14.60	15.27
170.	5.88	7.05	8.23	9.41	10.58	11.76	12.94	13.52
180.	5.24	6.29	7.34	8.39	9.44	10.49	11.54	12.06
190.	4.70	5.65	6.59	7.53	8.47	9.41	10.36	10.83
200.	4.25	5.09	5.94	6.79	7.64	8.50	9.34	9.77

When $45\sqrt{C_b} \leq (\ell_b/r_T) \leq 101\sqrt{C_b}$, $F_b = 33.3 - (16.33)(\ell_b/r_T)^2/10{,}000\ C_b$

When $(\ell_b/r_T) \geq 101\sqrt{C_b}$, $F_b = 170{,}000\ C_b/(\ell_b/r_T)^2$

Tables 3-6 and 3-7 present solutions for F_b by the three formulas for all values of l_b/r_T, $l_b d/A_f$ and C_b. They are also expressed in terms of moment factors (*MF*) for the different rolled shape beam sections in Col. (4), Col. (5), and Col. (6) of Tables 3-1 and 3-2 for use in computing beam resisting moments (M_{RX}).

Compression–Other Sections, Non-Compact or Non-Semi-Compact. The allowable compression stress in bending on the extreme fiber of all other sections is $F_b = 0.60\ F_y$ provided:

a. width-thickness ratios conform to limits given (1.9.1.2) and
b. for bending about the major axis, spacing of lateral bracing in the region of the compressive stress, l_b, does not exceed $76\ b_f/\sqrt{F_y}$. For Grade 36, $l_b \leqslant 12.67\ b_f$; for Grade 50, $l_b \leqslant 10.74\ b_f$ (1.5.1.4.6b).

Compression–Slender Unstiffened Compression Elements. Slender unstiffened compression elements have one free edge parallel to the direction of the compression stress (1.9.1.1), and have b/t ratios greater than $(95/\sqrt{F_y})$ for compression flanges of beams and angles or plates projecting from girders, and $(127/\sqrt{F_y})$ for stems of tees (1.9.1.2).

(*a*) *Beam Shapes, Plates and Angles in Flexural Compression.* The allowable compressive stress on slender elements consisting of flanges of beams and angles or plates projecting from beams is limited by two sets of criteria (1.9.1.2). F_b must not exceed $0.6\ F_y Q_s$ (Appendix C2) nor the two applicable values from three equations (Eqs. 1.5-6a; 1.5-6b; 1.5-7). In determining the b/t ratio, b is taken equal to one-half the flange width of rolled beam shapes, the flange width of channels, the distance from the free edge of plates to the first row of fasteners or welds, and the full width of legs of angles.

When

$$95/\sqrt{F_y} < b/t < 176/\sqrt{F_y},$$

$$\text{for Grade 36, } 15.83 < b/t < 29.3$$

$$\text{for Grade 50, } 13.43 < b/t < 24.9$$

$$Q_s = 1.415 - 0.00437\,(b/t)\sqrt{F_y} \;\ldots\ldots \text{(Appendix C2, Eq. C2-3).}$$

When

$$b/t \geqslant 176/\sqrt{F_y},$$

$$\text{for Grade 36, } b/t \geqslant 29.3$$

$$\text{for Grade 50, } b/t \geqslant 24.9$$

$$Q_s = 20{,}000/[\sqrt{F_y}(b/t)^2] \;\ldots\ldots \text{(Appendix C2, Eq. C2-4).}$$

See Table 3-8 for the allowable compression stress in bending for Grade 36 and Grade 50 compression flanges of beams and angles or plates projecting from beams.

(*b*) *Stems of Tees in Flexural Compression.* The allowable compressive stress on slender elements consisting of the stems of tees is limited similarly. F_b must not exceed $0.6\ F_y Q_s$ (Appendix C2) nor the two applicable values from three equations (Eqs. 1.5-6a; 1.5-6b; 1.5-7). In determining the b/t ratio of the stem of tees in compression over the full section, b is taken equal the full depth of the tee. The authors suggest for tees used alone as flexural members and not subject to uniform compression, that the width, b, of the compression flange stem can be taken as the distance from the neutral axis to the end of the stem. The remainder of the tee is in tension.

(Continued on page 134)

TABLE 3-8 Allowable Stress in Bending for Slender Unstiffened Compression Elements—Flanges of Beams and Angles or Plates Projecting from Beams; Grades 36 and 50

b/t	ALLOWABLE STRESS* (ksi) (Per Appendix C, Section C2)*			
	Grade 36		Grade 50	
	Q_s	$F_b = 0.60F_yQ_s$	Q_s	$F_b = 0.60F_yQ_s$
13.43	-	-	1.000	30.00
14	-	-	0.982	29.47
15	-	-	0.951	28.54
15.83	1.000	21.60	-	-
16	0.995	21.50	0.920	27.61
17	0.969	20.93	0.889	26.69
18	0.943	20.36	0.858	25.76
19	0.916	19.80	0.827	24.83
20	0.890	19.23	0.796	23.90
21	0.864	18.67	0.766	22.98
22	0.838	18.10	0.735	22.05
23	0.811	17.53	0.704	21.12
24	0.785	16.97	0.673	20.20
24.9	-	-	0.645	19.36
25	0.759	16.40	0.640	19.19
26	0.733	15.83	0.591	17.75
27	0.707	15.27	0.548	16.46
28	0.680	14.70	0.510	15.30
29	0.654	14.13	0.475	14.26
29.3	0.646	13.97	-	-
30	0.617	13.33	0.444	13.33
31	0.578	12.48	0.416	12.48
32	0.542	11.71	0.390	11.71
33	0.510	11.01	0.367	11.01
34	0.480	10.38	0.346	10.38
35	0.453	9.79	0.326	9.79
36	0.428	9.25	0.308	9.25
37	0.405	8.76	0.292	8.76
38	0.384	8.31	0.277	8.31
39	0.365	7.88	0.262	7.88
40	0.347	7.49	0.250	7.49
41	0.330	7.13	0.237	7.13
42	0.314	6.80	0.226	6.80
43	0.300	6.48	0.216	6.48
44	0.286	6.19	0.206	6.19
45	0.274	5.92	0.197	5.92

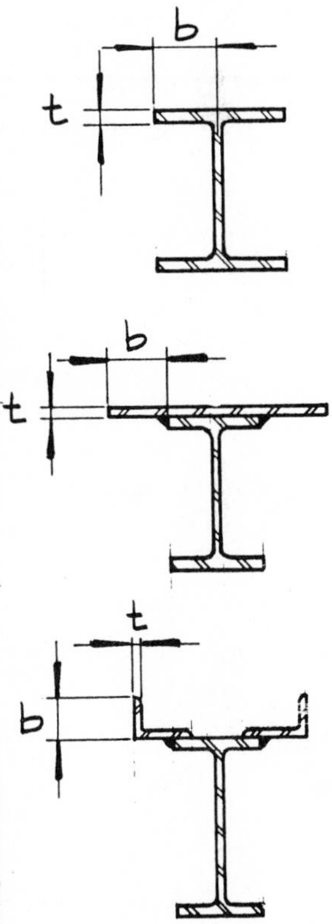

b
t
b
t
t
b

*Check that $F_b = 0.60\,F_yQ_s \leqslant$ applicable values in tables 3-6, -7. For $95/\sqrt{F_y} < (b/t) < 176/\sqrt{F_y}$, $Q_s = 1.415 - 0.00437\,(b/t)\sqrt{F_y}$. For $(b/t) > 176/\sqrt{F_y}$, $Q_s = 20{,}000/F_y(b/t)^2$.

When the stem of the tee in compression is within limits

$$127/\sqrt{F_y} < b/t < 176/\sqrt{F_y},$$

for Grade 36, $21.17 < b/t < 29.3$

for Grade 50, $17.96 < b/t < 24.9$

$$Q_s = 1.908 - 0.00715\,(b/t)\sqrt{F_y} \;\ldots\ldots\text{(Appendix C2; Eq. C2-5).}$$

When the b/t ratio exceeds the limit

$$b/t \geqslant 176/\sqrt{F_y},$$

$$Q_s = 20{,}000/[\sqrt{F_y}\,(b/t)^2] \;\ldots\ldots\text{(Appendix C2; Eq. C2-6).}$$

The allowable compression in bending for slender unstiffened stems of tees is shown in Table 3-9 for Grades 36 and 50. For tees under axial compression, see Table 4-6.

Channel or tee unstiffened slender compressive elements in bending must meet the following limitations (Appendix C2). (See Fig. 3-1.)

	Full Flange Width/Depth Ratio b_f/d	Flange Thickness to Web Thickness Ratio t_f/f_w
Channels	$0.25 \leqslant b_f/d \leqslant 0.50$	$2.0 \leqslant t_f/t_w \leqslant 3.0$
Rolled Tees	$b_f/d \geqslant 0.50$	$t_f/t_w \geqslant 1.10$
Built-up Tees	$b_f/d \geqslant 0.50$	$t_f/t_w \geqslant 1.25$

Compression–Slender Stiffened Compressive Elements. Slender "stiffened" compressive elements are those having lateral support along both edges, which are parallel to the direction of the compressive stress (1.9.2.1) and (b/t) ratios greater than the following limits (1.9.2.2):

$(b/t) > 238/\sqrt{F_y}$ for flanges of square and rectangular box sections of uniform thickness;

Grade 36, $b/t > 39.7$

Grade 50, $b/t > 33.7$

$b/t > 317/\sqrt{F_y}$ for cover plates perforated with access holes (Use full width and net area at largest hole for computation);

Grade 36, $b/t > 52.8$

Grade 50, $b/t > 44.8$

$b/t > 253\sqrt{F_y}$ for all other compressed elements;

Grade 36, $b/t > 42.2$

Grade 50, $b/t > 35.8$

The effect of slenderness, which tends to reduce compressive capacity, is taken into account for slender "stiffened" elements by use of a ficticious reduced width, b_e, and thereby a reduced cross-section considered effective for compression. This approach contrasts with that of using a reduced allowable stress as previously described for slender

(Continued on page 137)

TABLE 3-9 Allowable Stress in Bending for Slender Unstiffened Compression Elements—Stems of Tees; Grades 36 and 50.

b/t	ALLOWABLE STRESSES* (ksi) (Per Appendix C; Section C2)			
	Grade 36		Grade 50	
	Q_s	$F_b = 0.60\ F_yQ_s$	Q_s	$F_b = 0.60\ F_yQ_s$
17.96	-	-	1.000	30.00
18	-	-	0.997	29.93
19	-	-	0.947	28.42
20	-	-	0.896	26.90
21	-	-	0.846	25.38
21.17	1.000	21.60	-	-
22	0.964	20.82	0.795	23.87
23	0.921	19.90	0.745	22.35
24	0.878	18.97	0.694	20.83
24.9	-	-	0.649	19.47
25	0.835	18.04	0.640	19.19
26	0.792	17.12	0.591	17.75
27	0.749	16.19	0.548	16.46
28	0.706	15.26	0.510	15.30
29	0.663	14.34	0.475	14.26
29.3	0.651	14.06	-	-
30	0.617	13.33	0.444	13.33
31	0.578	12.48	0.416	12.48
32	0.542	11.71	0.390	11.71
33	0.510	11.01	0.367	11.01
34	0.480	10.38	0.346	10.38
35	0.453	9.79	0.326	9.79
36	0.428	9.25	0.308	9.25
37	0.405	8.76	0.292	8.76
38	0.384	8.31	0.277	8.31
39	0.365	7.88	0.262	7.88
40	0.347	7.49	0.250	7.49
41	0.330	7.13	0.237	7.13
42	0.314	6.80	0.226	6.80
43	0.300	6.48	0.216	6.48
44	0.286	6.19	0.206	6.19
45	0.274	5.92	0.197	5.92
46	0.262	5.67	0.189	5.67
47	0.251	5.43	0.181	5.43
48	0.241	5.20	0.173	5.20
49	0.231	4.99	0.166	4.99
50	0.222	4.79	0.160	4.79

t

b

t

b

*Check that $F_b = 0.60\ F_yQ_s \leqslant$ applicable values in Tables 3-6, -7. For $127/\sqrt{F_y} < (b/t) < 176/\sqrt{F_y}$, $Q_s = 1.908 - 0.00715\ (b/t)\sqrt{F_y}$. For $(b/t) \geqslant 176/\sqrt{F_y}$, $Q_s = 20{,}000/\sqrt{F_y}\ (b/t)^2$.

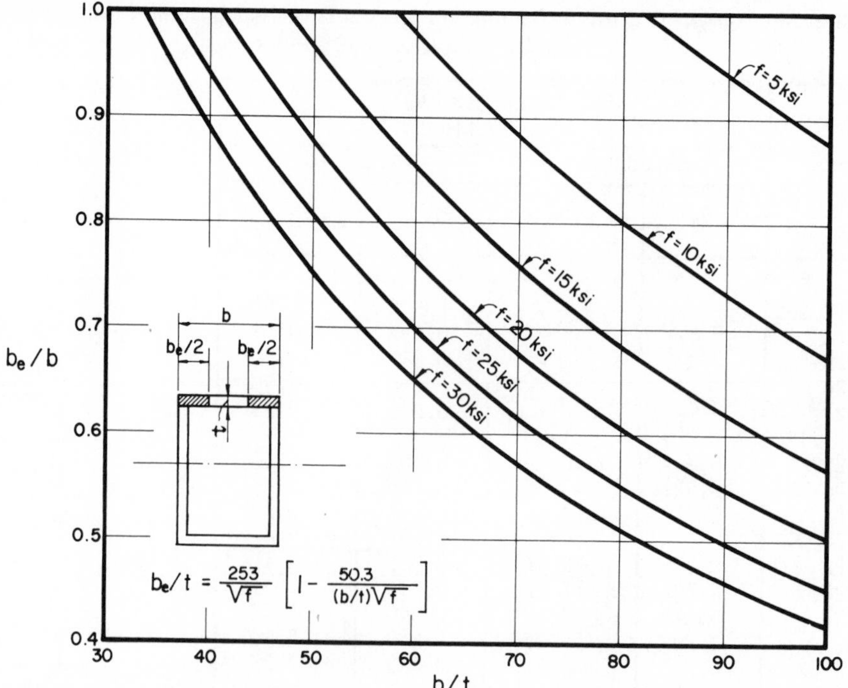

Fig. 3-6 Effective Width of Slender Stiffened Compression Elements of Square and Rectangular Box-Sections.

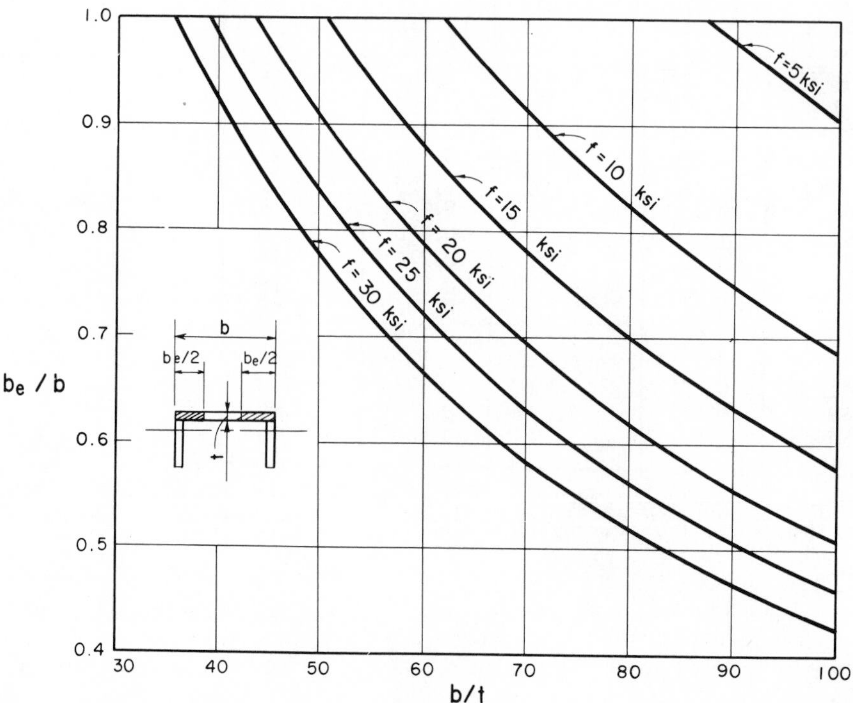

Fig. 3-7 Effective Width of Slender Stiffened Compression Elements Other Than Box-Sections.

"unstiffened" elements. Since the allowable stress here depends upon the unknown dimensions of a reduced cross-section, the design process becomes trial and review.

The reduced effective width, b_e, can be computed for any given width: thickness ratio, b/t, and assumed compressive stress, f, from the following equations from Appendix C:

$$b_e = (253t/\sqrt{f})\left[1 - \frac{50.3}{(b/t)\sqrt{f}}\right] \leqslant b$$ for the flanges of square and rectangular box sections of uniform thickness (C3-1).

$$b_e = (253t/\sqrt{f})\left[1 - \frac{44.3}{(b/t)\sqrt{f}}\right] \leqslant b$$ for other uniformly compressed elements (C3-2).

The computed value of the reduced effective width, b_e, can then be used to compute the moment of inertia of the reduced section and the compressive stress in bending. When the computed value of the compressive stress on bending agrees with the assumed value, the trial and review process is complete. Figure 3-6 and 3-7 can be helpful in determining the allowable compression stress in bending by this trial and error process. Note in Fig. 3-6 for square and rectangular sections of uniform thickness that for a compressive bending stress of 30 ksi, $b_e = b$ for b/t ratios less than 33.66.

Allowable Stresses in Shear

General. The average allowable shear stress, F_v, on the gross area (overall depth times thickness of web) of rolled sections and built-up members must not exceed 0.40 F_y (1.5.1.2).

$$\text{Grade 36}, f_v \leqslant 14.4 \text{ ksi}$$

$$\text{Grade 50}, f_v \leqslant 20.0 \text{ ksi}$$

(See "Exceptions for Built-up Girders" for the reduction in F_v with thin webs (1.10).)

Rigid Connections. Web shear stresses can be high within the boundaries of the rigid connection of two or more members whose webs lie in a common plane. Such webs should be reinforced when the web thickness, $t_w < 32M/A_{bc}\,F_y$, for

$$\text{Grade 36}, t_w < 0.889\,M/A_{bc};$$

$$\text{Grade 50}, t_w < 0.640\,M/A_{bc};$$

where M is the algebraic sum of the moments (unbalanced moment) in k-ft., directly across the connection and A_{bc} is the planar area of the connection web in square inches (Commentary 1.5.1.2).

Holes. The location of mechanical and electrical pipes, ducts, and conduit can be a problem in the structural design of beams and girders for buildings. Proper horizontal distribution of pipes and ducts through holes in beams can result in an integrated structural mechanical system allowing the use of ample depth of structure with greater economy, greater stiffness, and a smaller overall depth of the ceiling-floor sandwich.

General code provisions for beams with reinforced or non-reinforced holes have not been published by the AISC. No reduction in the allowable tension stress in bending or the allowable shear stress is required due to the interaction of such stresses for flexural members unless the members are designed on the basis of tension field action (1.10.7).

To determine if a beam web with holes needs to be reinforced requires a careful check of shear and bending stresses at the hole considering any secondary bending stresses caused by Vierendeel action above or below the hole.* If reinforcement is required on beam webs, it can be furnished by added web plates and supplementary flange plates or angles at the top and bottom of the hole.**

SPECIAL CASES

Tapered Members

Tapered members are excluded from the scope of this book. Special provisions applicable to problems of determining effective length, stiffness, and allowable stress have been included in the 1974 (Supplement No. 3) revision to the AISC Specifications (Appendix D).

Exceptions for Built-Up Girders

General. Built-up girders are deep homogeneous or hybrid flexural members that are used where the span, moment or shear are beyond the economical capability of rolled beam sections. Hybrid girders are defined as those with flanges of the same grade of steel and with the same cross-sectional area at any given section, but with different yield strength than that of the web (1.10.1).

Welded built-up plate girders have gained increasing importance because of the development of improved automatic submerged-arc welding techniques. Welded thin-web, off-center web, asymetrical, tapered, delta, and hybrid girders with or without built-in camber can now be produced economically.

Built-up girders with proper lateral support of the compression flange and with flanges continuously connected to the webs can be designed to meet either compact or semi-compact section criteria.

Exceptions are girders of A514 steel ($F_y = 90$ ksi) and hybrid girders (1.5.1.4), and tension field action girders (1.10.7).

Design criteria for thru-web built-up girders are based on the maximum load carrying capacity (ultimate strength) with a suitable factor of safety against inelastic post-buckling failure.

Proportions. As with rolled sections or cover-plated beams, welded plate girders must be proportioned by the moment of inertia of the gross section. No limit is placed on web bending stresses except for members subject to fatigue (1.7.1). No deduction need be made for holes in the tension or compression flange where total area of holes does not exceed 15 per cent of the gross flange area except for hybrid girders with an axial tension force equal to or greater than 0.15 F_y times the area of the gross section (1.10.1 and 1.14.3).

Web. When transverse web stiffeners are not provided, the maximum ratio of the clear distance between the flanges to the thickness of the web, h/t_w, is

$$h/t_w \leqslant 14{,}000/\sqrt{F_y\,(F_y + 16.5)} \ldots\ldots (1.10.2)$$

$$\text{Grade 36, } h/t_w \leqslant 322$$

$$\text{Grade 50, } t_w = \leqslant 243$$

*Cited references 6 and 8.
**Cited references 3, 9, and 10.

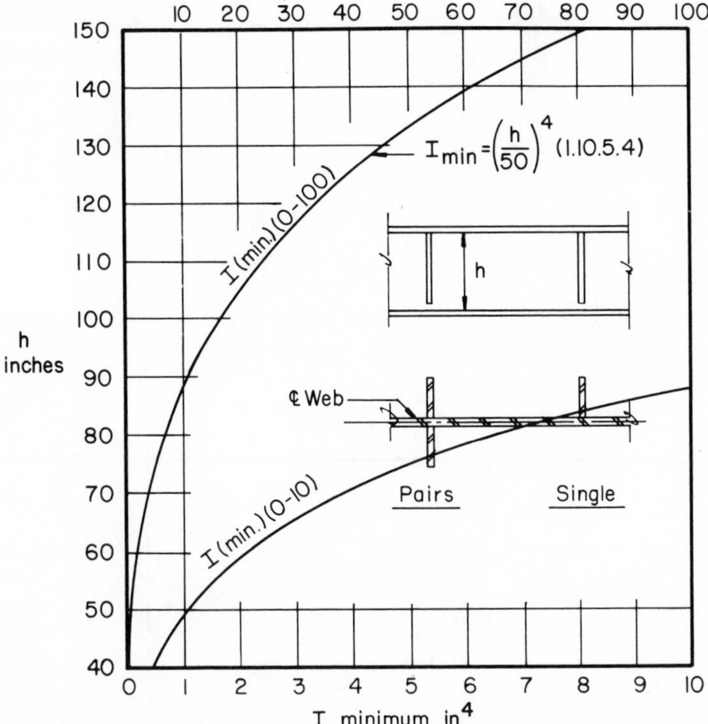

Fig. 3-8 Minimum Moment of Inertia of Intermediate Stiffeners (Pair or Single) About an Axis in the Web.

If transverse web stiffeners are provided at a minimum spacing of 1.5 times the girder depth, the maximum h/t_w ratio can be increased to

$$h/t_w \leqslant 2000/\sqrt{F_y} \text{ ...... (1.10.2);}$$

$$\text{Grade 36, } h/t_w \leqslant 333;$$

$$\text{Grade 50, } h/t_w \leqslant 283.$$

Stiffeners. See Figs. 3-8 and 3-9. Web stiffeners are not required when the following two criteria on proportions and stress are met (1.10.5.2):

1. $h/t_w < 260$ when $F_y < 46.2$ ksi
   $h/t_w < 14{,}000/\sqrt{F_y\,(F_y + 16.5)}$ when $F_y > 46.2$ ksi
   Grade 36, $h/t_w < 260$
   Grade 50, $h/t_w < 243$
2. $f_v \leqslant F_y C_v/2.89 \leqslant 0.40\,F_y$ ...... (Eq. 1.10-1)
   for Grade 36, $f_v \leqslant 12.46\ C_v \leqslant 14.4$ ksi
   for Grade 50, $f_r \leqslant 17.30\ C_v \leqslant 20.0$ ksi

where

$$C_v = 45{,}000\,k/[F_y\,(h/t_w)^2] \leqslant 0.8 \text{ ...... (1.10.5.2)}$$

$$C_v = [190/(h/t_w)]\sqrt{k/F_y} > 0.8 \text{ ...... (1.10.5.2)}$$

and

$$k = 4.00 + 5.34/(a/h)^2 \leqslant 1.0 \text{ ...... (1.10.5.2)}$$

$$k = 5.34 + 4.00/(a/h)^2 > 1.0 \text{ ...... (1.10.5.2)}$$

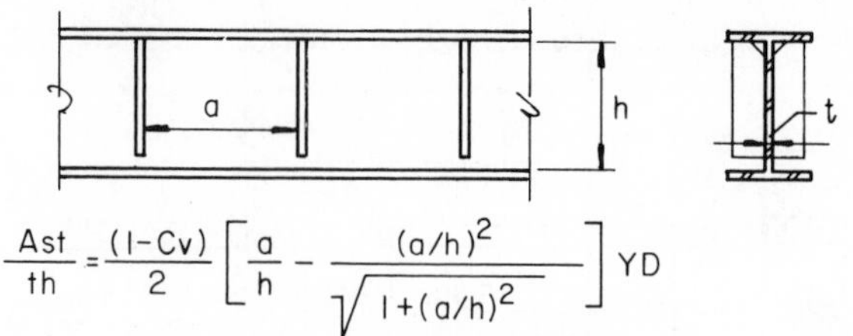

$$\frac{Ast}{th} = \frac{(1-Cv)}{2}\left[\frac{a}{h} - \frac{(a/h)^2}{\sqrt{1+(a/h)^2}}\right] YD$$

Y = Fy (Web) / Fy (Stiffener) = 1.0

D = 1.0 Stiffeners Furnished in Pairs

Ast = Area of One Pair of Intermediate Stiffeners

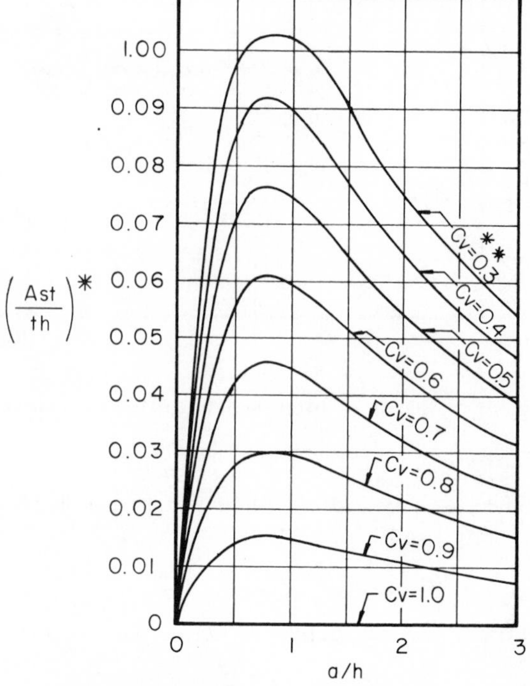

* For stiffeners furnished in pairs.
For single angle stiffeners multiply (Ast/th) by 1.8
For single plate stiffeners multiply (Ast/th) by 2.4

**C_v = Ratio of critical web stress to the shear yield stress of web material (1.10.5.2)

Fig. 3-9 Minimum Area of Intermediate Stiffeners.

For values of the allowable shear stress, F_v, with different h/t_w ratios, see Table 3-10 for tension field action girders and Table 3-11 for hybrid and non-tension field action girders. When the computed average shear stress, f_v, is less than the maximum allowable shear stress, F_v (listed in the right hand column of Table 3-11 under $a/h > 3.0$) and the h/t_w ratio is within specified limits, web reinforcement stiffeners are not required.

Tension field action may be considered in girders with vertical stiffeners strong enough to act as compression struts that will allow web membrane stresses due to shear forces, greater than those associated with the theoretical buckling load, to form diagonal tension fields similar in action to a Pratt truss. The consideration of tension field action is not permitted for hybrid girders (1.10.5.2); when the aspect ratio of the stiffener spacing,

(Continued on page 143)

TABLE 3-10 Allowable Shear Stress, F_v, for Tension Field Action Girders

	(h/t_w)	Intermediate Stiffener Aspect Ratio, (a/h)								
		0.50	0.75	1.00	1.25	1.50	1.75	2.00	2.50	3.00
Grade 36 (F_y = 36 ksi)	80							12.431	12.186	12.023
	90				12.393	12.061	11.798	11.588	11.279	11.068
	100			12.302	11.831	11.456	11.156	10.877	10.327	9.955
	110			11.080	11.370	10.749	10.236	9.830	9.233	8.822
	120		12.340	11.528	10.668	9.992	9.460	9.033	8.400	7.960
	130		12.057	10.973	10.091	9.403	8.855	8.413	7.753	7.289
	140		11.815	10.516	9.633	8.935	8.376	7.922	7.239	6.757
	150		11.507	10.148	9.263	8.558	7.989	7.525	6.824	6.327
	160	12.447	11.163	9.847	8.961	8.249	7.672	7.200	6.485	
	170	12.285	10.878	9.597	8.710	7.993	7.410	6.931		
	180	12.140	10.639	9.388	8.500	7.779	7.190	6.706		
	190	12.011	10.436	9.210	8.323	7.597	7.004			
	200	11.882	10.264	9.059	8.171	7.442				
	210	11.678	10.115	8.929	8.040	7.309				
	220	11.501	9.986	8.816	7.927					
	230	11.347	9.874	8.718	7.828					
	240	11.211	9.775	8.631						
	250	11.092	9.688	8.555						
	260	10.986	9.611	8.467						
	270	10.892	9.542							
	280	10.807	9.481							
	290	10.731	9.425							
	300	10.663	9.376							
	310	10.601								
	320	10.545								
	330	10.494								
Grade 50 (F_y = 50 ksi)	70						17.198	16.947	16.582	16.337
	80				16.853	16.370	15.981	15.669	15.208	14.890
	90			16.717	16.029	15.428	14.728	14.176	13.370	12.819
	100		17.232	16.109	15.017	14.082	13.348	12.760	11.891	11.287
	110		16.760	15.266	14.041	13.086	12.326	11.713	10.797	10.154
	120		16.366	14.527	13.300	12.329	11.550	10.917	9.965	9.292
	130		15.819	13.952	12.722	11.739	10.945	10.297	9.317	8.621
	140	17.172	15.297	13.495	12.264	11.272	10.466	9.806	8.803	8.089
	150	16.924	14.877	13.127	11.895	10.894	10.079	9.409	8.389	7.659
	160	16.707	14.533	12.825	11.593	10.586	9.762	9.084	8.049	
	170	16.492	14.248	12.576	11.342	10.330	9.500	8.815		
	180	16.164	14.009	12.366	11.132	10.115	9.280	8.590		
	190	15.887	13.806	12.189	10.954	9.934	9.094			
	200	15.650	13.634	12.038	10.802	9.779				
	210	15.446	13.485	11.908	10.672	9.646				
	220	15.269	13.356	11.795	10.559					
	230	15.115	13.244	11.696	10.460					
	240	14.979	13.145	11.610						
	250	14.860	13.058	11.534						
	260	14.754	12.981	11.466						
	270	14.659	12.912							
	280	14.575	12.851							

TABLE 3-11 Allowable Shear Stress, F_v, for Hybrid and Non-Tension Field Action Girders; Grades 36 and 50

	(h/t_w)	Intermediate Stiffeners Ratio, (a/h)									
		0.50	0.75	1.00	1.25	1.50	1.75	2.00	2.50	3.00	>3.00*
GRADE 50	50	20.0	20.0	20.0	20.0	20.0	20.0	20.0	20.0	20.0	20.0
	60	20.0	20.0	20.0	20.0	20.0	19.97	19.51	18.95	18.63	17.90
	70	20.0	20.0	20.0	18.67	17.72	17.12	16.72	16.24	15.97	15.35
	80	20.0	20.0	17.76	16.33	15.50	14.98	14.63	14.21	13.98	12.99
	90	20.0	18.97	15.95	14.52	13.68	12.78	12.19	11.50	11.12	10.27
	100	20.0	17.08	14.21	12.30	11.08	10.35	9.87	9.31	9.01	8.32
	110	20.0	15.52	12.02	10.17	9.16	8.56	8.16	7.70	7.44	6.87
	120	19.51	14.23	10.10	8.54	7.70	7.19	6.86	6.47	6.25	5.77
	130	18.01	12.43	8.61	7.28	6.56	6.12	5.84	5.51	5.33	4.92
	140	16.72	10.72	7.42	6.28	5.65	5.28	5.04	4.75	4.60	4.24
	150	15.61	9.34	6.46	5.47	4.93	4.60	4.39	4.14	4.00	3.70
	160	14.63	8.21	5.68	4.81	4.33	4.04	3.86	3.64	–	3.30
	170	13.66	7.27	5.03	4.26	3.83	3.58	3.42	–	–	2.88
	180	12.19	6.48	4.49	3.80	3.42	3.19	3.05	–	–	2.57
	190	10.94	5.82	4.03	3.41	3.07	2.87	–	–	–	2.30
	200	9.87	5.25	3.64	3.08	2.77	–	–	–	–	2.08
	210	8.95	4.76	3.30	2.79	2.51	–	–	–	–	1.89
	220	8.16	4.34	3.00	2.54	–	–	–	–	–	1.72
	230	7.46	3.97	2.75	2.33	–	–	–	–	–	1.57
	240	6.86	3.65	2.52	–	–	–	–	–	–	1.44
	250	6.32	3.36	2.33	–	–	–	–	–	–	1.33
	260	5.84	3.11	2.15	–	–	–	–	–	–	1.23
	270	5.42	2.88	–	–	–	–	–	–	–	–
	280	5.04	2.68	–	–	–	–	–	–	–	–
GRADE 36	50	14.40	14.40	14.40	14.40	14.40	14.40	14.40	14.40	14.40	14.40
	60	14.40	14.40	14.40	14.40	14.40	14.40	14.40	14.40	14.40	14.40
	70	14.40	14.40	14.40	14.40	14.40	14.40	14.19	13.78	13.55	13.02
	80	14.40	14.40	14.40	13.86	13.15	12.71	12.42	12.06	11.86	11.39
	90	14.40	14.40	13.39	12.32	11.69	11.30	11.04	10.72	10.54	10.13
	100	14.40	14.40	12.06	11.09	10.52	10.17	9.87	9.31	9.01	8.32
	110	14.40	13.17	10.96	10.08	9.16	8.55	8.16	7.70	7.44	6.87
	120	14.40	12.07	10.05	8.54	7.70	7.19	6.86	6.47	6.25	5.77
	130	14.40	11.15	8.61	7.28	6.56	6.12	5.84	5.51	5.33	4.92
	140	14.19	10.35	7.42	6.28	5.65	5.28	5.04	4.75	4.60	4.24
	150	13.24	9.34	6.46	5.47	4.93	4.60	4.39	4.14	4.00	3.70
	160	12.42	8.21	5.68	4.81	4.33	4.04	3.86	3.64	–	3.25
	170	11.69	7.27	5.03	4.26	3.83	3.58	3.42	–	–	2.89
	180	11.04	6.49	4.49	3.80	3.42	3.19	3.05	–	–	2.57
	190	10.46	5.82	4.03	3.41	3.07	2.87	–	–	–	2.30
	200	9.87	5.25	3.64	3.08	2.77	–	–	–	–	2.08
	210	8.95	4.76	3.30	2.79	2.51	–	–	–	–	1.89
	220	8.16	4.34	3.00	2.54	–	–	–	–	–	1.72
	230	7.46	3.97	2.75	2.33	–	–	–	–	–	1.57
	240	6.86	3.65	2.52	–	–	–	–	–	–	1.44
	250	6.32	3.36	2.33	–	–	–	–	–	–	1.33
	260	5.84	3.11	2.15	–	–	–	–	–	–	1.23
	270	5.42	2.88	–	–	–	–	–	–	–	–
	280	5.04	2.68	–	–	–	–	–	–	–	–
	290	4.70	2.50	–	–	–	–	–	–	–	–
	300	4.39	2.33	–	–	–	–	–	–	–	–
	310	4.11	–	–	–	–	–	–	–	–	–
	320	3.86	–	–	–	–	–	–	–	–	–
	330	3.63	–	–	–	–	–	–	–	–	–

*Stiffeners not required.

(a/h) is greater than 3 (Commentary 1.10.5); or when

$$F_v \geqslant 0.346\,F_y \ldots\ldots \text{(Commentary 1.10.5)}$$

$$F_v \geqslant 12.47 \text{ ksi with Grade 36}; F_v \geqslant 17.32 \text{ ksi with Grade 50.}$$

The spacing of web stiffeners, a, required for tension field action girders can be determined from the following expression for the allowable shear stress (.10.5.2):

$$F_v = \frac{F_y}{2.89}\left[C_v + \frac{1 - C_v}{1.15\sqrt{1 + a(h)^2}}\right]$$

$$\leqslant 0.40\,F_y \ldots\ldots \text{(Eq. 1.10-2)}$$

For solutions with F_y = 36 ksi and 50 ksi, see Table 3-10. Solutions for other values of F_y may be found in Appendix A, AISC Manual.

At end panels and at panels with large holes or adjacent thereto, the computed shearing stress with Grades 36 and 50 ksi must not exceed that shown in Table 3-11 (1.10.5.2; Eq. 1.10-1).

Composite Construction (Flexural)

General. Composite construction is a structural system in which a steel beam, or grider, and a reinforced concrete slab interact as a unit to resist flexural stresses and deflection due to bending (1.11.1). Composite beams of course also resist transient vibration (deflections) as a unit, but need not be fully composite as here defined to justify this assumed behavior. (See vibration check in Design Examples 3-6 and 3-11.) The steel beam may be a rolled shape or a built-up section. In simple spans the reinforced concrete slab acts as the compression flange of the composite steel-concrete beam, and part or all of the steel beam acts as the tension flange. Interaction of the concrete and the steel section requires transfer of the horizontal shear force between them. Mechanical steel shear connectors or full encasement of the steel beam in concrete cast integrally with the compression flange are considered to accomplish the transfer (1.11.1; 1.11.4).

The efficiency of composite beams can be increased by the use of an unsymmetrical steel section with more area in the tension flange than the compression flange. Unsymmetrical shapes can be made from built-up sections or rolled shapes with added coverplates. The authors recommend that cover plates be used only when the reduction in weight from that of a symmetrical section is greater than ten pounds per linear foot. A weight reduction of less than ten pounds per linear foot will usually not offset the added cost of attaching the cover plate with welds.

For assumed effective width of the concrete slab acting as a compression flange under various conditions, see Fig. 3-10. The limitations shown in Fig. 3-10 are based upon dimensions of slab, steel section, span, etc. (1.11.1).

The economy of composite construction depends upon a favorable cost comparison of the reduction of steel section weight and depth versus the addition of shear connectors, cover plates, or concrete encasement.

Composite-beams deflect from 30 to 50 percent less than the same sections as noncomposite beams. Since allowable stresses depend upon whether shoring is used, the designer must specify when shoring is required (1.11.2.1; 1.11.2.2). If shoring is not used, it is important to check the deflection of the steel section alone under construction loads. Usually, two-stage calculations for deflection will be necessary where shoring is not used; first, with the steel section only, and second with the composite section effective for full

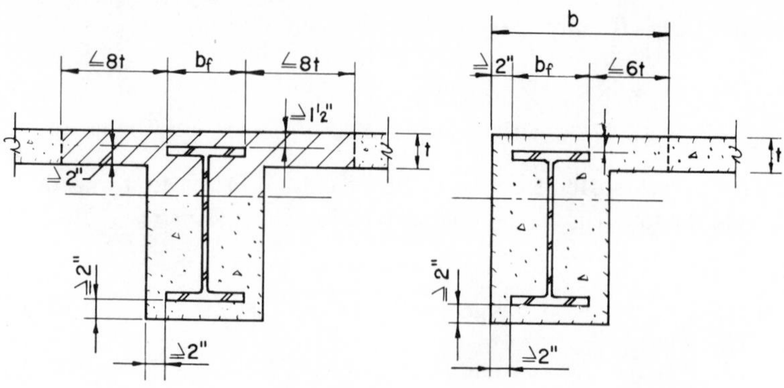

(a) Concrete Encasement

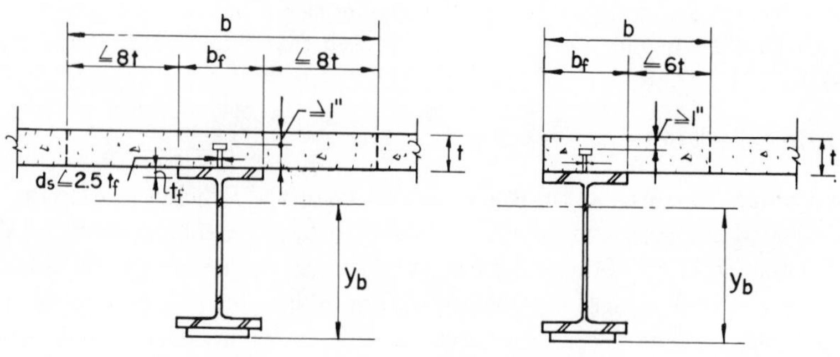

(b.) Mechanical Anchorage (Shear Connectors)

If not centered above web, $d_s \leqslant 2.5t_f$

Fig. 3-10 Composite Steel-Concrete Beams.

load. To minimize composite-beam deflection problems, the authors recommend the use of the following depth/span ratios (d_c/l) for fully stressed composite-beams where the depth, d_c, is measured from the top of the concrete compression flange to the bottom of the steel beam or cover plate. (See Fig. 3-11.)

Floors: $(d_c/l) = F_y/800$. Grade 36, $(d_c/l) = 1/22.2$
Grade 50, $(d_c/l) = 1/16$

Roofs: $(d_c/l) = F_y/1000$. Grade 36, $(d_c/l) = 1/27.7$
Grade 50, $(d_c/l) = 1/20$

Particularly when beams, designed as composite, support large open floor areas free of partitions or other damping sources, vibrations can be objectionable. To minimize human response to transient floor vibrations from pedestrian traffic in such areas, it is recommended that the steel beam depth-to-span ratio, $d/l \geqslant 0.05 = 1/20$ where the depth d is the depth of the steel beam alone (Commentary 1.13.2). (See Design Examples 3-6 and 3-11.)

Shear. The steel beam alone and its end connections must support the total end shear with no participation by the concrete whether it be total encasement or merely a slab (1.11.3).

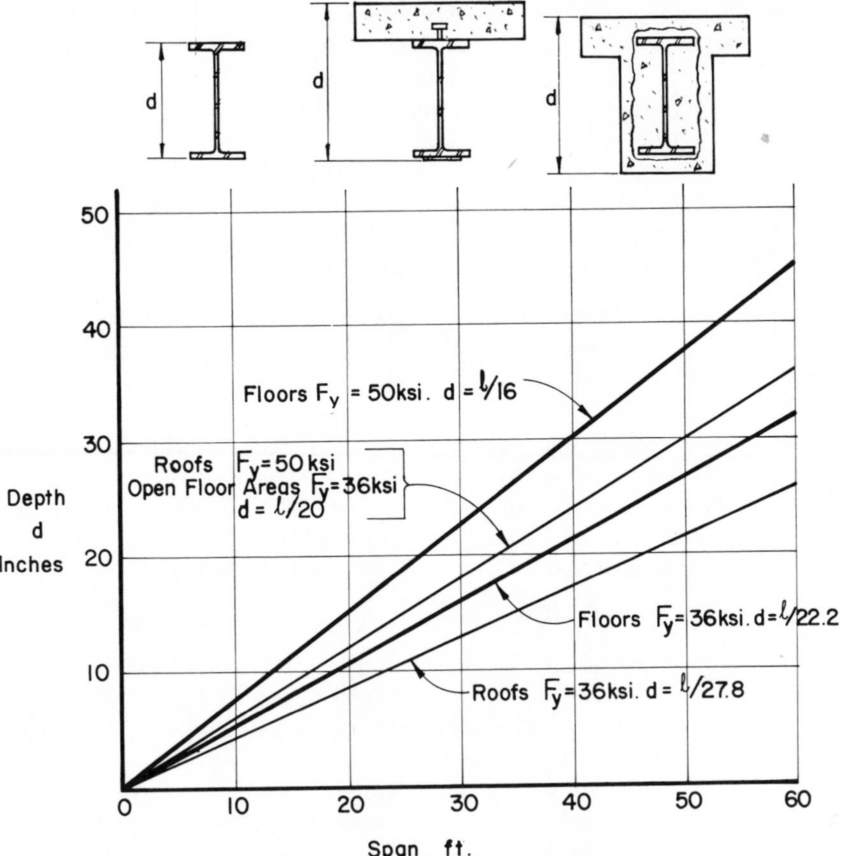

Fig. 3-11 Recommended Depth of Fully Stressed Beams.

Composite Concrete Encased Beams. Shear connectors are not required if the steel section is fully encased. The total horizontal shear is assumed to be transferred by natural bond (1.11.1). "Full concrete encasement" of the steel beam under this definition requires a minimum of 2 inches of concrete beyond the sides and soffit of the beam and $1\frac{1}{2}$ inches over the top of the beam. The slab must be a minimum of $3\frac{1}{2}$ inches deep at the edge of the minimum side encasement if it projects beyond minimum side encasement (1.11.1). (See Fig. 3-10.) Enclosing concrete must be laterally reinforced or confined all around the beam so as not to spall away, particularly across the tension flange (1.11.1). The authors recommend 4 × 4–W1.5 × W1.5 smooth welded wire fabric for deformed steel confinement reinforcement. This reinforcement will provide a steel ratio approximately equal to 0.0018 for the outer two inches of concrete encasement.

Prior to hardening of the concrete, the steel beam alone must support *all loads* applied at this stage of construction without exceeding the bending stress allowed for the particular steel section and its lateral support design condition (1.5.1). Alternatively, these loads may be supported on temporary shoring, and the stresses for this stage of loading may be taken as zero (1.11.2.1). After hardening of the concrete, the transformed composite section is considered to resist the stresses due to *all loads applied after concrete hardening*. The total bending stress in the steel beam, computed thus in two stages and combined, shall not exceed 0.66 F_y (1.11.2.1).

To avoid time consuming computations involving the composite transformed section, the designer can select a steel beam that alone will resist the positive moment produced by *all dead and live loads* without exceeding a bending stress of 0.76 F_y (1.11.2.1). In this case temporary shoring or proper lateral support of the compression flange must be provided until after the concrete encasement has hardened (1.11.2.1). The authors sugests this procedure for quick preliminary design only. Note that if deflections, ponding, or vibration effects are to be computed, the transformed section computations are still required. For these computations, E_c for the actual concrete weight is used (ACI 318-71; Section 8.3.1). Older codes related E_c only to f_c'.

Composite Beams with Shear Connectors. Design criteria for composite beams with shear connectors are based on the ultimate strength of the assembly but are set forth in terms of allowable flexural stresses and shear connector capacities for service loads (1.11.2.2; 1.11.4).*

Flexural tension stress in the concrete is neglected (1.11.2.2). Flexural compression stress in the concrete, f_c, must not exceed 0.45 f_c'. For all concretes, compute compression stress for the elastic section modulus of the full transformed composite section based on the modular ratio, $n = E_s/E_c$, for normal weight concrete (1.11.2.2). The modulus of elasticity of normal weight concrete, E_c, can be taken equal to

$$57{,}000\sqrt{f_c'} \ldots\ldots \text{(ACI 318-71)}.$$

Concrete stress and deflection must be computed for the total of additional dead load and live service loads applied after concrete hardens using 75 percent of the required concrete strength and for a corresponding modulus of elasticity, when temporary construction shores are not used (1.11.2.2).

The transformed composite section must have sufficient flexural strength to support all service dead and live loads without exceeding the bending stress allowed for the particular compact, semi-compact, or non-compact steel section and lateral support design condition (1.5.1.4). In regions of positive moment the steel section is exempt from the compact section requirement for the width/thickness ratio of unstiffened compression flanges ($b_f < 65/\sqrt{F_y}$, 1.5.1.4.1-b), from the compact section requirement for the width/thickness ratio of stiffened compression flanges ($b_f \leqslant 190/\sqrt{F_y}$, 1.5.1.4.1-c) and from the compact section requirement that limits the laterally unsupported length of the compression flange (1.5.1.4.1-e). The hardened reinforced concrete slab will provide adequate lateral support for the steel compression flange in regions of positive moment.

The magnitude of the section modulus of the transformed composite section is limited if "complete composite action" is not provided (1.11.2.2).

Complete Composite Design with Shear Connectors. The bending stress in the steel section can be based on the full transformed section modulus only when temporary construction shores are used and shear connectors are provided for complete composite action (1.11.2). "Complete composite action" is based on the ultimate strength concept of transferring an ultimate horizontal shear force between the concrete slab and steel section equal to the smaller of the ultimate compression force in the concrete slab, 0.85 $f_c'bt$, or the ultimate tension force in the steel beam, A_sF_y. A factor of safety of 2 for an allowable working stress is used to determine total load horizontal shear to be transferred between points of maximum positive moment and zero moment by shear connec-

*Cited Reference (2).

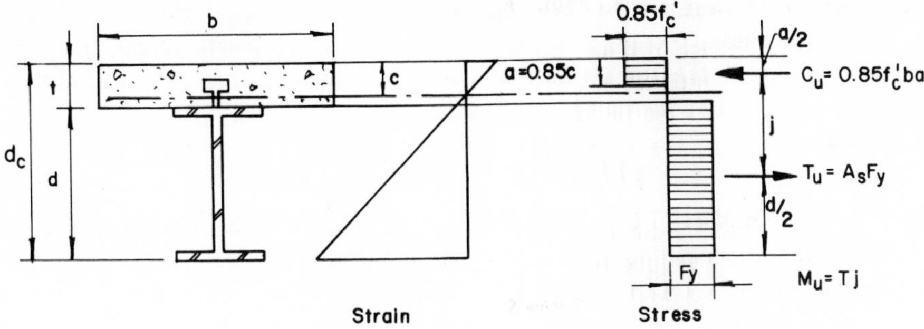

Neutral Axis in Flange

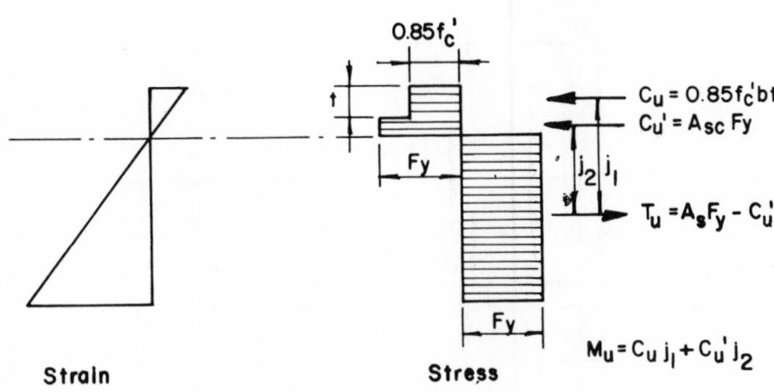

Neutral Axis Below Flange

Fig. 3-12 Composite Beam Stress Distribution at Ultimate Strength.

tors welded to the top flange of the steel beam and embedded in the concrete slab. The horizontal shear force, V_h, to be provided under total load is therefore taken as the smaller of:

$$\left.\begin{aligned} V_h &= 0.85 f_c' (bt/2) \ldots\ldots \text{(Eq. 1.11-3)} \\ V_h &= A_s F_y/2 \qquad \ldots\ldots \text{(Eq. 1.11-4)} \end{aligned}\right\} \ldots \text{(Fig. 3-12)}$$

When temporary construction shores are omitted, the magnitude of the section modulus of the transformed section for tension in the bottom of the beam, S_{tr}, is limited to insure that the bending stress in the steel beam under service loads will be well below the level of initial yielding (1.11.2.2; Commentary 1.11.2.2). For this condition,

$$S_{tr} \leqslant (1.35 + 0.35\, M_L/M_D)\, S_s \ldots\ldots \text{(Eq. 1.11-2)}$$

where S_s is the section modulus of the steel section alone, M_L is the moment caused by loads applied after the concrete compression strength has reached 75 percent of its design strength ($0.75\, f_c'$), and M_D is the moment caused by loads applied prior to this time.

Partial Composite Design with Shear Connectors. If the designer elects not to provide the number of shear connectors required for complete composite action, the effective section modulus of the transformed section for tension in the bottom of the beam, S_{eff}, is

$$S_{eff} = S_s + (V'_h/V_h)\,(S_{tr} - S_s) \;\ldots\ldots\; \text{(Eq. 1.11-1)}$$

where V_h is the smaller of $0.85\,f'_c bt/2$ or $A_sF_y/2$; and V'_h is the horizontal shear to be resisted by partial composite action. V'_h must not be less than $0.25\,V_h$ ...... (1.11.4). V'_h can be computed by rearranging Eq. 1.11-1:

$$V'_h = \frac{(S_{eff} - S_s)\,(V_h)}{(S_{tr} - S_s)} \geqslant 0.25\,V_h$$

The use of up to 25 percent fewer shear connectors than required for complete composite action will often reduce the effective section modulus only slightly. See the tabulated solutions for Eq. 1.11-1.

Effective Section Modulus, $\frac{S_{eff}}{S_s}$ S_x about bottom flange, steel only				
S_{tr}/S_s	V'_h/V_h			
	0.25	0.50	0.75	1.00
1.00	1.00	1.00	1.00	1.00
1.20	1.05	1.10	1.15	1.20
1.40	1.10	1.20	1.30	1.40
1.60	1.15	1.30	1.45	1.60
1.80	1.20	1.40	1.60	1.80
2.00	1.25	1.50	1.75	2.00
2.20	1.30	1.60	1.90	2.20
2.40	1.35	1.70	2.05	2.40
2.60	1.40	1.80	2.20	2.60
2.80	1.45	1.90	2.35	2.80
3.00	1.50	2.00	2.50	3.00

Partial composite action can result in deflections greater than computed using the moment of inertia of the transformed composite section, I_{tr}. Deflections for composite sections using partial composite action must therefore be computed using an effective moment of inertia, I_{eff}, that is a function of V'_h, V_h, I_{tr} and the moment of inertia of the steel section, I_s, where:

$$I_{eff} = I_s + \sqrt{V'_h/V_h}\,(I_{tr} - I_s) \;\ldots\ldots\; (1.11.4)$$

Shear Connectors. Headed steel studs are used almost exclusively today as shear connectors in composite construction. They have nearly replaced earlier connectors made from spirals, flexible channels, or hooked studs.

Headed studs are welded to the compression flange of the steel beam to transfer the horizontal shear between the beam and the concrete slab. Except for studs installed in the ribs of formed steel decks they must have a minimum lateral cover of one inch (1.11.4). (See Fig. 3-10.) To prevent stud welds from tearing out of thin beam flanges when not located over a web, the diameter must not be more than 2.5 times the beam flange thickness. Headed steel stud shear connectors can be spaced uniformly between locations of maximum positive moment and zero moment with a minimum spacing of $6(d_s)$ along the longitudinal axis of the beam and $4(d_s)$ transverse to the longitudinal axis. The maxim spacing of headed stud shear connectors is limited to eight times the total slab thickness (1.11.4)

The number of shear connectors required to resist the horizontal shear on each side of the maximum positive moment can be computed from V_h/q for complete composite action, and V_h'/q for partial composite action, where q is the allowable shear load for one connector as shown in the Specifications.

The anchorage value of headed stud shear connectors is proportional to their cross section area and is primarily influenced by the compressive strength and modulus of elasticity of the concrete. It has been shown that the ultimate strength of a headed steel stud shear connector, Q_u, can be predicted from the following empirical expression when $3 \leqslant f_c' \leqslant 5$ ksi, A_s is the cross-sectional area of the stud (in.^{2}), f_c' is the concrete compression strength (ksi), and E_c is the modulus of elasticity of the concrete (ksi):*

$$Q_u = 0.5\, A_s \sqrt{f_c' E_c}\ .$$

The values of the allowable horizontal shear load, q, for headed steel stud shear connectors are shown in Table 3-12. These values were obtained from this empirical formula using a factor of safety equal to 2 and from the AISC Specification (Tables 1.11.4 & 1.11.4A).

The AISC Specification allowable shear values were obtained from an ultimate strnegth formula developed for normal weight concrete using a factor of safety of 2.5:**

$$q = \frac{37.45\, A_s \sqrt{f_c'}}{2.5}$$

Allowable shear values for lightweight concrete $f_c' \leqslant 4$ ksi were obtained from the values for normal weight concrete multiplied by the coefficient in the Specification (Table 1.11.4A)

$$\text{Coeff.} = 0.00491\, w + 0.288$$

where w is the weight of the concrete in pcf.

The allowable shear values in Table 3-12 for these two different methods are within twelve percent of one another for $f_c' \leqslant 4$ ksi.

In the vicinity of concentrated loads the number of shear connectors, N_2, required between the concentrated load and the nearest point of zero moment must not be less than:

$$N_2 \geqslant \frac{N_1\,[(M\beta/M_{max}) - 1]}{\beta - 1} \quad \ldots\ldots \text{(Eq. 1.11-6)}$$

where

M = moment at concentrated load point $\leqslant M_{max}$

N_1 = number of shear connectors required between point of maximum and zero moment

$\beta = S_{tr}/S_s$ or S_{eff}/S_s as applicable.

*Cited Reference 4.

**Cited Reference 7.

TABLE 3-12 Allowable Horizontal Shear Load (q) for Headed Steel Stud Shear Connectors (kips)

Stud Diameter (in.)	f_c' (ksi)	Concrete Weight (pcf)									
		145		115		110		100		90	
		(1) AISC	(2) OSF	(1) AISC	(2) OSF	(1) AISC	(2) OSF	(1) AISC	(2) OSF	(1) AISC	(2) OSF
$\frac{1}{2}$	3	5.1	4.8	4.4	4.0	4.2	3.9	4.0	3.6	3.7	3.3
	3.5	5.5	5.4	4.7	4.5	4.5	4.4	4.3	4.1	4.0	3.7
	4	5.9	5.9	5.1	5.0	4.9	4.8	4.6	4.5	4.3	4.1
	5	5.9	7.0	5.7	5.9	5.5	5.7	5.1	5.3	4.8	5.0
$\frac{5}{8}$	3	8.0	7.5	6.9	6.3	6.6	6.1	6.2	5.6	5.8	5.2
	3.5	8.6	8.4	7.4	7.0	7.1	6.8	6.7	6.3	6.3	5.9
	4	9.2	9.2	7.9	7.8	7.6	7.5	7.1	7.0	6.7	6.4
	5	9.2	10.9	8.8	9.2	8.6	8.9	8.0	8.3	7.5	7.6
$\frac{3}{4}$	3	11.5	10.7	9.9	9.0	9.5	8.7	9.0	8.1	8.4	7.5
	3.5	12.5	12.1	10.8	10.1	10.4	9.8	9.7	9.1	9.1	8.4
	4	13.3	13.3	11.4	11.2	11.0	10.8	10.4	10.1	9.7	9.3
	5	13.3	15.7	12.8	13.2	12.4	12.8	11.6	11.9	10.9	11.0
$\frac{7}{8}$	3	15.6	14.6	13.4	12.3	12.9	11.9	12.1	11.1	11.4	10.2
	3.5	16.8	16.4	14.4	13.8	13.9	13.3	13.1	12.4	12.3	11.5
	4	18.0	18.1	15.4	15.2	14.9	14.7	14.0	13.7	13.1	12.7
	5	18.0	21.4	17.3	18.0	16.7	17.4	15.7	16.2	14.8	15.0

AISC - Per AISC Specification; Tables 1.11.4 & 1.11.4A.
OSF - Per Ollgaard, Slutter and Fisher. Factor of safety of 2; cited reference 7; $q = (1/2)(0.5\ A_s \sqrt{f_c' E_c})$.

Special Conditions. For use of longitudinal reinforcing bars as part of composite beams in negative bending, the authors recommend application limited to continuous construction where the steel beams are encased in concrete and end connections are not subject to moment reversals.

USE AND DERIVATIONS OF ROLLED SECTION SELECTION TABLES

General

As structural steel design has become more scientific, the basically simple problem of selecting a shape for a required resisting moment has become quite sophisticated. This sophistication is no problem when the selection is made by computer, but the need for a practicable manual solution remains. The designer with no access to a computer must prepare a final solution economically. The designer preparing preliminary selections for computer analysis needs a quick method to approximate an acceptable final design.

Under the earlier, more primitive AISC Specifications, the problem of converting a required resisting moment for a given unbraced length into the selection of an economical beam section consisted of a 60-second operation.*

*Cited Reference 1.

Operation:

(S) and (SL_u) are constants tabulated for all rolled shapes (on one page for 1948 shapes)
(1) Calculate $S' = M/20$, where M = required resisting moment in k-in.
(2) Calculate $S'L$, where L = unbraced length in feet
(3) Select from the table, the first section for which tabulated $SL_u \geqslant S'L$ and $S \geqslant S'$

Derivation:

Let F_b = allowable stress, ksi = $20\ L_u/L \leqslant 20$ ksi
$L_u = L$ where $Ld/bt = 600$
S = section modulus, in.^{3}
$M/F_b = S$; substituting for F_b,

$$S'L = SL_u$$

The principal complication of this simple process in the present AISC Specifications is that F_b = 20,000 psi has been replaced by nine variations, five of which involve complicated formulas to solve for F_b. The ground rules for selection of the proper F_b are spread over three pages of Specifications, plus explanations of same spread over four pages of reference, plus cross reference to numerous tables, research reports, etc. (This complication is about doubled in selecting a column with combined axial load and bending since it may bend about both axes. See Chapter 6.)

Table 3-1 (F_y = 36 ksi) and Table 3-2 (F_y = 50 ksi) are computer-prepared solutions for all nine conditions concentrated into one location for convenience in practical design. The authors' abbreviated interpretations of the ground rules for a given condition and shape, for which one (or two or three) of the nine conditions apply, appear as footnotes at the bottom of each page of the tables. It is admittedly not quite possible to match the 60-second time required in 1948. It may require about twice that time to select the solutions from the nine offered for application, plus about this time to apply them, for a total of three minutes. (The authors' goal was three minutes for routine design.)

Use of Beam Selection Tables

General. Two tables are provided; Table 3-1 for F_y = 36 ksi and Table 3-2 for F_y = 50 ksi. Let resisting moment = M_R. Subscripts x and y indicate axis about which moment is determined. Columns (1) through (6) are for bending about the major axis. Columns (1), (2), and (3) give the M_{Rx} value directly in k-ft. For most beam selections, one need merely read the value M_{Rx} in column (1) and the value L_c.

Where the tabulated $M_{Rx} \geqslant$ required M_R and $L_c \geqslant L_b$, the selection is completed if the tabulated dimension ratios are within the limits given in the footnotes. Where a value is shown for M_{Rx} in column (2), use this value as column (1) does not apply.

Where $L_c < L_b$ or the dimension ratios tabulated exceed the limits in the footnotes, the value of (l_b/r_T) must be computed. The footnotes show limits on (l_b/r_T) which determine whether the bending moment resistance is limited by the value in column (3) or must be computed with the moment factors (*MF*) of columns (4) and (5) or (5) and (6).

Entry Data.

1. Required moment of resistance, M_R, in k-ft.
2. The actual unbraced length of the compression flange. (Note that L_b is in feet and l_b in inches.)
3. C_b, the end condition coefficient. (See Table 6-2 or Fig. 3-2 giving values of C_b for various combinations of moments at successive bracing points.) The larger end moment, M_2, should be taken as M_R. $M_1 \leqslant M_2$. M_1 is negative for single curvature bending. If intermediate moments exceed M_2, C_b = 1.00. The lower the value of C_b, the lower the allowable compressive stress in bending becomes. C_b is always equal to or greater than 1.00.

Procedure–Bending About the Major Axis.

Step 1. Enter the correct table for F_y with M_R, L_b and C_b.

Step 2. Begin with the lightest section of the shape desired at a depth, $d = (F_y/1000)$ times the span for a roof member, or $(F_y/800)$ times the span for a floor member (Commentary 1.13.1).

Step 3. Read upward (toward the heavier sections) till the value in Col. (1), $M_{Rx} \geqslant$ required M_R. If the value in Col. (2) is not zero, Col. (1) does not apply and M_{Rx} is taken from Col. (2). Compare the tabulated L_c to L_b. If $L_c \geqslant L_b$, compare the tabulated dimension ratios $(b_f/2t_f)$ and (d/t_w) to the limits thereon (footnotes). If tabulated dimension ratios satisfy the limits, the section is adequate.*

Step 4. Where $L_c \leqslant L_b$ or the dimension ratio limits are not satisfied, compute (l_b/r_T) and compare its value to the limits shown which are based upon C_b. Column (3) gives the upper limit for M_{Rx} for these cases. For all channels bent about the major axis, use the value of M_{Rx} determined only from column (5) but not to exceed column (3). The footnotes in Table 3-1 for $F_y = 36$ ksi show:

if $(l_b/r_T) < 53.2\sqrt{C_b}$, use column (5) but $\leqslant$ column (3)
if $53.2\sqrt{C_b} \leqslant (l_b/r_T) \leqslant 101\sqrt{C_b}$, use the larger value of column (5) or (4) but $\leqslant$ column (3)
if $(l_b/r_T) > 101\sqrt{C_b}$, use the larger value of column (5) or (6) but $\leqslant$ column (3)

Note that column (3) offers a quick decision that the section is inadequate. If the value of M_{Rx} in column (3) is less than M_R, proceed upward in the table to another section.

Step 5. Columns (4), (5), and (6) tabulate moment factors (MF) for use in the three footnote equations yielding M_{Rx} in k-ft. Solve the appropriate pair of equations, (Cols. 4 and 5 or 5 and 6), using (MF) tabulated and functions of C_b and l_b. Note that l_b is in inches.

Procedure–Bending About the Minor Axis.

Step 1. Enter the correct table for F_y, with the required M_R and L_b.

Step 2. For doubly-symmetrical shapes, read M_{Ry} in column (7) or (8).

Step 3. For members where $(b_f/2t_f)$ exceeds limits 15.83 for $F_y = 36$ ksi or 13.44 for $F_y = 50$ ksi and for channels, read M_{Ry} in column (9).

Derivations of Beam Selection Tables (Tables 3-1 and 3-2)

The first six numbered columns solve for the resisting moment, M_{RX}, about the *x-x* (major) axis.

Column (1) Compact Sections.

$$M_{Rx} \text{ for } F_b = 0.66\,F_y \;\ldots\ldots\; (1.5.1.4.1)$$

applies to compact sections where $l_b \leqslant 76.0\,(b_f/\sqrt{F_y})$ and $l_b \leqslant 20{,}000/(d/A_f)F_y$. Note that in Tables 3-1 and 3-2 a value of $M_{Rx} = (\frac{0.66}{12}\,F_y\,S_x)$ is listed for all beams and that the user must compare his unbraced length in feet, L_b, to the tabulated values of L_c to determine if he has a compact section. For a compact section $L_b \leqslant L_c$.

Column (2) Semi-Compact Sections.

$$M_{Rx} \text{ for } F_b = F_y\,[0.79 - 0.002\,(b_f/2t_f)\sqrt{F_y}\,] \;\ldots\ldots\; (1.5.\text{-}5a)$$

Table 3-1 ($F_y = 36$ ksi); applies when $10.83 < (b_f/2t_f) \leqslant 15.83$; $(d/t_w) \leqslant 42.8$ and $L_b \leqslant L_c$.

Table 3-2 ($F_y = 50$ ksi); applies when $9.19 < (b_f/2t_f) \leqslant 13.44$; $(d/t_w) \leqslant 90.5$; and $L_b \leqslant L_c$. Values for $(b_f/2t_f)$ and (d/t_w) are tabulated under "Section Properties" in these tables.

Column (3) Non-Compact and Non-Semi-Compact Sections. M_{Rx} for $F_b = 0.60\,F_y$ applies to non-compact and non-semi-compact sections for tension (1.5.1.4.1) and also as an

*The limits for the dimension ratios are functions of the yield strength, F_y. These limits are different for Tables 3-1 and 3-2, and preclude direct conversion or comparison of resisting moment M_R from Table 3-1 to Table 3-2.

upper limit in compression for solutions using columns (4) and (5) or (5) and (6) . . . (1.5.1.4.6; Eqs. 1.5-6a, 1.5-6b, and 1.5-7). In addition it applies to all W-shapes not under columns (1), (2), (4), and (6).

Column (4) "Moment Factor" for Determining the Allowable Compression Stress in Bending.

$$F_b = \left[\frac{2}{3} - \frac{F_y (l/r_T)^2}{(1{,}530{,}000\, C_b)}\right] F_y \;\ldots\ldots \text{(Eq. 1.5-6a)}$$

Column (4) applies when $53.2\sqrt{C_b} \leqslant (l_b/r_T) \leqslant 119.0\sqrt{C_b}$ for $F_y = 36$ ksi and when $45.2\sqrt{C_b} \leqslant (l_b/r_T) \leqslant 101\sqrt{C_b}$ for $F_y = 50$ ksi. The tabulated moment factor, $MF = [F_y^2 S_x/[1836(r_T)^2]$. The resisting moment M_{Rx}, k-ft., can be determined using MF from column (4) as follows:

$$M_{Rx} = \left[\text{Column (1)} - \frac{(l_b)^2\,(MF)}{10{,}000\, C_b}\right] \leqslant \text{Column (3)}$$

Use the *larger* value for M_{Rx} as determined from columns (4) or (5), but do not exceed M_{Rx} of column (3). Column (4) is not applicable to channels.

Column (5) "Moment Factor" for Determining the Allowable Compression Stress in Bending.

$$F_b = \frac{12{,}000\, C_b}{l_b (d/A_f)} \;\ldots\ldots \text{(Eq. 1.5-7)}$$

Column (5) applies for all values of (l_b/r_T). The tabulated moment factor, $MF = S_x/(d/A_f)$. The resisting moment M_{Rx}, k-ft., can be determined using MF from column (5) as follows:

$$M_{Rx} = C_b (1000\, MF)/l_b \leqslant \text{Column (3)}.$$

Column (6) "Moment Factor" for Determining the Allowable Compression Stress in Bending.

$$F_b = \frac{170{,}000\, C_b}{(l_b/r_T)^2} \;\ldots\ldots \text{(Eq. 1.5-6b)}$$

Column (6) applies when $(l_b/r_T) > 119\sqrt{C_b}$ for $F_y = 36$ ksi; and when $(l_b/r_T) > 101\sqrt{C_b}$ for $F_y = 50$ ksi. The tabulated moment factor, $MF = 1{,}416{,}700 (r_T)^2$.

$$M_{Rx} = \frac{C_b (MF)}{(l_b)^2\,(100)} \leqslant \text{Column (3)}$$

Use the *larger* value of M_{Rx} from columns (6) or (5). Column (6) is not applicable to channels.

The last three numbered columns solve for the resisting moments, M_{Ry} about the *y-y* (minor) axis.

Column (7) Compact Sections.

$$M_{Ry} \text{ for } F_{by} = 0.75 F_y \;\ldots\ldots \text{(Eq. 1.5.1.4.3).}$$

The tabulated resisting moment in k-ft., $M_{Ry} = 0.75 F_y S_y/12$. Column (7) is applicable for doubly symmetrical I- and H-shapes in tension and compression when $(b_f/2t_f) \leqslant 10.83$ for $F_y = 36$ ksi, and $(b_f/2t_f) \leqslant 9.19$ for $F_y = 50$ ksi.

Column (8) Semi-Compact Sections.

$$M_{Ry} \text{ for } F_b = [1.075 - 0.005(b_f/2t_f)\sqrt{F_y}]\,F_y \ldots\ldots \text{(Eq. 1.5-5b)}$$

Column (8) is applicable for doubly symmetrical I- and H-shaped sections in tension and compression when $10.83 < (b_f/2t_f) < 15.83$ for $F_y = 36$ ksi, and when $9.19 < (b_f/2t_f) < 13.44$ for $F_y = 50$ ksi.

Column (9) Non-Compact and Non-Semi-Compact Sections. M_{Ry} for $F_b = 0.60F_y$. M_{Ry} (kip-ft.) $= 0.60F_yS_y/12$. For tension (1.5.1.4.5) compression (1.5.1.4.6b) in non-compact and non-semi-compact sections; for compression in sections not doubly symmetrical, and sections not loaded in the plane of their webs (1.5.1.4-6a).

SPECIAL CONSIDERATIONS

Deflection

Beams supporting roofs and floors must be designed "with due regard to the deflection produced by the design loads" (1.13.1). "Due regard" requires that beam deflections be limited. Deflection must not change behavior assumed for the structural frame so as to reduce strength assumed in design. Deflection and associated rotations must be compatible with design assumptions of the Type of Construction (1.2; 1.12.1; 1.12.2). Deflection must not cause structural damage or objectionable visible cracks to supported floors, roofs, partitions, ceiling, etc. Furthermore, "due regard" includes limiting deflection to maintain supported elements in serviceable as well as safe condition. These considerations are part of professional responsibility. Properly, the specifications do not attempt to set actual or computed deflection limits. The only actual limit prescribed is that (actual) live load deflection of beams and girders supporting plastered ceilings must not exceed a maximum value of 1/360 of the span (1.13.1).

It is recommended, as a guide for design, that the depth of fully stressed beams be limited (Commentary 1.13.1). For roof beams, a minimum depth $= F_y/1000$ times the span ($d = 0.036l$ for $F_y = 36$ ksi and $d = .050l$ for $F_y = 50$ ksi). For floor beams, a minimum depth $= F_y/800$ times the span ($d = 0.045l$ for $F_y = 36$ ksi and $d = 0.0625l$ for $F_y = 50$ ksi). (See Fig. 3-11.) If simple span or continuous members of less depth are used, the authors recommend that the maximum allowable bending stress be decreased in the same ratio as the depth is decreased from the recommended amount unless deflection calculations show that the predicted deflection will be serviceable (Commentary 1.13.1).

Vibration

Steel beams supporting large open floor areas free of partitions or other sources of damping must be designed with "due regard for vibration" (1.13.2). Depth of such beams should not be less than $\frac{1}{20}$ of the span in order to limit the motion that can cause perceptible vibrations disturbing to human comfort (Commentary 1.13.2).

Figure 3-13 can be used as a guide for checking beams and joists supporting large open floor areas. It is based on theoretical studies and physical data by Jack Wiss.* In the use of Fig. 3-13 it is suggested that the moment of inertia of steel joists be reduced fifteen percent since their shear rigidity is less than a solid plate.

---

*Cited Reference (13).

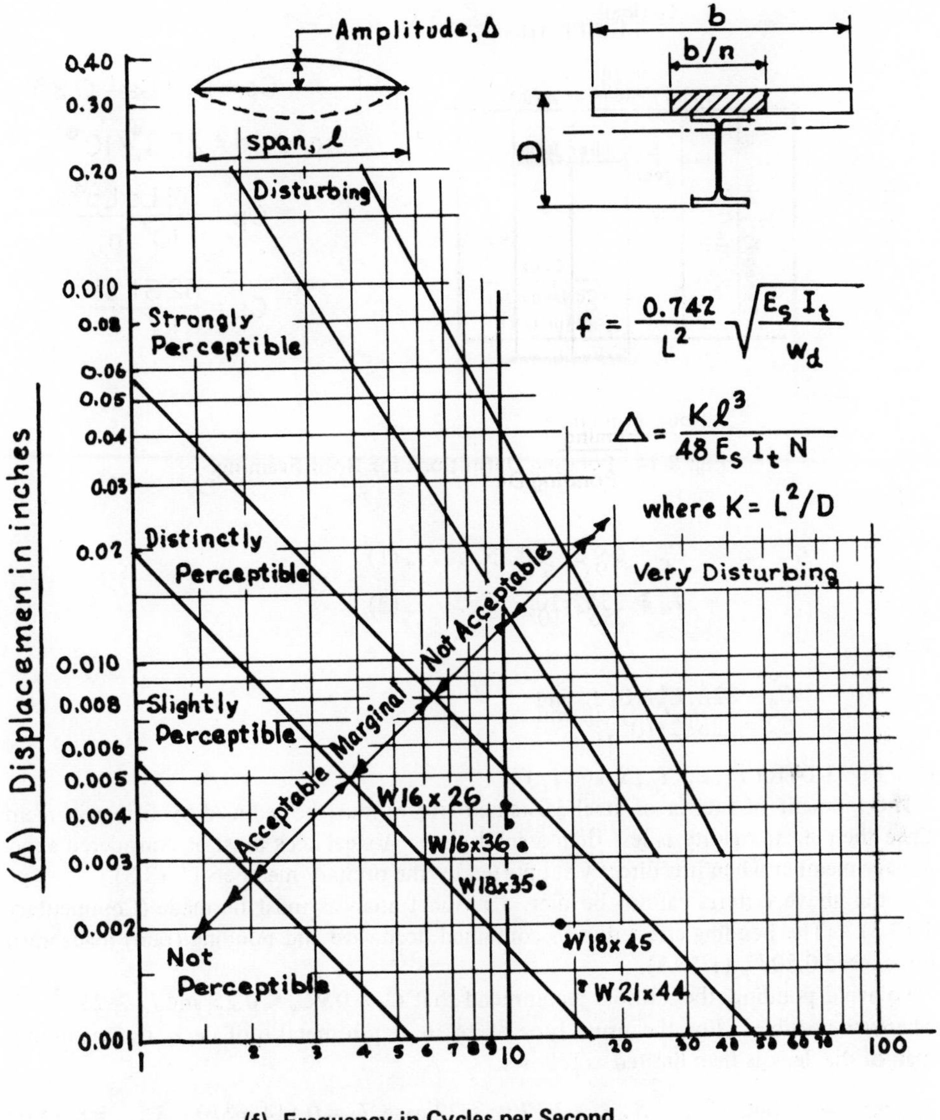

Fig. 3-13 Human Response for Evaluation of Foot Traffic Vibrations on Floors.

ℓ = span, in.	w = dead load, (lb./ft.)
L = span, ft.	N = number of beams in width $L/2$
D = depth, ft.	

Ponding

Ponding is the cumulative retention of water due to the progressive deflection of roof framing. If the roof framing is not provided with sufficient slope to drain, the roof system (deck, secondary and primary members) must be designed for stability under ponding conditions (1.13.3). Such a design may be based upon either a rational analysis or the following arbitrary limits:

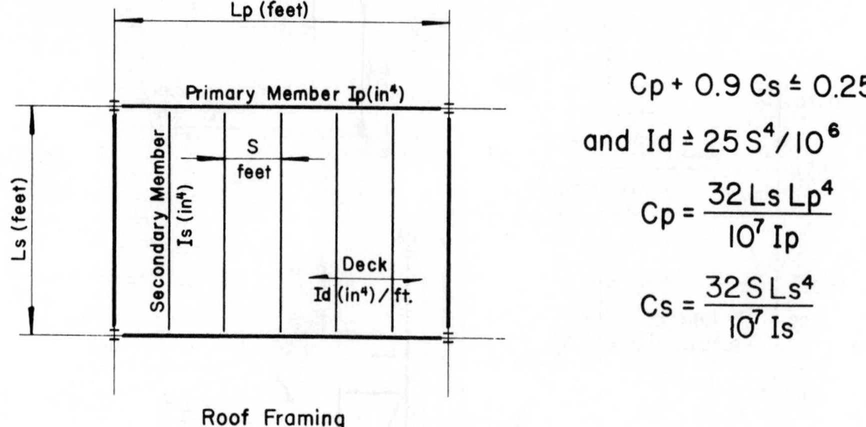

Fig. 3-14 Ponding Definitions for Roof Framing.

$$\left.\begin{array}{ll} C_p + 0.9\,C_s \leqslant 0.25 \ldots\ldots & (1) \\ I_d \geqslant 25S^4/10^6 \quad \ldots\ldots & (2) \end{array}\right\} \quad (1.13.3)$$

where

$$C_p = 32L_sL_p^4/10^7\,I_p \text{ and}$$
$$C_s = 32SL_s^4/10^7\,I_s$$

(See Fig. 3-14 for L_s, L_p, I_p, S, and I_s.)

The moment of inertia of steel joists and trusses must be reduced by fifteen percent since their shear rigidity is less than a solid plate. A steel deck must be considered a secondary member when it is directly supported by the primary members (1.13.3).

If the above criteria cannot be met, a rational analysis must be made (Commentary 1.13.3).* The bending stress due to combined dead, live and ponding (rain) loads must not exceed $0.80F_y$ (1.13.3).

To avoid ponding, the authors recommend that $C_p + 0.9C_s \leqslant 0.25$ and $I_d \geqslant 25S^4/10^6$ wherever possible. For the usual type of $1\frac{1}{2}$ in. depth metal roof deck, the maximum span of the deck is then limited as follows:

26 gage, $I_d = 0.066$ in^4/ft. $S_{max} = 7.17$ ft. 22 gage, $I_d = 0.114$ in^4/ft. $S_{max} = 8.22$ ft.
24 gage, $I_d = 0.091$ in^4/ft. $S_{max} = 7.77$ ft. 20 gage, $I_d = 0.137$ in^4/ft. $S_{max} = 8.60$ ft.

DESIGN EXAMPLES

BEAM EXAMPLE 3-1. SIMPLE SPAN ROOF BEAM—TYPE 2 CONSTRUCTION

Problem. Using ASTM Grade 36 steel, design the simple span roof beam of Example 1(b), Chapter 2.

Design Data. Refer to Example 1(b) and Fig. 3-15 for analysis data. The top flange of the beam is laterally braced by open web steel joist at points of concentrated loads. ($L_b = 6.0$ ft.)

The critical combination of moment and unbraced length of compression flange; $+M_R = 168.5$ k-ft.; $L_b = 6.0$ ft.; $M_1/M_2 = -1$ and $C_b = 1.0$ (single curvature). (See Fig. 3-15 and Table 6-2.)

*Cited Reference (5).

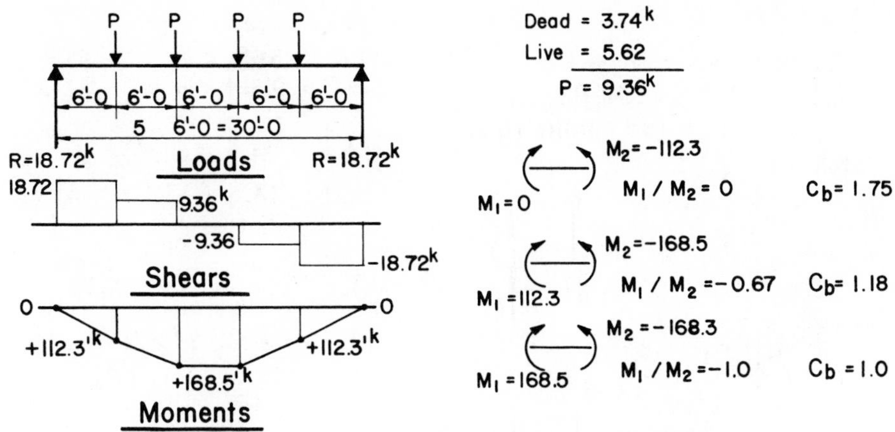

Fig. 3-15 Simple Span Roof Beam—Example 3-1.

Use of Beam Selection Table. For fully stressed roof beams, minimum recommended depth-span ratio = $F_y/1000$ (Commentary 1.13.1) = 36/1000 = 0.036; d(minimum) = (0.036)(360) = 12.96 inches.

Enter Table 3-1 beginning with W14 and proceed from the lightest to the heavier sections. The first section with $M_{Rx} > 168.5$ k-ft. in Col. (1) is W14 X 61. The second deeper section is W16 X 58; third deeper section is W18 X 50; and the fourth deeper section is W21 X 49.

Try W21 X 49 (minimum weight and maximum stiffness). Read $L_c = 6.88$ ft. $\geqslant L_b = 6.0$ ft. Compare the tabulated dimension ratios to the limits for Col. (1). See footnote (1). $(b_f/2t_f) = 6.12 < 10.83$; and $(d/t_w) = 56.57 < 106.7$. The section is *compact*; col. (1) is the correct resisting moment. $M_{Rx} = 184.7 > 168.5$ f-ft.; O.K. for flexure.

Shear. Check beam shear (1.5.1.2)

$$v = V/dt_w = (18.72)/[(20.82)(0.368)]$$

$$= 2.44 \text{ ksi O.K.} < 0.40 F_y = (0.40)(36) = 14.4 \text{ ksi.}$$

Deflection. Check beam live load deflection (1.13.1) using deflection coefficient from AISC Manual "Table of Equivalent Uniform Loads." Live load $P = 5.62$k; (see Fig. 3-15).

$$\Delta_L = \frac{(5)(4.80P)(1.01)\,l^3}{384\,EI}$$

$$= \frac{5(4.80)(5620)(1.01)(360)^3}{(384)(29{,}000{,}000)(971)}$$

$$= 0.587 \text{ in.} < l/360 = 1.00 \text{ in.; OK.}$$

BEAM EXAMPLE 3-2. EXTERIOR SPAN RIGID FRAME ROOF BEAM—TYPE 1 CONSTRUCTION

Problem. Using ASTM Grade 36 steel, design the exterior span rigid frame roof beam of Example 1(a), Chapter 2.

Design Data. Refer to Fig. 2-5(a) in Chapter 2 for analysis data for three span one-story rigid frame of Type 1 Construction. See Fig. 3-16 for $(D + L)$ moments and shears determined by an elastic analysis (STRESS) that considers elastic shortening of columns and

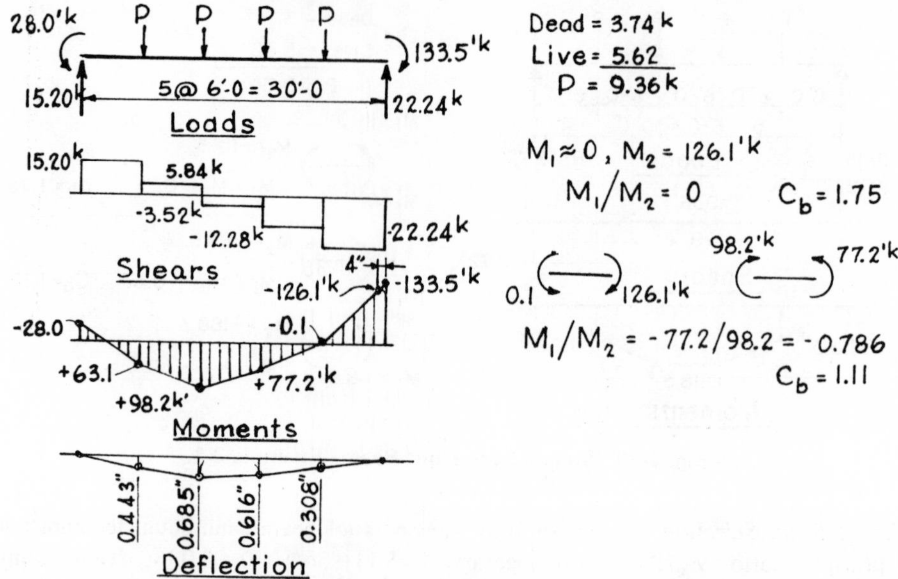

Fig. 3-16 Exterior Span, Rigid Frame Roof Beam—Example 3-2.

beams. As with all statically indeterminate structures, the beam and column sizes must first be assumed and then checked for conformance to the AISC Specifications. The top flange of the beam is braced by open web steel joists at points of concentrated loads ($L_b = 6.0$ ft.). The bottom flange of the beam is braced at the face of the column and at the point of inflection ($L_b = 5.68$ ft.).

Use of Beam Selection Table. The critical combination of moment and unbraced length of compression flange; $-M_R = 126.1$ k-ft.; $L_b = 5.68$; $M_1/M_2 = 0$ and $C_b = 1.75$ (single curvature). (See Fig. 3-16 and Table 6-2.) For fully stressed roof beams, minimum recommended depth-span ratio = F_y 1000 = 36/1000 = 0.036 (Commentary 1.13.1). This minimum depth ($d = 13$ in.) is not likely to result in economical weight. The authors recommend beginning routine design with a minimum $d = l/24$ for roof beams and $l/20$ for floor beams to avoid waste of design time. d (recommended) = 360/24 = 15 in.

Enter Table 3-1 beginning with W14 and proceed from the lightest to the heavier sections. The first section with $M_{Rx} > 126.1$ k-ft. in Col. (1) is W14 X 48. Repeat with W16, the first section is W16 X 40; with W18, the first section is W18 X 40.

Try W18 X 40 (minimum weight and maximum stiffness). Read maximum unbraced length, $L_c = 6.35$ ft.; actual unbraced lengths, $L_b = 5.68$ ft. for $-M$ and 6.00 for $+M$. Compare the tabulated dimension ratios $(b_f/2t_f) = 5.74 < 10.83$ and $(d/t_w) = 56.64 < 106.7$. The section is *compact* for both $-M$ and $+M$ conditions and therefore O.K. for flexure at $F_b = 0.66 F_y$. The tabulated M_{Rx} = 135.4 k-ft. > 126.1 k-ft. required.

Shear. Check beam shear (1.5.1.2)

$$v = V/dt_w = (22.24)/[(17.90)(0.316)]$$

$$= 3.93 \text{ ksi O.K.} < 0.40 F_y = 14.4 \text{ ksi.}$$

Deflection. Check beam live load deflection (1.13.1). Use simple span deflection coefficient from AISC Manual "Table of Equivalent Uniform Loads" and correct for end moments.

$$\Delta_L = \frac{(5)(4.80P)(1.01)l^3}{384\,EI} - \frac{(3)(M_L + M_R)l^2}{48\,EI}$$

$$= \frac{(5)(4.80)(5620)(1.01)(360)^3}{(384)(29{,}000{,}000)(612)} - \frac{(3)(28{,}000 + 133{,}500)(30/50)(360)^2}{(48)(29{,}000{,}000)(612)}$$

$$= 0.932 - 0.531$$

$$= 0.401 \text{ in.} < l/360 = 1.00 \text{ in. O.K.}$$

This value compares well with the computer value, see Fig. 3-16, $\Delta_L = (0.685) \cdot (5.62)/9.36 = 0.411$ in.

Wind. Shears and moments for $D + L + W$ are not critical because of the permitted one-third increase in allowable stresses with wind load (1.5.6). (See Example 1(a) in Chapter 2.)

BEAM EXAMPLE 3-3. INTERIOR SPAN—CANTILEVER- SUSPENDED-SPAN

Problem. Design the interior span roof beam of Fig. 2.5(e), Chapter 2. Use ASTM A572 Grade 50 steel.

Design Data. (See Fig. 3-17 for moments and shears.) The top flange is laterally braced in regions of positive moment by open web joists transmitting concentrated loads ($L_b = 6.0$ ft.). Regard the bottom flange as laterally braced in regions of negative moment by the connection at the face of the column and at points of inflection ($L_b = 4.17$ ft.) by change in stress.

The critical combination of moment and unbraced length of compression flange; $M_R = +84.25$ k-ft.; $L_b = 6.0$ ft.; $M_1/M_2 = -1$ and $C_b = 1.0$ for single curvature (See Fig. 3-17 and Table 6-2.)

Use of Beam Selection Table. Enter Col. (1) in Table 3-2 beginning with W18 (from Example 3) and proceed from the lightest to the heavier sections. The first section with $M_{Rx} > 84.2$ k-ft. in Col. (1) is W18 X 35. Note that the section is neither compact, Col. (1) nor semi-compact, Col. (2) because $L_b = 6.0$ ft. $> L_c = 4.84$ ft. Read Col. (3), $M_{Rx} = 144.7$ k-ft. $> M_R = 84.2$ k-ft., which suggests that a more shallow section might be satisfactory if the deflection is not excessive.

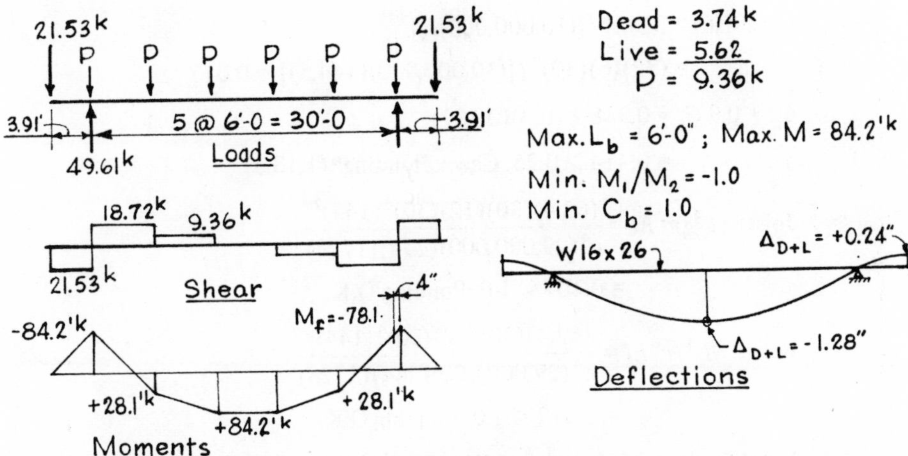

Fig. 3-17 Interior Span Roof Beam—Cantilever-Suspended-Span System—Example 3-3.

Re-enter Table 3-2 beginning with W16 and proceed from the lightest to the heavier sections. Read for W16 X 26, in Col. (1) $M_{Rx} = 105.3$. Note that this section is neither compact, (Col. (1), nor semi-compact, Col. (2) because $L_b = 6.0$ ft. $> L_c = 4.04$ ft. Read for Col. (3) $M_{Rx} = 95.7 > 84.2$. The value in Col. (3) is an upper limit.

Try W16 X 26. Read $r_T = 1.38$ and compute $(l_b/r_T) = 72/1.38 = 52.2$. $C_b = 1.0$; and $45.2\sqrt{C_b} < (l_b/r_T) < 101\sqrt{C_b}$. Use larger M_{Rx} from Col. (4) or Col. (5), but not to exceed the value in Col. (3).

Col. (4)

$$\text{Read } MF = 27.384$$

$$\text{Compute } M_{Rx} = [\text{Col. (1)} - (MF)l_b^2/(10{,}000\ C_b)] \leqslant \text{Col. (3)}$$

$$= \{105.3 - (27.384)(72)^2/[(10{,}000)(1)]\} \leqslant 95.7$$

$$84.2 \text{ k-ft.} < 91.1 \text{ k-ft.} < 95.7 \quad \text{O.K. for flexure.}$$

Col. (5)

$$\text{Read } MF = 4.64$$

$$\text{Compute } M_{Rx} = [1000\ (MF)\ C_b/l_b] \leqslant \text{Col. (3)}$$

$$= [(1000)(4.64)(1)/(72)] \leqslant 95.7$$

$$= 64.4 \text{ k-ft.} < 95.7$$

M_{Rx} from Col. (4) = 91.1 k-ft. > 84.25 k-ft. The section is OK for flexure.

Shear. Check *W16* X *26* for shear.

$$v = V/dt_w = (21.53/[(15.65)(0.25)]$$

$$= 5.50 \text{ ksi O.K.} < 0.40 F_y = (0.40)(50) = 20 \text{ ksi.}$$

Deflection. Check *W16* X *26* for live load deflection (1.13.1). (See Fig. 3.)

$$\Delta_L = (1.28)(5.62)/(9.36) = 0.77 \text{ in.} < l/360 = 1.0 \text{ in.; O.K.}$$

Ponding. Check deck and W16 X 26 for ponding (1.13.3). Primary member W16 X 26 ($I_p = 244$ in.^{4}); secondary member 20H6 ($I_s = 170 \times 0.85 = 144.5$ in.^{4}); $L_p = 30$ ft.; $L_s =$ 30 ft.; $S = 6$ ft.; 22 gage metal deck ($I_d = 0.114$ in.^{4}/ft.). (See Fig. 3-14.)

Joist and Beam

$$C_p = 32L_sL_p^4/[(10{,}000{,}000)I_p]$$

$$= (32)(30)(30)^4/[(10{,}000{,}000)(244)] = 0.0318$$

$$C_s(\text{20H6}) = 32SL_s^4/[(10{,}000{,}000)I_s]$$

$$= (32)(6)(30)^4/[(10{,}000{,}000)(144.5)] = 0.017$$

$$C_p + 0.9\ C_s = 0.318 + (0.9)(0.107)$$

$$= 0.414 > 0.25. \text{ Check Ponding* (1.13.3)}$$

$$\text{Joist: } \gamma L^4/\pi^2 EI = \frac{(62.4)(30)(12)(30)^4(144)^2}{\pi^4(29{,}000{,}000)(244)(1728)}$$

$$= 0.107 < 1.0. \text{ Ponding O.K.}$$

$$\text{Beam: } \gamma L^4/\pi^4 EI = \frac{(62.4)(30)(12)(30)^4(144)^2}{\pi^4(29{,}000{,}000)(244)(1728)}$$

$$= 0.317 < 1.0. \text{ Ponding O.K.}$$

$$\text{Metal Deck: } = 25S^4/10^6 = (25)(6)^4/10^6 = 0.0324$$

$$= I_d(\text{22 gage}) = 0.114 \text{ in.}^4/\text{ft.} > 0.0324 \text{ in.}^4\text{; OK.}$$

***Cited reference 14.**

BEAM EXAMPLE 3-4. SUSPENDED EXTERIOR SPAN, CANTILEVER-AND-SUSPENDED-SPAN

Problem. Design exterior span roof beam of Example 1(e) in Chapter 2. Use ASTM A572 Grade 50 steel.

Design Data. (See Figure 3-18 for moments and shears.) The top flange of the beam is laterally braced by open web steel joist at points of concentrated loads (L_b = 6.0 ft.).

The critical combination of moment and unbraced compression flange; $+M_R$ = 134.8 k-ft.; L_b = 6.0 ft.; M_1/M_2 = 0.875 and C_b = 1.148 for single curvature. (See Table 6-2.)

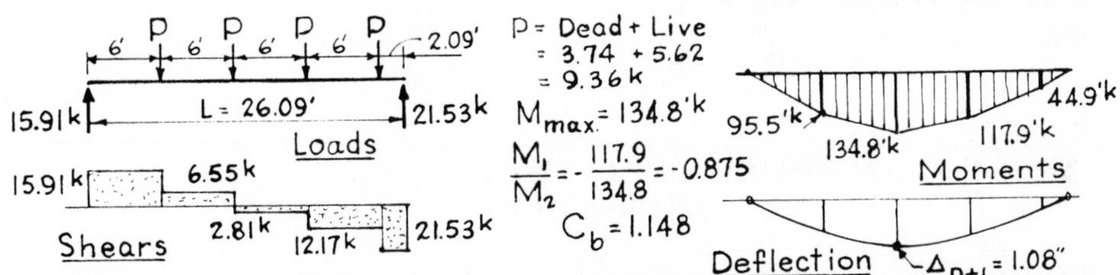

Fig. 3-18 Exterior Span Roof Beam—Cantilever-Suspended-Span System—Example 3-4.

Use of Beam Selection Table. For fully stressed roof beams, minimum recommended depth-span ratio = $F_y/1000$ = 50/1000 = 0.05 (Commentary 1.13.1). Minimum d = (0.05)(26.09)(12) = 15.65. Enter Table 3-2 beginning with W16 and proceed from the lightest to the heavier sections. The first section with $M_{Rx} > 134.8$ k-ft. in Col. (1) is W16 X 36, and the second higher section is W18 X 35.

Try W18 X 35 (minimum weight and maximum stiffness). Note that L_b = 6.0 ft. > L_c = 4.84 ft. showing that W18 X 35 is neither compact nor semi-compact. Read r_T = 1.510 and compute

$$(l_b/r_T) = (6.0)(12)/1.510 = 47.7.$$

Because

$$(l_b/r_T) < 45.2\sqrt{C_b} = 45.2\sqrt{1.148} = 48.4$$

use value of M_{Rx} from Col. (5) but not to exceed the value from Col. (3). Since $(l_b/r_T) < 45\sqrt{C_b}$, Cols. (4) and (6) are not applicable. (See footnotes, Table 3-2.)

Col. (5)

$$\text{Read } MF = 8.41$$

$$\text{Compute } M_{Rx} = 1000\,(MF)\,C_b/l_b \leqslant \text{Col. (3)}$$

$$= (1000)(8.41)(1.148)/72 \leqslant \text{Col. (3)}$$

$$= 134.1 \text{ k-ft.} \leqslant \text{Col. (3)} = 144.7$$

M_{Rx} = 134.1 k-ft. ≈ 134.8 k-ft. The section is OK. for flexure.

Shear. Check W18 X 35 for shear.

$$v = V/dt_w = (21.53)/[(17.71)(0.098)]$$

$$= 4.08 \text{ ksi} < 0.40F_y = (0.40)(50) = 20 \text{ ksi; OK.}$$

Deflection. Check beam live load deflection (1.13.1). (See Fig. 3-16.)

$$\Delta_L = (0.96)(5.62)/9.36 = 0.58 \text{ in.} < l/360 = 1.0 \text{ in.; OK.}$$

BEAM EXAMPLE 3-5. SIMPLE SPAN OFFICE FLOOR BEAM–TYPE 2 CONSTRUCTION

Problem. Design non-composite Type 2 Construction simple span floor beam using ASTM Grade 36 steel.

Design Data. Refer to Fig. 2-8, Chapter 2, for floor layout. Typical floor beams supporting a metal deck and LWA concrete slab span 25′-0″ and are spaced 8′-4″ on centers. Floor dead and live loads are as follows:

LWA concrete (110 pcf) slab	=	41 psf
Partition allowance	=	20
Ceiling and mechanical allowance	=	10
Beam	=	4
Dead	=	75 psf

Reduced live load
(50 psf) [100 - (0.08)(8.33)(25)]/100 = 41 psf
$(D + L)$ = 116 psf

The top flange is continuously braced laterally by the slab ($L_b = 0$). $M_R = WL^2/8 = (0.116)(8.33)(25)^2/8 = 75.4$ k-ft.

Use of Beam Selection Table. For fully stressed floor beams; minimum recommended depth-span ratio = $F_y/800 = 36/800 = 0.045$; d (minimum = (0.045)(25)(12) = 13.5 in. (See Fig. 3-11.)

Enter Table 3-1 beginning with W14 and proceed from the lighter to the heavier sections. The first section with $M_{Rx} > 75.4$ k-ft. in Col. (1) is W14 X 30; the second higher section is W16 X 26.

Try W16 X 26. Compare the tabulated dimension ratios $(b_f/2t_f) = 7.97 < 10.83$ and $(d/t_w) = 62.6 < 106.7$. The section is compact and therefore OK for flexure. $M_{Rx} = 75.8 > 75.4$ k-ft. from Column (1).

Shear. Check beam shear (1.5.1.2).

$$v = V/dt_w = (0.116)(8.33)(12.5)/[(15.65)(0.25)]$$
$$= 12.08/3.91$$
$$= 3.09 \text{ ksi} < 0.40F_y = (0.4)(36) = 14.4 \text{ ksi; OK.}$$

Deflection. Check beam live load deflection (1.13.1).

$$\Delta_L = 5wl^4/(384EI)$$
$$= (5)(41)(8.33)(300)^4/[(12)(384)(29{,}000{,}000)(300)]$$
$$= 0.345 \text{ in.} < l/360 = 300/360 = 0.833 \text{ in. OK.}$$

BEAM EXAMPLE 3-6. VIBRATION.

Check beam in Example 3-5 for transient vibrations in open areas without partitions. If the partitions are present on the floor area supported by the beam, they will usually provide sufficient dampening to reduce transient vibrations to a level that is tolerable for human comfort. (See Fig. 3-13 and 3-19.) Even though the beam is not a composite beam, the authors recommend that it be considered to act compositely for vibration

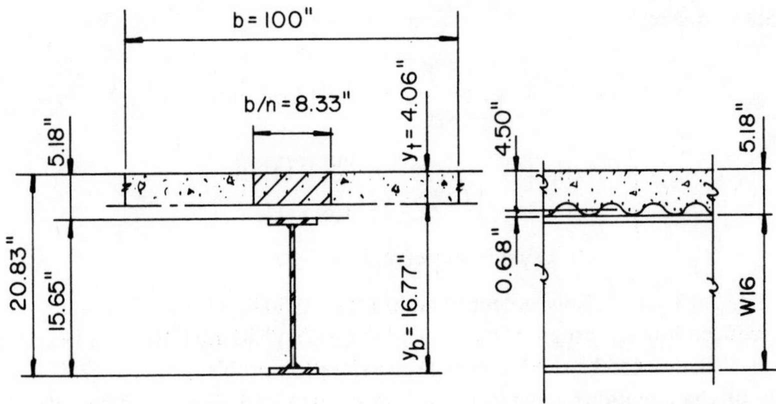

Fig. 3-19 Vibration Check—Simple Span Floor Beam—Example 3-6.

calculations and that the flange width be taken equal to the beam spacing. $f_c' = 4{,}000$ psi; $w = 110$ pcf.

$$n = E_s/E_c = 29{,}000{,}000/2{,}407{,}870^* = 12$$

$$8.33(y_t)^2/2 = (7.67)\left(20.83 - \frac{15.65}{2} - y_t\right)$$

$$(y_t)^2 + 1.84y_t - 23.95 = 0$$

$$y_t = 4.06 \text{ in.}; y_b = 20.83 - 4.06 = 16.77 \text{ in.}$$

$$I_{tr} = (8.33)(4.06)^3/3 \qquad = \quad 186$$

$$I_x \text{ of } W16 \times 26 \qquad = \quad 300$$

$$(7.67)(16.77 - 15.65/2)^2 = \underline{\ 614\ }$$

$$1100 \text{ in.}^4$$

Assume

$$N = (l/2)/(\text{beam spacing})$$

$$(25/2)/8.33$$

$$= 1.5$$

Computed displacement,

$$\Delta = \frac{Kl^3}{48\,EIN} = \frac{L^2 l^3}{D(48)EIN} \text{ (See Fig. 3-13.)}$$

$$= \frac{(25)^2\,(300)^3}{(1.736)(48)(29{,}000{,}000)(1100)(1.50)}$$

$$= 0.0042 \text{ in.}$$

---

*Cited Reference (12).

Calculate frequency,

$$f = \frac{0.742}{L^2}\sqrt{\frac{E_s I}{w_D}} \text{ using dead load of 55 psf.}$$

$$= \frac{(0.742)}{(25)^2}\sqrt{\frac{(29{,}000{,}000)(1100)}{(55)(8.33)}}$$

$$= 9.9 \text{ cycles per second}$$

Enter Fig. 3-13 with displacement (amplitude) = 0.0042 in. and frequency = 9.9 cps and read reduced human response that is DISTINCTLY PERCEPTIBLE. A heavier and/or deeper rolled section can be used to reduce the vibration problem.

A study of the equations and plotted curves in Fig. 3-13 shows that the effect of the transformed moment of inertia *reduces* amplitude proportionally to ($I_t r$) and *increases* frequency proportionally to its square root to ($\sqrt{I_{tr}}$). Due to the slope of the curves, merely using a heavier section may not improve human responses much. The amplitude, however, is reduced also proportionally to the *depth* (of the transformed section). Thus, to improve transient vibration effects upon human response, an increase in I_{tr} accomplished by an increase in depth is most effective. Several trials will illustrate this difficulty.

Trial Size	yb in.	I_{tr} in.^{4}	Δ in.	f cps	Response Range
W16 × 26	16.75	1100	0.0042	9.9	0.75 in "Distinctly Perceptible"
W16 × 36	16.39	1485	0.0032	11.5	0.67 in "Distinctly Perceptible"
W18 × 35	18.10	1698	0.0025	12.3	0.55 in "Distinctly Perceptible"
W18 × 45	17.70	2142	0.0020	13.8	0.42 in "Distinctly Perceptible"
W21 × 44	20.18	2603	0.0014	15.2	0.20 in "Distinctly Perceptible;" close to "Slightly Perceptible"–and Acceptable.

Note the slow convergence for increases in moment of inertia for the section without increases in depth.

BEAM EXAMPLE 3-7. EXTERIOR SPAN RIGID FRAME FLOOR BEAM—10th STORY OF 25 STORY MOMENT RESISTING RIGID FRAME

Problem. Design rigid frame floor beam using ASTM Grade 36 steel.

Design Data. Refer to Fig. 2-6, 2-7, 2-8 and 2-10(a) in Chapter 2, "Analysis and Design," for analysis data. See Fig. 3-20 for summary of gravity and wind moments.

The top flange of the beam is laterally braced by the reinforced concrete slab ($L_b = 0$). Because the compression in the beam due to axial load is small ($f_a = 5.77/18 = 0.32$ ksi) and the ratio $f_a/F_a = 0.32/16.17 = 0.02 < 0.15$, the authors would consider the bottom flange to be laterally braced at the point of inflection and at the face of the column. $L_b = 8.23 - 0.58 = 7.65$ ft. (See Fig. 3-20.)

The critical combination of moment and unbraced compression flange length is maximum $-M$ for Dead + Live + Wind (see Fig. 3-20); $-M_R = (-307.2)(3/4) = -230.4$ k-ft. The combined design moment was reduced to compensate for the one-third increase in allowable bending stress with wind (1.5.6). $L_b = 7.65$ ft.; $M_1/M_2 = 0$; $C_b = 1.75$ for single curvature. Enter Table 3-1 with maximum moment of $(D + L)$ or $0.75(D + L + W)$.

Use of Beam Selection Table. For fully stressed floor beams; minimum recommended depth-span ratio = $F_y/800$ (Commentary 1.13.1). $F_y/800 = 36/800 = 0.0625$. Min d = (0.05)(30)(12) = 18 in. Enter Col. (1) of Table 3-1 beginning with W18 and proceed

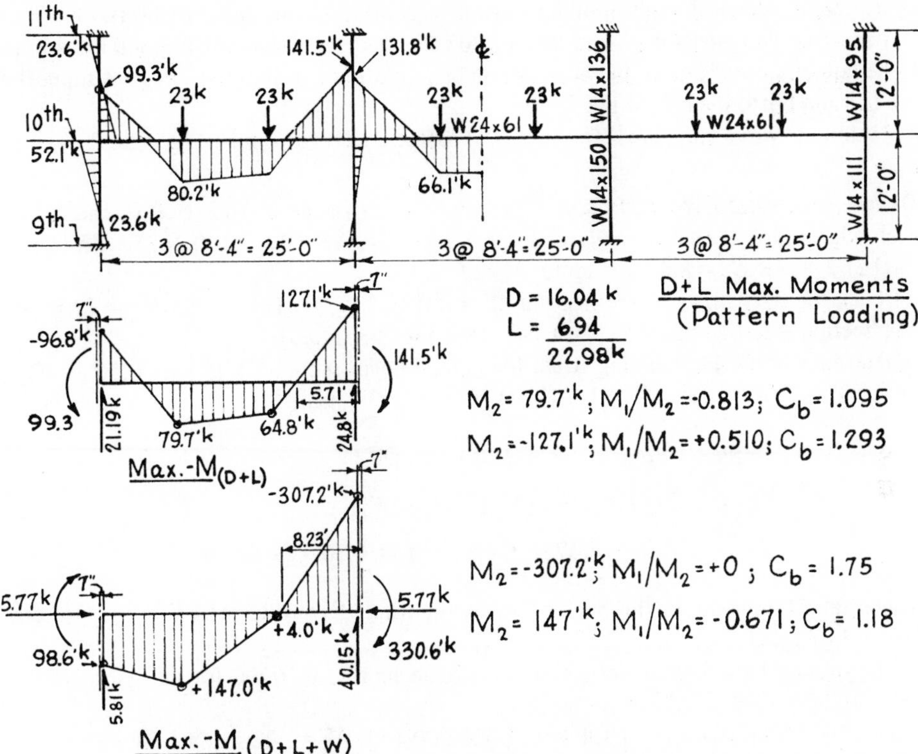

Fig. 3-20 Exterior Span Floor Beam—10th Floor (of 25)—Type 1 Example 3-7.

from the lightest to the heavier sections. The first section with $M_{Rx} > 230.4$ k-ft. is $W18 \times 64$; the second, deeper section is $W21 \times 62$; the third deeper section is $W24 \times 61$.

Try W24 $\times$ *61*. (Minimum weight and maximum stiffness.) $L_b = 7.65 > L_c = 7.41$ and section is neither compact nor semi-compact. Read $r_T = 1.74$ and compute $(l_b/r_T) = (7.65)(12)/1.74 = 52.8$. Because $(l_b/r_T) < 53.2\sqrt{C_b} = 53.2\sqrt{1.75} = 70.4$, use value of M_{Rx} from Col. (5) but not to exceed the value from Col. (3). Since $(l_b/r_T) < 53.2\sqrt{C_b}$, Col's. (4) and (6) are not applicable. (See footnotes Table 3-1.)

Col. (5)

$$\text{Read } MF = 22.74. \quad l_b = 7.65 \times 12 = 91.8$$

$$\text{Compute } M_{Rx} = 1000\,(MF)\,C_b/l_b \leqslant \text{Col. (3)} = 233.9$$

$$= (1000)(22.74)(1.75)/91.8 \leqslant 233.9$$

$$= 433.5 \text{ k-ft.} > 233.9. \text{ Use } M_{Rx} = 233.9$$

$$M_{Rx} = 233.9 > 230.4 \text{ k-ft. OK for flexure.}$$

Shear. Check $W\,24 \times 61$ for shear.

$$v = V/dt_w = (41.5)(3/4)/[(23.72)(0.419)]$$

$$= 3.1 \text{ ksi} < 0.4F_y = 0.4 \times 36 = 14.4; \text{OK for shear.}$$

BEAM EXAMPLE 3-8. WIND MULLION—TYPE 2 CONSTRUCTION.

Problem. Design a simple span wind mullion using ASTM Grade 36 steel with a maximum width of 2 inches and a maximum deflection of $l/200$.

Design Data. Vertical wind mullions have a span of 28 ft., are spaced 10.0 ft. on centers and must be designed for a wind load of 20 psf. The compression flange will be assumed to be laterally unsupported between points of support (L_b = 28.0 ft.). Neglect support at the top and bottom edges of the glass.

Maximum wind moment, $M_w = (0.02)(10)(28)^2/8 = 19.6$ k-ft.; $L_b = 28.0$ ft.; and $C_b = 1$.

Design Box-Section Wind Mullion. The compression flange is connected to the webs at each edge and is therefore "stiffened" (1.9.2.1). Try TS 10 × 2 × 0.1875 with $(l_b/b) = 28(12)/2 = 168$, $S_x = 8.58$ in.^{3} and $I_x = 42.9$ in.^{4}.

Check to determine if the section is compact. $l_b/b = 168 > 1200/F_y = 1200/36 = 33$. Section is not compact (1.5.1.4.1-f). (See Fig. 3-5, note 1.)

Determine allowable bending stress for non-compact section. Use $F_b = 0.60\,F_y$ if $d \leqslant 6b$ and $(b - 2t)/t < 238/\sqrt{F_y}$ (1.5.1.4.4). (See Fig. 3-5, note 2.)

$$d = 10 \text{ in.} < 6b = 6(2) = 12 \text{ in. OK.}$$

$$(b - 2t)/t = [2 - 2(0.1875)]/0.1875$$

$$= 8.67 < 238/F_y = 238/\sqrt{36} = 39.67 \text{ OK.}$$

Allowing a $\frac{1}{3}$ increase in the bending stress for wind, $F_b = 1.33(0.60F_y) = 1.33(0.60) \cdot (36) = 28.7$ ksi (1.5.6).

The authors believe that deflection should be limited to $l/200$ for mullions. $l/200 = 28(12)/200 = 1.68$ in. Check deflection.

$\Delta = (5)(20/12)(10)(336)^4/[(384)((29{,}000{,}000)(42.9)] = 2.22$ in. > 1.68 in. Try TS 10 × 2 × 0.25; $\Delta = 2.22(42.9/59.4) = 1.60$ in. < 1.68 in. OK.

Design Solid Rectangular Wind Mullion. Try bar 12 × 1 with lateral support at midspan ($L_b = 28/2 = 14$ ft.). $l_b/b = 14(12)/1 = 168$, $S_x = 1(12)^2/6 = 24$ in.^{3} and $I_x = 1(12)^3/12 = 133$ in.^{4}

$d = 12$ in. $> 10b = 10(1) = 10$ and $l_b/b = 168/1 = 168 > 2500/F_y = 2500/36 = 69.4$. Section is non-compact and the allowable bending stress, F_b, can be determined from Fig. 3-3 and Fig. 6-5.

Enter Fig. 3-3 with $b/d = 1/12 = 0.083$ and read $(l_b/r_y)_{\text{equiv.}}/\sqrt{l_b/b} = 8.3$. Compute $(l_b/r_y)_{\text{equiv.}} = 8.3\sqrt{l_b/b} = 8.3\sqrt{168/1} = 107.6$. From Fig. 6-5 read $F_a = 11.99$ ksi (AISC Eq. 1.5-1). Allowing a $\frac{1}{3}$ increase in bending stress for wind compute $F_b = (1.33)(11.99) = 15.95$ ksi.

Check if bending stress $f_b \leqslant F_b = 15.95$ ksi. $f_b = M_w/S_x = (19.6)(12)/24 = 9.8$ ksi $<$ 15.95 ksi OK.

Check deflection.

$$\Delta = (5)(20/12)(10)(336)^4/[(384)(29{,}000{,}000)(144)]$$

$$= 0.66 \text{ in.} = l/509 < l/240 \text{ OK.}$$

PLATE GIRDER DESIGN EXAMPLES—GENERAL

As observed for rolled sections, under "Use and Derivations of Rolled Section Selection Tables," the design of built-up girders has also become sophisticated. The complications in establishing the allowable stresses for a built-up girder preclude simple classification as in Tables 3-1, -2 for rolled sections. In addition to the nine variations involved in determining allowable stresses, built-up girder designs involve various buckling limits, "tension field" action, and hybrid sections. For tension field action girders an added complication of relating the allowable bending tension stress to shear stress is introduced (1.10.7).

The allowable bending stress is limited to a maximum of 0.60 F_y for hybrid girders (1.5.1.4.1, 1.5.1.4.2, and 1.5.1.4.3) and tension field action girders when combined shear and tension stress are considered (1.10.7). These additional complications plus the fact that a built-up section is a custom design for a particular application make it impossible to prepare semi-automatic solutions like those in Tables 3-1 and 3-2.

Examples 3-9 and 3-10 show the trial-and-error design procedures required at each step of built-up girder design. If the examples seem simple, consider that only the trial sizes that proved successful are shown. These examples utilized the simple case where top flange bracing is continuous. In evaluating weight comparisons between the examples, conclusions are probably valid only for this simple case.

PLATE GIRDER EXAMPLE 3-9. WELDED TENSION FIELD ACTION PLATE GIRDER—TYPE 2 CONSTRUCTION

Problem. Design a simple span welded tension field action plate girder for Type 2 construction. Use ASTM Grade 36 steel. Span is 60 ft. and maximum depth is limited to 60 in.

Design Data. The girder supports a uniform load of 2.0 kips per foot and two concentrated loads of 100 kips at the third points of the span. A reinforced concrete floor slab provides continuous lateral support for the compression flange (l_b = 0). (See Fig. 3-21 for design moments and shears.)

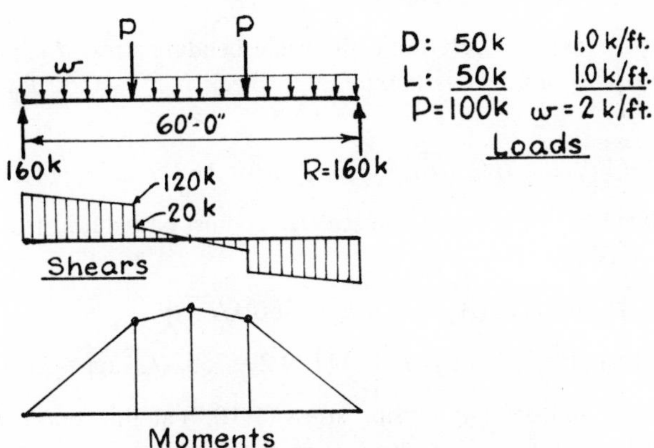

Fig. 3-21 Built-Up Girder Shears and Moments—Examples 3-9 and 3-10.

Preliminary Web Design. Try a 56 in. tension field action web and assume intermediate stiffener aspect ratio $(a/h) \leq 1.0$. Refer to Table 3-10; for $(a/h) = 1.0$ and $(h/t_w) = 180$; $F_v = 9.388$ ksi. Try web plate 56 × 5/16; $h/t_w = 56/0.3125 = 179 < 180$; OK.

$$f_v = V/A_w = 160/[(56)(0.3125)] = 9.14 \text{ ksi}$$

$$< 9.39 \text{ ksi; OK for shear.}$$

Preliminary Flange Design. Assume 1.5 in. thick flanges with the allowable stress in bending based on combined shear and tension stress (1.10.7) at the location of the concentrated loads;

$$F_b = \left(0.825 - 0.375 \frac{f_v}{F_v}\right) F_y \leq 0.60 F_y \ldots\ldots \text{(Eq. 1.10-7)}$$

$$= [0.825 - (0.375)(9.14)(120/160)/9.39]\,(36) \leq (0.60)(36)$$

$$= 19.84 < 21.6 \text{ ksi.}$$

Required flange area; $A_f = M/[(h + t_f)F_b] - A_w/6$

$$= (2800)(12)/[(56 + 1.5)(19.84)] - 17.5/6$$

$$= 29.45 - 2.92$$

$$= 26.53 \text{ in.}^2$$

Try 18 in. × 1 $\frac{1}{2}$ in. flange plates; $A_f = 27.0$ in.$^2 > 26.53$ in.^{2} OK.

Check width/thickness ratio of unstiffened compression flange; limit $(b_f/2t_f) \leqslant 95/\sqrt{F_y} = 15.83$ (1.9.1.2); $(b_f/2t_f) = 18/[(2)(1.5)] = 6.0 < 15.83$; OK; (1.10.3).

Preliminary Girder Cross-Section Properties. (See Fig. 3-22 for elevation and section of girder.) Cross-section properties of the girder are as follows:

$$A_w = 17.5 \text{ in.}^2$$

$$A_f = 27.0 \text{ in.}^2$$

$$A_w/A_f = 17.5/27.0 = 0.648$$

$$h/t_w = 56/0.3125 = 179.2$$

$$I_x = (0.3125)(56)^3/12 + [(18)(1.5)^3/12 + (27.0)(28.75)^2](2)$$

$$= 49{,}218 \text{ in.}^4$$

$$S_x = 49{,}218/29.5 = 1{,}668 \text{ in.}^3$$

Allowable Bending Stress, F_b. Determine allowable bending stress, F_b, at center of span where shear is zero and moment is maximum. Check if $F_b \leqslant 0.60\ F_y = (0.60)(36) = 21.6$ ksi.

$$(h/t_w) = 179.2 > 760/\sqrt{F_b} = 760/\sqrt{21.6} = 163.5 \ldots\ldots (1.10.6)$$

The allowable flange bending compression stress, F_b, must be reduced from $F_b = 21.6$ ksi to F'_b (1.10.6; Eq. 1.10-5):

$$F'_b = F_b\ [1 - 0.0005\ (A_w A_f)\ (h/t_w - 760/\sqrt{F_b})]$$

$$= (21.6)\ [1 - (0.0005)(0.686)(179.2 - 760/\sqrt{21.6}] = 21.5 \text{ ksi.}$$

For combined shear and bending tension stress (1.10.7) at the concentrated loads the allowable stress in bending; $F_b = 19.84$ ksi (see preliminary flange design).

Computed Bending Stress, f_b.

At center of span:

$$f_b = M/S_x = (2900)(12)/1668 = 20.9 \text{ ksi} < 21.5 \text{ ksi OK.}$$

At concentrated loads:

$$f_b = M/S_x = (2800)(12)/1668 = 20.1 \text{ ksi} \approx 19.84 \text{ ksi OK. (within 1.3 percent).}$$

End Panel Stiffener Spacing Requirements. Compute the maximum shearing stress; $f_v = V/A_w = 160/17.5 = 9.14$ ksi.

The maximum spacing, a, of end panel stiffeners can be estimated from the equation: $a = 11{,}000\ t\sqrt{1/1000}/\sqrt{f_v} = 348\,t/\sqrt{f_v} \ldots\ldots$ (Commentary 1.10.5). $a = (348)(0.3125)/\sqrt{9.14} = 36.0$ in.

Compute the maximum allowable shearing stress, $F_v \ldots\ldots$ (1.10.5.2). Eq. 1.10-1). Compare with f_v for an assumed end panel stiffener spacing, $a = 36.0$ in. For $a = 36.0$ in.; $(a/h) = 36/56 = 0.643$; $k = 4 + 5.34/(a/h)^2 = 4 + 5.34/(0.643)^2 = 16.91$.

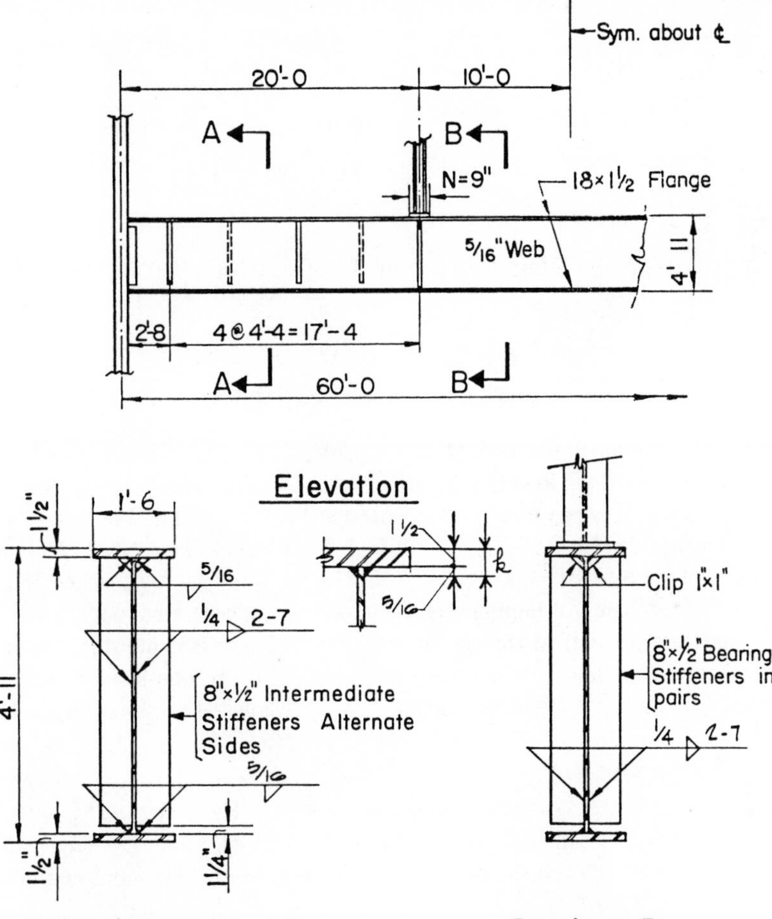

Fig. 3-22 Welded Tension Field Action Plate Girder Details—Example 3-9.

$C_v = 45{,}000\, k/[F_y\, (h/t_w)^2] = (45{,}000)\,(16.91)/[(36)\,(179.2)^2]$
$= 0.658;$
$F_v = F_y C_v/2.89 = (36)\,(0.658)/2.89 = 8.20$ ksi.
$f_v = 9.14 > F_v = 8.20$; N.G.

Try a reduced stiffener spacing $a = 32$ in. For $a = 32$ in.;

$(a/h) = 32/56 = 0.571;$
$k = 4 + 5.34/(0.571)^2 = 20.38;$
$C_v = (45{,}000)\,(20.38)/[(36)\,(179.2)^2] = 0.793;$

and

$$F_v = (36)\,(0.793)/2.89 = 9.88 > f_v = 9.14 \text{ ksi; OK for shear.}$$

Use end panel stiffener spacing of $a = 32$ in.

Intermediate Stiffener Spacing Requirements. The clear distance between end panel stiffeners and first interior concentrated load = (20) (12) − 32 = 208 in. Try $a = 208/4 =$

52 in. and $a/h \leqslant 1.0$ as assumed in the preliminary design of the tension field action web. For $a = 52$ in.; $(a/h) = 52/56 = 0.928$; $k = 4 + 5.34/(a/h)^2 = 4 + 5.34/(0.928)^2 = 10.20$; $C_v = (45{,}000)(10.20)/[(36)(179.2)^2] = 0.397$; and

$$F_v = [F_y/289]\ \{C_v + (1 - C_v)/[1.15\sqrt{1 + (a/h)^2}]\} \ldots\ldots (1.10.5.2, \text{Eq. } 1.10.2)$$
$$= [(36/2.89)]\ \{0.397 + (1 - 0.397)/[1.15\sqrt{1 + (0.928)^2}]\}$$
$$= 9.73 \text{ ksi}$$

F_v can also be obtained by interpolation from Table 3-10 for $(a/h) = 0.928$ and $(h/t_w) = 179.2$.

$$f_v = V/A_w = [160 - (2)(2.67)]/17.5$$
$$= 8.84 \text{ ksi} < F_v = 9.73 \text{ ksi; OK.}$$

Space three intermediate stiffeners at 52 in. centers, between the end panel stiffeners and the interior concentrated load bearing stiffeners. (See Fig. 3-22.)

Maximum distance between interior concentrated loads = 240 in. For $a/h = 240/56 = 4.28$; $h = 5.34 + 4/(4.28)^2 = 5.56$; $C_v = 45{,}000\,(5.56)/[(36)(179.2)^2] = 0.216$; and $F_v = (36)(0.216)/2.89 = 2.69$ ksi. $f_v = V/A_w = 20/17.5 = 1.14$ ksi $< F_v = 2.69$ ksi; OK. Intermediate stiffeners are not required between interior concentrated loads.

The fact that intermediate stiffeners are not required between interior concentrated loads can also be ascertained quickly from Table 3-11 for non-tension field action girders, where for $(h/t_w) < 180$ and $(a/h) > 3$, $F_v > 2.566 > 1.14$ ksi; OK; (1.10.10.1; 1.10.10.2).

Intermediate Stiffener Size. Intermediate stiffeners must extend approximately to the edge of the flanges (1.10.5). The width/thickness ratio, $(w/t) \leqslant 95/\sqrt{F_y} = 95/\sqrt{36} = 15.83$ (1.9.1.2). For an 8 in. stiffener width, $t = w/15.83 = 0.505$. Try single stiffeners 8 × 1/2 in. on alternate sides of the web. Check minimum area and minimum moment of inertia requirements (1.10.5.4).

Minimum area of intermediate stiffeners can be obtained from Fig. 3-9 or calculated from 1.10.5.4 Eq. 1.10-3.

$$A_{st} = [(1 - C_v)/2]\ [(a/h) - (a/h)^2/\sqrt{1 + (a/h)^2}]\ YDht.$$

For $Y = 1$; $D = 2.4$ (single plate stiffeners); $t = 0.3125$; $a = 52$ in.; $h = 56$ in.; $(a/h) = 52/56 = 0.929$; $(h/t_w) = 56/0.3125 = 179.2$; $k = 4 + 5.34/(a/h)^2 = 4 + 5.34/(0.929)^2 = 10.19$;

$$C_v = 45{,}000k/[E_y(h/t)^2] \ldots\ldots (1.10.5.2)$$
$$= 45{,}000\,(10.19)/[(36)/(179.2)^2]$$
$$= 0.397$$

$$A_{st} = \left[\frac{(1 - 0.397)}{2}\right]\left[0.929 - \frac{(0.929)^2}{\sqrt{1 + (0.929)^2}}\right](1)(2.4)(56)(0.3125)$$
$$= 3.76 \text{ in.}^2$$

A_{st} for single plate stiffeners 8″ × 1/2″ = 4.00 > 3.76 in.^{2}; OK.

Minimum moment of inertia of intermediate stiffeners can be obtained from Fig. 3-8 or calculated from I (min.) $= (h/50)^4 = 1.574$ (1.10.5.4). Moment of inertia of single plate stiffeners about an axis in the plane of the web = $0.50(8)^3/12 + 4(4.156)^2 = 21.3 + 69.1 = 90.4$ in.$^4 > (56/50)^4 = 1.57$ in.^{4} OK.

Use $8 \times \frac{1}{2}$ in. single plate intermediate stiffeners on alternate sides of the web as shown in Fig. 3-22.

Bearing Stiffener Requirements. Check whether bearing stiffeners are required at the concentrated load points. Note bearing stiffeners must fit closely to the flanges, through or to which the load or reaction is delivered (1.10.5.1). Intermediate stiffeners may be cut short ($4t$) from the tension flange; single intermediate stiffeners must be attached to the compressive flange to resist uplift, not merely fitted for compressive bearing (1.10.5.4; 1.5.3.4)

If $R \leqslant 0.75\ F_y t\ (N + 2k)$, bearing stiffeners are not required (1.10.10.1; Eq. 10.10-8).

$t = 0.3125$ in.
$N = 9$ in. (length of bearing parallel to web)
$k = 1.50 + 0.3125 = 1.8125$ in. (See Fig. 3-22).
$R = 100k$ (See Fig. 3-21).
$0.75\ F_y t\ (N + 2k) = (0.75)(36)(0.3125)(9 + 1.8125)$
$= 91.2k < R = 100\ k$

Bearing stiffeners are required at the concentrated load points.

Design a pair of bearing stiffeners under each 100 kip concentrated column load (1.10.5.1). Try 2 - $8 \times 1/2$ in. plates (the same size as the intermediate stiffeners).

$w/t = 8/0.50 = 16$ OK ≈ 15.83 (1.9.1.2). (See Fig. 3-23.)
$A_{st} = (16)(1/2) + (7.81)(0.3125) = 10.44$ in.^{2}
$I_x = (0.5)(16.31)^3/12 = 180.8$ in.^{4}
$r_x = \sqrt{I_x/A_{st}} = \sqrt{(180.8)/10.44}$
$= 4.16$ in.

Check the axial load capacity of a pair of plate stiffeners acting with a part of the web as a column. The unbraced length of the stiffener web column can be taken equal to $\frac{3}{4}$ of the depth of the web (1.10.5.1). The stiffeners are braced in the direction parallel to the web. From Fig. 6-5, for $k = 1$; $kl/r = (1)(3/4)(56)/4.16 = 10.10$. Read $F_a = 21.15$ ksi $\leqslant$ 21.6 . . . . . . (1.5.1.3.4). $f_a = P/A = 100/10.44 = 9.58$ ksi < 21.15 ksi; OK.

Check bearing stress on stiffener area beyond fillet weld connecting web to flange (1.10.5.1); $f_p = (100)/[(8\text{-}1.0)(0.50)(2)] = 14.28$ ksi $< F_p = 0.90\ F_y = (0.90)(36) =$ 32.4 ksi (1.5.1.5.1) OK.

Use $8 \times 1/2$ in. bearing stiffeners in pairs under each concentrated load with a tight fit to top and bottom flanges.

Deflection.

$$\Delta_{D+L} = (5)(200/12)(720)^4/[(384)/29{,}000{,}000)(49{,}218)]$$
$$+ (100{,}000)(240)\,[(3)(720)^2 - (4)(240)^2]/[(24)(29{,}000{,}000)(49{,}218)]$$
$$= 0.408 + 0.928$$
$$= 1.336 \text{ in.} = d/538$$

($L = D$, See Fig. 3-21). $\Delta_L = 1.336/2 = 0.668$ in. $< l/360$.

$l/360 = 2.00$ in.; OK for plastered ceilings (1.13.1).

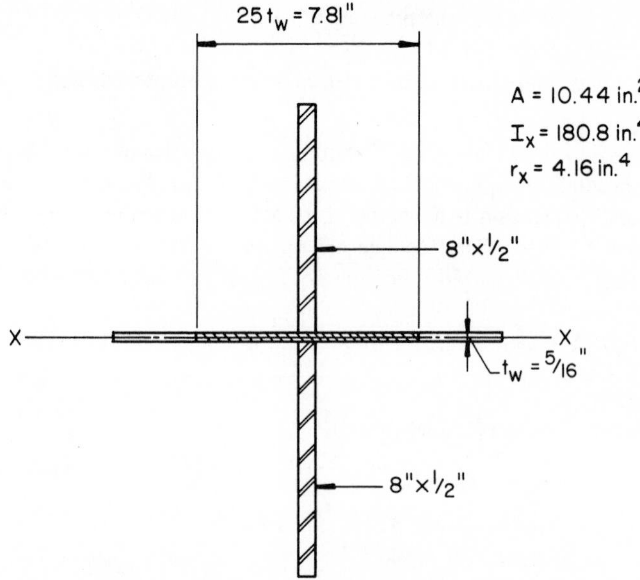

Fig. 3-23 Intermediate and Bearing Stiffeners—Example 3-9.

Calculated Weight.

Flanges; (1.5) (18) (2) (3.4) (60)		11,016
Web; (56) (0.3125) (3.4) (60)		3,570
Stiffeners;		750
Intermediate (8) (0.5) (8) (3.4) (54.75)/12	= 496	
Bearing (8) (0.5) (4) (3.4) (56)/12	= 254	
	Total	15,336 lb.

The weight of this tension field action welded plate girder made from A36 steel is 31 percent more than the similar hybrid girder of design example 3-10.

PLATE GIRDER EXAMPLE 3-10. HYBRID WELDED PLATE GIRDER—TYPE 2 CONSTRUCTION

Problem. Redesign the girder of Example 3-9 as a welded hybrid plate girder. Use ASTM Grade 36 web and stiffener steel and Grade 50 flanges with the same maximum depth of 60 in. as in Example 3-9.

Design Data. Same as Example 3-9. (See Fig. 3-21 for design moments and shears.)

Preliminary Web Design. Try a 56 in. web and assume intermediate stiffener aspect ratio $a/h = 0.6$. Refer to Table 3-11; for $(a/h) = 0.6$ and $(h/t_w) = 180$; $F_v = 9.21$ ksi (by interpolation). Try web plate 56 X 5/16; $(h/t_w) = 56/0.3125 = 179.2 < 180$ OK.

$$f_v = V/A_w = 160/[(56)(0.3125)] = 9.14 \text{ ksi} < 9.21; \text{OK.}$$

Preliminary Flange Design. Assume 1.5 in. thick flanges (1.10.3).

$$M/[(h + t_f)F_b] - A_w/6 = (2900)(12)/[(56 + 1.5)(30)] - 17.5/6 = 17.26 \text{ in.}^2$$

Try 12 X $1\frac{1}{2}$ flange plates; $A = 18.0$ in.^{2}

Check width thickness ratio of unstiffened compression flange:

$$(b_f/2t_f) \leqslant 95/\sqrt{F_y} = 13.43 \text{ (1.9.1.2):}$$

$$(b_f/2t_f) = 12/[(2)(1.5)] = 4.0 < 13.43; \text{OK.}$$

Preliminary Girder Cross-Section. Girder cross-section properties are as follows:

$$A_w = 17.5 \text{ in.}^2; A_f = 18 \text{ in.}^2; A_w/A_f = 17/(5/8) = 0.972$$
$$h/t_w = 56/0.3125 = 179.2;$$
$$I_x = \frac{(0.3125)(56)^3}{12} + \left[\frac{(12)(1.5)^3}{12} + (18)(28.75)^2\right](2) = 34{,}366 \text{ in.}^4$$
$$S_x = 34{,}366/29.5 = 1164 \text{ in.}^3$$

Determine Allowable Bending Stress, F_b. For hybrid girders the allowable stress in bending is limited to a maximum value of $0.60\,F_y = 0.60(50) = 30$ ksi. For $l_b = 0$ check if $F_b \leqslant 30$ ksi. $h/t_w = 179.2 > 760/\sqrt{F_b} = 760/\sqrt{30} = 107.5$ (1.10.6). Allowable flange bending compression stress, F_b, must be reduced to F_b' (1.10.6; Eq. 1.10-5):

$$F_b' = F_b\ [1 - 0.0005\,(A_w/A_f)\,(h/t_w - 760/\sqrt{F_b})]$$
$$= 30\ [1 - 0.0005\,(0.972)\,(179.2 - 760/\sqrt{30})]$$
$$= 29.4 \text{ ksi}$$

Check Maximum Bending Stress.

$$f_b = MS_x = (2900)(12)/1164 = 29.8 \text{ ksi} \approx F_b' = 29.4 \text{ ksi; OK.}$$

End Panel Stiffener Requirements. Compute the maximum shearing stress; $f_v = V/A_w = 160/17.5 = 9.14$ ksi.

The spacing of end panel stiffeners is not based on tension field action and can therefore be the same as in Design Example 3-9, viz., $a = 32$ in.

Intermediate Stiffener Spacing Requirements. Clear distance between end panel stiffeners and first interior concentrated load = (20) (12) - 32 = 208″. Try $a = 208/6 \approx 34$ in. and $a/h = 34/56 = 0.607 \approx 0.60$ as assumed in the preliminary design. For $a = 34$ in.; $(a/h) = 34/56 = 0.607$; $k = 4 + 5.34/(a/h)^2 = 4 + 5.34/(0.607)^2 = 18.49$;

$$C_v = 45{,}000\,h/[F_y\,(h/t_w)^2] = (45{,}000)(18.49)/[(36)(179.2)^2]$$
$$= 0.719;$$

and

$$F_v = F_y C_v/2.89 = (36)(0.719)/2.89$$
$$= 8.95 \text{ ksi (1.10.5.2 Eq. 1.10-1).}$$

F_v can also be obtained by interpolation from Fig. 3-11 for $(a/h) = 0.607$ and $(h/t_w) = 179.2$.

$$f_v = V/A_w = [160 - (2)(2.67)]/17.5$$
$$= 8.84 \text{ ksi} < F_v = 8.95 \text{ ksi; OK.}$$

Space intermediate stiffeners, 2 spaces @ 34 in. and 4 spaces @ 35 in. centers, from the first (end panel) stiffener (in lieu of 4 spaces @ 52 in. centers as shown in Fig. 3-22 for tension field action intermediate stiffeners).

Between interior concentrated loads $f_v = V/A_w = 20/17.5 = 1.14$ ksi. From Table 3-11 for $(h/t_w) = 179.2$ and $(a/h) = 240/56 > 3$, determine that intermediate stiffeners are not required because $F_v \approx 2.566 > f_v = 1.14$ ksi.

Intermediate Stiffener Size. Intermediate stiffeners must extend approximately to the edge of the flanges (1.10.5). The width/thickness ratio, $(w/t) \leqslant 95/\sqrt{F_y} = 95/\sqrt{36} = 15.83$ (1.9.1.2). For a $5\frac{1}{2}$ in. stiffener width, $t_{min} = w/15.83 = 5.5/15.83 = 0.374$. Try single stiffeners $5\frac{1}{2} \times \frac{9}{16}$ in. on alternate sides of the web. Check minimum area and minimum moment of inertia requirements (1.10.5.4).

Minimum area of intermediate stiffeners can be read from Fig. 3-9 or calculated (1.10.5.4; Eq. 1.10-3).

$$A_{st} = [(1 - C_v)/2] \; [(a/h) - (a/h)^2/\sqrt{1 + (a/h)^2}\,] \; YD\,ht.$$

For $Y = 1$; $D = 2.4$ for single plate stiffeners; $t = 0.3125$; $a = 34$ in.; $h = 56$ in.; $(a/h) = 34/56 = 0.607$; $(h/t_w) = 56/0.3125 = 179.2$; $k = 4 + 5.34/(a/h)^2 = 4 + 5.34/0.607 = 12.80$;

$$C_v = 45{,}000\, k/[F_y\, (h/t)^2] = (45{,}000)\,(12.80)/[(36)\,(179.2)^2]$$

$$= 0.498 \; (1.10.5.2)$$

$$A_{st} \geqslant \left[1 - \frac{0.498}{2}\right] \left[0.607 - \frac{(0.607)^2}{\sqrt{1 + (0.607)^2}}\right] (1)\,(2.4)\,(56)\,(0.3125)$$

$$\geqslant 3.08 \text{ in.}^2 \;\ldots\ldots\; (\text{Eq. } 1.10\text{-}3)$$

Use single plates $5\frac{1}{2} \times \frac{9}{16} = 3.09$ in.$^2 > 3.08$; OK.

Minimum moment of inertia of intermediate stiffeners can be read from Fig. 3-8 or calculated (1.10.5.4).

$$I_{(\text{Min})} = (h/50)^4 = (56/50)^4 = 1.574 \text{ in.}^4$$

Moment of inertia of single plate stiffeners about an axis parallel to the web,

$$(0.5625)\,(5.3)^3/12 + (3.09)\,(3.03)^2 = 36.2 \text{ in.}^4 > 1.574; \text{OK}.$$

Use $5\frac{1}{2} \times \frac{9}{16}$ in. single plate stiffeners on alternate sides of the web; attach to top flange; stop 2 in. above bottom flange (1.10.5.4).

Bearing Stiffener Requirements. Design a pair of bearing stiffeners under each 100 kip concentrated column load (1.10.5.1). Try 2–$5\frac{1}{2} \times \frac{9}{16}$ in. plates which are the same size as the intermediate stiffeners; $(w/t) = 5.5/0.5625 = 9.78 < 95\sqrt{F_y} = 15.83$ (1.9.1.2).

Check the axial load capacity of a pair of plate stiffeners acting with a part of the web as a column as shown in Fig. 3-22. The unbraced length of the stiffener web column can be taken equal to $\frac{3}{4}$ of the depth of the web (1.10.5.1). The stiffeners are braced in the direction parallel to the web. From Fig. 6-5 for $k = 1$ and $(kl/r) = (1)\,(3/4) \cdot (56)/3.30 = 12.72$ the allowable compressive stress $F_a = 21.01$ ksi. $f_a = P/A = 100/6.19 = 16.16$ ksi < 21.01 ksi; OK.

Check bearing stress on stiffener area beyond fillet weld connecting web to flange (1.10.5.1).

$$f_p = (100)/[(5.5\text{-}1.0)\,(0.5625)\,(2)] = 19.75 \text{ ksi.}$$

$$F_p = 0.90\, F_y = (0.9)\,(36) = 32.4 \text{ ksi} > 19.75 \text{ OK } (1.5.1.5.1).$$

Use $5\frac{1}{2} \times \frac{9}{16}$ in. bearing stiffeners in pairs under each concentrated load with a tight fit to top and bottom flanges.

Deflection.

$$\Delta_{D+L} = (5)\,(2/12)\,(720)^4/[(384)\,(29{,}000)\,(34{,}336)]$$

$$+ (100)\,(240)\,[(3)\,(720)^2 - (4)\,(240)^2]/[(24)\,(29{,}000)\,(34{,}336)]$$

$$= 0.585 + 1.329 = 1.914 = l/376$$

Δ_L ($L = D$ See Fig. 3-21) $= 1.914/2 = 0.957$ in. $< l/360 = 2.00$ in.; OK for plastered ceilings (1.13.1).

Calculated Weight.

Flanges; (1.5) (12) (2) (3.4) (60)		7,344
Web; (36) (0.3125) (3.4) (60)		3,570
Stiffeners;		772
Intermediate (5.5) (0.5625) (12) (3.4) (54.75)/12	= 576	
Bearing (5.5) (0.5625) (4) (3.4) (56)/12	= 196	
		Total = 11,686 lb.

The weight of this hybrid welded plate girder is 23 percent less than the similar tension field action girder of Design Example 3-9. The savings are in the weight of the flanges. Tension and braced compression elements are, of course, ideal for taking advantage of higher yield point grades of steel. In these examples the tension field action theory with its more complicated calculations resulted in no saving in weight, probably due to the proportions of girder depth and span.

BEAM EXAMPLE 3-11. COMPOSITE STEEL–CONCRETE BEAM WITHOUT TEMPORARY SHORING.

Problem. Design a simple span composite steel-concrete (Type 2 Construction) floor beam using A572 Grade 50 steel, normal weight concrete ($f_c' = 3500$ psi), using headed steel stud shear connectors. Temporary construction shores are not to be required, but deflection under construction dead load will be limited to $l/600$. The beam supports an open sales floor. Compare the cost of this design with that for a composite concrete encased beam.

Design Data. (Refer to Fig. 3-24.) 50 ft. span beams are spaced 16′-8″ on centers. Service dead and live floor loads are as follows:

5″ reinforced concrete slab	63 psf
Steel beam	8 psf
CONSTRUCTION DEAD LOAD (*CDL*)	71 psf
Ceiling and mechanical	10 psf
Dead Load (*D*)	81 psf
Live Load (*L*)	200 psf
DEAD LOAD + LIVE LOAD (*D* + *L*)	281 psf

The top flange of the beam is laterally supported by attached formwork during construction and by the slab after the concrete has hardened. Design moments at various load conditions are:

$$M_{CDL} = (0.071)(16.67)(50)^2/8 = 369.9 \text{ k-ft.}$$

$$M_{(f_c'=0.75\, f_c')} = (0.010 + 0.200)(16.67)(50)^2/8 = 1{,}094.0 \text{ k-ft.}$$

$$M_{(D+L)} = (0.281)(16.67)(50)^2/8 = 1{,}463.8 \text{ k-ft.}$$

$$M_L = (0.200)(16.67)(50)^2/8 = 1{,}041.9 \text{ k-ft.}$$

Preliminary Design.

1. Compute the elastic section modulus of the steel beam, S_s, required for construction loads;

$$S_s = M_{CDL}/(0.66F_y)$$

$$= \frac{(369.9)(12)}{(0.66)(50)} = 134.5 \text{ in.}^3$$

(Continued on page 177)

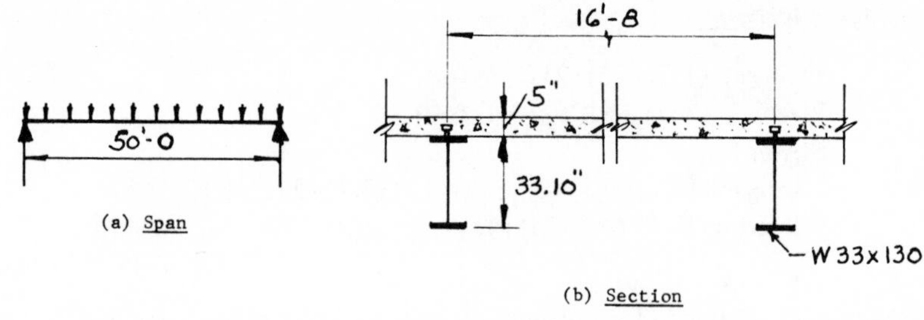

(a) Span

(b) Section

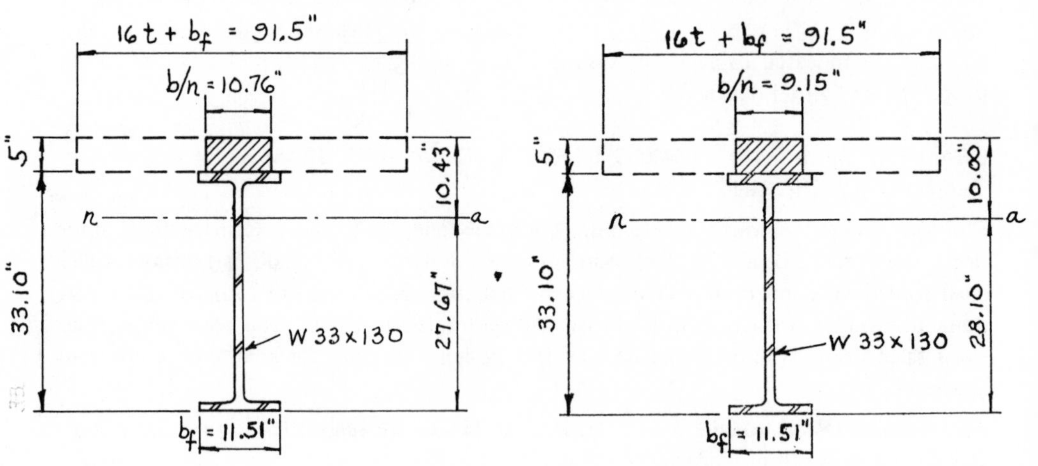

f'_c = (0.75)(3,500) = 2,625 psi

E_s = 29,000,000 psi

E_c 2,920,000 psi

$n = E_s/E_c = 10$

	A (in.)^2	y (in.)	Ay(in.)^3
W 33 x 130	38.3	16.55	633
Flange	45.7	35.6	1,627
Σ	84.0		2,360

W33 x 130 = 6,710

$(38.3)(11.55)^2$ = 5,109

$(9.15)(5)^3/12$ = 95

$(45.7)(7.50)^2$ = 2,571

I_{tr} = 14,485

(d) Construction Load Section Properties

f'_c = 3,500 psi

E_s = 29,000,000 psi

E_c = 3,372,000 psi

$n = E_s/E_c = 8.5$

	A (in.)^2	y (in.)	Ay (in.)^3
W 33 x 130	38.3	16.55	633
Flange	53.8	35.60	1,915
Σ	92.1		2,548

W 33 x 130 I = 6,710

$(38.3)(11.12)^2$ = 4,736

$(10.76)(5)^3/12$ = 112

$(53.8)(7.93)^2$ = 3,378

I_{tr} = 14,941 in.^{4}

S_{tr} = 14,941/27.67 = 540 in.^{3}

(c) Service Load Section Properties

Fig. 3-24 Simple Span Composite Steel-Concrete Beam Section Properties for Flexure—Example 3-11.

2. Compute the elastic section modulus of the transformed composite section, S_{tr}, required for total dead and live loads. Assume a compact steel section and complete composite action.

$$S_{tr} = M_{(D+L)}/(0.66F_y)$$
$$= \frac{(1463.8)(12)}{(0.66)(50)} = 532.3 \text{ in.}^3$$

3. Compute minimum depth of section. Minimum depth of steel section recommended to provide acceptable human response to walking vibrations; $d = l/20 = (50) \cdot (12)/20 = 30$ in. Minimum total depth (slab plus beam) recommended for fully stressed beams to minimize deflection problems; $d_c = l/16 = (50)(12)/16 = 37.5$ in. (Commentary 13.1). Try W33; $d = 33$ in. $> l/20 = 30$ in. OK; $d_c = 33 + 5 = 38$ in. $> l/16 = 37.5$ in. OK.

4. Select preliminary trial section from AISC table of properties for composite beams. These tables are based on $f_c' = 3000$ psi but can be used for preliminary design. Try W33 × 130 without bottom cover plate; $S_{tr} = 538$ in.$^3 > 532.3$ in.^{3} OK; $S_s = 406$ in.$^3 >$ 134.5 in.^{3} OK.

5. Check the trial section to determine that S_{tr} required is less than the maximum allowed for construction without temporary shores (1.11.2.2). From Fig. 3-24c $S_{tr} = 540$ in.^{3} for $f_c' = 3500$ psi.

$$540 \leqslant [1.35 + 0.35\, M_{(f_c = 0.75\, f_c')}/M_{CDL}]\, S_s \text{ ...... (Eq. 1.11-2)}$$
$$\leqslant [1.35 + (0.35)(1094.0)/369.9]\; 406$$
$$\leqslant 968 \text{ in.}^3 \text{ OK.}$$

Steel Stress. Compute the bending stress, f_b, in the bottom flange of the steel beam for a total load moment $M_{(D+L)} = 1463.8$ k-ft. Section properties of the composite steel-concrete beam are shown in Fig. 3-24c; section modulus of the transformed section referred to the bottom flange of the steel beam, $S_{tr} = 540$ in.^{3}

$$f_b = (1463.8)(12)/540$$
$$= 32.5 \text{ ksi} < 0.66F_y = (0.66)(50) = 33 \text{ ksi OK.}$$

Concrete Stress. Compute f_c for construction without temporary shores and for the moment due to loads applied after the concrete has obtained 75 percent of its design strength, $M_{(f_c = 0.75\, f_c')} = 1094.0$ k-ft.; $n = 10$; $y_t = 10.00$ in.; and $I_{tr} = 14{,}485$ in.^{4} (See Fig. 3-24d).

$$f_c = Mc/(nI_{tr})$$
$$= (1094.0)(12)(10.0)/[(10)(14{,}485)]$$
$$= 0.906 \text{ ksi} < 0.45(0.75)f_c' = 0.45(0.75)(3.5) = 1.181 \text{ ksi OK.}$$

Shear Connectors for Full Composite Action. Try $\frac{7}{8}$″ diameter × $3\frac{1}{2}$ in. steel headed stud shear connectors with diameter $d_s = 0.875$ and height $h = 3.5$ in.;

$$d_s = 0.875 \text{ in.} < 2.5\, t_f (1.11.4). \quad (2.5)(0.855) = 2.14 \text{ in. OK. any location}$$

Determine the allowable shear load, q, for each stud from Table 3-12. For $f_c' = 3.5$ ksi and $w = 145$ pcf, $q = 16.4$ kips.

Compute the number of studs (N_s) required between the location of maximum moment and zero moment, as the *smaller* of the amounts required to develop the total steel section in tension or the concrete flange area in compression:

W 33 X 130

$$N_s = A_s F_y/(2q) = (38.3)(50)/[(2)(16.4)]$$
$$= 58.4 \text{ say } 59 \text{ studs}$$

Concrete (5" X 91.5")

$$N_s = 0.85 f'_c A_c/(2q) = (0.85)(3.5)(5)(91.5)/[(2)(16.4)]$$
$$= 41.5 \text{ say } 42 \text{ studs (governs)}$$

Full composite action requires a total of (42) (2) = 84 = $\frac{7}{8}$" dia. X $3\frac{1}{2}$ in. headed steel studs equally spaced throughout the full length of the beam.

Shear Connectors for Partial Composite Action. Compute the total horizontal shear force, V_h, required for full composite action as the smaller of the amounts required to develop the total steel section in tension or the concrete flange area in compression;

W 33 X 130

$$V_h = A_s F_y/2 \ldots\ldots \text{(Eq. 1.11-4)}$$
$$= 38.3(50)/2 = 957.5 \text{ kips}$$

Concrete (5" X 91.5")

$$V_h = 0.85 f'_c A_c/2 \ldots\ldots \text{(Eq. 1.11.3)}$$
$$= (0.85)(3.5)(5)(91.5)/2 = 680.5 \text{ kips (governs)}$$

Using the smaller value for V_h from above, compute total horizontal shear force, V'_h, to be resisted by connectors for partial composite action;

$$V'_h = [(S_{eff} - S_s)/(S_{tr} - S_s)]\, V_h \ldots\ldots \text{(Eq. 1.11-1)}$$
$$= [(532.3 - 359)/(540 - 359)]\, 680.5$$
$$= (173.3/181)\, 680.5$$
$$= 651.6 \text{ kips} > \frac{0.85}{4} f'_c A_c/2 = \frac{680.5}{4} = 170.1 \ldots\ldots (1.11.4)$$

Compute the number of $\frac{7}{8}$" dia. X $3\frac{1}{2}$" steel headed stud shear connectors required between the location of maximum moment and zero moment for partial composite action;

$$N_s = V'_h/q = 651.6/16.4$$
$$= 39.7 \text{ say } 40 \text{ studs}$$

Partial composite action requires a total of (40)(2) = 80 - $\frac{7}{8}$" dia. X $3\frac{1}{2}$" headed studs equally spaced throughout the full length of the beam, which is two less than required for full composite action.

Use 80 $\frac{7}{8}$" X $3\frac{1}{2}$" headed studs spaced at 7.5 in. throughout length of beam.

Deflection. Check deflection for construction dead load;

$$\Delta_{(CDL)} = \frac{(5)(71)(16.67)(50)(600)^3}{(384)(29{,}000{,}000)(6710)}$$

$$= 0.855 \text{ in.} < l/600 = 1 \text{ in. OK.}$$

Check additional deflection for superimposed ceiling dead load ($D = 10$ psf) and live load ($L = 200$ psf) based on $I_{tr} = 14{,}941$ in.^{4} (See Fig. 3-24.)

$$\Delta_L = \frac{(5)(210)(16.67)(50)(600)^3}{(384)(29{,}000{,}000)(14{,}941)}$$

$$= 1.136 \text{ in.} < l/360 = 1.667 \text{ in. OK} \;\ldots\ldots (1.13.1)$$

Vibration. (See Fig. 3-13 and 3-25.) Compute frequency (f) for the composite beam for a dead load ($D = 81$ psf) using the moment of inertia of the transformed composite steel section based on a width of concrete slab equal to the beam spacing.

$$f = (0.742/L^2)\sqrt{E_s I_{tr}/w_d}$$

$$= [(0.742/(50)^2]\sqrt{(29{,}000{,}000)(17{,}339)/[(81)(16.67)]}$$

$$= 5.72 \text{ cycles per second}$$

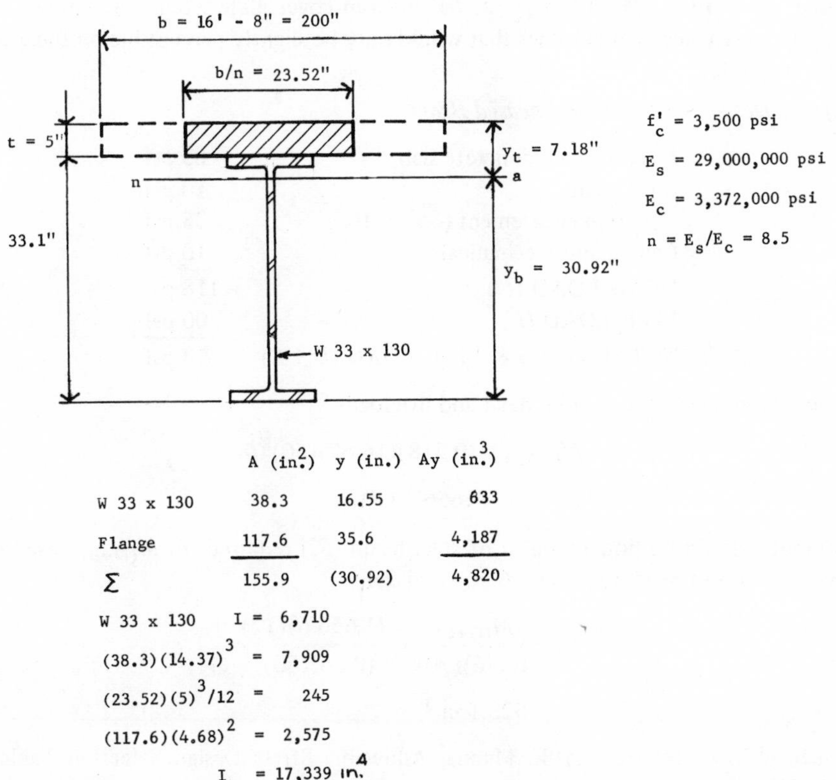

	A (in.^{2})	y (in.)	Ay (in.^{3})
W 33 x 130	38.3	16.55	633
Flange	117.6	35.6	4,187
Σ	155.9	(30.92)	4,820

W 33 x 130 I = 6,710

$(38.3)(14.37)^3$ = 7,909

$(23.52)(5)^3/12$ = 245

$(117.6)(4.68)^2$ = 2,575

I_{tr} = 17,339 in.^{4}

Fig. 3-25 Simple Span Composite Steel-Concrete Beam Section Properties for Vibration—Example 3-11.

Assume that the effective number of beams (N) affected by transient walking vibrations is equal to those within a width equal to one-half the length of the span. (See Fig. 3-13.)

$$N = (L/2)/(\text{Beam spacing in feet})$$
$$= (50/2)/16.67$$
$$= 1.5 \text{ beams affected}$$

Compute the displacement (Δ) of the composite steel beam.

$$\Delta = \frac{Kl^3}{48 E_s I_{tr} N} \text{ ...... (Fig. 3-13)}$$

where

$$K = L^2/D = 50^2/3.175 = 787$$

$$= \frac{(787)(600)^3}{(48)(29{,}000{,}000)(17{,}339)(1.5)} = 0.00469 \text{ in.}$$

To determine probable extent of "human response," enter Fig. 3-13 with a frequency of 5.72 cps and a displacement (amplitude) of 0.00469. The intersection of these values on the graph of human response to walking vibrations is in the lower area of marginal acceptance, and walking vibrations can be expected to be distinctly perceptible.

If acceptable values are required a deeper beam with a greater moment of inertia must be used. A W 36 X 135 with a $1\frac{1}{2}''$ X 10″ bottom cover plate would be required to produce completely acceptable values that would only be slightly perceptible on the graph of human response.

Alternate Design for Concrete Encased Beam

5″ reinforced concrete slab	63 psf
Steel beam	10 psf
Concrete encasement (35″ X 16″)	35 psf
Ceiling and mechanical	10 psf
DEAD LOAD (D)	118 psf
LIVE LOAD (L)	200 psf
TOTAL DEAD & LIVE LOAD ($D + L$)	318 psf

Compute moment due to total dead and live loads.

$$M_{(D+L)} = (0.318)(16.67)(50)^2/8$$
$$= 1656.6 \text{ k-ft.}$$

Compute elastic section modulus of steel beam (S_s) required for $M_{(D+L)}$ based on an allowable stress in bending $f_b = 0.76 F_y$ (1.11.2.1)

$$S_s = \frac{M_{(D+L)}}{(0.76)(50)} = \frac{(1{,}656.6)(12)}{(0.76)(50)}$$
$$= 523.1 \text{ in.}^3$$

Select W 36 X 160 from AISC Manual Allowable Stress Design Selection Table $S_s = 542 > 523.1$ in.^{3} OK.

Cost Comparison. Compare cost of composite concrete encased beam with cost of non-encased beam using headed steel stud shear connectors.

Non-Encased Beam

W 33 X 130: 130 lbs. @ \$0.30/lb.	= \$39.00
1.6 - $\frac{7}{8}''$ dia. X $3\frac{1}{2}''$ studs/ft. @ 1.00	= 1.60
Spray-on Fire Resistance, 8.67 SF @ \$0.75/SF	= 6.50
Total	= \$47.10/ft.

Encased Beam

W 36 X 160: 160 lbs. @ \$0.30/lb.	= \$48.00
Concrete stem 0.144 CY @ \$50/CY	= 7.20
Forms 7.16 SF @ \$2.25/SF	= 16.11
Welded Wire Fabric 817 lbs. @ \$0.30/lb.	= 2.61
Total	= \$73.92/ft.

The economy of composite construction using headed steel stud shear connectors is obvious.

REFERENCES CITED–BEAMS

1. "Economical Selection of Beams," W. P. Stewart; March 18, 1948; *Engineering News-Record.*
2. "Tentative Recommendations for the Design and Construction of Composite Beams and Girders for Buildings," Progress Report of Joint ASCE-ACI Committee on Composite Construction; *Structural Journal*, Dec. 1960; ASCE.
3. "Reinforcement Requirements for Girder Web Openings," E. P. Segner, Jr.: **90**, ST3; June 1964; ASCE.
4. "Flexural Strength of Steel-Concrete Composite Beams," R. G. Slutter and G. C. Driscoll; **91**, ST2; April, 1965; ASCE.
5. "Ponding of Two-Way Roof Systems," F. J. Marino; July, 1966; **3**, No. 3, *Engineering Journal*; AISC.
6. "Design of Beams with Web Openings," United States Steel, ADUSS 27-3500-02, April 1968.
7. "Shear Strength of Stud Connectors in Lightweight and Normal Weight Concrete," J. G. Ollgaard, R. G. Slutter, and J. W. Fisher; April, 1971; **8**, No. 2, *Engineering Journal*, AISC.
8. "Recommended Design Procedures for Beams with Web Openings," J. E. Power; Oct. 1971; **8**, No. 4; *Engineering Journal*; AISC.
9. "Simplified Plastic Analysis for Reinforced Web Holes," R. G. Redwood; October 1971; **8**, No. 4; *Engineering Journal*; AISC.
10. "Tables for Plastic Design of Beams with Rectangular Holes," R. G. Redwood; Jan. 1972; **9**, No. 1; *Engineering Journal*; AISC.
11. "Moment-Rotation Curves for Locally Buckling Beams," J. J. Climenhaga and R. P. Johnson; **99**, ST6; June 1972; ASCE.
12. *Structural Lightweight Concrete Design*, 1974; supplement to *CRSI Handbook*; CRSI.
13. "Human Perception of Transient Vibrations," J. F. Wiss and R. A. Parmalee; **100**, ST4; April, 1974; ASCE.
14. "Failure of Simply Supported Flat Roofs by Ponding of Rain," James Chinn; April 1965; No. 2; Engineering Journal; AISC.

SELECTED ADDITIONAL REFERENCES–BEAMS

1. "Failure of Simply-Supported Flat Roofs by Ponding of Rain," J. Chinn; April 1965; **2**, No. 2, *Engineering Journal*; AISC.
2. "Vibration of Steel Joist-Concrete Slab Floors," K. H. Lenzen; July, 1966; **3**, No. 3, *Engineering Journal*, AISC.

3. "An Investigation of the Effective Slab Width for Composite Construction," T. A. Hagood, Jr., L. Guthrie, and P. G. Hoadley; Jan. 1968; **5**, No. 1, *Engineering Journal*; AISC.
4. "Design of Unstiffened Girders," A. L. Abolitz; Jan. 1969; **6**, No. 1; *Engineering Journal*; AISC.
5. "A Unified Approach to the Elastic Lateral Buckling of Beams," D. A. Nethercot and K. C. Rockey; July 1972; **9**, No. 4; *Engineering Journal*; AISC.
6. "Fast Check for Ponding," L. B. Burgett; Jan. 1973; **10**, No. 1; *Engineering Journal*; AISC.
7. "Welded Connections under Combined Shear and Moment," J. L. Dawe and G. L. Kulak; **100**, ST4; April, 1974; ASCE.

4

JOISTS AND TRUSSES

JOISTS

General

Steel joists are possibly the most commonly used flexural member in steel building construction. Joists are steel trusses, standardized in accordance with two specifications:

(1) "Standard Specifications for Open Web Steel Joists, J-Series and H-Series;" *SJI and AISC*, Oct. 1, 1974; and

(2) "Standard Specification for Longspan Steel Joists, LJ-Series and LH-Series, and Deep Longspan Steel Joists, DLJ-Series and DLH-Series;" *SJI and AISC*; Oct. 1, 1974.

Standard Steel Joists

In ordinary applications, the Engineer* does not design steel joists, but merely selects the standard type and size required for the job load and span. Designation of a standard type and size consists simply of a number (nominal depth in inches), the letter or letters representing the series type, and the number representing design load capacity within the series. The letter "J" represents the use of material with yield strength, $F_y = 36$ ksi throughout; the letter "H", $F_y = 50$ ksi for the chords, and, at the supplier's option, $F_y = 36$ or 50 ksi for the web sections. For example, "20H6" is a 20 in. deep joist, $F_y = 50$ ksi (for deflection) and load capacity such that the maximum resisting moment = 406 kip-in. and the end reaction capacity = 5100 lbs.

Standard Limitations

The following limitations, restrictions, and conditions of usage comprise the essential factors permitting standardization of steel joists as a special class in the general category of steel trusses:

1. Chords are essentially parallel. (In the Longspan and the Deep Longspan series the top chord is provided with a standard slope of $\frac{1}{8}$ in./ft. if desired and the nominal depth is then that at the midspan.) For the J- and H-series, camber may be provided at the

*The practicing Engineer responsible for the structural design of a particular construction project using joists.

manufacturer's option; for other series, it is standard. (J- and H-parallel chord, 2, 4.7; LJ-, LH-, DLJ-, DLH-103.4c, 103.6, 104.3.)

2. Allowable tensile stresses are 22,000 psi and 30,000 psi for F_y = 36 ksi and 50 ksi respectively (J- and H-4.2; LJ-, LH-, DLJ-, and DLH-103.2).

3. Joists are designed for gravity loads as simple span trusses (J- and H-4.1; LJ-, LH-, DLJ-, and DLH-103.1).

4. Joists are designed for uniform loadings applied to the top chords (J- and H-4.1; LJ-, LH-, DLJ-, and DLH-103.1).

5. For design the top chord is considered to be fully braced against lateral movement under design loads by the floor or roof deck construction (J- and H-4.1; LJ-, LH-, DLJ-, and DLH-103.1).

6. Joist construction includes minimum standard bridging and end anchorages. The bridging, anchorage of bridging, and end anchorages are required to be installed before the application of construction loads. The standard bridging of specified types, with minimum sizes, and within maximum spacing limits is designed to provide temporary bracing of the top chord and to hold the member in its proper position against lateral movement (J- and H-5.4 and 5.5; LJ-, LH-, DLJ-, and DLH-104.5, and 104.6).

7. End connection to steel columns for lateral stability is required (J- and H-5.6b; LJ-, LH-, DLJ-, and DLH-104.7b). These connections may be designed and specified as special Type 2 construction for resisting wind load only (AISC-1.2).

8. Joists may be utilized as either primary or secondary members. "Primary" in the structural sense of forming a part of the primary structural system resisting lateral displacement. Secondary if used only as flexural members supporting gravity loads between members of a primary bracing system.

9. Joists in the J- and H-Series and the LJ- and LH-Series may be used for the direct support of either floors or roofs in buildings. Joists in the DLJ- and DLH-Series are suitable for the direct support of roofs only (J-, H-, 2; LJ-, LH-, DLJ-, and DLH-101).

10. Joists are to be specified by the Engineer as manufactured in accordance with the Standard Specifications and according to an SJI approved design.

In practice where the above ten limitations are applicable, the Engineer simply indicates the designation for the standard joist required, and particular job conditions for the exact span, end anchorage, anchorage condition for the lines of bridging and type required, connections to columns, type of floor or roof construction, slope of joist, or pitch of top chords, etc. The requirement that the manufacture be in accordance with AISC Specifications and the design approved by SJI is usually included in the job specifications. The Engineer may also specify submission of design calculations for his approval.

Span, Depth, and Load Limitations

The limitations of span, depth, and load for standard joist designs are given in the AISC Manual.

Special Conditions or Requirements

There are frequently special job requirements or conditions not possible to standardize or to accommodate by standardized designs. The Specifications for the use of joists in these cases require that the design features not covered by joist specifications should conform to the AISC *Specification for the Design, Fabrication, and Erection of Structural Steel for Buildings* or the *Specification for the Design of Cold-Formed Steel Structural Members* (J-, H-, 4.1; and LJ-, LH-, DLJ-, and DLH-103.1). Even for these special conditions, the Engineer is not usually required to design steel joists. Since the different manufacturers produce standard joists using different materials (hot-rolled or cold-formed; webs designed for F_y = 36 or 50 ksi for H-joists), different sections, and connection details, a single design might restrict competitive bidding. Common practice for maximum efficiency in design is to indicate the special conditions, such as headers, concentrated loads, unusual camber or top chord pitch, etc., and to require that the bidders submit designs for approval (1.1.1; 1.1.2).

Designation for Special Requirements

In designating these special requirements, the Engineer need only show the dimensions, locations, and any special features, such as the profile for a variable depth cantilever. (See Fig. 4-1(a).) For suspended concentrated loads such as heavy piping, light conveyors, etc. the load should be stated and identified as dead, live, or impact, and located in plan so that the supplier can lay out the web system for the special design required. (See Fig. 4-1(b).) "Pitch" is a term referring to the deviation from horizontal for the top chord only, usually for roof joists. For all standard joists (LJ-, LH-, DLJ-, and DLH-) maximum pitch is $\frac{1}{8}$ in. per foot. Greater pitch can of course be supplied, but requires special design. The Engineer must indicate the amount and direction of pitch, and whether one-way or two-way. "Slope" is the deviation from horizontal for an entire parallel chord joist. Maximum slope is $\frac{1}{2}$ in. per foot for all standard joists. A greater slope will require special design. (See Fig. 4-1(c).) See text, Standard Load Tables.

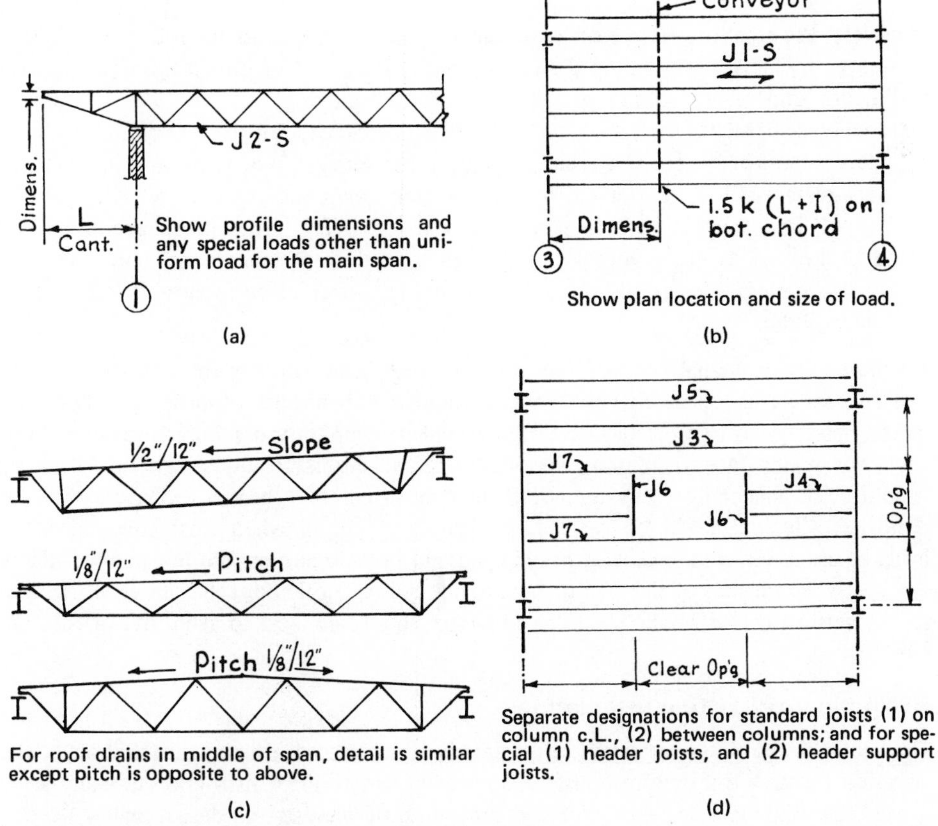

Fig. 4-1 Designation of Special Conditions for Joists. (a) Cantilever Spans. (b) Concentrated Loads on Bottom Chord. (c) Slope and Camber. (c) Headers and Concentrated Loads Resulting.

The usual condition requiring headers is an opening. The headers should be shown as solid beams or special joists. The header support is usually provided by adjacent full span joists, and the concentrated load from the header requires special design of these joists. The Engineer must show location in plan and the size of the opening so that the special support joist may be laid out for the concentrated load. If some line loads or concentrated point loads must be supported by the header in addition to the interrupted joist(s) such as a wall, skylight frame, etc., the type, amount, and location or distribution of such loads must be shown. (See Fig. 4-1(d).)

Show profile dimensions and any special loads other than uniform load for the main span.

(*a*) *Cantilever Spans*. Fig. 4-1.

Show plan location and size of load.

(*b*) *Concentrated Loads on Bottom Chord*. Fig. 4-1.

For roof drains in middle of span, detail is similar except pitch is opposite to above.

(*c*) *Slope and Pitch–Standard Joists*. Fig. 4-1.

Separate designations for standard joists (1) on column c.L., (2) between columns; and for special (1) header joists, (2) header support joists.

(*d*) *Headers and Concentrated Loads Resulting*. Fig. 4-1.

Inspection of Joist Materials and Erection

Materials. Pre-erection inspection requirements upon delivery to the job site are rather limited (1.1.2). Approved shop drawings are the main tool for this inspection. Length and depth dimensions should match those on the drawings, and the joists should be reasonably straight. Mill tests on the material and affidavits for the fabrication from the suppliers should be on file. A visual inspection for damage in handling and for evidence of inadequate welds is desirable. If random, infrequent inadequacies in the welds are discovered, field correction before erection will save delay. Such inadequacies may include broken welds, poor appearance of welds, welds too small or too short, etc. If systematic inadequacies are discovered too extensive for quick correction, field load tests may be required.

Erection. The principal erection features requiring inspection are the field connections and positioning of joists. Specified bridging and end anchorages of joists should be completed before construction loads are applied unless temporary bracing is provided. Bottom chord extensions to columns, masonry anchors, all bolts in end anchors, and bridging should be in place before any temporary bracing of columns, beams, or masonry walls is removed. Joists should be parallel, straight, and plumb when final connections of bridging are completed; misfit of bridging should be field corrected if not plumb. Where any special conditions of the joist design are not symmetrical about the center of the span or the vertical center of the joist, the inspector should be alert to avoid mis-orientation end-for-end.

Evaluation of Existing Installations

Occasionally, an Engineer may be required to review an existing installation. The occasion for such a review may arise at any time from during construction to many years later. (See Example 1 for the detailed procedure of analysis and design review.) Such a review is simple, but the determination of existing conditions-member sizes, grade, adequacy of welds, bracing, bridging, value of connections, etc. may require much investigation. Guided by results of a thorough review based upon the best field data available to select an appropriate load, the final assessment of existing load capacity is best determined by a load test when the theoretical studies show probable doubt of adequate capacity. When the purpose of the evaluation is to accommodate some special concentrated load like a heavy pipe, light conveyor, skylight opening, etc. and member sizes and conditions are adequate, only the connections are in question. Check design if available or use field inspection to check dimensions of welds for connections of web members where specified arbitrary minimum shear controls design. (See Table 4-4.) Note that a number of members would be capable of resisting larger load, particularly for load reversal from tension to compression, and also if welds were increased to suit the new load conditions.

Typical Design

A typical manufacturer's computer-prepared design which would be submitted for the review and approval of the Engineer follows. The example includes material used, yield point specified, and design data (furnished by the Engineer on the structural drawings and in the job specifications), layout of the joist profile, stress analysis, sections of all members, and details of the connections (welding) . . . . . . (1.1.2).

EXAMPLE 1.

The Engineer has specified *48 LH 17* steel joists meeting AISC Specifications for a clear span of 84′-4″ to support a roof with a total load, $w = 646$ plf. The Engineer's details show 4 in. bearing at each end, underslung, parallel chord joists. See Fig. 4-2.

The following (computer-prepared) set of calculations for design is typical of those which might be supplied by the joist supplier for the approval of the Engineer.*

JOIST 48 LH 17

Overall length = 85.000 ft.; working length = 85.000 ft. Right end–underslung; bearing length = 4.00 in.; bearing depth = 5.00 in. Left end–underslung; bearing length = 4.00 in.; bearing depth = 5.00 in. Panel lengths: left end = 78.00 in.; 9 panels @ 96.00 in.; right end = 78.00 in. Chords parallel; nominal depth = 48.5 in.; $w = 646$ plf. End reactions, $R = 0.5 \times 85 \times 646 = 27{,}455$ lbs.

This design in accordance with Standard Specifications for Longspan Steel Joists, LJ- and LH-Series, adopted by the American Institute of Steel Construction, Inc., and the Steel Joist Institute, Oct. 1, 1974.

$F_y = 50{,}000$ psi; $F_t = 30{,}000$ psi; for locations and magnitudes of chord forces used in the chord section, see "Stress Analysis," Table 4-2. See also Fig. 4-3.

Top chord Q-factor = 1.000; at nominal depth, $I = 6{,}563$ in.^{4} Approximate uniform load at deflection = $L/360$,

$$w = EI/(1.15 \times 675 \times L^3) = 399 \text{ plf}^{**}$$

Maximum spacing between rows of bridging = 170 × 1.688/12 = 23.91 ft. (103.4 - a)
> 21.0 ft. (104.5 - b)

*Courtesy of the CECO CORPORATION, Chicago, Illinois.

**Note that deflection computed as for a solid web beam with $I = 6{,}563$ in.^{4} has been increased by 15 per cent (factor 1.15) to allow for the truss system web effect. The standard Load Tables provided in 1972 showed a live load, $w = 403$ at $L/360$ deflection agreeing closely with the value here computed. In 1974, this value was *reduced* to $w = 346$ corresponding to an increased factor of 1.34.

TABLE 4-1 Chord Sections and Properties

Member	Nominal Section	Area (sq. in.)	$\bar{y}$ (in.)	c (in.)	r_{x-x} (in.)	I_{x-x} (in.^{4})	I_{y-y} (in.^{4})	r_{y-y} (in.)
Top chord	2s 4 × 4 × $\frac{7}{16}$	6.610	1.161	2.839	1.226	9.938	18.844	1.688
Bot chord	2s 4 × 4 × $\frac{3}{8}$	5.719	1.138			8.717		

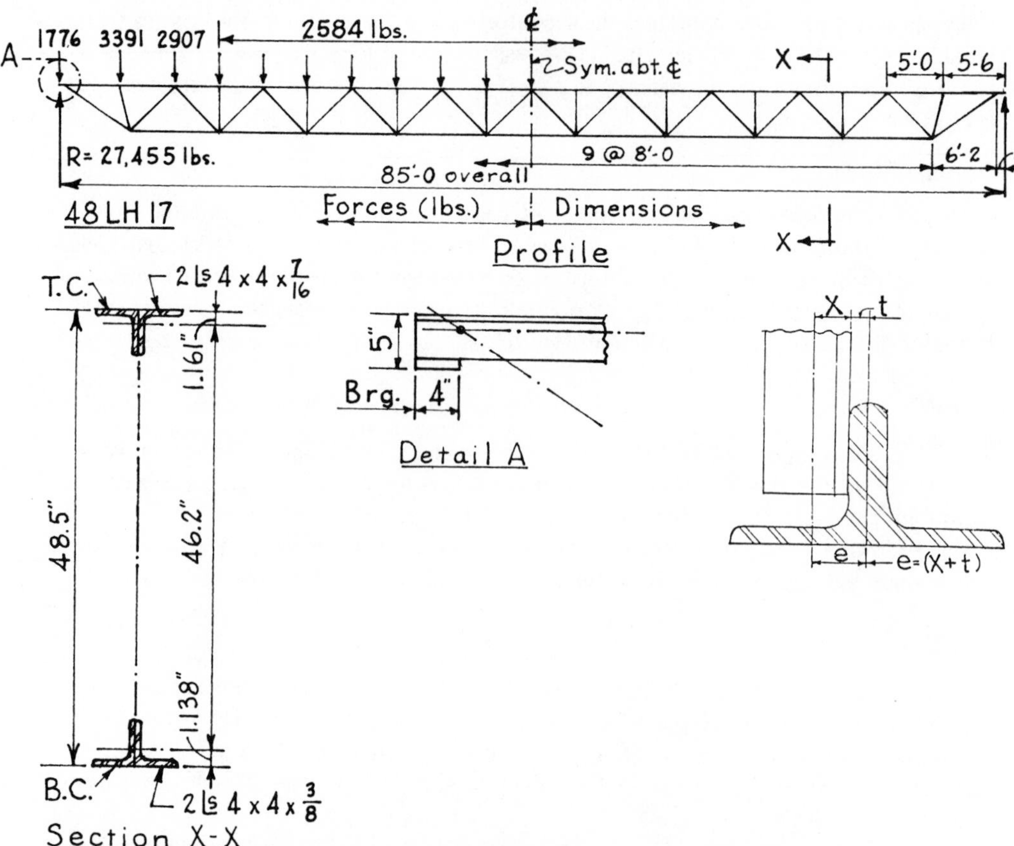

Fig. 4-2 Joist Design Review Example—Profile, Loads, and Dimensions.

Bot. chord:

Maximum force = 151,533 lbs.

Maximum stress = 151,533/5.719 = 26,496 psi

Top chord–interior panel:

Maximum force = 150,191 lbs.;

Maximum stress = 150,191/6.610 = −22,722 psi; $L/r = 39.15$;

$F_a = 25{,}941$ psi; $F_e' = 97{,}230$ psi

At panel point, $M = wL^2/12 = 10{,}336$ lb.-in.

$$f_b = Mc/I_{xx} = (10{,}336)(2.839)/9.938 = 2{,}953 \text{ psi}$$

$$f_a + f_b = 22{,}722 + 2{,}953 = 25{,}675 < 25{,}941 \text{ psi}$$

At mid-panel, $M = wL^2/24 = 5{,}168$ lb.-in.

$$f_b = M\bar{y}/I_{xx} = 604 \text{ psi}; \; C_m = 1 - 0.4 f_a/F_e' = 0.907$$

$$\frac{f_a}{F_a} + \frac{C_m f_b}{F_b(1 - f_a/F_e')} = 0.876 + 0.024 = 0.900 < 1.00$$

Top chord–end panel:

Maximum force = 43,352 lbs.; $f_a = 43{,}352/6.610 = 6{,}559$ psi

The bending moment equations are based upon the top chord as a continuous member. $L/r = 53.83$; $F_a = 23{,}730$ psi; $F_e' = 51{,}427$ psi.

At panel point, $M = wL^2/9.436 = 24,851$ lb.-in.*

$$f_b = Mc/I_{xx} = 7{,}100 \text{ psi}$$

$$f_a + f_b = 6{,}559 + 7{,}100 = 13{,}659 \text{ psi} < 23{,}730 \text{ psi}$$

At mid-panel, $M = wL^2/12.084 = 19{,}406$ lb.-in.*

$$f_b = M\bar{y}/I_{xx} = 2{,}266 \text{ psi}; C_m = 1 - 0.3 f_a/F_e' = 0.962$$

$$\frac{f_a}{F_a} + \frac{C_m f_b}{F_b(1 - f_a/F_e')} = 0.276 + 0.083 = 0.360 < 1.00$$

STRESS ANALYSIS AS A PIN-CONNECTED TRUSS

For loads and reactions, see Fig. 4-2. The designation of web members and panel points is shown in Fig. 4-3. The complete stress analysis and results are shown in Table 4-2, for full specified uniform loading. It will be noted that for any uniform loading (shear equal zero at mid-span) the stress in diagonal web members at or near the center is very low. If designed only for the full specified uniform loading, these members could be overloaded by partial span loadings. To provide for partial span loading many designers routinely perform a second stress analysis for trusses and girders with one-half span fully loaded and the other half with dead load only. The dead load of the joists and the bridging alone is usually small compared to the total design load, and so the critical partial loading is most probable during construction before any other dead load is applied. The AISC Specifications provide for such a partial load, conservatively neglecting the dead load of the joists and bridging, by requiring that the vertical shear used for design of web members in the longspan joists shall not be less than 25 percent of the (total) end reaction under full specified loading (LJ-, LH-, DLJ-, and DLH-series 103.4b); in the open web joists, 50 percent of the maximum end reaction (J- and H-series 4.4b). In Table 4-2 it will be noted that the minimum shear requirement controlled design for web members C4, T5, and C5.

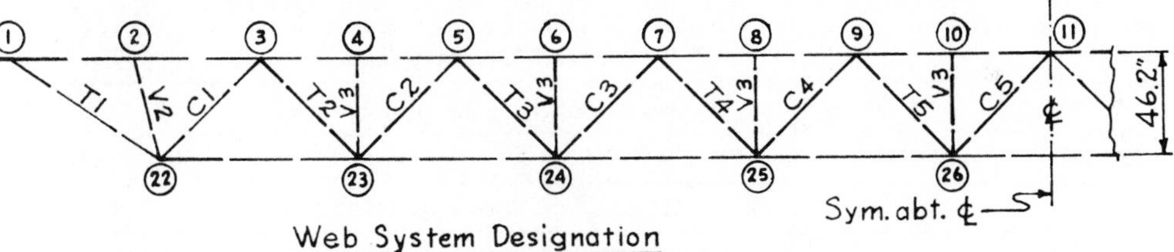

Fig. 4-3 Joist Example—Web System Designation.

SELECTION OF SECTIONS FOR WEB MEMBERS

The sections, section properties, and effective design lengths used are shown in Table 4-3. The results of investigation of these sections by the design formulas prescribed in the Standard Specifications for Longspan Steel Joists are summarized in Table 4-4.

*These moment coefficients reproduce computer analysis moments.

TABLE 4-2 Stress Analysis as a Pin-Connected Truss for Full Load.

Member	Web lengths (in.)	Forces in Members (lbs.) (+) Tension (-) Compression	
T1	90.66	(90.66/46.2)(27,455 - 1776)	= + 50,386
TC	1 to 2	(78/90.66)(50,386)	=– 43,352
V2	47.73	(47.73/46.2)(3,391)	= - 3,504
TC	2 to 3	-43,352 + (12/46.2)(3,391)	= - 42,471
C1	66.62	(-27.455 + 1,776 + 3,391)(66.62/46.2)	= - 32,138
BC	22-23	(43,352) - (12/46.2)(3,391) + (32,138)(48/66.62)	= + 65,626
T2	66.62	[(32,138)(46.2/66.62) - 2,907](66.62/46.2)	= + 27,946
V3	46.2		= - 2,584
C2	66.62	[(-27,946)(46.2/66.62) + 2,584] (66.62/46.2)	= - 24,220
BC	23-24	65,626 + (27,946 + 24,220)(48/66.62)	= +103,211
TC	3 to 5	-42,471 - (32,138 + 27,946)(48/66.662)	= - 85,761
T3	66.62	[(24,220)(46.2/66.62) - 2,584] (66.62/46.2)	= + 20,494
TC	5 to 7	-85,761 - (24,220 + 20,494)(48/66.62)	= -117,976
C3	66.62	[(-20,494)(46.2/66.62) + (2,584)] (66.62/46.2)	= - 16,768
BC	24-25	103,211 + (20,494 + 16,768)(48/66.62)	= +130,057
T4	66.62	[(16.768)(46.2/66.62) - 2,584] (66.62/46.2)	= + 13,041
TC	7 to 9	-117,976 - (16,768 + 13,041)(48/66.62)	= -139,453
C4	66.62	[-(13,041)(46.2/66.62) + 2,584] (66.62/46.2)	= - 9,315*
BC	25-26	130,057 + (13,041 + 9,315)(48/66.62)	= +146,164
T5	66.62	[(9,315)(46.2/66.62) - 2,584] (66.62/46.2)	= + 5,589*
C5	-66.62	[(-5,589)(46.2/66.62) + 2,584] (66.62/46.2)	= - 1,863*
TC	9 to 11	-139,453 - (9,315 + 5,589)(48/66.62)	= -150,191
BC	26-27	146,164 + (5,589 + 1,863)(48/66.62)	= +151,533

*Minimum force = (0.25 × 27,455)(66.62/46.2) = 9,898 lbs.

TABLE 4-3 Web Members—Sections Used and Properties

Member	Equal Leg Angles		Area (sq. in.)	r (in.)	$\bar{x}$ (in.)	I (in.^{4})	Design l(in.)
	No.	Leg × Thickness					
T1	2 ⌊s	2.00 × 0.231	1.741	0.611	0.585	0.65	90.66
V2	1 ⌊	1.75 × 0.165	0.550	0.540	0.498	0.16	47.73
C1	2 ⌊s	2.50 × 0.228	2.176	0.773	0.709	1.30	66.62
T2	2 ⌊s	1.75 × 0.142	0.954	0.543	0.490	0.28	66.62
V3	1 ⌊	1.75 × 0.142	0.477	0.543	0.490	0.14	46.20
C2	2 ⌊s	2.50 × 0.228	2.176	0.773	0.709	1.30	66.62
T3	1 ⌊	2.00 × 0.182	0.695	0.394	0.567	0.27	66.62
C3	2 ⌊s	2.00 × 0.182	1.390	0.618	0.567	0.53	66.62
T4	1 ⌊	1.75 × 0.142	0.477	0.346	0.490	0.14	66.62
C4	1 ⌊	2.50 × 0.304	1.428	0.762	0.737	0.83	66.62
T5	1 ⌊	1.50 × 0.138	0.395	0.295	0.426	0.08	66.62
C5	1 ⌊	2.50 × 0.304	1.428	0.762	0.737	0.83	66.62

Approvals

The problem of reviewing computer designs has become routine.* It is, of course, not limited to steel joist designs submitted by suppliers. The principle that a computer will perform repetitions endlessly of the same routine is well known, and saves much manual review. When the program used has been previously reviewed completely, further review can be limited to check that the input used was correctly taken off the job specifications and drawings. Thus, design calculations prepared by the same supplier for a number of similar joist designs need be checked only for the correct load, span, F_y, etc. after one design using the same program has been fully reviewed and approved.

Chords. In this example, Table 4-2 has been organized as the designer's review, and would not be repeated. All stresses computed therein were within one pound of the computer results, so that this portion of the program can be considered fully checked. The computer output for chord member analysis and design showed complete calculations and is easily checked.

Computer data.

T.C.–2*Ls*4 X 4 X 0.437 X 85′-00; bearing weld 5″; panel point welds 3″.

B.C.–2*Ls*4 X 4 X 0.375 X 73′-00; bearing weld 5″; panel point welds 5″.

The chords can be considered braced at the panel points as double-angle members, and to the ends of the panel point welds as single angle members, unbraced between the welds (LJ-, LH-, DLJ-, and DLH-103.4e). Note also that fillers (or welds) are required at the middle of each top chord interior panel points (maximum spacing = 24 in.) and in all double angle web members in compression (for joists more than 28 in. deep)

T.C. ⅃L Interior panel; $l = 48''$; $l/r = 39.15$

L Interior panel; $l = 48 - 3 = 45''$

⅃L End panel; $l = 66''$; $l/r = 53.83$

L End panel; $l = 66 - 3 = 63''$

Webs. The web members in tension, (+) forces, are easily checked in one operation.

$$f_a = P/A \leqslant 30{,}000 \text{ psi}$$

The double angle compression members are also easily checked since the eccentricity can be neglected when less than the distance from the c.g. of the chord to the top of the chord (LJ-, LH-, DLJ-, DLH-103.5 e). Thus, as shown, the check consists of

$$f_a/F_a \leqslant 1.00$$

where F_a is based upon the l/r, using l as the panel-point-to-panel-point length (LJ-, LH-, DLJ-, and DLH- 103.3).

The eccentricity of the single angle tension members can also be neglected. For single angle compression members, the eccentricity e = the sum of the thickness of the top or bottom chord angle (whichever is the larger) plus the distance x for the web angle. (See sketch, Fig. 4-2.) Check member $V2$:

$$e = 0.437'' + 0.498'' = 0.935''$$
$$M = (-3{,}504)(0.935) = 3{,}276 \text{ lb.-in.}$$
$$f_b = Mc/I = (3{,}276)(0.498)/0.16$$
$$= 10{,}197 \text{ psi} \approx 10{,}178 \text{ psi OK}$$

*Cited Reference (1).

TABLE 4-4 Web Members—Design Investigation

Member	Design Axial Force (lbs.)	l/r	f_a (psi)	F_a (psi)	f_b (psi)	F'_e (psi)	$\frac{f_a}{F_a} + \frac{C_m f_b}{\left(1 + \frac{f_a}{F'_e}\right) 30}$	= total 1.000	Required Weld Length (in.)‡
T1	50,386	148.27	28,936	30,000	–	–	0.000 + 0.000	–	7.34
V2	– 3,504	88.39	– 6,368	–17,259	10,178	–19,070	0.369 + 0.509	0.878	1.43
C1	–32,138	86.24	–14,769	–17,656	–	–	0.000 + 0.000	–	4.75
T2	27,946	122.64	29,304	30,000	–	–	0.000 + 0.000	–	6.63
V3	– 2,584	85.05	– 5,419	–17,491	8,341	–20,601	0.310 + 0.397	0.707	1.23
C2	–24,220	86.24	–11,130	–17,656	–	–	0.000 + 0.000	–	3.58
T3	20,494	169.06	29,493	30,000	–	–	0.000 + 0.000	–	7.58
C3	–16,768	107.79	–12,065	–12,825	–	–	0.000 + 0.000	–	3.10
T4	13,041	192.68	27,350	30,000	–	–	0.000 + 0.000	–	6.18
C4	– 9,898*	87.40	– 6,933	–17,468	10,314	–19,504	0.397 + 0.533	0.930	2.67
T5	9,898*	225.47	25,060	30,000	–	–	0.000 + 0.000	–	4.83
C5	– 9,898*	87.40	– 6,933	–17,468	10,314	–19,504	0.397 + 0.533	0.930	2.67

‡Required to develop the design forces tabulated at 21,000 psi (103.2 d).

*Based upon 25 percent of the end reaction (103.4 b).

TABLE 4-5 Review of Specified Weld Lengths for Web Member Connections

Member	Axial Force (lbs.)	(0.5) (0.6) F_yA (lbs.)	(0.5) F_aA (lbs.)	Axial Force 2 (Eq. 1.6-1a)	Strength of Weld (lbs./in.)	0.3 F_yA	Required length of Weld total per member, (in.) 0.5 F_aA	by M'f'r'	0.5 × Force (Eq. 1.6-1a)
T1	+50,386	26,115	–	–	3,400	14.7	–	14.68	–
V2	– 3,504	8,250	4,750	1,995	2,450	3.4	1.94	1.43	0.81
C1	–32,138	32,640	18,775	–	3,385	9.6	9.5	9.5	–
T2	+27,946	14,310	–	–	2,110	13.3	–	13.3	–
V3	– 2,584	7,155	4,170	1,827	2,110	3.4	1.94	1.23	0.87
C2	–24,220	14,310	8,420	–	3,385	7.2	7.2	7.16	–
T3	+20,494	10,425	–	–	2,702	7.6	–	7.58	–
V3	– 2,584	7,155	4,170	1,827	2,110	3.4	1.94	1.23	0.87
C3	–16,768	20,850	8,915	–	2,702	7.8	6.2	6.2	–
T4	+13,041	7,155	–	–	2,110	6.2	–	6.18	–
V3	– 2,584	7,155	4,170	1,827	2,110	3.4	1.94	1.23	0.87
C4	– 9,898*	21,420	12,600	5,322	4,515	4.8	2.80	2.67	1.20
T5	+ 9,898*	5,925	–	–	2,050	4.8	–	4.83	–
V3	– 2,584	7,155	4,170	1,827	2,110	3.5	1.94	1.23	0.87
C5	– 9,898*	21,420	12,600	5,322	4,515	4.8	2.80	2.67	1.20

*Based upon 25 percent of the total end reaction (103.4-b).

Connections. The computer-prepared design, Table 4-4, includes a tabulation of the weld lengths required. The tabulated numbers are the lengths in inches required at each end of each angle (for double angle members multiply by two). The lengths shown are based upon the axial forces from the stress analysis or specified minimum shears and allowable weld stresses of 21,000 psi (LJ-, LH-, DLJ-, and DLH- 103.2 d).

Since there are several arbitrary but conservative Specification requirements concerning connections, differing for openweb and longspan steel joists, it is prudent to review the proposed connection welds to ensure that the programmer's interpretation is satisfactory to the reviewer. In this example, *48 LH 17*, the minimum vertical shear for design of the web system is 25 percent of the total (full load) end reaction (LJ-, LH-, DLJ-, and DLH- 103.4 b). Joint connections must develop the design stress, but not less than 50 percent of the "allowable strength of the member" (LJ-, LH-, DLJ-, and DLH- 103.5).

The ambiguous phrase "allowable strength of the member" can be justifiably interpreted in a descending order of conservatism as:

(a) the *cross-sectional* capacity at allowable stress; for tension or compression, equal to ($0.60\, F_y A$);
(b) the *member* capacity for type of axial stress developed under full uniform load considering the unbraced length; for tension, the net section at allowable stress, ($0.60\, F_y A$); for compression, ($F_a A$);
(c) the *member* capacity for the type of axial stress developed under full uniform load considering the unbraced length and the eccentricity, if any. For single-angle compression members with eccentricity in bending out of the plane of the joist (see Fig. 4-2), this capacity will be the design load multiplied by the reciprocal of the solution to AISC Eq. (1.6-1a) using the design load. (See Table 4-4, member C5: $-9{,}898 \times 1.00/0.930 = 10{,}643$ lbs.)

It will be noted that the third interpretation controls none of the connecting weld lengths in this design example. See Table 4-5. This case would arise only where a member were made with an oversize section of more than double the required design capacity. It is, of course, logical to design so that the connection is adequate to develop strength in excess of the buckling strength to avoid a sudden brittle failure mode, but it is illogical somewhat to penalize a supplier for using oversize members by excessive weld requirements.

In this case it will be noted that the manufacturer simply provided weld lengths based upon controlling design loads, whether from analyses or the specified arbitrary minima. Since the design example utilized a variety of efficient sections closely fitting the design loads, this solution closely satisfies interpretations (a) and (b) for all tensile members, and (b) for double-angle compression members ($e = 0$), and easily exceeds (c) for the single-angle compression members.

The specification of unrelated arbitrary criteria for both minimum strength of web members and their connections, complicated by ambiguity, could result in illogical applications. In applying such rules the user should recognize that the *intent* of the Specifications is simply to provide a uniform safety factor logically throughout the entire assembly, and that excess overdesign of individual elements is illogical.

For open-web joists (J- and H-series) the minimum connection strength requirements for web members is phrased differently: "Joint connections shall be capable of withstanding the forces due to an ultimate load equal to at least two times the design load shown in the applicable Standard Load Table," (J- and H- 4.5 b). This requirement is somewhat more obscure except as applied to a load test. The allowable stress on welds is specific

(AISC 1.5.3; LJ-, LH-, DLJ-, and DLH- 103.2 d). The ultimate strength of welds is not specified except that a factor of 1.7 may be applied for plastic design (AISC 2.8). If 1.7 is considered a safe overload factor for welds, the conservative interpretation for minimum strength in open-web joist connections would be to provide 2.0/1.7 = 1.18 times the weld length required for the forces due to the design load at the allowable weld stresses.

TRUSSES

Trusses in general are flexural members, planar or three dimensional (space frames), made up of separate members designed principally for axial loads. In most trusses, the members are laid out in triangular patterns; the primary stress analysis is based upon axial stresses only, in all members as though pin-connected at each end. End connections are made symmetrical about gravity axes of members or are designed to resist unavoidable eccentricities in layout. For very large, non-pin-connected trusses, a secondary analysis of bending moments created by axial lnegth changes in members with rigid end connections is usually performed. Joists are small trusses within very definite limitations of design, load condition, etc. (See "Joists.") Vierendeel trusses are a special category in which web members, usually verticals only, are designed for shear, moment, and axial load. Vierendeel trusses usually are designed with parallel chords, and square or rectangular pattern of web and chord members, although occasionally with a top chord steeply pitched in one or both directions.

The AISC Specifications provide comprehensive provisions for analysis of all types of frames and for design of tension ties, compressive members with or without shear or moment, and connections thereof. These provisions are adequate for analysis and design of all the component members in any truss, and consequently few provisions apply to truss design per se. Perhaps, the most important provisions applicable most frequently in truss design relate to use of structural tees, angle, or double angle members; slenderness effects upon compressive capacity for these shapes or elements thereof considered "stiffened" or "unstiffened." The most efficient explanation of these provisions is by example; citing pertinent provisions as they appear to control at various stages in an actual truss design. (See Example 4-2 following.) Note that it became necessary to create an original design aid for use of structural tees (Table 4-6) in the course of this example for design of the top chord (laterally braced continuously). See also Chapter 6—Columns for a brief example from Table 4-6 for design of a structural tee strut as a column, braced at ends only.

DESIGN EXAMPLE 4-2. SIMPLE SPAN WELDED ROOF TRUSS—GRADE 50 CHORDS AND GRADE 36 WEB

Problem. Design simple span welded Pratt roof truss with a maximum depth of approximately $\frac{1}{13}$ of the span using Grade 50 tee-section chords and Grade 36 web members. Span is 180 feet. Trusses are spaced 15 ft. on centers and support $7\frac{1}{2}$ in. deep long-span deck on top flange of top chord tee.

Design Data. The roof truss supports a dead load of 25 psf and a roof live load of 25 psf. See Fig. 4-4 for design loads, member axial forces, and top chord bending moments.

Bottom Chord Design. See Fig. 4-4. Maximum axial tension force = 222 kips. Assume bottom chord bracing at quarter points of span. Maximum recommended (l/r) of bottom chord = 240 (1.8.4).

Allowable tension stress; $F_t = 0.60\ F_y = (0.60)(50) = 30$ ksi (1.5.1.1). Minimum area of bottom chord, $A = 222/30 = 7.40$ in.^{2}

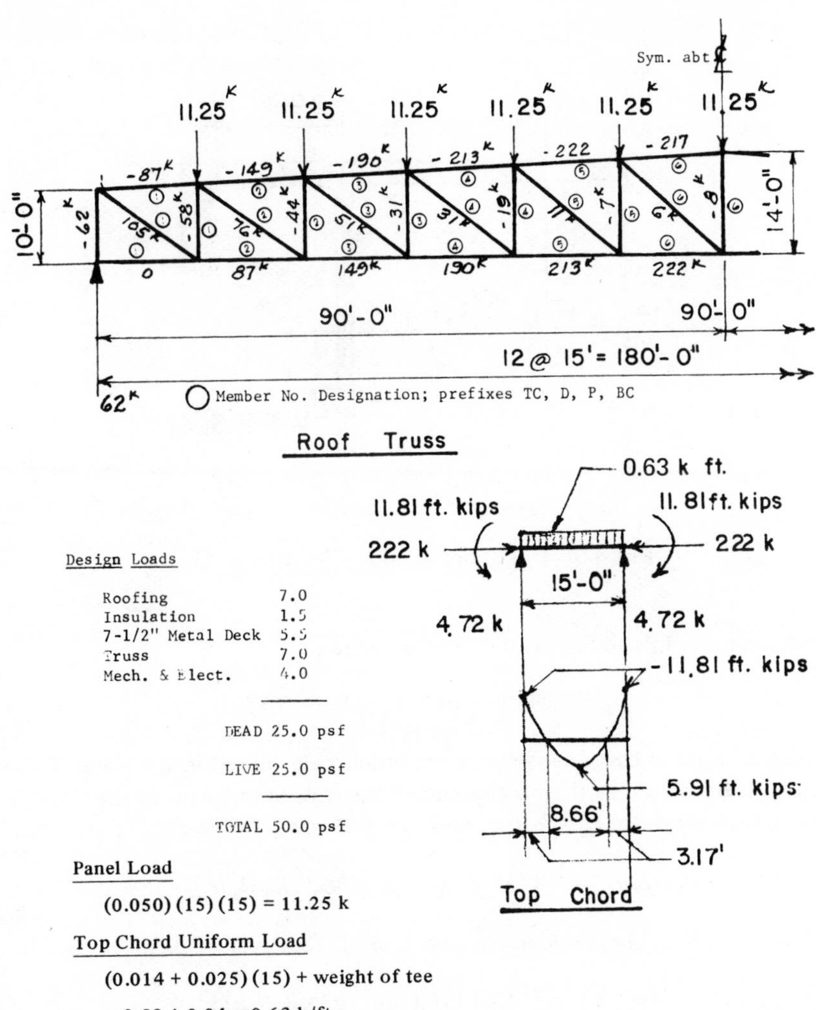

Fig. 4-4 Roof Truss Design Loads, Axial Forces, and Top Chord Secondary Bending Moments.

Try *WT 6 X 26.5*

$$A = 7.80 \text{ in.}^2;\ r_x = 1.51 \text{ in.};\ r_y = 2.48 \text{ in.}$$

$$(l_y/r_y) = [(180)(12)/4]/2.48 = 218 < 240 \text{ OK.}$$

$$(l_x/r_x) = (15)(12)/1.51 = 119 < 240, \text{OK.}$$

$$f_t = 222/7.80 = 28.5 \text{ ksi} < F_t = 30 \text{ ksi, OK.}$$

Top Chord Design. (See Fig. 4-4.) Maximum axial compression force = 222 kips. Negative bending moment at panel points; $-M_x = wl^2/12 = -11.81$ k-ft. Positive bending moment midspan; $+M_x = wl^2/24 = 5.91$ k-ft.

Try WT 10.5 X 41. See Fig. 4-5.

In the region of positive bending moment the metal deck as a diaphram will provide continuous lateral support for the top compression flange of the tee, $l_b = 0$. In regions of negative moment the compression stem of the tee can be considered laterally supported

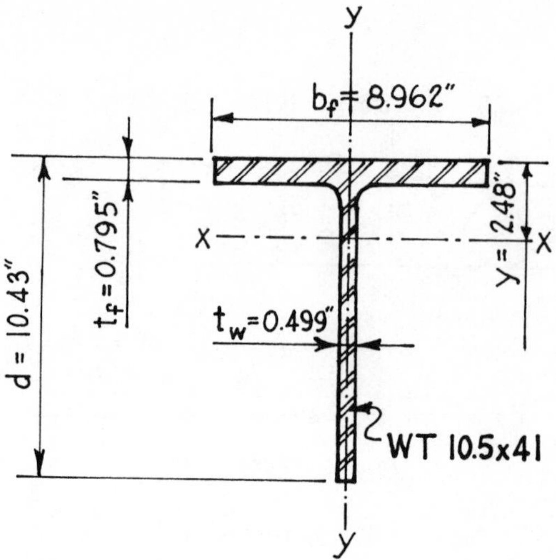

Fig. 4-5 Top Chord Section—WT 10.5 X 41.

at the truss panel points and at the points of inflection:

$$l_b = 0.2113\,l = (0.2113)(15) = 3.17 \text{ ft.}$$

For axial load the top chord tee section is continuously braced in the plane of the roof deck ($l_y = 0$) and braced at the panel points of the truss in the plane of the truss ($l_x = 15$ ft.). Slenderness ratios for $r_x = 3.10$ in. and $r_y = 1.99$ are as follows:

$$K_x(l_x/r_x) = (1)(15)(12)/3.10 = 58.1; K_y(l_y/r_y) = 0.$$

Check WT 10.5 X 41 for combined stresses (1.6.1). (See Fig. 4-5.)

$$A = 12.1 \text{ in.}^2; I_x = 116.0 \text{ in.}^4; b_f = 8.962 \text{ in.};$$

$$d = 10.43 \text{ in.}; t_w = 0.499 \text{ in.; and } t_f = 0.795 \text{ in.}$$

(1) *Compute axial and bending compressive stresses:*

$f_a = P/A = 222/12.1 = 18.35$ ksi.

f_b (top flange at midspan) $= M_x c/I_x = (5.91)(12)(2.48)/116.0 = 1.52$ ksi.

f_b (bottom of stem at panel point) $= M_x c/I_x = (11.81)(12)(7.95)/116.0 = 9.71$ ksi.

(2) *Check section properties of tee for compression stress:*
$(b/t) = (d/t_w) = 10.43/0.499 = 20.9 > 127/\sqrt{F_y} = 127/\sqrt{50} = 17.96$, and full depth of stem is not effective (1.9.1.2).

(3) *Check profile properties of tee*

$$(b_f/d) = 8.962/10.43 = 0.859 \text{ OK} > 0.5 \text{ (Appendix C, Table C1);}$$

$$(t_f/t_w) = 0.795/0.499 = 1.593 \text{ OK} > 1.10 \text{ (Appendix C, Table C1).}$$

$$Q_a = 1.00 \text{ for unstiffened compression elements (Com. 1.9).}$$

(4) *Compute compressive stress reduction factor for stem of tee* (Appendix C, Section C2, Eq. C2-5)

$$Q_s = 1.908 - 0.00715(b/t)\sqrt{F_y}$$
$$= 1.908 - 0.00715(d/t_w)\sqrt{F_y}$$
$$= 1.908 - (0.00715)(10.43/0.499)\sqrt{50}$$
$$= 0.851$$

which checks the value given in Table 4-6 for structural tees limited by width-thickness ratios.

For negative bending only, the stem of the tee is in compression and the width-thickness ratio can be based on the depth of the stem below the neutral axis. $(b/t) = (d/t_w) = (10.43 - 2.48)/0.499 = 15.93 < 127/\sqrt{F_y} = 127/\sqrt{50} = 17.96$; OK (1.9.1.2). The stem of the tee is fully effective in bending.

(5) *Compute modified column slenderness ratio,* C_c'. This ratio divides elastic buckling from inelastic buckling and is modified by Q_s to account for the reduced effective depth of the stem of the tee (Appendix C, Section C5). (See Table 4-6.)

$$C_c' = \sqrt{2(\pi)^2 E/(Q_a Q_s F_y)} = \sqrt{2\pi^2(29{,}000)/[(0.851)(1.000)(50)]} = 116.0$$

(6) *Compute allowable axial compression stress permitted in the absence of bending* (Appendix C, Section C5, Eq. C5-1);

$$F_a = Q_s Q_a F_y \left[\frac{1 - \dfrac{(Kl/r)^2}{2C_c'^2}}{5/3 + (3/8)(Kl/r)/C_c' - (1/8)(Kl/r)^3/(C_c')^3} \right]$$

$$= (0.851)(1.000)(50) \left[\frac{1 - \dfrac{(58.1)^2}{(2)(116.0)^2}}{5/3 + (3/8)(58.1)/116.0 - (1/8)(58.1)^3/(116.0)^3} \right]$$

$$= (0.851)(1.000)(50)(0.875)/1.839$$
$$= 20.25 \text{ ksi.}$$

(7) *Compute Euler stress divided by factor of safety (1.6.1);*

$$F_e' = 12\pi^2 E/[(23)(Kl_x/r_x)^2]$$
$$= 12\pi^2(29{,}000)/[(23)(58.1)^2]$$
$$= 44.2 \text{ ksi.}$$

(8) *Compute the column curvature coefficient that is applied to the bending term of the interaction formula for combined stresses* (Commentary 1.6.1; Table C1.6.1.2, Category C).

$$C_m = 1 - 0.4 f_a/F_e' = 1 - (0.4)(18.3)/44.2 = 0.834$$

(9) *Compute allowable bending stress permitted in absence of axial force.*

In the region of positive bending moment the top chord tee section qualifies as a compact section (1.5.1.4.1).

(1.5.1.4.1a) . . . . . . flange continuously connected to the web

(1.5.1.4.1b) . . . . . . $b_f/(2t_f) = 8.962/[(2)(0.795)] = 5.64 < 65/\sqrt{50} = 9.18$

(1.5.1.4.1d) . . . . . . $(d/t_w) = 10.43/0.499 = 20.9 < 257/\sqrt{50} = 36.3$

(1.5.1.4.1e) . . . . . . $l_b = 0$

$\therefore\ F_b = 0.66\,F_y = 0.66\,(50) = 33.0$ ksi.

In the region of negative bending moment the tee section also qualifies as a compact section (1.5.1.4.1). For this moment condition the stem of the tee section is in compression and the area of the compression flange consists of that part of the stem of the tee below the neutral axis. The unbraced length of the compression flange is the distance from the panel point of the truss to the point of inflection of the beam (l_b = 38 in.).

$$l_b = 38 \text{ in.} < 76 b_f/\sqrt{F_y} \ldots\ldots (1.5.1.4.1e).$$

$$< (76)(10.43 - 2.48)/\sqrt{50} = 85 \text{ in. OK}$$

$$l_b = 38 \text{ in.} < 20{,}000/[(d/A_f)F_y]$$

$$< \frac{20{,}000/50}{10.43/[(10.43 - 2.48)(0.499)]}$$

$$< 152 \text{ in. OK.}$$

$$\therefore\ F_b = 0.66\,F_y = 0.66\,(50) = 33.0 \text{ ksi.}$$

(10) *Compute the sum of the combined stress ratios at mid-span* (1.6.1; Eq. 1.6-1a).

$$f_a/F_a + C_m f_b/[(1 - f_a/F_e')F_b] = \frac{18.35}{20.25} + \frac{(0.834)(1.52)}{(1 - 18.35/44.2)(33)}$$

$$= 0.906 + 0.066$$

$$= 0.972 < 1.000 \text{ OK}$$

(11) *Compute the sum of the combined stress ratios at truss panel point* (1.6.1; Eq. 1.6-1b).

$$f_a/(0.60F_y) + f_b/F_b = 18.35/[(0.60)(50)] + 9.71/33.0$$

$$= 0.612 + 0.294$$

$$= 0.906 < 1.00 \ldots\ldots \text{OK.}$$

End Panel Diagonal Design. (see Fig. 4-6.). Maximum axial tension force = 105 kips. Maximum recommended $(Kl/r) = 240$ . . . . . . (1.8.4).

Allowable tension stress; $F_t = 0.60\,F_y = (0.60)(36) = 21.6$ ksi (1.5.1.1) = 22.0 ksi (1.5). Some confusion exists in these Specification provisions. Use $F_t = 22$ ksi. Minimum area of diagonal, $A = 105/22 = 4.77$ in.^{2}

Try *2 Ls 5 × 3 × 5/16 with long legs back to back*

$A = 4.80$ in.^{2}; $r_x = 1.61$ in.; $r_y = 1.26$ in.; $(Kl/r_y) = (18.03)(12)/1.26 = 171.7 < 240$ OK. $f_t = 105/4.80 = 21.8$ ksi $< F_t = 22.0$ ksi. OK.

First Interior Vertical Design. Maximum axial compression force = 58 kips. See Fig. 4-7.

Try *2 Ls 4 × $3\frac{1}{2}$ × $\frac{5}{16}$*

$A = 4.49$ in.^{2}; $r_x = 1.26$ in.; and $r_y = 1.59$ in. Maximum slenderness ratio is $(Kl/r_x) = (1)(10.33)(12)/1.26 = 98.4$

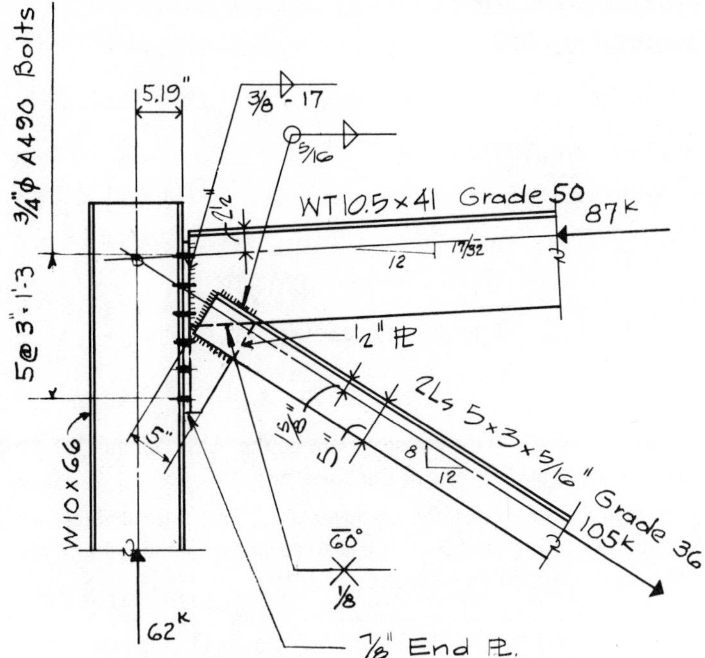

Fig. 4-6 Connection of Roof Truss and Column.

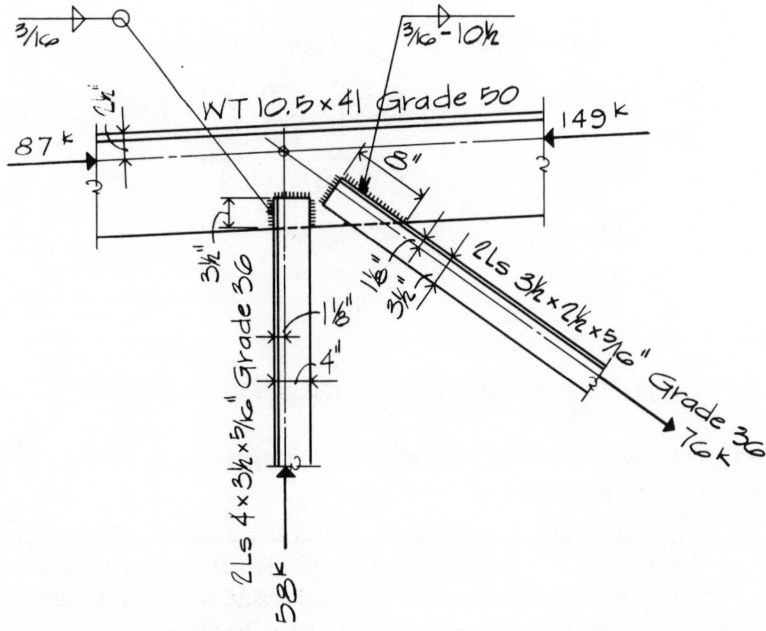

Fig. 4-7 Connection of Roof Truss, Top Chord, and Web Members.

Check 2 Ls 4 × $3\frac{1}{2}$ × $\frac{5}{16}$ for compression (1.5.1.3.1):

(1) *Section properties of 2 Ls 4 × $3\frac{1}{2}$ × $\frac{5}{16}$*

$$(b/t) = 4/(5/16) = 12.8 > 76/\sqrt{F_y} = 76/\sqrt{36} = 12.67.$$

The double angles are not fully effective (1.9.1.2).

(2) *Compute compressive stress reduction factor for 2 Ls 4 × $3\frac{1}{2}$ × $\frac{5}{16}$* (Appendix C, Section C2, Eq. C2-1).

$$\begin{aligned} Q_s &= 1.34 - 0.00447\,(b/_t)\sqrt{F_y} \\ &= 1.34 - (0.00447)\,(12.8)\,\sqrt{36} \\ &= 0.997 \end{aligned}$$

= the value given in the AISC Manual of properties for separated double angles limited by width-thickness ratios.

(3) *Compute modified column slenderness ratio, C'_c.* This ratio divides elastic buckling from inelastic buckling and is modified by Q_s to account for the reduced effective width of the legs of the angles (Appendix C, Section C5).

$$\begin{aligned} C'_c &= \sqrt{2\pi^2 E/(Q_a Q_s F_y)} \\ &= \sqrt{2\pi^2 (29{,}000)/[(1.000)\,(0.997)\,(36)]} \\ &= 126.3 \end{aligned}$$

(4) *Compute allowable axial compression stress* (Appendix C, Section C5, Eq. C5-1)

$$F_a = \frac{Q_s Q_a F_y \left[1 - \dfrac{(Kl/r)^2}{2(C'_c)^2}\right]}{[5/3 + (3/8)\,(kl/r)/C'_c - (1/8)\,(kl/r)^3/(C'_c)^3]}$$

$$= \frac{(1.000)\,(0.997)\,(36)\left[1 - \dfrac{(98.4)^2}{(2)\,(126.3)^2}\right]}{5/3 + (3/8)\,(98.4)/(126.3) - (1/8)\,(98.4)^3/(126.3)^3}$$

$$= (1.000)\,(0.997)\,(36)\,(0.698)/1.899$$

$$= 13.16 \text{ ksi.}$$

(5) *Compute compressive stress:*

$$f_a = P/A = 58/4.49 = 12.92 \text{ ksi} < F_a = 13.16 \text{ ksi OK.}$$

DESIGN EXAMPLE 4-3. TYPICAL ROOF TRUSS CONNECTION DETAIL OF TOP CHORD AND WEB MEMBERS.

Problem. Detail typical welded connection of vertical and diagonal web members to top chord of truss shown in Fig. 4-7. The designer is ordinarily not required to detail the size, length and placement of welds for such connections because they must be shown on the shop drawings (1.1.2). He need only show the truss configuration, the type of connections desired and the forces in all members as shown in Fig. 4-7 . . . . . . (1.1.1). A typical connection for the truss in Fig. 4-7 is detailed here to demonstrate specification requirements for the detailer.

Design data. See Figure 4-7 for member sizes, grade of steel, and axial forces. Use E70XX electrodes (1.5.3).

Vertical Member Connection to Top Chord. Member size (2 *Ls* $4 \times 3\frac{1}{2} \times \frac{5}{16}$) Compression force = 58 kips. The maximum fillet weld size along the edges of members $\frac{1}{4}''$ or more in thickness is $\frac{1}{16}$ in. less than the thickness of the material (1.17.6-2). For 2 *Ls* $4 \times 3\frac{1}{2} \times \frac{5}{16}$, the maximum fillet weld size that can be used along the edge of the outstanding legs is $\frac{5}{16} - \frac{1}{16} = \frac{1}{4}$ in. (1.17.6.2). The minimum fillet weld size allowed is based on the thickness of the thicker part joined (1.17.5). The 0.499 in. thickness of the stem of the top chord, WT 10.5 × 41, requires a minimum fillet weld size of $\frac{3}{16}$ in. (Table 1.17.5).

Try a $\frac{3}{16}$ in. fillet weld. The working strength of a $\frac{3}{16}$ in. fillet weld can be obtained from Table 4-7 or computed. Allowable shearing stress on the throat (1.14.7) of the fillet welds is 21 ksi (Table 1.5.3).

$$\tfrac{3}{16}\text{ in. weld: } (21.0)(3/16)(1/\sqrt{2}) = 2.78 \text{ kips/in.}$$

Total length of $\frac{3}{16}$ in. fillet weld required = 58/2.78 = 20.9 in. Use $\frac{3}{16}$ in. fillet weld. Extend angles 3.50 in. on the stem of the tee as shown in Fig. 4-7. Allowing for the loss in length of the weld that occurs at ends (1.14.7), we develop an effective weld length of (2) [(3.5) (2) + 4 − (2) (3/16)] = 21.2 in. > 20.9 in. OK.

TABLE 4-7 Fillet Weld Strength (Table 1.5.3)*

Size Inches	Strength Kips per inch
$\frac{1}{8}$	1.86
$\frac{3}{16}$	2.78
$\frac{1}{4}$	3.71
$\frac{5}{16}$	4.64
$\frac{3}{8}$	5.57
$\frac{7}{16}$	6.50
$\frac{1}{2}$	7.42
$\frac{9}{16}$	8.35
$\frac{5}{8}$	9.28
$\frac{3}{4}$	11.14
$\frac{7}{8}$	12.99
1	14.84

*AWS A5.1 or A5.5 E70XX electrodes for A36 and A53 Grade B matching base metal.

AWS A5.17, F7X-EXXX flux electrode combination for A242 and A375 matching base metal.

AWS A5.18, E705-X or E70U-1 electrodes for A501, A529, A570 Grade E, A572 Grades 42 to 60, and A5888.

It is not necessary to locate the welds for end connections of double angle members (1.15.3) in such a manner as to balance the forces about the neutral axis unless they are subject to a repeated variation in stress (1.15.3; 1.7).

Diagonal Member Connection to Top Chord. Member is (2 *Ls* $3\frac{1}{2} \times 2\frac{1}{2} \times \frac{5}{16}$) Tension force = 76 kips. Minimum fillet weld size along edge of outstanding leg is based upon the thickness of the thicker part joined = $\frac{3}{16}$ in. (1.17.5). Maximum size fillet is $\frac{5}{16} - \frac{1}{16} = \frac{1}{4}$ in. (1.17.6.2).

Total length of $\frac{1}{4}$ in. weld required = 76/3.71 = 20.5 in. Extend angles 8.0 in. on the stem of the tee as shown in Fig. 4-7 which will provide an effective weld length in excess of the 20.5 inches required.

End Diagonal Member Connection to Top Chord (Member D ①). Weld small $\frac{1}{2}$ in. gusset plate to stem of WT 10.5 × 41 top chord with an AISC prequalified manual shielded metal-arc double-vee groove joint of limited thickness and grind smooth (1.17.2). (See Fig. 4-6.) End diagonal (2 *Ls* ⅃L 5 × 3 × $\frac{5}{16}$) tension force = 105 kips. Maximum fillet weld along edges of diagonal = $\frac{5}{16} - (\frac{1}{16}) = \frac{1}{4}$ in. (1.17.6.2). Minimum fillet weld size (1.17.5) based on the thickness of the thicker part joined ($\frac{1}{2}$" tee stem) = $\frac{3}{16}$ in. (Table 1.17.5). Try a $\frac{1}{4}$ in. fillet weld; working strength from Table 4-7 = 3.71 kips/in. Total length of $\frac{1}{4}$ in. fillet weld required = 105/3.71 = 28.3 in. Extend D-① 2 *Ls* ⅃L 5 × 3 × $\frac{5}{16}$ 5 inches onto stem of tee and gusset plate as shown in Fig. 4-6. Allowing for the loss in weld that occurs at corners and ends (as conservative practice) the effective weld length of 2 [5 + 5 + 5 - ($\frac{1}{4}$) (2)] = 29.0 in. > 28.3 in., OK.

Truss Connection to Column. Field bolt truss to W10 × 49 column using A490 field bolts in a friction-type end plate connection. (See Fig. 4-6.) This connection can be laid out so that no moment is created in the truss connection or no moment in the column. If moment is eliminated in the truss connection, the simple span vertical shear will be delivered at the face of the column. The resulting moment in the column will be 0.5 dR. Unless this moment is balanced, as by a similar truss load on the opposite column face, it will usually be preferable to lay out the connection for zero moment to the column. In order to avoid an eccentric load on the column, locate end diagonal such that its workline intersects the workline of the top chord tee at the centerline of the column. This layout will subject the bolts connecting the truss to the column to an axial shear force of 62 kips and a moment of (62 k) (5.19 in.) = 322 k-in. (See Fig. 4-6.)

No exact method exists for locating the neutral axis to determine the tension stress in the bolts for this type of connection. The neutral axis is located somewhere below the centroid of the bolt pattern. It is conservative common practice to assume its location conveniently at the centroid of the bolts. Assume 12-$\frac{3}{4}$" dia. A490 bolts (2 rows of 6 each at 3 in. centers).

Number of bolts: 12; let y = lever arm to neutral axis

$$\Sigma y^2 \text{ of bolt pattern} = (1.5)^2 + (4.5)^2 + (7.5)^2 = 157.5 \text{ in.}^2$$

Compute nominal tension force in each top bolt:

$$F = Mc/\Sigma y^2 = (322)(7.5)/[(157.5)(2)] = 7.66 \text{ kips}$$

Compute the increase in the tension force (Q) in the A490 top bolts due to prying action.* (See Fig. 4-8.) Assume $\frac{7}{8}$ in. thick Grade 50 end plate thickness (t_f).

---

*Cited reference (2)

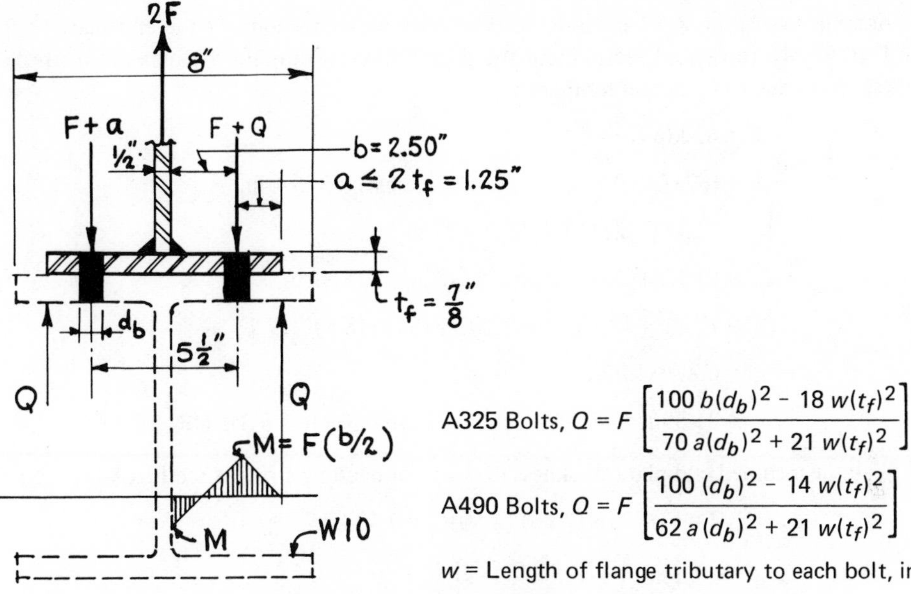

Fig. 4-8 Bolt Tension Prying Action.

$$
\begin{aligned}
Q &= F\left[\frac{100\,(d_b)^2 - (14)w(t_f)^2}{62a(d_b)^2 + 21w(t_f)^2}\right] \\
&= (7.66)\left[\frac{(100)\,(0.75)^2 - (14)\,(3)\,(0.875)^2}{(62)\,(1.25)\,(0.75)^2 + (21)\,(3)\,(0.875)^2}\right] \\
&= (7.66)\,(0.262) \\
&= 2.01 \text{ kips}
\end{aligned}
$$

Compute the total tension force $(Q + F)$ in each top bolt:

$$F + Q = 7.66 + 2.01 = 9.67 \text{ kips}$$

Compute the tension stress in each top bolt:

$$
\begin{aligned}
f_t &= (Q + F)/A_b = 9.67/0.4418 \\
&= 21.9 \text{ ksi} < 54 \text{ ksi } (1.5.2.1) \text{ OK.}
\end{aligned}
$$

Compute the allowable shear stress (F_v) in the $\frac{3}{4}''$ dia. A490 bolts for the minimum specified pretension load (T_b) of 35 kips (AISC Specification for Structural Joints–Table 3 Fastener Tension).

$$
\begin{aligned}
F_v &= 20\,(1 - f_t A_b/T_b) \ldots\ldots (1.6.3) \\
&= 20\,[1 - (21.9)\,(0.4418)/35] \\
&= 14.47 \text{ ksi OK.}
\end{aligned}
$$

Compute the shear stress (f_v) in the bolts:

$$
\begin{aligned}
f_v &= V/A_b = 62/[(12)\,(0.4418)] \\
&= 11.69 \text{ ksi} < F_v = 14.47 \text{ ksi OK.}
\end{aligned}
$$

Assume two $\frac{3}{8}$ in. X 17 in. long vertical fillet welds to connect the end plate to the WT 10.5 X 41 top chord section and the gusset plate. Compute the maximum stress in these welds due to shear and moment:

$$V = 62 \text{ kips}$$

$$M = (62)(5.19 + 0.875 \text{ plate thickness}) = 376 \text{ k-in.}$$

$$A_w = (0.375)(2)(17)/\sqrt{2} = 9.02 \text{ in.}^2$$

$$I_w = (0.375)(2)(17)^3/[(12)(\sqrt{2})] = 217.1 \text{ in.}^4$$

$$f_v = V/A_w + Mc/I_w = 62/9.02 + (376)(8.5)/217.1$$

$$= 6.87 + 14.72$$

$$= 21.59 \text{ ksi} \approx 21 \text{ ksi} \ldots\ldots (1.14.7, \text{Table } 1.5.3). \text{ OK.}$$

Check the assumed end plate thickness of $\frac{7}{8}$ in. for bending stress. (See Fig. 4-8.)

$$M = F(b/2) = (7.66)(2.50)/2 = 9.58 \text{ k-in.}$$

$$S = [w - (0.75 + 0.0625)]\,(t_f)^2/6$$

$$= (3 - 0.8125)(0.875)^2/6$$

$$= 0.279 \text{ in.}^3$$

$$f_b = M/S = 9.58/0.279$$

$$= 34.3 \text{ ksi} < 0.75\,F_y = 0.75\,(50) = 37.5 \text{ ksi OK.}$$

DESIGN EXAMPLE 4.4 USE OF STRUCTURAL-TEE AXIAL COMPRESSION LOAD TABLES

Problem. Design top chord of simple span truss using an ASTM A572 (Grade 50) structural tee section.

Design Data. Assume pin connected truss with $K = 1$ (Commentary 1.8). Maximum axial compression load is 467 kips and the unbraced length about both the major and the minor axis is 20 ft. ($K_x l_x = K_y l_y = 20$ ft.)

*Use of Structural-Tee Axial Compression Load Tables**

Enter Table 4-6 with $K_x l_x = K_y l_y = 20$ and select a structural tee section that is structurally sufficient for an axial compression load $P_m = 467$ kips with an unbraced length of 20 ft.

It is evident from the tabulation of structural tees that are acceptable in Table 4-6 that the WT 12 X 80 is the lightest and most economical section. This tee section also has sufficient stem depth for welding of web member angles. The design selection using Table 4-6 is now complete.

Check Structural-Tee Axial Compression Load Tables. Using a manual computation, check the WT 12 X 80 top chord section to show the design computations avoided and resulting design time savings through the use of Table 4-6. (See Figure 4-9.)

$$b/t = d/t_w = 12.36/0.656 = 18.84$$

$$b/t > 127/\sqrt{F_y} = 127/\sqrt{50} = 17.96 \ldots\ldots (\text{Appendix C, Sect. C2})$$

---

*Cited Reference (3)

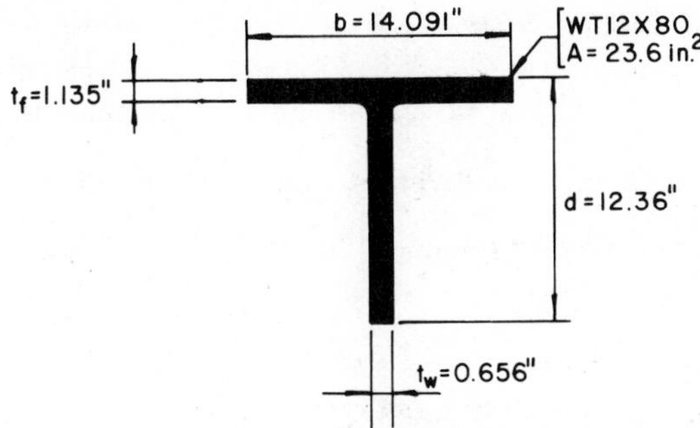

Fig. 4-9 Top Chord Section.

$$b/t < 176/\sqrt{F_y} = 176/\sqrt{50} = 24.89 \ldots\ldots \text{(Appendix C, Sect. C2)}$$

$$Q_s = 1.908 - 0.00715(b/t)\sqrt{F_y} \ldots\ldots \text{(Appendix C, Eq. C2-5)}$$

$$= 1.908 - (0.00715)(18.84)(\sqrt{50})$$

$$= 0.955 \text{ (See Table 4-6; } Q_s = 0.96)$$

$$Q_a = \frac{\text{Effective Area}}{\text{Actual Area}} \ldots\ldots \text{(Appendix C, Sect. C4)}$$

$$= 1.000$$

The allowable axial compression stress, F_{at}, for reduced section structural tees with $Kl/r < C_c' = \sqrt{(2\pi^2 E)/(Q_s Q_a F_y)}$ (Appendix C, Sect. C5) is a function of numerous variables, $F[Q_s, Q_a, F_y, (Kl/r, E]$

$$C_c' = \sqrt{\frac{2\pi^2(29{,}000)}{(0.955)(1.000)(50)}} = 109.5$$

$$Kl_y/r_y = (1)(240)/3.35 = 71.64 < C_c'$$

$$F_{at} = \frac{Q_s Q_a \left[1 - \dfrac{(Kl/r)^2}{2(C_c')^2}\right] F_y}{\dfrac{5}{3} + \dfrac{3(Kl/r)}{8\,C_c'} - \dfrac{(Kl/r)^3}{8(C_c')^3}} \ldots\ldots \text{(Appendix C, Sect. C5, Eq. C5-1)}$$

$$= \frac{(0.955)(1.000)\left[1 - \dfrac{(71.64)^2}{2(109.5)^2}\right](50)}{\dfrac{5}{3} + \dfrac{(3)(71.64)}{(8)(109.5)} - \dfrac{(71.64)^3}{(8)(109.5)^3}}$$

$$P_m = 19.99\,(23.6)$$

$$= 472 \text{ kips (See Table 4-6; } P_m = 472 \text{ kips)}$$

It is evident from this example that Table 4-6 provides a one step instant solution for the allowable axial load on structural tees and avoids the laborious computations necessary for a manual solution.

CITED REFERENCES

1. "Integrated System for Production of Joists," A. M. Lount and S. H. Simmonds, April 1975, ST4, ASCE.
2. "Behavior of Bolts in Tee-Connections Subject to Prying Action," *Research Series No. 353*, University of Illinois, Sept. 1969.
3. "Slender Unstiffened Axially Loaded Compression Members," R. Coba and R. Prokop, *Structural Thesis*, Univ. of Illinois, Chicago, 1976.

SELECTED REFERENCES–JOISTS AND TRUSSES

1. "Buckling of Steel Angle and Tee Struts," J. B. Kennedy and M. K. S. Murty; **98**, No. ST11; November, 1972; ASCE.
2. "Composite Open-Web Steel Joists," R. H. R. Tide and T. V. Galambos; January, 1970; **7**, No. 1; *Engineering Journal*, AISC.
3. "Ultimate Capacity of Single Bolted Angle Connections," J. B. Kennedy and G. R. Sinclair; August, 1969; **95**, No. ST8; ASCE.

STRUCTURAL TEES AS MAIN COMPRESSION MEMBERS

Table 4-6, following, presents tabulated allowable axial compression loads on structural tees. Two good reasons for including this table are: (1) the data are not available elsewhere, and (2) manual computation for tees is not only tedious, but also extremely tricky. The path of computations through cross-references in the Specifications, Appendix C, and the Commentary is tortuous. A simplified version of this path, with non-relevant, dead-end cross-references excluded, is shown below. See also Example 4-4, page 204.

Many tee stems are "slender" elements even in Grade 36; some not, become slender in Grade 50; and, of course, even more in the higher strength grades. There are two basic types of "slender" elements, stiffened and unstiffened. Combinations of the two become formidable problems, not in arithmetic, but in interpretation of the Specification requirements. Instead of three direct self-contained procedures, the applicable requirements combine, separate, and finally combine into one massive formula (C5-1), with limits thereon referenced back to the beginning. The real problem of axial load plus bending on combined stiffened and unstiffened elements is truly unique (C6).

SLENDER COMPRESSION ELEMENTS

Types: (1) stiffened, (2) unstiffened, and (3) combinations.

Typical Stiffened Elements: square and rectangular box sections: channel webs in compression, etc.

Typical Unstiffened Elements: angles, tees, channel flanges, web stiffeners, etc.

Typical Combinations: built-up shapes with projecting cover plates or angles as on box sections, rib-stiffened plates (twin-tees), etc.

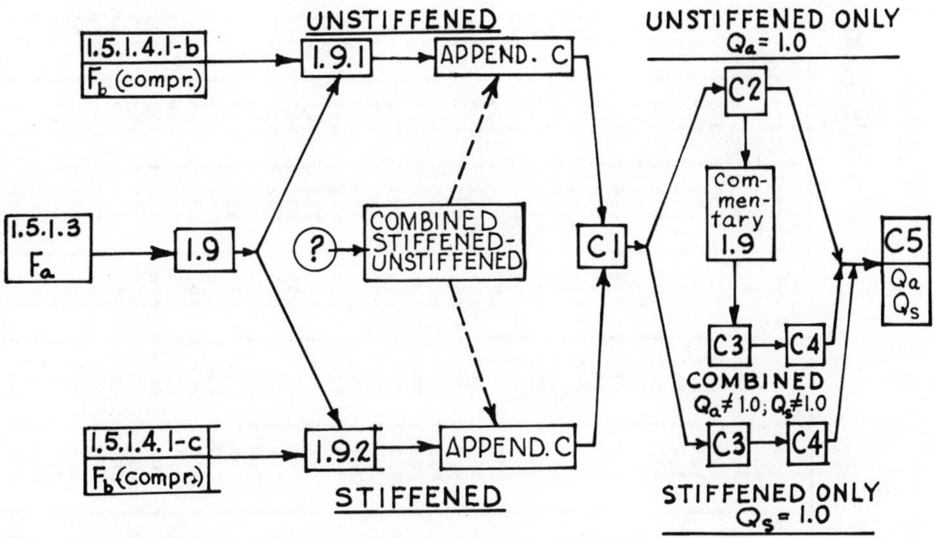

TABLE 4-6

STRUCTURAL TEES—MAIN MEMBERS

Allowable Axial Compression Load, P_m (Kips)

F_y 36 Ksi

Effective Length $K\ell$ in ft., x-x Axis

Size	WT18	WT18	WT18	WT18	WT18	WT18	WT18	WT18	WT18	WT18	WT18	WT16.5	WT16.5	WT16.5	WT16.5	WT16.5	WT16.5
Weight	150	140	130	122.5	115	97	91	85	80	75	67.5	120	110	100	76	70.5	x65
0	953	890	813	735	654	551	484	414	369	323	271	762	690	582	375	327	284
1	949	886	810	732	651	549	482	413	368	322	270	759	687	580	374	326	283
2	944	882	806	729	648	546	480	411	367	321	269	755	683	577	372	325	282
3	940	878	802	725	645	544	478	410	365	320	268	751	680	574	371	323	281
4	935	873	798	722	642	542	476	408	364	319	267	747	676	571	369	322	280
5	930	869	794	718	639	539	474	406	362	317	266	742	672	568	367	321	279
6	925	864	789	714	636	537	472	404	361	316	265	738	668	564	365	319	277
7	919	858	785	710	632	534	469	402	359	315	264	733	663	561	363	317	276
8	913	853	780	706	628	531	467	400	357	313	263	727	658	557	361	316	275
9	907	847	775	701	625	528	464	398	356	312	262	722	653	553	359	314	273
10	901	841	769	696	620	525	462	396	354	310	261	716	648	549	357	312	272
11	894	835	764	692	616	522	459	394	352	309	259	710	643	545	355	310	270
12	888	829	758	687	612	518	456	392	350	307	258	704	638	540	353	308	269
13	881	822	752	681	608	515	454	390	348	305	257	698	632	536	350	306	267
14	873	816	746	676	603	512	451	387	346	304	255	692	626	531	348	304	265
15	866	809	740	671	598	508	448	385	344	302	254	685	620	526	345	302	264
16	859	802	734	665	593	504	444	382	342	300	253	678	614	521	343	300	262
17	851	794	727	659	589	501	441	380	340	298	251	671	608	516	340	298	260
18	843	787	720	653	583	497	438	377	337	297	250	664	601	511	337	296	258
19	835	779	713	647	578	493	435	374	335	295	248	656	595	506	335	293	256
20	826	771	706	641	573	489	431	372	333	293	247	649	588	500	332	291	254
21	818	763	699	635	567	485	428	369	330	291	245	641	581	495	329	289	252
22	809	755	692	628	562	480	424	366	328	289	243	633	574	489	326	286	250
23	800	747	684	621	556	476	421	363	325	287	242	625	567	483	323	284	248
24	791	738	676	615	550	472	417	360	323	284	240	617	559	477	320	281	246
25	782	729	668	608	544	467	413	357	320	282	238	608	551	471	317	278	244
26	772	720	660	601	538	463	409	354	318	280	237	599	544	465	314	276	242
27	762	711	652	593	532	458	405	351	315	278	235	591	536	458	310	273	239
28	752	702	644	586	526	453	401	347	312	276	233	582	528	452	307	270	237
29	742	693	635	579	520	448	397	344	309	273	231	572	520	445	304	268	235
30	732	683	627	571	513	443	393	341	306	271	229	563	511	439	300	265	232
31	722	673	618	563	506	438	389	337	304	268	227	554	503	432	297	262	230
32	711	663	609	556	500	433	385	334	301	266	226	544	494	425	293	259	228
33	701	653	600	548	493	428	380	331	298	264	224	534	485	418	289	256	225
34	690	643	591	540	486	423	376	327	295	261	222	524	476	411	286	253	223
35	679	633	581	531	479	418	372	323	292	259	220	514	467	404	282	250	220

Effective Length, $K\ell$, in feet — Y-Y Axis																	
0	953	890	813	735	654	551	484	414	369	323	271	762	690	582	375	327	284
1	947	885	808	731	650	546	480	411	366	321	269	758	686	579	372	325	282
2	941	879	803	726	646	541	475	407	363	318	266	753	681	575	369	322	279
3	934	873	798	721	642	535	470	403	359	315	264	747	676	571	365	318	277
4	927	866	791	716	637	529	465	399	356	312	261	741	670	566	361	315	274
5	920	859	785	710	632	522	459	394	352	308	258	734	664	562	356	311	270
6	912	851	778	704	627	514	453	389	347	304	255	728	658	556	352	307	267
7	903	843	771	697	621	507	446	384	342	300	252	720	652	551	346	303	263
8	894	835	763	691	615	498	439	378	337	296	248	713	645	545	341	298	259
9	885	826	755	683	609	489	432	372	332	292	245	705	637	540	335	293	255
10	875	817	747	676	603	480	424	366	327	287	241	696	630	533	329	288	251
11	865	808	738	668	596	471	416	359	321	282	237	688	622	527	323	283	246
12	855	798	729	660	589	461	408	352	315	277	232	679	614	520	317	277	241
13	844	787	719	652	582	450	399	345	309	272	228	669	605	513	310	271	236
14	832	777	710	643	574	439	390	338	302	266	223	660	596	506	303	265	231
15	821	766	699	635	567	428	380	330	296	261	219	649	587	499	296	259	226
16	809	755	689	625	559	417	371	322	289	255	214	639	578	491	288	253	221
17	796	743	678	616	551	405	361	314	282	249	209	628	568	483	280	246	215
18	784	731	667	606	542	392	350	305	274	243	204	617	558	475	272	239	209
19	771	719	656	596	534	380	339	297	267	236	198	606	548	467	264	232	203
20	757	706	645	586	525	367	328	288	259	230	193	595	537	458	256	225	197
21	744	693	633	576	516	353	317	279	251	223	187	583	526	450	247	218	191
22	729	680	621	565	507	339	306	269	243	216	181	571	515	441	238	210	184
23	715	666	608	554	497	325	294	259	235	209	176	558	504	432	229	202	177
24	700	653	595	543	488	310	281	250	226	202	169	545	492	422	219	194	170
25	685	638	582	531	478	295	269	239	217	194	163	532	481	413	210	186	163
26	670	624	569	520	468	280	256	229	208	187	157	519	468	403	200	178	156
27	654	609	555	508	457	264	242	218	199	179	150	506	456	393	190	169	149
28	638	594	542	495	447	248	229	207	189	171	144	492	443	383	179	160	141
29	622	579	527	483	436	231	215	196	179	162	137	477	431	372	168	151	133
30	605	563	513	470	425	216	201	184	169	154	130	463	417	361	157	142	125
31	588	547	498	457	414	202	188	173	159	145	122	448	404	351	147	133	117
32	571	531	483	444	403	190	176	162	149	137	115	433	390	340	138	124	110
33	553	514	468	431	391	178	166	152	140	128	108	418	376	328	130	117	104
34	535	497	452	417	379	168	156	144	132	121	102	402	362	317	123	110	98
35	517	480	436	403	367	159	148	135	125	114	96	386	347	305	116	104	92
36	498	462	420	389	355	150	139	128	118	108	91	370	332	293	109	98	87
37	479	444	404	374	342	142	132	121	112	102	86	353	317	281	104	93	82
38	459	425	387	359	330	135	125	115	106	97	82	336	302	268	98	88	78
39	439	407	369	344	317	128	119	109	101	92	78	319	286	255	93	84	74
40	419	388	352	329	303	121	113	104	96	87		303	272	243	89	80	
41	399	369	335	313	290	116	108	99	91	83		289	259	231	84		
42	380	352	319	298	276	110	102	94				275	247	220			
43	363	335	305	285	264							262	236	210			
Q_s =	1.00	1.00	0.99	0.94	0.90	0.89	0.84	0.77	0.72	0.68	0.63	1.00	0.99	0.92	0.78	0.73	0.69

TABLE 4-6 (*Continued*)

STRUCTURAL TEES—MAIN MEMBERS

Allowable Axial Compression Load, P_m (Kips)

F_y 36 Ksi

Size	WT16.5	WT15	WT15	WT15	WT15	WT15	WT15	WT15	WT15	WT13.5	WT13.5	WT13.5	WT13.5	WT13.5	WT13.5	WT13.5	WT12
Weight	59	105	95	86	66	62	62	54	49.5	88.5	80	72.5	57	51	47	42	80
Effective Length $K\ell$ in ft. — x-x Axis																	
0	237	667	604	510	357	315	283	255	217	564	510	438	320	255	217	180	510
1	236	664	601	508	356	314	282	253	217	560	507	435	319	254	216	179	506
2	235	660	597	505	354	312	281	252	216	557	504	433	317	253	215	178	503
3	234	656	594	502	352	311	279	251	215	553	500	430	315	251	214	178	499
4	233	652	590	499	350	309	278	250	214	549	496	427	313	250	213	177	494
5	232	647	586	495	348	307	277	249	213	545	492	423	311	248	211	176	489
6	231	643	581	492	346	306	275	247	212	540	488	420	309	247	210	175	484
7	230	638	577	488	344	304	273	246	210	535	484	416	306	245	209	174	479
8	229	632	572	484	342	302	272	244	209	530	479	412	304	243	207	172	474
9	228	627	567	480	339	300	270	243	208	524	474	408	301	241	206	171	468
10	227	621	562	476	337	298	268	241	207	519	469	404	299	239	204	170	461
11	226	615	556	472	334	295	266	240	205	513	464	399	296	237	202	169	455
12	224	609	551	467	331	293	264	238	204	507	458	395	293	235	201	167	448
13	223	603	545	463	329	291	262	236	203	500	452	390	290	233	199	166	441
14	222	596	539	458	326	288	260	234	201	494	446	385	287	231	197	165	434
15	220	589	533	453	323	286	258	232	200	487	440	380	284	228	195	163	427
16	219	582	526	448	320	283	256	230	198	480	434	375	280	226	193	162	419
17	218	575	520	442	317	281	253	228	196	473	427	369	277	223	191	160	411
18	216	567	513	437	314	278	251	226	195	466	421	364	274	221	189	158	403
19	215	560	506	431	310	275	249	224	193	458	414	358	270	218	187	157	395
20	213	552	499	426	307	272	246	222	191	450	407	352	266	215	185	155	386
21	212	544	492	420	303	269	244	220	190	442	400	346	263	213	183	153	378
22	210	536	484	414	300	267	241	218	188	434	392	340	259	210	181	152	369
23	208	527	476	408	296	264	239	216	186	426	385	333	255	207	178	150	359
24	207	519	469	401	293	260	236	213	184	417	377	327	251	204	176	148	350
25	205	510	461	395	289	257	233	211	182	408	369	320	247	201	174	146	340
26	203	501	453	389	285	254	230	208	180	400	361	314	243	198	171	144	330
27	202	492	444	382	281	251	228	206	178	390	352	307	238	195	169	142	320
28	200	483	436	375	277	247	225	203	176	381	344	300	234	192	166	140	310
29	198	473	427	368	273	244	222	201	174	372	335	293	230	189	163	138	299
30	196	464	418	361	269	241	219	198	172	362	327	285	225	185	161	136	288
31	194	454	409	354	265	237	216	196	170	352	318	278	221	182	158	134	277
32	193	444	400	347	261	234	213	193	168	342	309	270	216	178	155	132	266
33	191	434	391	339	256	230	210	190	166	332	299	263	211	175	153	130	254
34	189	423	382	332	252	226	206	188	163	321	290	255	206	171	150	128	243
35	187	413	372	324	248	223	203	185	161	311	280	247	202	168	147	126	231

Effective Length, $K\ell$, in feet — Y-Y Axis																	
0	237	667	604	510	357	315	283	255	217	564	510	438	320	255	217	180	510
1	235	663	600	507	354	312	281	252	216	560	506	435	317	253	215	178	506
2	233	658	596	503	350	309	278	250	213	556	502	431	313	250	213	177	503
3	231	653	591	500	346	305	274	247	211	551	498	428	309	247	210	175	498
4	228	648	586	495	341	301	271	243	208	546	493	424	305	244	207	172	494
5	225	642	580	491	336	297	267	240	205	540	488	420	300	240	204	170	489
6	223	635	575	486	330	292	263	236	202	534	483	415	295	236	201	167	484
7	220	629	569	481	324	287	258	232	198	528	477	411	289	232	198	164	479
8	216	622	562	476	318	282	253	228	195	521	471	406	283	227	194	161	473
9	213	614	555	470	312	276	248	223	191	515	465	400	277	222	190	158	467
10	209	607	548	465	305	270	243	218	187	507	458	395	270	217	186	155	461
11	206	599	541	459	297	264	237	213	183	500	451	389	263	212	182	151	454
12	202	590	533	452	290	257	232	208	178	492	444	383	256	207	177	147	447
13	198	581	525	446	282	250	226	203	174	484	437	377	248	201	172	144	440
14	194	572	517	439	274	243	219	197	169	475	429	371	240	195	168	140	433
15	189	563	509	432	265	236	213	191	164	467	421	364	232	189	163	136	425
16	185	554	500	425	256	229	206	185	159	458	413	357	223	183	157	131	417
17	180	544	491	418	247	221	199	179	154	448	405	350	214	176	152	127	409
18	176	533	481	410	238	213	192	172	148	439	396	343	205	169	146	122	401
19	171	523	472	402	228	204	184	166	143	429	387	335	196	162	141	118	392
20	166	512	462	394	218	196	177	159	137	419	378	328	186	155	135	113	384
21	161	501	452	386	207	187	169	152	131	409	368	320	176	147	129	108	375
22	155	490	442	378	197	178	161	144	125	398	358	312	165	140	122	103	366
23	150	478	431	369	186	169	152	137	119	387	349	303	155	132	116	97	356
24	144	467	420	360	174	159	144	129	112	376	338	295	144	123	109	92	346
25	139	454	409	351	163	149	135	121	105	364	328	286	132	115	102	86	336
26	133	442	398	342	151	139	126	113	99	353	317	277	122	106	95	81	326
27	127	429	386	332	140	129	117	105	92	341	306	268	114	99	88	75	316
28	121	416	374	323	130	120	108	97	85	328	295	259	106	92	82	70	305
29	114	403	362	313	121	112	101	91	79	316	284	250	98	85	76	65	295
30	108	390	350	303	113	104	94	85	74	303	272	240	92	80	71	61	283
31	101	376	337	293	106	98	89	79	69	290	260	230	86	75	67	57	272
32	95	362	324	282	99	92	83	74	65	277	248	220	81	70	63	53	261
33	89	347	311	271	94	86	78	70	61	263	235	209	76	66	59	50	249
34	84	333	297	260	88	81	74	66	58	249	222	199	72	62	56	47	237
35	79	318	284	249	83	77	69	62		235	210	188	68	59	53		224
36	75	302	270	238	79	72	66			222	198	178	64				212
37	71	287	255	226	74	69				210	188	168					201
38	67	272	242	215						199	178	159					190
39		258	230	204						189	169	151					181
40		245	219	194						180	161	144					172
41		234	208	184						171	153	137					163
42		223	198	176						163	146	130					156
43		212	189	168						156	139	124					149
Q_s =	0.63	1.00	1.00	0.93	0.85	0.80	0.77	0.74	0.69	1.00	1.00	0.95	0.88	0.79	0.73	0.67	1.00

TABLE 4-6 (*Continued*)

STRUCTURAL TEES—MAIN MEMBERS
Allowable Axial Compression Load, P_m (Kips)

F_y 36 Ksi

Size	WT12	WT12	WT12	WT12	WT12	WT12	WT12	WT12	WT12	WT12	WT12	WT10.5	WT10.5	WT10.5	WT10.5	WT10.5	WT10.5
Weight	72.5	65	60	55	50	47	42	38	34	30.5	27.5	71	63.5	56	48	41	36.5
Effective Length $K\ell$ in ft. — x-x Axis																	
0	462	409	370	312	259	268	217	179	148	135	110	451	404	356	305	261	210
1	459	406	368	310	257	267	216	178	147	134	110	448	401	354	302	259	209
2	456	403	365	308	256	265	214	177	146	133	109	444	397	350	300	257	207
3	452	400	363	306	254	263	213	176	146	132	108	440	393	347	297	255	205
4	448	396	359	303	252	261	211	175	145	132	108	435	389	343	294	252	203
5	444	393	356	301	250	259	210	174	144	131	107	430	384	339	291	250	201
6	439	389	353	298	248	257	208	173	143	130	106	424	379	335	288	247	199
7	434	385	349	295	246	254	206	171	142	129	106	419	374	330	284	244	197
8	429	380	345	292	243	252	205	170	141	128	105	412	369	325	280	240	195
9	424	375	341	289	241	249	203	168	139	127	104	406	363	320	276	237	192
10	418	371	337	285	238	247	201	167	138	126	103	399	357	314	272	233	190
11	413	365	333	282	235	244	198	165	137	125	102	392	350	309	268	230	187
12	407	360	328	278	232	241	196	163	136	123	102	385	343	303	263	226	184
13	400	355	323	274	229	238	194	162	134	122	101	377	336	297	259	222	181
14	394	349	318	270	226	235	192	160	133	121	100	369	329	290	254	217	178
15	387	343	313	266	223	232	189	158	131	120	99	361	322	284	249	213	175
16	380	337	308	262	220	228	187	156	130	118	98	352	314	277	244	209	172
17	373	331	303	258	217	225	184	154	128	117	97	344	306	270	238	204	169
18	366	325	297	253	213	221	181	152	127	116	96	335	298	262	233	199	165
19	358	318	292	249	210	218	179	150	125	114	95	325	289	255	227	194	162
20	350	311	286	244	206	214	176	148	123	113	93	316	281	247	221	189	158
21	342	304	280	239	202	210	173	146	122	111	92	306	272	239	215	184	155
22	334	297	274	234	199	206	170	143	120	110	91	296	263	231	209	179	151
23	326	290	267	229	195	202	167	141	118	108	90	285	253	223	203	173	147
24	317	282	261	224	191	198	164	139	116	107	89	275	244	214	196	168	143
25	308	275	254	219	187	194	161	136	115	105	87	264	234	205	190	162	139
26	299	267	248	214	182	190	158	134	113	103	86	253	223	196	183	156	135
27	290	259	241	208	178	185	155	131	111	101	85	241	213	187	176	150	131
28	281	251	234	203	174	181	151	129	109	100	84	229	202	177	169	144	126
29	271	242	227	197	170	176	148	126	107	98	82	217	191	168	162	138	122
30	261	234	219	191	165	172	144	124	105	96	81	205	180	158	154	131	117
31	251	225	212	185	160	167	141	121	103	94	79	192	169	148	147	124	113
32	241	216	204	179	156	162	137	118	101	92	78	180	158	139	139	118	108
33	231	207	196	173	151	157	134	115	98	91	77	170	149	130	131	111	103
34	220	198	188	167	146	152	130	112	96	89	75	160	140	123	123	104	98
35	209	188	180	160	141	147	126	110	94	87	74	151	132	116	116	98	93

Effective Length, KL, in feet — Y-Y Axis																	
0	462	409	370	312	259	268	217	179	148	135	110	451	404	356	305	261	210
1	459	406	367	309	257	265	215	177	147	133	109	448	401	354	301	258	207
2	456	403	364	307	255	262	212	175	145	130	107	445	398	351	297	255	204
3	452	399	360	304	252	258	209	173	143	128	105	441	394	348	292	251	201
4	448	396	356	301	250	253	206	170	141	125	102	436	390	344	287	246	197
5	443	392	352	297	247	249	202	167	138	121	100	432	386	341	281	241	193
6	439	388	347	293	244	244	198	164	136	118	97	427	382	337	275	236	188
7	434	383	342	289	241	238	194	161	133	114	94	422	377	332	269	230	183
8	428	379	337	285	238	232	189	157	130	110	90	416	372	328	262	224	178
9	423	374	331	281	234	226	184	153	127	105	87	410	367	323	254	217	172
10	417	369	326	276	231	219	179	149	124	101	83	404	361	318	247	211	166
11	411	363	319	271	227	213	174	145	120	96	79	398	356	313	238	203	160
12	405	358	313	266	223	205	169	141	117	91	75	391	350	308	230	196	154
13	398	352	306	261	219	198	163	136	113	85	71	384	343	302	221	188	147
14	392	346	300	255	215	190	157	132	109	80	66	377	337	297	211	180	140
15	385	340	293	250	210	182	151	127	105	74	62	370	330	291	202	171	132
16	377	333	285	244	206	173	144	122	101	68	57	362	324	284	192	162	125
17	370	327	278	238	201	165	138	116	97	61	52	355	316	278	181	153	117
18	362	320	270	232	196	156	131	111	93	55	46	346	309	272	170	144	108
19	355	313	262	225	191	146	124	105	88	49	42	338	302	265	159	134	100
20	346	306	254	219	186	136	116	100	83	44	38	330	294	258	147	123	91
21	338	298	245	212	181	126	109	94	79	40	34	321	286	251	135	113	82
22	330	291	236	205	175	116	101	87	74	37	31	312	278	244	123	103	75
23	321	283	227	198	170	106	92	81	68			303	270	236	113	94	69
24	312	275	218	191	164	97	85	74	63			293	261	229	104	86	63
25	303	267	209	183	158	90	78	69	58			284	252	221	95	80	58
26	294	259	199	176	152	83	72	63	54			274	243	213	88	74	54
27	284	250	189	168	146	77	67	59	50			264	234	205	82	68	50
28	274	242	179	160	140	72	62	55	46			253	225	196	76	63	46
29	264	233	168	151	134	67	58	51	43			243	215	188	71	59	43
30	254	224	158	143	127	62	54	48	40			232	205	179	66	55	40
31	244	214	148	134	120	58	51	45	38			221	195	170	62	52	
32	233	205	139	126	113	55	48					209	185	160	58	49	
33	222	195	130	118	107							197	174	151	55	46	
34	211	185	123	112	100							186	164	142			
35	200	175	116	105	95							176	155	134			
36	189	165	109	99	90							166	147	127			
37	179	156	104	94	85							157	139	120			
38	169	148	98	89	80							149	132	114			
39	161	141	93	85	76							141	125	108			
40	153	134	89	81	73							134	119	103			
41	146	127	84	77	69							128	113	98			
42	139	121	80	73	66							122	108	93			
43	132	116	77	70	63							116	103	89			
Q_s =	1.00	0.99	0.97	0.89	0.81	0.90	0.81	0.74	0.69	0.69	0.63	1.00	1.00	1.00	1.00	1.00	0.91

TABLE 4-6 *(Continued)*

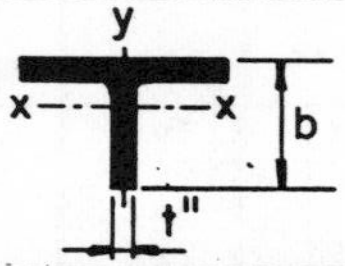

STRUCTURAL TEES—MAIN MEMBERS

Allowable Axial Compression Load, P_m (Kips)

F_y 36 Ksi

Size	WT10.5	WT10.5	WT10.5	WT10.5	WT10.5	WT9	WT9	WT9	WT9	WT9	WT9	WT9	WT9	WT9	WT9	WT9	WT9
Weight	34	31	27.5	24.5	22	57	52.5	48	42.5	38.5	35	32	30	27.5	25	22.5	20
Effective Length $K\ell$ in ft. x-x Axis																	
0	184	155	126	108	88	363	333	305	270	246	222	195	185	160	132	109	88
1	183	154	125	107	88	360	330	302	268	244	220	193	183	158	131	108	88
2	182	153	124	107	87	356	326	298	265	242	218	191	181	157	130	107	87
3	180	151	123	106	87	351	322	295	262	239	216	189	180	155	129	106	86
4	179	150	123	105	86	347	318	291	259	236	213	187	177	154	127	105	85
5	177	149	122	104	85	342	313	287	255	233	210	184	175	152	126	104	85
6	175	148	120	104	85	336	308	282	251	229	207	181	173	150	124	103	84
7	173	146	119	103	84	330	303	277	247	225	203	178	170	148	123	102	83
8	171	144	118	102	83	324	297	272	243	221	200	175	168	145	121	100	82
9	169	143	117	101	83	318	291	266	238	217	196	172	165	143	119	99	81
10	167	141	116	100	82	311	285	260	234	213	192	169	162	140	117	97	80
11	165	139	114	99	81	303	278	254	229	208	188	165	158	138	115	96	78
12	163	138	113	97	80	296	271	248	223	203	183	161	155	135	113	94	77
13	160	136	111	96	79	288	264	241	218	198	179	157	152	132	111	93	76
14	158	134	110	95	78	280	256	234	212	193	174	153	148	129	109	91	75
15	155	132	108	94	77	271	248	227	206	188	169	149	145	126	106	89	73
16	152	129	107	92	76	263	240	220	200	182	164	145	141	123	104	87	72
17	150	127	105	91	75	254	232	212	194	176	159	140	137	120	101	85	70
18	147	125	103	90	74	244	223	204	188	170	153	136	133	117	99	83	69
19	144	123	102	88	73	235	214	196	181	164	148	131	129	113	96	81	68
20	141	120	100	87	72	225	205	187	174	158	142	126	124	110	93	79	66
21	138	118	98	85	71	215	196	179	167	151	136	121	120	106	91	77	64
22	135	115	96	84	70	204	186	170	160	145	130	116	115	102	88	75	63
23	131	113	94	82	69	193	176	160	152	138	124	111	111	98	85	73	61
24	128	110	92	81	67	182	166	151	145	131	117	105	106	95	82	70	59
25	125	108	90	79	66	170	155	141	137	123	111	100	101	90	79	68	58
26	121	105	88	77	65	159	144	131	129	116	104	94	96	86	75	65	56
27	118	102	86	76	64	147	134	121	120	108	97	88	91	82	72	63	54
28	114	99	84	74	62	137	124	113	112	100	90	82	85	78	69	60	52
29	110	96	82	72	61	127	116	105	104	94	84	76	80	73	65	58	50
30	106	94	80	70	60	119	108	98	97	87	78	71	75	69	62	55	48
31	103	91	77	69	58	112	101	92	91	82	73	67	70	64	58	52	46
32	99	87	75	67	57	105	95	86	86	77	69	63	66	60	55	50	44
33	95	84	73	65	56	98	89	81	80	72	65	59	62	57	51	47	42
34	90	81	70	63	54	93	84	77	76	68	61	55	58	53	48	44	40
35	86	78	68	61	53	87	80	72	72	64	58	52	55	50	46	41	38

Effective Length, $K\ell$, in feet — Y-Y Axis																
0	184	155	126	108	88	363	333	305	270	246	222	195	185	160	132	109	88
1	182	153	124	106	87	360	330	302	267	243	220	192	182	158	130	108	87
2	179	151	123	104	85	357	327	299	263	240	217	190	179	155	128	106	85
3	176	148	121	102	83	353	324	296	259	236	214	187	176	152	126	104	83
4	173	146	119	100	81	349	320	293	255	232	210	184	172	149	123	102	81
5	170	143	116	97	79	345	316	289	250	228	206	180	167	145	121	100	79
6	166	140	114	94	77	341	312	285	245	223	201	176	163	141	118	97	76
7	161	136	111	90	74	336	308	281	239	218	196	172	158	137	114	95	73
8	157	133	108	87	71	331	303	277	233	212	191	168	152	132	111	92	70
9	152	129	105	83	68	325	298	272	226	206	186	163	146	128	107	89	67
10	147	125	102	79	65	320	293	268	220	200	180	158	140	123	103	86	63
11	142	121	99	75	62	314	287	263	213	193	174	153	134	117	99	82	60
12	137	116	95	70	58	308	282	257	205	187	168	148	127	112	94	79	56
13	131	111	92	65	54	301	276	252	197	179	161	142	120	106	90	75	52
14	125	107	88	61	50	295	270	246	189	172	154	136	113	100	85	71	48
15	119	102	84	55	46	288	263	240	181	164	147	130	105	93	80	67	43
16	112	96	80	50	42	281	257	234	172	156	140	124	97	86	75	63	38
17	106	91	76	44	38	274	250	228	163	148	132	117	89	79	69	59	34
18	99	85	71	40	33	266	243	222	154	139	124	110	80	72	63	54	30
19	91	80	67	36	30	258	236	215	144	130	116	103	72	65	58	50	27
20	84	74	62	32	27	251	229	208	134	121	108	96	65	59	52	45	25
21	76	67	57	29	25	242	221	201	123	111	99	89	59	53	47	41	22
22	69	61	52			234	213	194	113	102	90	81	53	48	43	37	
23	64	56	48			225	205	187	103	93	82	74	49	44	39	34	
24	58	51	44			217	197	179	95	85	76	68	45	41	36	31	
25	54	47	40			207	189	171	87	79	70	63	41	37	33	29	
26	50	44	37			198	180	163	81	73	64	58	38	35	31	27	
27	46	41	34			189	171	155	75	67	60	54	35	32	29		
28	43	38	32			179	162	147	69	63	56	50					
29	40	35				169	153	138	65	59	52	47					
30						158	143	129	61	55	48	43					
31						148	134	121	57	51	45	41					
32						139	126	114	53	48	43	38					
33						131	118	107	50	45	40	36					
34						123	111	101	47								
35						116	105	95									
36						110	99	90									
37						104	94	85									
38						99	89	81									
39						94	85	76									
40						89	81	73									
41						85	77	69									
42						81	73	66									
43						77	70	63									
Q_s =	0.85	0.78	0.72	0.69	0.63	1.00	1.00	1.00	1.00	1.00	1.00	0.96	0.97	0.91	0.83	0.76	0.69

TABLE 4-6 *(Continued)*

STRUCTURAL TEES—MAIN MEMBERS

Allowable Axial Compression Load, P_m (Kips)

F_y 36 Ksi

Size	WT9 X	WT8 X	WT8 X	WT8 X	WT8 X	WT8 X	WT8 X	WT8 X	WT8 X	WT8 X	WT8 X	WT8 X	WT8 X	WT7 X	WT7 X	WT7 X	WT7
Weight	17.5	48	44	39	35.5	32	29	25	22.5	20	18	15.5	13	365	332.5	302.5	275
Effective Length Kℓ in ft. — x-x Axis																	
0	70	305	279	248	227	203	184	157	130	100	88	66	47	2311	2112	1922	1747
1	70	301	276	246	224	201	182	156	129	99	88	66	47	2291	2093	1904	1730
2	69	297	272	243	222	199	180	154	128	98	87	65	46	2268	2071	1883	1710
3	69	293	268	240	219	196	178	152	126	97	86	65	46	2243	2047	1860	1688
4	68	288	264	236	216	193	175	150	124	96	85	64	46	2215	2020	1835	1664
5	67	283	259	232	212	190	172	148	123	95	84	63	45	2185	1991	1807	1637
6	67	277	254	228	208	186	169	145	121	94	83	62	45	2153	1960	1777	1609
7	66	271	248	224	204	183	166	143	119	92	81	62	44	2119	1927	1745	1578
8	65	265	242	219	200	179	162	140	116	91	80	61	44	2083	1891	1711	1546
9	65	258	236	214	195	175	158	137	114	89	79	60	43	2044	1854	1676	1511
10	64	251	230	208	190	170	154	134	112	87	77	59	42	2004	1815	1638	1475
11	63	243	223	203	185	166	150	131	109	85	76	58	42	1962	1774	1599	1438
12	62	236	216	197	179	161	146	127	106	84	74	57	41	1918	1731	1558	1398
13	61	227	208	191	174	156	141	124	104	82	72	56	40	1872	1687	1515	1357
14	60	219	200	184	168	151	136	120	101	80	71	55	40	1824	1640	1470	1314
15	59	210	192	178	162	145	131	116	98	78	69	53	39	1775	1592	1424	1270
16	58	201	184	171	156	140	126	112	95	75	67	52	38	1724	1543	1376	1224
17	57	191	175	164	149	134	121	108	92	73	65	51	37	1672	1491	1327	1176
18	56	181	166	157	143	128	115	104	88	71	63	50	37	1617	1438	1276	1127
19	55	171	157	149	136	122	110	100	85	68	61	48	36	1562	1384	1223	1076
20	54	160	147	141	129	115	104	95	81	66	59	47	35	1504	1327	1169	1023
21	53	149	137	133	121	109	98	90	78	63	57	46	34	1445	1269	1113	969
22	52	138	126	125	113	102	92	86	74	61	55	44	33	1383	1209	1055	913
23	50	127	116	117	106	95	85	81	70	58	53	43	32	1321	1148	995	855
24	49	116	106	108	97	87	78	76	66	56	50	41	31	1256	1084	934	796
25	48	107	98	99	90	80	72	70	62	53	48	40	31	1190	1019	871	735
26	47	99	91	92	83	74	67	65	58	50	46	38	30	1121	952	806	680
27	45	92	84	85	77	69	62	60	54	47	43	36	29	1051	883	748	630
28	44	85	78	79	72	64	58	56	50	44	41	35	28	979	822	695	586
29	43	80	73	74	67	60	54	52	47	41	38	33	27	913	766	648	546
30	41	74	68	69	62	56	50	49	44	38	35	31	25	853	716	606	510
31	40	70	64	65	58	52	47	46	41	36	33	29	24	799	670	567	478
32	38	65	60	61	55	49	44	43	38	34	31	28	23	750	629	532	449
33	37	61	56	57	52	46	41	40	36	32	29	26	22	705	591	500	422
34	35	58	53	54	49	43	39	38	34	30	28	24	21	664	557	471	397
35	34	55	50	51	46	41	37	36	32	28	26	23	20	627	526	445	375

Effective Length, $K\ell$, in feet (Y-Y Axis)																	
0	70	305	279	248	227	203	184	157	130	100	88	66	47	2311	2112	1922	1747
1	69	302	276	246	224	201	182	155	128	99	87	65	46	2300	2102	1913	1739
2	68	299	274	242	221	198	179	152	126	97	86	64	45	2288	2091	1903	1729
3	66	296	271	238	217	195	177	149	123	96	84	62	44	2276	2080	1892	1719
4	64	293	268	234	214	191	173	145	120	93	82	60	43	2262	2067	1880	1709
5	63	289	265	229	209	187	170	141	117	91	80	58	42	2248	2054	1868	1697
6	61	285	261	224	205	183	166	137	114	89	78	56	40	2233	2040	1855	1685
7	58	281	257	219	200	179	162	132	110	86	76	54	39	2217	2025	1842	1673
8	56	277	253	213	194	174	157	127	106	83	73	51	37	2201	2010	1827	1660
9	53	272	249	207	189	169	153	122	102	80	70	49	35	2184	1994	1812	1646
10	51	268	244	201	183	163	148	116	97	77	67	46	33	2166	1977	1797	1632
11	48	263	240	194	176	158	142	110	92	74	64	43	31	2148	1960	1781	1617
12	45	257	235	187	170	152	137	103	87	70	61	40	29	2128	1942	1764	1601
13	42	252	230	180	163	145	131	97	82	67	58	36	27	2109	1924	1747	1585
14	39	246	225	172	156	139	125	90	77	63	55	33	25	2088	1904	1729	1569
15	35	240	219	164	149	132	119	82	71	59	51	29	22	2067	1885	1711	1552
16	32	234	214	156	141	125	113	75	65	55	47	25	20	2045	1865	1692	1534
17	28	228	208	147	133	118	106	67	59	50	43	22	17	2023	1844	1673	1516
18	25	222	202	138	125	110	99	60	52	46	39	20	15	2000	1822	1653	1498
19	22	215	196	129	116	102	92	53	47	41	35	18		1976	1800	1632	1479
20	20	208	189	119	107	94	85	48	42	37	32			1952	1778	1611	1460
21		201	183	109	98	86	77	44	38	34	29			1928	1755	1590	1440
22		194	176	100	89	78	70	40	35	31	26			1903	1731	1568	1419
23		187	169	91	82	72	64	36	32	28	24			1877	1707	1546	1399
24		179	162	84	75	66	59	33	29	26	22			1850	1682	1523	1377
25		171	155	77	69	61	54	31	27	24	20			1824	1657	1499	1356
26		163	147	71	64	56	50	29	25					1796	1632	1475	1334
27		155	140	66	59	52	47							1768	1606	1451	1311
28		147	132	61	55	48	43							1740	1579	1426	1288
29		138	124	57	51	45	40							1711	1552	1401	1265
30		129	116	54	48	42	38							1681	1524	1375	1241
31		121	108	50	45	39	35							1651	1496	1349	1216
32		114	102	47	42	37	33							1621	1467	1322	1192
33		107	96	44	40									1589	1438	1295	1166
34		101	90											1558	1409	1268	1141
35		95	85											1526	1379	1240	1115
36		90	80											1493	1348	1211	1088
37		85	76											1460	1317	1182	1061
38		81	72											1426	1285	1153	1034
39		76	68											1392	1253	1123	1006
40		73	65											1357	1220	1092	978
41		69	62											1321	1187	1061	949
42		66	59											1285	1153	1030	920
43		63	56											1249	1119	998	890
Q_s =	0.63	1.00	1.00	1.00	1.00	1.00	1.00	0.99	0.91	0.79	0.77	0.67	0.57	1.00	1.00	1.00	1.00

TABLE 4-6 *(Continued)*

STRUCTURAL TEES—MAIN MEMBERS

Allowable Axial Compression Load, P_m (Kips)

F_y 36 Ksi

Effective Length $K\ell$ in ft. — x–x Axis

Size	WT7	WT7	WT7	WT7	WT7	WT7	WT7	WT7	WT7	WT7	WT7	WT7	WT7	WT7	WT7	WT7	WT7
Weight	250	227.5	213	199	185	171	160	157	143.5	132	123	118.5	114	109.5	105.5	101	96.5
0	1588	1445	1352	1264	1175	1086	1017	998	912	838	782	752	724	696	670	642	613
1	1571	1430	1337	1250	1162	1074	1006	986	900	828	772	742	714	686	661	633	605
2	1553	1412	1320	1233	1146	1059	993	972	887	815	760	731	703	676	651	623	596
3	1532	1392	1301	1215	1129	1043	978	956	873	801	747	718	691	664	639	612	585
4	1508	1370	1280	1195	1109	1024	962	939	856	786	732	704	677	650	626	599	572
5	1483	1346	1257	1172	1088	1004	944	920	838	769	716	688	662	635	612	585	559
6	1456	1320	1232	1148	1065	982	925	899	819	750	698	671	645	619	596	570	544
7	1427	1292	1205	1123	1041	959	905	877	798	730	679	652	627	602	579	554	528
8	1396	1263	1176	1095	1014	934	883	853	775	709	659	633	608	583	561	536	511
9	1363	1232	1146	1066	987	907	859	828	752	687	638	612	588	563	542	518	493
10	1328	1199	1114	1036	957	880	835	801	727	663	615	590	567	542	522	498	474
11	1292	1164	1081	1003	927	850	809	773	700	638	591	567	544	521	501	478	454
12	1254	1128	1046	970	894	820	782	744	673	612	566	543	521	498	479	456	433
13	1215	1091	1010	935	861	788	754	714	644	585	540	517	496	474	456	434	411
14	1174	1051	972	898	826	754	725	682	614	556	513	491	471	448	432	410	388
15	1132	1011	932	860	790	720	695	649	583	527	484	463	444	422	407	385	365
16	1087	969	891	821	752	684	663	614	550	496	455	435	416	395	380	360	340
17	1042	925	849	780	713	646	631	579	517	464	424	405	387	367	353	333	314
18	994	880	805	738	672	608	597	541	482	430	392	374	357	337	325	305	287
19	946	833	760	694	630	567	562	503	445	395	359	341	325	306	295	276	258
20	895	785	712	649	586	526	526	463	407	359	325	309	294	277	266	249	233
21	843	734	664	602	541	482	488	422	370	326	294	280	267	251	241	226	212
22	789	683	613	553	495	440	449	384	337	297	268	255	243	229	220	206	193
23	733	629	562	506	453	402	411	351	308	272	246	233	222	209	201	189	176
24	676	578	516	464	416	370	378	323	283	250	225	214	204	192	185	173	162
25	623	532	476	428	383	341	348	297	261	230	208	198	188	177	170	160	149
26	576	492	440	396	354	315	322	275	241	213	192	183	174	164	158	148	138
27	534	456	408	367	328	292	298	255	224	197	178	169	161	152	146	137	128
28	497	424	379	341	305	271	277	237	208	183	166	158	150	141	136	127	119
29	463	396	354	318	285	253	259	221	194	171	154	147	140	132	127	119	111
30	433	370	330	297	266	236	242	207	181	160	144	137	131	123	118		
31	405	346	309	278	249	221	226	193	170	150							
32	380	325	290	261	234	208	212	182	159								
33	357	306	273	246	220	195	200										
34	337	288	257	231	207		188										
35	318	272	243				178										

Effective Length, $K\ell$, in feet — Y-Y Axis																	
0	1588	1445	1352	1264	1175	1086	1017	998	912	838	782	752	724	696	670	642	613
1	1580	1438	1345	1257	1169	1081	1012	993	907	834	778	748	720	692	666	638	610
2	1571	1430	1338	1250	1162	1075	1006	987	901	829	773	743	715	688	662	634	606
3	1562	1421	1330	1242	1155	1068	1000	981	896	823	768	738	711	683	658	630	602
4	1552	1412	1321	1234	1147	1061	993	974	890	818	763	733	706	678	653	626	598
5	1541	1402	1312	1226	1139	1053	986	967	883	812	757	728	701	673	648	621	594
6	1530	1392	1302	1216	1131	1045	978	960	876	805	751	722	695	668	643	616	589
7	1519	1381	1292	1207	1122	1037	970	952	869	799	745	716	689	662	637	611	584
8	1506	1370	1281	1197	1112	1028	961	943	861	792	738	710	683	656	632	605	578
9	1494	1358	1270	1186	1102	1019	953	935	853	784	731	703	676	650	626	599	573
10	1481	1346	1259	1175	1092	1009	943	926	845	776	724	696	670	643	619	593	567
11	1467	1333	1246	1164	1081	999	934	917	837	769	717	689	663	637	613	587	561
12	1452	1320	1234	1152	1070	989	924	907	828	760	709	681	656	630	606	580	555
13	1438	1306	1221	1140	1059	978	913	897	818	752	701	673	648	622	599	574	548
14	1422	1292	1208	1127	1047	967	903	887	809	743	692	665	640	615	592	567	542
15	1407	1277	1194	1114	1034	955	892	876	799	734	684	657	632	607	584	559	535
16	1391	1262	1180	1101	1022	943	880	865	789	724	675	649	624	599	576	552	528
17	1374	1247	1165	1087	1009	931	869	854	778	714	666	640	615	591	569	544	520
18	1357	1231	1150	1073	995	919	857	842	768	705	656	631	607	582	560	537	513
19	1339	1215	1135	1058	982	906	844	830	757	694	647	621	598	574	552	528	505
20	1321	1198	1119	1043	968	893	832	818	745	684	637	612	589	565	543	520	497
21	1303	1181	1103	1028	953	879	819	805	734	673	627	602	579	556	535	512	489
22	1284	1163	1086	1013	939	866	806	792	722	662	617	592	570	546	526	503	481
23	1265	1145	1069	997	923	852	792	779	710	651	606	582	560	537	516	494	472
24	1245	1127	1052	980	908	837	778	766	697	639	595	572	550	527	507	485	463
25	1225	1109	1034	964	892	823	764	752	685	627	584	561	539	517	497	476	454
26	1204	1089	1016	947	876	808	750	738	672	615	573	550	529	507	487	466	445
27	1183	1070	998	929	860	792	735	724	659	603	561	539	518	497	477	457	436
28	1162	1050	979	911	843	777	720	709	645	591	550	528	507	486	467	447	427
29	1140	1030	960	893	826	761	705	695	632	578	538	516	496	475	457	437	417
30	1118	1009	940	875	809	744	689	679	618	565	525	504	485	464	446	427	407
31	1096	988	921	856	791	728	673	664	603	552	513	492	473	453	435	416	397
32	1073	967	900	837	773	711	657	648	589	538	500	480	461	441	424	405	387
33	1049	945	880	818	755	694	641	632	574	525	487	468	449	430	413	395	377
34	1026	923	859	798	736	677	624	616	559	511	474	455	437	418	401	384	366
35	1001	901	838	778	717	659	607	600	544	496	461	442	424	406	390	372	355
36	977	878	816	757	698	641	590	583	528	482	447	429	412	393	378	361	344
37	952	855	794	737	678	622	572	566	512	467	433	415	399	381	366	349	333
38	926	831	772	715	658	604	554	548	496	452	419	402	385	368	353	337	321
39	900	807	749	694	638	585	535	530	480	437	405	388	372	355	341	325	310
40	874	782	726	672	617	565	517	512	463	421	390	374	358	342	328	313	298
41	847	757	702	650	596	545	498	494	446	405	375	359	344	328	315	300	286
42	820	732	678	627	575	525	478	475	429	389	360	345	330	315	302	288	274
43	792	706	654	604	553	505	459	456	411	373	345	330	316	301	288	275	261
Q_s =	1.00	1.00	1.00	1.00	1.00	1.00	1.00	1.00	1.00	1.00	1.00	1.00	1.00	1.00	1.00	1.00	1.00

TABLE 4-6 *(Continued)*

STRUCTURAL TEES—MAIN MEMBERS

Allowable Axial Compression Load, P_m (Kips)

F_y 36 Ksi

Size	WT7	WT7	WT7	WT7	WT7	WT7	WT7	WT7	WT7	WT7	WT7	WT7	WT7	WT7	WT7	WT7	WT7
Weight	92	88	83.5	79	75	71	68	63.5	59.5	55.5	51.5	47.5	43.5	42	39	37	34
Effective Length $K\ell$ in ft. x-x Axis: 0	583	559	529	501	475	451	432	404	378	352	326	302	276	268	248	235	216
1	575	552	522	494	469	445	426	398	373	347	322	298	272	264	245	232	213
2	566	543	514	486	461	438	419	392	366	341	316	293	268	260	241	229	210
3	555	533	504	477	452	429	411	384	359	334	310	287	262	255	236	225	206
4	544	521	493	466	442	420	402	375	351	327	302	280	256	249	231	220	202
5	530	509	481	455	431	409	392	366	342	318	295	273	249	243	226	215	197
6	516	495	468	442	419	398	381	356	332	309	286	265	242	237	219	210	192
7	501	480	454	429	406	386	370	345	322	300	277	257	234	229	213	204	187
8	484	465	439	414	392	373	357	333	311	289	267	247	225	222	206	197	181
9	467	448	423	399	377	359	344	320	299	278	256	238	216	214	198	191	175
10	449	430	406	383	362	344	330	307	286	266	245	227	207	205	190	184	168
11	429	412	388	366	346	328	315	293	273	254	233	216	197	196	182	176	162
12	409	392	370	348	329	312	300	278	259	241	221	205	186	187	173	168	155
13	388	372	351	329	311	295	284	263	245	227	208	193	175	177	164	160	147
14	366	351	330	309	292	277	267	247	230	213	195	181	163	166	154	152	139
15	343	329	309	289	272	259	249	230	214	198	181	168	151	155	144	143	131
16	318	305	287	268	252	239	231	212	197	182	166	154	138	144	134	134	123
17	293	281	264	245	231	219	212	194	180	166	151	140	125	132	123	124	114
18	267	256	240	222	208	198	192	175	162	149	135	125	112	120	111	114	105
19	240	230	216	199	187	178	172	157	145	134	121	112	100	108	100	104	95
20	217	208	195	180	169	160	155	142	131	121	109	101	90	97	90	94	86
21	197	189	176	163	153	145	141	129	119	109	99	92	82	88	82	85	78
22	179	172	161	149	139	132	128	117	108	100	90	84	75	80	75	77	71
23	164	157	147	136	128	121	117	107	99	91	83	77	68	74	68	71	65
24	151	144	135	125	117	111	108	98	91	84	76	70	63	68	63	65	60
25	139	133	124	115	108	103	99	91	84	77	70	65	58	62	58	60	55
26	128	123	115	107	100	95	92	84	78	71	65	60	53	58	53	55	51
27	119	114	107	99	93	88	85	78	72	66	60	56	50	53	50	51	47
28	111	106	99	92	86	82	79	72	67	62				50	46	48	44
29	103	99	93													45	41
30																42	38
31																	
32																	
33																	
34																	
35																	

Effective Length, $K\ell$, in feet — Y-Y Axis																	
0	583	559	529	501	475	451	432	404	378	352	326	302	276	268	248	235	216
1	580	556	526	498	473	449	429	402	376	350	324	301	275	266	247	233	214
2	576	553	523	495	470	446	427	399	373	348	322	299	273	264	244	231	212
3	573	549	520	492	466	443	424	396	371	345	320	296	271	261	242	228	209
4	569	545	516	488	463	440	420	393	368	342	317	294	269	258	240	225	206
5	564	541	512	485	460	436	417	390	365	340	314	292	267	256	237	222	203
6	560	537	508	481	456	433	413	386	361	336	312	289	264	252	234	218	200
7	555	532	503	476	452	429	409	382	358	333	309	286	262	249	231	214	196
8	550	527	499	472	448	425	405	379	354	330	305	283	259	246	228	210	193
9	545	522	494	467	443	421	401	375	350	326	302	280	256	242	224	206	189
10	539	517	489	463	439	416	396	370	346	323	299	277	253	238	221	202	185
11	533	511	483	458	434	412	392	366	342	319	295	273	250	234	217	197	180
12	527	505	478	452	429	407	387	361	338	315	291	270	247	230	213	192	176
13	521	499	472	447	424	402	382	357	334	310	287	266	243	226	209	187	171
14	515	493	466	441	418	397	376	352	329	306	283	263	240	221	205	182	166
15	508	487	460	436	413	392	371	347	324	302	279	259	236	217	200	176	161
16	501	480	454	430	407	387	365	341	319	297	275	255	233	212	196	170	156
17	494	474	448	424	402	381	360	336	314	292	271	251	229	207	191	165	150
18	487	467	441	417	396	375	354	330	309	287	266	246	225	202	187	159	145
19	480	459	434	411	389	369	348	325	304	282	261	242	221	197	182	152	139
20	472	452	427	404	383	363	341	319	298	277	257	238	217	191	177	146	133
21	464	445	420	398	377	357	335	313	293	272	252	233	213	186	171	139	127
22	456	437	413	391	370	351	329	307	287	266	247	228	209	180	166	132	120
23	448	429	406	384	363	344	322	301	281	261	241	224	204	174	161	125	114
24	440	421	398	376	356	338	315	294	275	255	236	219	200	168	155	118	107
25	432	413	390	369	349	331	308	288	269	250	231	214	195	162	149	111	100
26	423	405	382	361	342	324	301	281	262	244	225	209	190	156	143	103	93
27	414	396	374	354	335	317	294	274	256	238	220	203	185	149	137	95	86
28	405	387	366	346	327	310	286	267	249	231	214	198	181	143	131	89	80
29	396	378	357	338	320	303	279	260	243	225	208	192	176	136	125	83	75
30	386	369	349	330	312	295	271	253	236	219	202	187	170	129	118	77	70
31	377	360	340	321	304	288	263	245	229	212	196	181	165	122	112	72	65
32	367	351	331	313	296	280	255	238	222	205	190	175	160	115	105	68	61
33	357	341	322	304	288	272	247	230	214	199	183	169	154	108	99	64	58
34	347	331	313	295	279	264	238	222	207	192	177	163	149	101	93	60	54
35	337	321	303	286	271	256	230	214	200	184	170	157	143	96	88	57	51
36	326	311	293	277	262	247	221	206	192	177	164	151	138	90	83	54	48
37	316	301	284	268	253	239	212	197	184	170	157	145	132	86	78	51	46
38	305	290	274	258	244	230	203	189	176	162	150	138	126	81	74	48	43
39	294	279	263	248	235	221	194	180	168	155	142	131	119	77	71	46	41
40	282	269	253	239	225	212	184	171	160	147	135	125	114	73	67	43	39
41	271	257	242	228	216	203	175	163	152	140	129	119	108	70	64	41	
42	259	246	232	218	206	194	167	155	145	133	123	113	103	66	61		
43	247	235	221	208	196	185	159	148	138	127	117	108	98	63	58		
Q_s =	1.00	1.00	1.00	1.00	1.00	1.00	1.00	1.00	1.00	1.00	1.00	1.00	1.00	1.00	1.00	1.00	1.00

TABLE 4-6 *(Continued)*

STRUCTURAL TEES—MAIN MEMBERS

Allowable Axial Compression Load, P_m (Kips)

F_y 36 Ksi

Size	WT7	WT7	WT7	WT7	WT7	WT7	WT7	WT7	WT7	WT6	WT6	WT6	WT6	WT6	WT6	WT6	WT6
Weight	30.5	26.5	24	21.5	19	17	15	13	11	95	80.5	66.5	60	53	49.5	46	42.5
Effective Length $K\ell$ in ft. (x-x Axis)																	
0	194	168	152	130	113	93	77	61	44	603	512	423	382	337	315	292	270
1	191	166	151	129	112	92	76	61	43	594	504	417	376	332	310	287	266
2	188	164	148	127	111	91	75	60	43	584	495	409	369	325	304	281	260
3	185	161	146	125	109	90	74	59	43	572	485	400	361	318	297	275	254
4	181	158	143	122	107	88	73	59	42	559	474	390	352	310	289	267	247
5	177	154	140	120	105	87	72	58	41	545	461	379	342	300	281	259	240
6	172	151	136	117	103	85	71	57	41	529	447	367	331	290	271	250	231
7	167	147	133	114	101	83	69	56	40	512	432	354	319	279	261	241	222
8	162	142	129	111	98	82	68	55	40	494	416	339	307	268	250	231	212
9	156	138	125	107	96	80	66	54	39	475	398	324	293	255	238	220	202
10	150	133	120	104	93	78	65	52	38	454	380	309	279	242	226	208	191
11	144	128	116	100	90	75	63	51	37	433	361	292	264	228	213	196	180
12	138	123	111	96	87	73	61	50	37	410	341	274	248	214	199	183	167
13	131	117	106	92	84	71	59	49	36	387	320	256	231	199	185	169	155
14	124	111	101	87	81	68	58	47	35	362	298	237	214	182	169	155	141
15	116	105	95	83	78	66	56	46	34	336	275	216	195	166	153	140	127
16	109	99	90	78	74	63	54	44	33	309	251	195	176	148	137	125	112
17	101	92	84	73	70	60	51	43	32	281	226	173	157	131	121	110	100
18	92	86	78	68	67	57	49	41	31	252	202	155	140	117	108	99	89
19	83	79	71	63	63	54	47	39	30	226	181	139	125	105	97	88	80
20	75	71	65	57	59	51	45	38	29	204	163	125	113	95	87	80	72
21	68	65	59	52	55	48	42	36	28	185	148	114	103	86	79	72	65
22	62	59	53	47	50	45	40	34	27	169	135	104	93	78	72	66	59
23	57	54	49	43	46	41	37	32	26	154	123	95	86	72	66	60	54
24	52	50	45	40	42	38	34	30	24	142	113	87	79	66	61	55	50
25	48	46	41	37	39	35	32	28	23	131	104	80	72	61	56	51	
26	45	42	38	34	36	32	29	26	22	121	97	74	67				
27	41	39	35	31	33	30	27	24	21	112	90						
28	38	36	33	29	31	28	25	23	19								
29	36	34	31	27	29	26	24	21	18								
30		32	29	25	27	24	22	20	17								
31		30	27	24	25	23	21	19	16								
32					24	21	19	17	15								
33					22	20	18	16	14								
34					21	19	17	15	13								
35								15	12								

Effective Length, $K\ell$, in feet Y-Y Axis																	
0	194	168	152	130	113	93	77	61	44	603	512	423	382	337	315	292	270
1	192	166	151	129	112	92	76	60	43	598	508	420	380	335	313	289	268
2	190	164	148	127	110	90	74	59	42	594	504	417	376	332	310	287	266
3	188	161	146	125	107	88	73	57	41	589	500	413	373	329	308	284	263
4	185	158	143	123	105	86	71	55	39	583	495	409	369	326	305	282	261
5	182	155	140	120	102	84	69	53	38	577	490	405	365	322	301	279	258
6	179	151	137	117	98	81	67	50	36	571	485	400	361	318	298	275	255
7	176	147	133	114	95	79	65	48	35	564	479	395	357	314	294	272	252
8	173	143	129	111	91	76	63	45	33	557	473	390	352	310	290	268	248
9	169	139	125	108	87	73	60	42	31	550	466	385	347	306	286	264	245
10	165	134	121	104	83	69	58	38	28	542	459	379	342	301	282	260	241
11	161	129	117	100	79	66	55	35	26	534	452	373	337	296	277	256	237
12	157	124	112	96	74	62	52	31	24	526	445	367	331	291	272	252	233
13	153	119	107	92	69	59	49	27	21	517	438	361	325	286	267	247	228
14	149	113	102	88	64	55	46	24	19	508	430	354	319	281	262	242	224
15	144	107	97	84	59	50	42	21	16	498	421	347	313	275	257	237	220
16	139	101	91	79	53	46	39	18	14	489	413	340	306	269	251	232	215
17	134	95	85	74	48	42	35	16	13	479	404	333	300	263	246	227	210
18	129	88	79	69	42	37	31			469	395	325	293	257	240	222	205
19	124	81	73	64	38	33	28			458	386	318	286	251	234	216	200
20	119	74	67	58	34	30	25			447	377	310	278	244	228	210	194
21	113	68	61	53	31	27	23			436	367	302	271	238	222	205	189
22	107	62	55	48	28	25	21			425	357	293	263	231	215	199	183
23	101	56	50	44	26	23	19			413	347	285	255	224	209	192	178
24	95	52	46	41	24	21	18			401	337	276	247	217	202	186	172
25	89	48	43	37	22	19				389	326	267	239	209	195	180	166
26	83	44	40	35						376	315	257	230	202	188	173	160
27	77	41	37	32						363	304	248	222	194	180	166	153
28	71	38	34	30						350	292	238	213	186	173	159	147
29	66	35	32	28						337	281	228	204	178	165	152	140
30	62	33	30	26						323	269	218	194	170	157	145	133
31	58	31	28	24						309	256	208	185	161	149	137	126
32	55									294	244	197	175	153	141	130	119
33	51									279	231	186	165	144	133	122	112
34	48									264	218	176	156	135	125	115	106
35	46									249	205	166	147	128	118	108	100
36	43									236	194	157	139	121	112	102	.94
37	41									223	184	148	131	114	106	97	89
38	39									212	174	141	125	108	100	92	85
39	37									201	165	133	118	103	95	87	80
40	35									191	157	127	112	98	90	83	76
41										182	150	121	107	93	86	79	73
42										173	143	115	102	89	82	75	69
43										165	136	110	97	85	78	72	66
44										158	130	105	93	81	75	69	63
Q_s =	1.00	1.00	1.00	0.96	0.94	0.86	0.81	0.74	0.63	1.00	1.00	1.00	1.00	1.00	1.00	1.00	1.00

TABLE 4-6 *(Continued)*

STRUCTURAL TEES—MAIN MEMBERS

Allowable Axial Compression Load, P_m (Kips)

F_y 36 Ksi

Effective Length $K\ell$ in ft. — x-x Axis

Size	WT6	WT6	WT6	WT6	WT6	WT6	WT6	WT6	WT6	WT6	WT6	WT6	WT6	WT6	WT6	WT5	WT5
Weight	39.5	36	32.5	29	26.5	25	22.5	20	18	15.5	13.5	11	9.5	8.25	7	56	50
0	251	229	206	184	168	159	143	127	114	92	71	62	49	41	27	356	318
1	247	225	203	181	166	157	141	125	113	91	70	62	48	41	27	350	312
2	242	221	199	178	163	154	138	123	111	89	69	61	48	40	27	342	304
3	236	215	194	173	159	150	135	120	109	87	68	60	47	40	26	332	295
4	229	210	189	169	155	147	132	117	106	86	66	59	46	39	26	321	286
5	222	203	183	164	150	143	128	114	104	83	65	58	45	39	26	309	275
6	215	196	176	158	145	138	124	110	101	81	63	56	44	38	25	296	263
7	206	188	169	152	139	133	120	106	98	79	62	55	43	37	25	282	250
8	197	180	161	145	133	128	115	102	94	76	60	54	42	36	24	266	236
9	188	171	153	138	127	123	110	97	91	73	58	52	41	35	24	250	221
10	177	161	145	130	120	117	105	92	87	70	56	50	40	34	23	233	205
11	167	152	136	123	113	111	99	87	83	67	54	49	39	33	23	214	188
12	155	141	126	114	106	105	94	82	79	64	51	47	38	32	22	195	170
13	144	130	116	106	98	98	87	76	74	61	49	45	36	31	22	174	151
14	131	119	106	96	90	91	81	71	70	57	46	43	35	30	21	152	131
15	118	107	95	87	81	83	74	64	65	54	44	41	33	29	20	133	115
16	104	94	84	77	72	76	67	58	60	50	41	39	32	28	20	116	101
17	92	83	74	68	64	68	60	51	54	46	38	37	30	26	19	103	89
18	82	74	66	61	57	60	54	46	49	42	36	35	29	25	18	92	80
19	74	67	59	54	51	54	48	41	44	37	32	32	27	24	17	83	71
20	67	60	54	49	46	49	43	37	40	34	29	30	25	22	17	75	64
21	61	55	49	45	42	44	39	34	36	31	27	27	24	21	16	68	58
22	55	50	44	41	38	40	36	31	33	28	24	25	22	19	15		
23	50	46	40	37	35	37	33	28	30	26	22	23	20	18	14		
24	46	42	37	34	32	34	30	26	28	23	20	21	18	16	13		
25					30	31	28	24	25	22	19	19	17	15	13		
26						29	26		23	20	17	18	16	14	12		
27									22	19	16	17	15	13	11		
28									20	17	15	15	14	12	10		
29												14	13	11	9		
30												13	12	10	9		
31												13	11	10	8		
32														9			
33																	

Effective Length, $K\ell$, in feet — Y-Y Axis																	
0	251	229	206	184	168	159	143	127	114	92	71	62	49	41	27	356	318
1	249	227	205	183	167	157	141	126	113	90	70	61	47	40	27	353	315
2	247	225	203	181	165	155	139	124	111	89	69	58	46	39	26	350	312
3	244	223	201	179	163	152	137	122	108	87	67	55	43	37	24	346	308
4	242	221	199	176	161	150	134	120	105	85	66	52	41	34	23	342	304
5	239	219	197	174	159	146	132	117	102	82	64	48	38	32	22	337	300
6	236	216	194	171	156	143	129	114	99	80	62	44	35	29	20	332	296
7	233	213	192	168	153	140	125	111	95	77	60	40	32	26	18	327	291
8	230	210	189	165	150	136	122	108	91	74	58	35	28	22	16	322	286
9	227	207	186	162	147	132	118	105	87	71	56	30	24	18	14	316	281
10	223	204	183	158	144	127	114	102	83	68	53	24	20	15	12	310	276
11	220	200	180	155	141	123	110	98	78	64	51	20	16	12	10	304	270
12	216	197	177	151	137	118	106	94	73	60	48	17	14	10	8	297	264
13	212	193	174	147	134	113	101	90	68	56	45	14	12			290	258
14	208	189	170	143	130	108	97	86	63	52	42	12				283	251
15	203	186	167	139	126	103	92	82	57	48	39					275	245
16	199	181	163	134	122	97	87	77	52	44	36					268	238
17	194	177	159	130	118	92	82	73	46	39	33					260	231
18	190	173	155	125	113	86	76	68	41	35	29					252	223
19	185	169	151	120	109	80	70	63	37	31	26					243	216
20	180	164	147	115	104	73	65	57	33	28	24					235	208
21	175	159	143	110	100	66	59	52	30	25	22					226	200
22	169	155	139	105	95	61	53	47	27	23	20					216	191
23	164	150	134	100	90	55	49	43	25	21	18					207	183
24	159	145	129	94	84	51	45	40	23	20	17					197	174
25	153	139	125	88	79	47	41	37	21	18	15					187	165
26	147	134	120	82	74	43	38	34								177	156
27	141	129	115	76	68	40	35	32								166	146
28	135	123	110	71	63	37	33	29								156	137
29	129	117	105	66	59	35	31	27								145	127
30	123	111	99	62	55	33	29	26								136	119
31	116	105	94	58	52	31	27	24								127	111
32	109	99	88	54	49	29	25	22								119	105
33	103	93	83	51	46											112	98
34	97	88	78	48	43											106	93
35	91	83	74	45	41											100	87
36	86	78	70	43	38											94	83
37	82	74	66	41	36											89	78
38	77	70	63	39	34											84	74
39	74	67	59	37	33											80	70
40	70	63	56	35	31											76	67
41	67	60	54	33	30											73	64
42	63	58	51													69	61
43	61	55	49													66	58
Q_s =	1.00	1.00	1.00	1.00	1.00	1.00	1.00	1.00	1.00	0.93	0.83	0.89	0.81	0.79	0.61	1.00	1.00

TABLE 4-6 (*Continued*)

STRUCTURAL TEES—MAIN MEMBERS

Allowable Axial Compression Load, P_m (Kips)

F_y 36 Ksi

Size	WT5	WT5	WT5	WT5	WT5	WT5	WT5	WT5	WT5	WT5	WT5	WT5	WT5	WT5	WT5	WT5	WT5
Weight	44.5	38.5	36	33	30	27	24.5	22.5	19.5	16.5	14.5	12.5	10.5	9.5	8.5	7.5	5.75
Effective Length $K\ell$ in ft. x-x Axis 0	283	244	229	210	191	172	156	143	124	105	92	79	67	61	54	46	27
1	278	239	224	205	187	168	152	140	122	103	91	78	66	60	53	46	27
2	271	233	219	200	182	163	148	137	119	100	89	76	64	59	52	45	26
3	263	226	212	194	176	158	143	133	115	97	86	74	63	57	51	44	26
4	254	218	204	187	170	152	138	128	111	94	84	72	61	56	50	43	25
5	244	209	196	179	162	145	132	123	106	90	81	70	59	54	48	42	25
6	233	200	187	170	154	138	125	117	101	86	78	67	57	52	47	40	24
7	221	189	177	161	146	130	117	111	96	82	74	64	54	51	45	39	23
8	208	177	166	150	136	121	109	104	90	77	71	61	52	48	43	37	23
9	195	165	154	139	126	112	101	97	84	72	67	57	49	46	41	36	22
10	180	152	142	128	116	102	92	89	77	66	63	54	46	44	39	34	21
11	165	138	129	116	104	92	82	81	70	60	58	50	43	42	37	32	20
12	148	124	115	102	92	80	72	73	63	54	54	46	40	39	35	30	19
13	131	108	100	89	79	69	62	63	55	48	49	42	37	36	32	28	18
14	114	93	86	76	68	59	53	55	47	41	44	37	33	33	30	26	17
15	99	81	75	67	60	52	46	48	41	36	39	33	29	30	27	24	16
16	87	72	66	58	52	46	41	42	36	32	34	29	26	27	24	22	15
17	77	63	58	52	46	40	36	37	32	28	30	26	23	24	22	19	14
18	69	57	52	46	41	36	32	33	29	25	27	23	20	22	19	17	13
19	62	51	47	41	37	32	29	30	26	22	24	20	18	19	17	16	12
20	56	46	42	37	34			27	23	20	22	18	16	18	16	14	11
21	50									18	20	17	15	16	14	13	10
22											18	15	14	14	13	12	9
23											16	14	12	13	12	11	8
24														12	11	10	8
25														11	10	9	7
26																8	6
27																	
28																	
29																	
30																	
31																	
32																	
33																	
34																	
35																	

Effective Length, $K\ell$, in feet — Y-Y Axis																	
0	283	244	229	210	191	172	156	143	124	105	92	79	67	61	54	46	27
1	280	242	227	208	189	170	154	141	123	103	91	78	66	59	52	45	26
2	278	239	225	206	187	168	153	139	121	102	89	76	64	57	50	43	25
3	275	237	222	203	185	166	151	137	119	100	86	74	62	54	47	41	24
4	271	234	219	201	183	164	149	135	117	98	84	72	60	50	44	38	23
5	268	231	216	198	180	162	147	132	114	96	81	69	58	47	41	35	21
6	264	227	213	195	177	159	144	129	112	94	78	67	56	42	37	31	19
7	259	223	210	192	174	157	142	126	109	92	74	64	53	38	32	27	17
8	255	220	206	188	171	154	139	123	106	89	70	60	50	33	28	23	15
9	250	215	202	185	168	151	137	119	103	87	66	57	47	27	23	18	13
10	245	211	198	181	165	148	134	115	100	84	62	53	44	22	18	15	11
11	240	207	194	177	161	145	131	111	96	81	58	49	40	18	15	12	9
12	235	202	189	173	157	141	128	107	93	78	53	45	37	15	13	10	8
13	229	197	185	169	153	138	125	103	89	74	48	41	33	13	11	9	6
14	223	192	180	164	149	134	121	99	85	71	43	36	29	11	9		
15	217	187	175	160	145	130	118	94	81	67	37	32	25				
16	211	181	170	155	141	126	114	89	77	64	33	28	22				
17	205	175	164	150	136	122	110	84	72	60	29	25	19				
18	198	170	159	145	132	118	106	79	68	56	26	22	17				
19	191	164	153	140	127	114	102	74	63	52	23	20	16				
20	184	157	147	134	122	109	98	68	58	47	21	18	14				
21	177	151	141	129	117	105	94	62	53	43	19	16	13				
22	169	144	135	123	112	100	90	57	48	39	17	15					
23	162	138	129	117	106	95	85	52	44	36							
24	154	131	122	111	101	90	81	48	41	33							
25	146	124	115	105	95	85	76	44	37	30							
26	137	116	108	98	89	79	71	41	35	28							
27	129	109	101	92	83	74	66	38	32	26							
28	120	101	94	85	77	69	61	35	30	24							
29	112	94	88	80	72	64	57	33	28	23							
30	104	88	82	74	67	60	54	31	26	21							
31	98	82	77	70	63	56	50	29	24	20							
32	92	77	72	65	59	53	47	27	23	18							
33	86	73	68	61	56	50	44	25									
34	81	69	64	58	52	47	42										
35	77	65	60	55	49	44	39										
36	73	61	57	52	47	42	37										
37	69	58	54	49	44	39	35										
38	65	55	51	46	42	37	33										
39	62	52	48	44	40	35	32										
40	59	50	46	42	38	34	30										
41	56	47	44	40	36	32	29										
42	53	45	42	38	34	31	27										
43	51	43	40														
Q_s	1.00	1.00	1.00	1.00	1.00	1.00	1.00	1.00	1.00	1.00	1.00	1.00	1.00	1.00	1.00	0.98	0.73

TABLE 4-6 *(Continued)*

STRUCTURAL TEES—MAIN MEMBERS

Allowable Axial Compression Load, P_m (Kips)

F_y 36 Ksi

Size	WT4	WT4	WT4	WT4	WT4	WT4	WT4	WT4	WT4	WT4	WT4	WT4	WT4	WT3	WT3	WT3	WT3
Weight	33.5	29	24	20	17.5	15.5	14	12	10	8.5	7.5	6.5	5	12.5	10	7.75	8
Effective Length $K\ell$ in ft. x–x Axis																	
0	213	184	152	127	111	98	89	76	64	54	48	41	29	79	64	49	51
1	208	180	149	124	108	96	87	74	62	53	47	41	29	77	61	48	49
2	201	174	144	120	105	93	84	72	61	51	46	40	28	73	58	45	47
3	193	167	137	115	100	89	80	69	58	50	44	38	27	69	55	43	45
4	184	159	130	109	95	84	76	65	56	47	43	37	26	64	51	40	42
5	174	150	123	102	89	79	72	62	53	45	41	35	25	58	46	36	38
6	163	140	114	95	82	73	67	57	50	43	39	34	24	51	41	32	35
7	151	129	105	87	75	67	62	53	47	40	37	32	23	44	35	28	31
8	138	118	94	79	68	60	56	48	43	37	34	30	21	37	28	23	26
9	124	105	83	70	59	53	50	42	40	34	32	28	20	29	22	18	21
10	109	92	71	60	50	45	43	36	36	30	29	26	18	24	18	15	17
11	93	78	59	50	42	37	36	30	31	27	26	23	17	19	15	12	14
12	78	65	50	42	35	31	30	25	27	23	23	21	15	16	13	10	12
13	67	56	42	36	30	26	26	22	23	20	20	18	13	14		9	10
14	57	48	37	31	26	23	22	19	20	17	17	15	11				
15	50	42	32	27	22	20	19	16	17	15	15	13	10				
16	44	37	28	23	20	17	17	14	15	13	13	12	9				
17	39	32							13	11	12	10	8				
18									12	10	11	9	7				
19											9	8	6				
20											9	8	6				
21																	
22																	
23																	
24																	
25																	
26																	
27																	
28																	
29																	
30																	
31																	
32																	
33																	
34																	
35																	

Effective Length, $K\ell$, in feet — Y-Y Axis																	
0	213	184	152	127	111	98	89	76	64	54	48	41	29	79	64	49	51
1	210	182	151	126	110	97	87	75	62	53	47	40	28	78	62	48	50
2	208	180	149	124	108	96	86	74	61	52	45	38	27	77	61	47	48
3	205	177	147	122	107	95	84	72	59	50	42	36	26	75	60	46	46
4	201	174	144	120	105	93	82	70	57	48	40	34	24	73	58	45	43
5	198	171	141	117	103	91	80	68	55	46	37	31	22	71	56	44	41
6	194	167	138	115	101	89	77	66	52	44	34	28	20	68	55	42	38
7	189	164	135	112	98	87	75	64	49	41	30	25	18	66	52	40	34
8	185	160	132	109	96	85	72	62	46	39	26	21	16	63	50	38	31
9	180	155	128	106	93	82	69	59	43	36	22	17	13	60	48	37	27
10	175	151	125	103	90	80	66	56	40	33	18	14	11	57	45	34	23
11	170	146	121	100	87	77	62	53	36	30	15	12	9	54	43	32	19
12	164	141	117	96	84	74	59	50	32	26	12	10	8	50	40	30	16
13	158	136	112	93	81	71	55	47	28	23	10	8	6	47	37	28	14
14	152	131	108	89	77	68	51	44	24	20	9	7		43	34	25	12
15	146	125	103	85	74	65	47	40	21	17				39	31	22	10
16	139	120	98	81	70	62	43	37	19	15				35	27	20	9
17	132	114	93	76	67	58	39	33	17	13				31	24	17	
18	125	108	88	72	63	55	35	29	15	12				27	21	16	
19	118	101	83	67	58	51	31	26	13	11				25	19	14	
20	111	95	77	62	54	47	28	24	12	10				22	17	13	
21	103	88	71	57	50	43	25	22						20	16	11	
22	95	81	65	52	45	39	23	20						18	14	10	
23	87	74	60	48	42	36	21	18						17	13	10	
24	80	68	55	44	38	33	19	16						15	12	9	
25	73	62	51	41	35	31	18	15						14	11		
26	68	58	47	38	33	28	17	14									
27	63	54	43	35	30	26											
28	59	50	40	32	28	24											
29	55	46	38	30	26	23											
30	51	43	35	28	24	21											
31	48	41	33	26	23	20											
32	45	38	31	25	21	19											
33	42	36	29	23	20	18											
34	40	34	27														
35	37																
36																	
37																	
38																	
39																	
40																	
41																	
42																	
43																	
Q_s =	1.00	1.00	1.00	1.00	1.00	1.00	1.00	1.00	1.00	1.00	1.00	1.00	0.91	1.00	1.00	1.00	1.00

TABLE 4-6 *(Continued)*

STRUCTURAL TEES—MAIN MEMBERS

Allowable Axial Compression Load, P_m (Kips)

F_y 36 Ksi

Effective Length $K\ell$ in ft. — x-x Axis

Size	WT3	WT3	WT2.5	WT2.5	WT2	MT7	MT6	MT5	MT5	MT4	MT4	MT4	MT4	MT4	MT4	MT3	MT3
Weight	6	4.25	9.25	8	6.5	8.6	5.9	11.45	4.5	18.85	17.5	16.3	9.25	3.25	2.75	16.87	10
0	38	27	59	51	41	27	18	73	16	120	109	103	59	13	13	107	64
1	37	26	56	48	39	27	18	71	15	117	106	101	57	13	13	104	61
2	36	25	52	45	36	27	18	70	15	113	103	97	56	13	12	99	58
3	34	24	47	41	32	27	18	68	15	108	99	93	54	12	12	93	54
4	32	22	42	36	27	26	17	66	15	103	94	88	51	12	12	86	49
5	29	20	35	30	21	26	17	64	15	97	89	83	49	12	11	78	44
6	26	19	28	24	15	26	17	61	14	90	83	77	46	12	11	69	38
7	23	16	21	18	11	26	17	58	14	83	77	71	43	11	10	59	32
8	20	14	16	14	8	25	16	55	14	75	70	64	39	11	10	49	25
9	17	12	13	11		25	16	52	13	67	62	56	36	11	9	39	20
10	13	9	10			25	16	49	13	58	54	48	32	10	8	31	16
11	11	8				24	16	45	13	48	46	40	28	10	8	26	13
12	9	6				24	15	42	12	41	39	33	23	9	7	22	11
13	8	6				23	15	38	12	35	33	28	20	9	6		
14	7	5				23	15	34	11	30	28	25	17	8	5		
15						23	14	29	11	26	25	21	15	8	5		
16						22	14	26	10	23	22	19	13	7	4		
17						22	14	23	10		19		12	6	4		
18						21	13	20	10				10	6	3		
19						21	13	18	9					5			
20						20	12	17	8					5			
21						20	12	15	8					4			
22						19	11	14	7					4			
23						19	11		7								
24						18	11		6								
25						17	10		6								
26						17	10		5								
27						16	9										
28						16	9										
29						15	8										
30						14	8										
31						14	7										
32						13	7										
33						12											
34						12											
35						11											

Effective Length, $K\ell$, in feet — Y-Y Axis																	
0	38	27	59	51	41	27	18	73	16	120	109	103	59	13	13	107	64
1	37	26	58	50	40	27	17	71	15	118	107	102	57	12	12	105	62
2	36	25	56	49	39	26	17	69	14	116	106	101	56	11	11	103	61
3	34	24	55	47	37	25	16	67	13	115	104	99	54	10	9	101	59
4	32	22	53	45	35	24	14	65	12	112	102	97	52	9	7	98	58
5	30	21	51	44	33	22	13	62	10	110	100	95	49	7	5	95	56
6	27	19	48	42	31	21	11	59	8	107	97	93	46	5	4	91	54
7	25	17	46	39	28	19	10	56	6	105	95	90	43	4		88	51
8	22	15	43	37	26	17	8	52		102	92	88	40			84	49
9	19	13	40	35	23	16		48		98	89	85	37			80	46
10	15	10	37	32	19	14		44		95	86	82	33			75	43
11	13	8	34	29	16	11		40		92	82	79	29			71	40
12	11	7	31	26	14	10		35		88	79	76	25			66	37
13	9	6	27	23	12			31		84	75	72	21			60	34
14	8	5	24	20	10			26		80	71	69	18			55	30
15	7		21	17	9			23		76	67	65	16			49	27
16			18	15	8			20		72	63	61	14			43	23
17			16	13				18		67	59	57	12			38	21
18			14	12				16		62	55	53	11			34	18
19			13	11				14		57	50	49				31	17
20			12	10				13		52	45	44				28	15
21			10							48	41	40				25	14
22										43	37	37				23	12
23										40	34	34				21	11
24										36	31	31				19	
25										34	29	28					
26										31	27	26					
27										29	25	24					
28										27	23	23					
29										25	22	21					
30										23	20	20					
31										22		18					
32																	
33																	
34																	
35																	
36																	
37																	
38																	
39																	
40																	
41																	
42																	
43																	
Q_s =	1.00	1.00	1.00	1.00	1.00	0.50	0.48	1.00	0.55	1.00	1.00	1.00	1.00	0.63	0.74	1.00	1.00

TABLE 4-6 *(Continued)*

STRUCTURAL TEES—MAIN MEMBERS

Allowable Axial Compression Load, P_m (Kips)

F_y 36 Ksi

Size	MT3	MT2.5	MT2	MT2	MT2	ST12	ST12	ST12	ST12	ST12	ST10	ST10	ST10	ST10	ST9	ST7.5	ST7.5
Weight	2.2	9.45	8.15	6.9	6.5	60	52.95	50	45	39.95	47.5	42.5	37.5	32.7	27.35	25	21.45
Effective Length $K\ell$ in ft. x-x Axis																	
0	11	60	52	44	41	380	337	318	285	224	302	270	238	208	174	159	136
1	11	57	49	41	39	378	335	316	283	223	300	268	236	206	172	157	135
2	10	54	44	38	35	375	333	314	282	221	298	266	234	205	171	155	133
3	10	49	39	33	30	373	330	311	280	220	295	263	232	203	169	153	132
4	9	43	32	27	24	370	327	309	277	218	292	261	230	201	167	151	130
5	9	37	24	21	18	367	325	307	275	216	289	258	227	199	165	149	127
6	8	30	17	15	12	363	322	304	273	215	286	255	225	196	163	146	125
7	8	22	12	11	9	360	318	301	270	213	282	252	222	194	160	143	123
8	7	17	10	8		356	315	298	267	211	279	248	219	191	158	141	120
9	6	13				352	311	295	265	209	275	245	216	188	155	137	117
10	6	11				348	308	292	262	207	270	241	213	186	152	134	114
11	5					344	304	288	259	204	266	237	210	183	149	131	111
12	4					340	300	285	255	202	262	233	206	179	146	127	108
13	4					335	296	281	252	199	257	229	202	176	143	123	105
14	3					330	291	277	249	197	252	224	199	173	140	120	101
15	3					325	287	274	245	194	247	220	195	169	137	116	98
16						320	282	270	241	192	242	215	191	166	133	111	94
17						315	278	265	238	189	237	210	187	162	129	107	90
18						310	273	261	234	186	231	205	183	158	126	103	86
19						305	268	257	230	183	226	200	178	154	122	98	82
20						299	263	252	226	180	220	195	174	150	118	93	78
21						293	257	248	222	177	214	189	169	146	114	89	74
22						287	252	243	217	174	208	184	165	142	110	83	69
23						281	246	238	213	171	202	178	160	138	105	78	64
24						275	241	233	209	168	195	172	155	133	101	73	60
25						269	235	228	204	164	189	166	150	129	96	67	55
26						263	229	223	199	161	182	160	145	124	92	62	51
27						256	223	218	194	157	175	154	139	119	87	58	47
28						249	217	213	190	154	168	147	134	114	82	54	44
29						242	211	207	185	150	161	141	128	109	77	50	41
30						236	204	202	180	146	154	134	123	104	72	47	38
31						228	198	196	174	143	146	127	117	99	68	44	36
32						221	191	190	169	139	139	120	111	94	63	41	34
33						214	184	184	164	135	131	113	105	88	60	39	32
34						206	177	178	158	131	123	106	99	83	56	36	30
35						198	170	172	153	127	116	100	93	78	53	34	28

Effective Length, $K\ell$, in feet — Y-Y Axis																	
0	11	60	52	44	41	380	337	318	285	224	302	270	238	208	174	159	136
1	10	59	50	43	40	374	332	311	280	220	297	265	233	203	170	155	133
2	9	57	49	41	38	367	326	304	273	215	290	259	226	198	165	150	129
3	7	55	47	39	37	359	319	295	265	210	282	252	219	192	159	144	124
4	5	53	44	37	35	350	311	285	256	204	273	244	210	184	153	137	119
5	3	51	41	35	32	339	302	273	247	197	263	236	200	176	146	129	112
6		48	38	32	30	328	292	261	236	190	252	226	190	167	138	121	105
7		46	35	29	27	316	282	247	224	182	240	216	178	157	129	112	98
8		42	31	26	24	303	271	233	212	173	227	205	165	147	120	102	90
9		39	27	22	21	289	259	217	198	164	213	193	152	136	110	91	81
10		36	23	19	17	274	246	201	184	155	198	180	138	124	99	79	72
11		32	19	15	14	259	233	183	169	144	183	167	123	111	87	67	62
12		28	16	13	12	243	219	165	153	134	167	153	106	98	75	56	52
13		24	14	11	10	226	205	145	136	122	149	138	91	84	64	48	44
14		21	12	9	9	208	190	125	118	110	131	122	78	72	55	41	38
15		18	10	8	8	189	174	109	103	98	114	107	68	63	48	36	33
16		16	9			169	157	96	90	86	100	94	60	55	42	32	29
17		14				150	140	85	80	76	89	83	53	49	37	28	26
18		13				134	125	76	71	68	79	74	47	44	33		
19		11				120	112	68	64	61	71	66	43	39			
20						108	101	61	58	55	64	60					
21						98	92	56	52	50	58	54					
22						89	83			45	53	50					
23						82	76										
24						75	70										
25						69	65										
26							60										
27																	
28																	
29																	
30																	
31																	
32																	
33																	
34																	
35																	
36																	
37																	
38																	
39																	
40																	
41																	
42																	
43																	
Q_s =	0.78	1.00	1.00	1.00	1.00	1.00	1.00	1.00	1.00	0.88	1.00	1.00	1.00	1.00	1.00	1.00	1.00

TABLE 4-6 *(Continued)*

STRUCTURAL TEES—MAIN MEMBERS
Allowable Axial Compression Load, P_m (Kips)

F_y 36 Ksi

Size	ST6	ST6	ST6	ST5	ST4	ST3.5	ST3	ST2.5	ST2	ST1.5
Weight	20.4	17.5	15.9	12.7	9.2	7.65	6.25	5	3.85	2.85
Effective Length Kℓ in ft. — x-x Axis										
0	130	111	101	81	58	49	40	32	24	18
1	128	110	100	79	57	47	38	30	23	17
2	126	108	98	78	55	46	37	29	21	14
3	124	106	96	76	54	44	35	27	19	11
4	121	104	94	74	51	42	32	24	16	8
5	118	101	92	71	49	39	30	21	13	5
6	115	99	90	69	46	36	27	18	9	3
7	112	96	87	66	43	33	24	14	7	
8	108	93	84	63	40	30	20	11	5	
9	104	90	81	60	37	26	16	9		
10	100	87	78	56	33	23	13	7		
11	96	83	75	53	29	19	11	6		
12	92	80	71	49	25	16	9			
13	87	76	68	45	22	13	8			
14	82	72	64	41	19	12				
15	77	68	60	36	16	10				
16	72	63	56	32	14	9				
17	66	59	52	28	13					
18	61	54	47	25	11					
19	55	49	43	23						
20	49	45	38	20						
21	45	40	35	18						
22	41	37	32	17						
23	37	34	29	15						
24	34	31	27	14						
25	32	29	25							
26	29	26	23							
27	27	24	21							
28	25	23	20							
29	23	21	18							
30		20								
31										
32										
33										

Effective Length, $K\ell$, in feet — Y-Y Axis																
0	130	111	101	81	58	49	40	32	24	18						
1	127	108	99	78	56	47	38	30	23	17						
2	123	104	95	76	54	45	36	29	22	16						
3	118	100	91	72	51	42	33	26	19	14						
4	113	95	87	68	48	39	30	23	17	12						
5	107	89	82	64	44	35	27	20	14	9						
6	100	83	76	59	39	31	23	17	11	7						
7	93	76	70	54	35	26	19	13	8	5						
8	85	68	63	48	29	21	15	10	6	4						
9	76	60	56	42	24	17	12	8	5							
10	67	51	48	35	19	14	9	6								
11	58	42	40	29	16	11	8									
12	49	36	34	24	13	10										
13	41	30	29	21	11											
14	36	26	25	18												
15	31	23	22	16												
16	27	20	19													
17	24															
18																
19																
20																
21																
22																
23																
24																
25																
26																
27																
28																
29																
30																
31																
32																
33																
34																
35																
36																
37																
38																
39																
40																
41																
42																
Q_s =	1.00	1.00	1.00	1.00	1.00	1.00	1.00	1.00	1.00	1.00	1.00	1.00	1.00	1.00	1.00	1.00

TABLE 4-6 *(Continued)*

STRUCTURAL TEES—MAIN MEMBERS

Allowable Axial Compression Load, P_m (Kips)

F_y 50 ksi

Size	WT18	WT18	WT18	WT18	WT18	WT18	WT18	WT18	WT18	WT18	WT18	WT16.5	WT16.5	WT16.5	WT16.5	WT16.5	WT16.5
Weight	150	140	130	122.5	115	97	91	85	80	75	67.5	120	110	100	76	70.5	x65
Effective Length Kℓ in ft. x-x Axis 0	1227	1071	941	834	725	609	515	424	372	322	271	939	798	653	386	330	284
1	1221	1066	937	831	722	607	513	423	371	321	270	934	794	650	384	329	283
2	1214	1061	932	827	719	604	511	421	369	320	269	929	790	646	383	328	282
3	1208	1055	927	823	715	601	508	419	368	319	268	923	785	643	381	326	280
4	1200	1049	922	818	711	599	506	418	366	317	267	917	780	639	379	325	279
5	1193	1042	917	814	708	596	504	416	365	316	266	911	775	635	377	323	278
6	1185	1036	911	809	703	593	501	414	363	315	265	904	770	631	375	322	277
7	1176	1029	905	804	699	589	499	412	362	314	264	897	764	627	373	320	275
8	1167	1021	899	798	695	586	496	410	360	312	263	890	758	622	371	318	274
9	1158	1013	892	793	690	582	493	408	358	311	262	882	752	617	369	317	273
10	1149	1005	885	787	685	579	490	406	356	309	261	874	746	612	367	315	271
11	1139	997	878	781	681	575	488	403	354	308	259	866	739	607	365	313	270
12	1128	988	871	775	675	571	484	401	353	306	258	857	732	602	362	311	268
13	1118	980	863	768	670	567	481	399	351	304	257	848	725	597	360	309	266
14	1107	970	856	762	665	563	478	396	348	303	255	839	717	591	357	307	265
15	1095	961	848	755	659	559	475	394	346	301	254	830	710	585	355	305	263
16	1084	951	839	748	653	554	471	391	344	299	253	820	702	579	352	303	261
17	1072	941	831	741	647	550	468	388	342	297	251	810	694	573	349	300	259
18	1059	931	822	733	641	545	464	385	340	296	250	799	685	566	346	298	258
19	1047	920	813	726	635	541	460	383	337	294	248	789	677	560	343	296	256
20	1034	909	804	718	629	536	457	380	335	292	247	778	668	553	340	293	254
21	1020	898	795	710	622	531	453	377	333	290	245	766	659	546	337	291	252
22	1007	887	785	702	615	526	449	374	330	288	243	755	650	539	334	288	250
23	993	875	776	694	609	521	445	371	327	286	242	743	640	532	331	286	248
24	979	863	766	685	602	515	440	368	325	284	240	731	631	525	328	283	246
25	964	851	755	677	595	510	436	365	322	281	238	719	621	518	324	281	243
26	950	839	745	668	587	505	432	361	320	279	237	707	611	510	321	278	241
27	934	826	734	659	580	499	428	358	317	277	235	694	601	502	318	275	239
28	919	814	724	650	573	493	423	355	314	275	233	681	590	494	314	272	237
29	903	801	713	640	565	488	419	351	311	272	231	668	580	486	311	270	234
30	888	787	702	631	557	482	414	348	308	270	229	654	569	478	307	267	232
31	871	774	690	621	549	476	409	344	305	268	227	641	558	470	303	264	230
32	855	760	679	612	541	470	405	341	303	265	226	627	547	461	300	261	227
33	838	746	667	602	533	464	400	337	300	263	224	612	535	453	296	258	225
34	821	732	655	591	525	457	395	333	296	260	222	598	524	444	292	255	222
35	804	717	643	581	516	451	390	330	293	258	220	583	512	435	288	251	220

Effective Length, $K\ell$, in feet — Y-Y Axis																	
0	1227	1071	941	834	725	609	515	424	372	322	271	939	798	653	386	330	284
1	1218	1064	935	829	720	604	510	421	369	320	269	933	793	649	382	327	281
2	1209	1056	928	824	716	597	505	417	366	317	266	925	787	644	379	324	279
3	1199	1048	921	817	711	591	500	413	362	314	264	918	780	639	375	321	276
4	1189	1039	913	811	705	583	494	408	358	311	261	909	773	634	371	318	273
5	1177	1029	905	804	699	575	488	403	354	307	258	900	766	628	366	314	270
6	1165	1019	896	796	693	566	481	398	350	303	255	890	758	622	361	310	266
7	1152	1008	887	788	686	557	473	392	345	299	252	880	750	615	356	305	263
8	1139	997	877	780	679	547	465	386	340	295	248	869	741	608	350	300	259
9	1124	985	867	771	672	537	457	380	334	291	245	858	731	601	344	296	255
10	1109	972	856	762	664	526	449	373	329	286	241	846	722	594	338	290	250
11	1094	959	845	752	656	514	440	367	323	281	237	833	711	586	331	285	246
12	1078	946	833	742	648	502	430	359	317	276	232	820	701	578	324	279	241
13	1061	932	821	732	639	490	420	352	311	271	228	807	690	569	317	273	236
14	1043	917	808	721	631	477	410	344	304	266	223	793	679	561	310	267	231
15	1025	902	796	710	621	464	400	336	298	260	219	779	667	552	302	261	226
16	1007	886	782	699	612	450	389	328	291	254	214	764	655	542	294	255	220
17	987	870	768	687	602	436	378	320	283	248	209	748	642	533	286	248	215
18	968	854	754	675	592	421	366	311	276	242	204	733	629	523	278	241	209
19	947	837	740	662	582	406	354	302	268	236	198	716	616	513	269	234	203
20	927	819	725	650	571	390	342	292	260	229	193	700	602	502	260	227	197
21	905	801	709	637	561	374	329	283	252	223	187	683	589	492	251	219	190
22	883	783	694	623	550	357	316	273	244	216	181	665	574	481	242	211	184
23	861	764	678	610	538	340	303	263	236	209	176	647	560	469	232	203	177
24	838	745	661	596	527	323	289	253	227	201	169	629	545	458	222	195	170
25	814	725	644	581	515	304	275	242	218	194	163	610	529	446	212	187	163
26	790	705	627	567	503	286	261	231	209	186	157	591	514	434	202	178	156
27	765	685	610	552	490	267	246	220	199	179	150	571	497	422	191	169	149
28	740	664	592	536	478	248	230	208	190	171	144	551	481	409	180	160	141
29	714	642	573	521	465	231	215	197	180	162	137	530	464	397	169	151	133
30	688	620	554	505	452	216	201	184	170	154	130	509	447	384	157	142	125
31	661	598	535	488	438	202	188	173	159	145	122	487	430	370	147	133	117
32	634	575	516	472	424	190	176	162	149	137	115	465	412	357	138	124	110
33	605	552	496	455	410	178	166	152	140	128	108	442	393	343	130	117	104
34	577	528	475	438	396	168	156	144	132	121	102	419	374	328	123	110	98
35	547	503	454	420	382	159	148	135	125	114	96	396	355	314	116	104	92
36	518	478	433	402	367	150	139	128	118	108	91	374	336	299	109	98	87
37	490	453	411	383	352	142	132	121	112	102	86	354	318	284	104	93	82
38	465	430	390	365	336	135	125	115	106	97	82	336	302	269	98	88	78
39	441	408	370	346	321	128	119	109	101	92	78	319	286	255	93	84	74
40	419	388	352	329	305	121	113	104	96	87		303	272	243	89	80	
41	399	369	335	313	290	116	108	99	91	83		289	259	231	84		
42	380	352	319	298	276	110	102	94				275	247	220			
43	363	335	305	285	264							262	236	210			
Q_s =	0.93	0.87	0.82	0.77	0.71	0.71	0.64	0.57	0.53	0.49	0.45	0.89	0.82	0.74	0.57	0.53	0.49

TABLE 4-6 *(Continued)*

STRUCTURAL TEES—MAIN MEMBERS
Allowable Axial Compression Load, P_m (Kips)

F_y 50 Ksi

Size	WT16.5	WT15	WT15	WT15	WT15	WT15	WT15	WT15	WT15	WT13.5	WT13.5	WT13.5	WT13.5	WT13.5	WT13.5	WT13.5	WT12
Weight	59	105	95	86	66	62	62	54	49.5	88.5	80	72.5	57	51	47	42	80
Effective Length $K\ell$ in ft. — x-x Axis																	
0	237	850	702	576	387	328	290	258	217	750	613	498	353	264	219	179	678
1	236	845	698	572	385	327	289	257	216	745	610	495	351	263	218	179	673
2	235	840	694	569	383	325	287	256	215	739	605	492	349	262	217	178	667
3	234	834	690	566	381	324	286	255	215	733	601	488	347	260	216	177	660
4	233	828	685	562	379	322	285	253	214	727	596	484	344	259	215	176	653
5	232	821	679	558	376	320	283	252	212	719	590	480	342	257	213	175	646
6	231	814	674	553	374	318	281	251	211	712	584	476	339	255	212	174	638
7	230	806	668	549	372	316	280	249	210	704	578	471	336	253	210	173	629
8	229	798	662	544	369	314	278	248	209	696	572	466	333	251	209	172	620
9	228	790	655	539	366	312	276	246	208	687	565	461	330	249	207	171	611
10	227	781	648	534	363	310	274	244	206	678	558	456	327	247	206	169	601
11	226	772	641	529	360	307	272	243	205	668	551	450	324	245	204	168	590
12	224	763	634	523	357	305	270	241	204	658	543	444	321	243	202	167	579
13	223	753	627	517	354	302	268	239	202	648	535	439	317	240	201	165	568
14	222	743	619	511	351	300	266	237	201	637	527	432	313	238	199	164	556
15	220	733	611	505	347	297	264	235	199	626	519	426	310	235	197	162	544
16	219	722	602	499	344	294	261	233	198	615	510	419	306	233	195	161	531
17	218	711	594	492	340	291	259	231	196	603	501	413	302	230	193	159	518
18	216	700	585	486	337	289	257	229	195	591	492	406	298	228	191	158	505
19	215	688	576	479	333	286	254	227	193	578	482	398	293	225	189	156	491
20	213	676	567	472	329	283	252	225	191	565	472	391	289	222	186	155	472
21	212	664	557	465	325	279	249	223	189	552	462	384	285	219	184	153	462
22	210	651	548	457	321	276	246	220	188	539	452	376	280	216	182	151	448
23	208	639	538	450	317	273	244	218	186	525	442	368	276	213	180	149	432
24	207	626	527	442	313	270	241	216	184	511	431	360	271	210	177	148	416
25	205	612	517	434	309	266	238	213	182	496	420	352	266	207	175	146	400
26	203	599	506	426	304	263	235	211	180	482	409	343	261	203	172	144	384
27	202	585	496	418	300	259	232	208	178	467	397	335	256	200	170	142	367
28	200	570	485	410	295	256	229	206	176	451	385	326	251	197	167	140	349
29	198	556	473	401	291	252	226	203	174	435	373	317	246	193	165	138	332
30	196	541	462	392	286	248	223	200	172	419	361	308	240	190	162	136	313
31	194	526	450	384	281	245	220	198	170	402	349	298	235	186	159	134	294
32	193	510	438	374	276	241	217	195	168	386	336	289	230	183	156	132	276
33	191	495	426	365	271	237	213	192	165	368	323	279	224	179	154	130	260
34	189	479	414	356	266	233	210	189	163	351	309	269	218	175	151	128	245
35	187	462	401	346	261	229	207	186	161	332	296	259	212	171	148	125	231

Effective Length, $K\ell$, in feet — Y-Y Axis																	
0	237	850	702	576	387	328	290	258	217	750	613	498	353	264	219	179	678
1	235	844	697	572	383	325	287	256	215	744	609	494	349	262	217	178	673
2	233	837	692	567	378	321	284	253	213	737	603	490	345	259	215	176	667
3	231	829	686	563	374	318	281	250	211	729	598	486	340	255	212	174	660
4	228	821	679	558	368	313	277	246	208	721	591	481	335	252	209	172	653
5	225	812	673	552	362	309	273	243	205	712	585	476	329	248	206	169	645
6	223	803	665	546	356	303	269	239	202	703	577	470	323	244	203	166	637
7	220	793	657	540	349	298	264	235	198	693	569	464	316	239	199	164	628
8	216	782	649	534	342	292	259	230	195	682	561	458	309	234	195	161	619
9	213	771	640	527	334	286	254	226	191	671	553	452	301	229	192	157	609
10	209	759	631	520	327	280	248	221	187	659	544	445	293	224	187	154	599
11	206	747	621	513	318	273	242	216	183	647	534	438	285	218	183	151	588
12	202	734	611	505	309	266	236	210	178	634	524	430	276	213	178	147	577
13	198	721	601	497	300	259	230	205	174	621	514	422	267	207	174	143	566
14	194	707	590	489	291	251	223	199	169	607	503	414	258	200	169	139	554
15	189	693	579	480	281	244	217	193	164	593	492	406	248	194	164	135	541
16	185	679	568	471	271	236	210	187	159	578	481	398	238	187	158	131	528
17	180	664	556	462	260	227	202	180	154	562	469	389	228	180	153	127	515
18	176	648	544	453	250	218	195	174	148	547	457	380	217	173	147	122	501
19	171	632	532	443	238	210	187	167	143	530	445	370	205	165	141	117	487
20	166	616	519	433	227	200	179	160	137	514	432	361	194	158	135	113	473
21	161	599	506	423	215	191	171	153	131	497	419	351	182	150	129	108	458
22	155	581	492	413	202	181	162	145	125	479	406	341	169	141	123	103	443
23	150	564	478	402	190	171	154	137	119	461	392	330	156	133	116	97	427
24	144	546	464	391	176	161	145	129	112	442	377	320	144	124	109	92	411
25	139	527	450	380	163	150	135	121	105	423	363	309	132	115	102	86	394
26	133	508	435	368	151	139	126	113	99	403	348	297	122	106	95	81	377
27	127	488	419	357	140	129	117	105	92	383	333	286	114	99	88	75	360
28	121	468	404	345	130	120	108	97	85	363	317	274	106	92	82	70	342
29	114	448	388	332	121	112	101	91	79	341	301	262	98	85	76	65	324
30	108	427	372	320	113	104	94	85	74	320	284	250	92	80	71	61	305
31	101	405	355	307	106	98	89	79	69	299	267	237	86	75	67	57	286
32	95	383	338	294	99	92	83	74	65	281	251	224	81	70	63	53	268
33	89	360	320	281	94	86	78	70	61	264	236	211	76	66	59	50	252
34	84	340	302	267	88	81	74	66	58	249	222	199	72	62	56	47	238
35	79	320	285	253	83	77	69	62		235	210	188	68	59	53		224
36	75	303	270	239	79	72	66			222	198	178	64				212
37	71	287	255	226	74	69				210	188	168					201
38	67	272	242	215						199	178	159					190
39		258	230	204						189	169	151					181
40		245	219	194						180	161	144					172
41		234	208	184						171	153	137					163
42		223	198	176						163	146	130					156
43		212	189	168						156	139	124					149
Q_s =	0.45	0.92	0.84	0.76	0.66	0.60	0.57	0.54	0.50	0.96	0.87	0.78	0.70	0.59	0.53	0.48	0.96

TABLE 4-6 (*Continued*)

STRUCTURAL TEES—MAIN MEMBERS

Allowable Axial Compression Load, P_m (Kips)

F_y 50 Ksi

Effective Length $K\ell$ in ft. — x-x Axis

Size	WT12	WT12	WT12	WT12	WT12	WT12	WT12	WT12	WT12	WT12	WT12	WT10.5	WT10.5	WT10.5	WT10.5	WT10.5	WT10.5
Weight	72.5	65	60	55	50	47	42	38	34	30.5	27.5	71	63.5	56	48	41	36.5
0	573	473	425	345	271	298	227	182	148	135	110	627	557	446	414	309	234
1	568	470	422	343	269	296	226	181	147	134	110	621	552	442	410	307	233
2	564	466	419	340	268	294	224	180	146	133	109	615	546	438	406	304	231
3	558	462	416	338	266	292	223	179	145	132	108	607	539	433	402	301	229
4	553	457	412	335	264	290	221	177	144	132	108	599	532	427	397	297	227
5	547	453	408	332	261	287	220	176	143	131	107	590	524	421	392	294	224
6	540	448	403	328	259	285	218	175	142	130	106	581	516	415	386	290	222
7	533	442	399	325	257	282	216	173	141	129	106	571	507	408	380	286	219
8	526	436	394	321	254	279	214	172	140	128	105	560	497	401	374	282	216
9	518	431	389	317	251	276	212	170	139	127	104	549	487	393	367	277	213
10	510	424	384	313	248	272	209	169	138	126	103	537	476	385	360	272	210
11	502	418	378	309	245	269	207	167	137	125	102	524	465	377	353	267	207
12	493	411	372	305	242	266	205	165	135	123	102	511	453	368	345	262	204
13	484	404	366	300	239	262	202	164	134	122	101	498	441	359	337	256	200
14	474	397	360	296	236	258	200	162	133	121	100	483	428	350	329	251	196
15	465	389	353	291	233	254	197	160	131	120	99	469	415	340	320	245	193
16	455	381	347	286	229	250	194	158	130	118	98	454	402	330	311	239	189
17	444	373	340	281	225	246	192	156	128	117	97	438	388	319	302	233	185
18	434	365	333	276	222	242	189	154	127	116	96	422	373	308	293	226	181
19	423	357	326	270	218	238	186	152	125	114	95	405	358	297	283	220	177
20	411	348	318	265	214	233	183	149	123	113	93	388	343	285	273	213	172
21	400	339	311	259	210	229	180	147	122	111	92	370	327	274	262	206	168
22	388	330	303	253	206	224	177	145	120	110	91	352	310	261	252	199	163
23	375	320	295	247	202	219	173	142	118	108	90	333	293	249	241	191	158
24	363	311	287	241	197	214	170	140	116	106	89	314	276	236	230	184	154
25	350	301	278	235	193	209	167	138	114	105	87	294	258	223	218	176	149
26	337	290	269	228	188	204	163	135	113	103	86	273	240	209	206	168	144
27	323	280	261	222	184	199	160	133	111	101	85	253	222	195	194	160	139
28	309	269	252	215	179	193	156	130	109	100	84	236	207	181	182	151	133
29	295	259	242	208	174	188	152	127	107	98	82	220	193	169	169	143	128
30	281	247	233	201	169	182	149	125	105	96	81	205	180	158	158	134	122
31	266	236	223	194	164	177	145	122	103	94	79	192	169	148	148	125	117
32	251	224	213	187	159	171	141	119	100	92	78	180	158	139	139	118	111
33	236	212	203	179	154	165	137	116	98	90	77	170	149	130	131	111	105
34	222	200	193	172	149	159	133	113	96	89	75	160	140	123	123	104	99
35	209	189	183	164	143	153	129	110	94	87	74	151	132	116	116	98	93

Effective Length, Kℓ, in feet — Y-Y Axis																	
0	573	473	425	345	271	298	227	182	148	135	110	627	557	446	414	309	234
1	568	469	422	342	269	295	225	180	146	133	109	622	552	443	408	305	231
2	563	466	417	339	267	290	222	178	144	130	107	616	547	439	401	300	228
3	558	461	413	335	264	286	218	175	143	128	105	609	541	434	394	295	223
4	552	457	408	332	261	281	215	172	140	125	102	601	534	429	385	289	219
5	546	452	402	327	258	275	211	169	138	121	100	594	527	424	375	282	214
6	539	446	396	323	255	269	207	166	135	118	97	585	520	418	365	275	208
7	532	441	390	318	252	262	202	163	133	114	94	576	511	412	354	268	202
8	524	435	383	313	248	255	197	159	130	110	90	566	503	405	342	259	196
9	516	428	376	308	245	248	192	155	127	105	87	556	494	398	329	251	189
10	508	422	369	303	241	240	187	151	124	101	83	545	484	391	316	242	182
11	499	415	361	297	236	231	181	147	120	96	79	534	474	383	302	232	174
12	490	408	353	291	232	223	175	142	117	91	75	523	464	376	287	222	166
13	481	400	345	285	228	214	169	138	113	85	71	510	453	367	272	211	158
14	471	393	336	278	223	204	162	133	109	80	66	498	442	359	256	200	150
15	461	385	327	271	218	195	155	128	105	74	62	485	430	350	239	189	141
16	451	376	318	264	213	184	148	123	101	68	57	471	418	341	222	177	131
17	440	368	308	257	208	174	141	117	97	61	52	457	406	332	203	164	121
18	429	359	298	250	203	163	134	112	93	55	46	443	393	322	184	151	111
19	417	350	288	242	198	151	126	106	88	49	42	428	380	312	165	138	101
20	406	341	277	234	192	139	118	100	83	44	38	413	366	302	149	124	91
21	394	331	266	226	186	127	110	94	78	40	34	397	352	291	135	113	82
22	381	321	255	218	181	116	101	87	73	37	31	381	338	280	123	103	75
23	369	311	244	209	175	106	92	81	68			365	323	269	113	94	69
24	355	301	232	201	168	97	85	74	63			348	308	258	104	86	63
25	342	291	220	192	162	90	78	69	58			330	292	246	95	80	58
26	328	280	207	182	156	83	72	63	54			312	276	234	88	74	54
27	314	269	194	173	149	77	67	59	50			293	259	221	82	68	50
28	300	257	181	163	142	72	62	55	46			274	242	209	76	63	46
29	285	246	169	153	135	67	58	51	43			256	226	196	71	59	43
30	270	234	158	143	128	62	54	48	40			239	211	183	66	55	40
31	254	222	148	134	121	58	51	45	38			224	198	171	62	52	
32	239	209	139	126	113	55	48					210	186	161	58	49	
33	225	197	130	118	107							197	174	151	55	46	
34	212	185	123	112	100							186	164	142			
35	200	175	116	105	95							176	155	134			
36	189	165	109	99	90							166	147	127			
37	179	156	104	94	85							157	139	120			
38	169	148	98	89	80							149	132	114			
39	161	141	93	85	76							141	125	108			
40	153	134	89	81	73							134	119	103			
41	146	127	84	77	69							128	113	98			
42	139	121	80	73	66							122	108	93			
43	132	116	77	70	63							116	103	89			
Q_s =	0.89	0.82	0.80	0.71	0.61	0:72	0.61	0.54	0.49	0.50	0.45	1.00	0.99	0.90	0.98	0.85	0.73

TABLE 4-6 (*Continued*)

STRUCTURAL TEES—MAIN MEMBERS

Allowable Axial Compression Load, P_m (Kips)

F_y 50 Ksi

Size	WT10.5	WT10.5	WT10.5	WT10.5	WT10.5	WT9	WT9	WT9	WT9	WT9	WT9	WT9	WT9	WT9	WT9	WT9	WT9
Weight	34	31	27.5	24.5	22	57	52.5	48	42.5	38.5	35	32	30	27.5	25	22.5	20
0	199	160	127	108	88	504	462	423	375	322	269	222	212	179	140	111	88
1	198	159	126	107	88	498	457	418	371	319	267	220	210	177	139	111	88
2	196	157	125	107	87	492	451	413	367	315	264	218	208	175	138	110	87
3	195	156	124	106	87	485	444	407	361	311	260	215	206	173	136	109	86
4	193	155	123	105	86	477	437	400	356	306	256	212	203	171	135	108	85
5	191	154	122	104	85	468	429	392	350	301	252	209	200	169	133	106	84
6	189	152	121	104	85	458	420	384	343	295	248	206	197	167	132	105	84
7	187	151	120	103	84	448	410	375	336	290	243	202	194	164	130	104	83
8	185	149	119	102	83	437	400	366	328	283	238	198	190	161	128	102	82
9	182	147	118	101	83	425	390	356	320	277	233	194	187	159	126	101	81
10	180	145	116	100	82	413	378	346	312	270	227	190	183	155	124	99	79
11	177	144	115	99	81	401	367	335	303	262	221	185	179	152	121	98	78
12	175	142	113	97	80	387	354	324	294	254	215	181	175	149	119	96	77
13	172	140	112	96	79	373	341	312	284	246	209	176	171	146	117	94	76
14	169	137	111	95	78	359	328	300	274	238	202	171	166	142	114	93	75
15	166	135	109	94	77	344	314	287	263	229	195	165	161	138	112	91	73
16	163	133	107	92	76	328	300	274	253	220	188	160	156	135	109	89	72
17	160	131	106	91	75	312	285	260	241	211	181	154	151	131	106	87	70
18	156	128	104	90	74	295	269	245	230	202	173	148	146	126	103	85	69
19	153	126	102	88	73	278	253	231	218	192	166	142	141	122	100	83	68
20	150	123	100	87	72	259	236	215	205	181	157	136	135	118	97	80	66
21	146	121	99	85	71	241	219	199	193	171	149	130	130	114	94	78	64
22	142	118	97	84	70	221	201	183	179	160	140	123	124	109	91	76	63
23	139	116	95	82	69	203	184	167	166	148	131	116	118	104	88	74	61
24	135	113	93	81	67	186	169	154	152	137	122	109	112	99	84	71	59
25	131	110	91	79	66	171	156	142	140	126	113	102	106	94	81	69	58
26	127	107	89	77	65	159	144	131	130	116	104	95	99	89	77	66	56
27	123	104	86	76	64	147	134	121	120	108	97	88	92	84	74	64	54
28	119	101	84	74	62	137	124	113	112	100	90	82	86	79	70	61	52
29	114	98	82	72	61	127	116	105	104	94	84	76	80	73	66	58	50
30	110	95	80	70	60	119	108	98	97	87	78	71	75	69	62	55	48
31	106	92	78	69	58	112	101	92	91	82	73	67	70	64	58	53	46
32	101	89	75	67	57	105	95	86	86	77	69	63	66	60	55	50	44
33	96	85	73	65	56	98	89	81	80	72	65	59	62	57	51	47	42
34	92	82	70	63	54	93	84	77	76	68	61	55	58	53	48	44	40
35	87	78	68	61	53	87	80	72	72	64	58	52	55	50	46	41	38

Effective Length $K\ell$ in ft. — x-x Axis

Effective Length, $K\ell$, in feet Y-Y Axis																	
0	199	160	127	108	88	504	462	423	375	322	269	222	212	179	140	111	88
1	197	158	125	106	87	499	458	419	370	318	266	220	209	176	138	110	87
2	194	155	124	104	85	494	453	414	364	313	262	216	205	173	136	108	85
3	190	153	122	102	83	488	447	409	357	307	257	213	201	169	134	106	83
4	187	150	119	100	81	481	441	403	349	301	252	209	196	165	131	104	81
5	183	147	117	97	79	474	434	397	341	294	246	204	190	161	127	102	78
6	178	144	115	94	77	466	427	390	331	286	240	199	184	156	124	99	76
7	173	140	112	90	74	457	419	383	321	278	233	194	178	151	120	97	73
8	168	136	109	87	71	449	411	376	310	269	226	189	171	146	116	94	70
9	163	132	106	83	68	439	402	368	299	259	219	183	164	140	112	90	67
10	157	128	103	79	65	429	393	359	287	249	211	177	156	134	108	87	63
11	151	124	99	75	62	419	383	350	274	239	202	170	148	127	103	84	60
12	145	119	96	70	58	408	373	341	261	228	194	163	139	120	98	80	56
13	138	114	92	65	54	397	363	331	247	216	184	156	130	113	93	76	52
14	131	109	88	61	50	385	352	321	233	204	175	149	121	106	88	72	48
15	124	104	84	55	46	373	341	311	218	192	165	141	111	98	82	68	43
16	117	98	80	50	42	361	329	300	202	179	154	133	100	90	76	64	38
17	109	92	76	44	38	348	317	289	185	165	144	125	89	81	70	59	34
18	101	86	71	40	33	334	305	278	168	151	132	116	80	72	64	55	30
19	93	80	67	36	30	320	292	266	151	136	121	107	72	65	58	50	27
20	84	74	62	32	27	306	279	253	136	123	109	98	65	59	52	45	25
21	76	67	57	29	25	291	265	241	124	112	99	89	59	53	47	41	22
22	69	61	52			276	251	228	113	102	90	81	53	48	43	37	
23	64	56	48			260	236	214	103	93	82	74	49	44	39	34	
24	58	51	44			244	221	200	95	85	76	68	45	41	36	31	
25	54	47	40			228	206	186	87	79	70	63	41	37	33	29	
26	50	44	37			211	191	172	81	73	64	58	38	35	31	27	
27	46	41	34			195	177	160	75	67	60	54	35	32	29		
28	43	38	32			182	164	148	69	63	56	50					
29	40	35				169	153	138	65	59	52	47					
30						158	143	129	61	55	48	43					
31						148	134	121	57	51	45	41					
32						139	126	114	53	48	43	38					
33						131	118	107	50	45	40	36					
34						123	111	101	47								
35						116	105	95									
36						110	99	90									
37						104	94	85									
38						99	89	81									
39						94	85	76									
40						89	81	73									
41						85	77	69									
42						81	73	66									
43						77	70	63									
$Q_s =$	0.66	0.58	0.52	0.50	0.45	1.00	1.00	1.00	1.00	0.94	0.87	0.79	0.80	0.74	0.63	0.56	0.50

TABLE 4-6 (*Continued*)

STRUCTURAL TEES—MAIN MEMBERS

Allowable Axial Compression Load, P_m (Kips)

F_y 50 Ksi

Size	WT9	WT8	WT8	WT8	WT8	WT8	WT8	WT8	WT8	WT8	WT8	WT8	WT8	WT7	WT7	WT7	WT7
Weight	17.5	48	44	39	35.5	32	29	25	22.5	20	18	15.5	13	365	332.5	302.5	275
0	70	423	387	345	315	280	236	182	145	104	91	66	47	3210	2934	2670	2427
1	70	418	382	341	311	277	233	181	144	103	90	66	47	3177	2902	2640	2398
2	69	411	376	336	307	273	230	178	142	102	89	65	46	3138	2865	2605	2365
3	69	404	369	330	302	268	226	176	140	101	88	64	46	3095	2824	2565	2327
4	68	395	362	324	296	263	222	173	138	99	87	64	46	3047	2777	2521	2285
5	67	386	353	317	290	258	218	170	136	98	86	63	45	2996	2727	2473	2239
6	67	376	344	310	283	252	213	167	134	97	85	62	45	2939	2673	2421	2189
7	66	366	335	302	276	245	208	164	131	95	83	61	44	2879	2614	2365	2135
8	65	354	324	294	268	239	202	160	129	93	82	61	44	2815	2552	2305	2078
9	65	342	313	285	260	231	196	156	126	92	81	60	43	2748	2487	2242	2018
10	64	330	302	275	251	224	190	152	123	90	79	59	42	2677	2417	2176	1954
11	63	316	289	265	242	216	184	148	120	88	77	58	42	2602	2345	2106	1887
12	62	302	277	255	232	207	177	144	117	86	76	57	41	2524	2269	2033	1816
13	61	288	263	244	222	198	170	139	113	84	74	56	40	2443	2190	1957	1743
14	60	272	249	233	212	189	163	134	110	82	72	54	40	2359	2107	1878	1667
15	59	256	234	221	201	179	155	129	106	80	70	53	39	2271	2022	1795	1587
16	58	240	219	209	190	170	147	124	103	77	69	52	38	2180	1933	1710	1505
17	57	222	203	196	178	159	139	119	99	75	67	51	37	2086	1841	1621	1419
18	56	204	187	183	166	148	130	113	95	73	65	50	37	1988	1745	1529	1330
19	55	185	170	169	153	137	121	108	91	70	62	48	36	1888	1646	1433	1237
20	54	167	153	155	140	126	112	102	86	67	60	47	35	1783	1544	1334	1141
21	53	152	139	141	127	114	102	96	82	65	58	45	34	1676	1438	1231	1042
22	52	138	127	128	116	104	93	89	77	62	56	44	33	1564	1329	1126	949
23	50	127	116	117	106	95	85	83	73	59	53	43	32	1449	1218	1030	868
24	49	116	106	108	97	87	78	76	68	56	51	41	31	1332	1118	946	798
25	48	107	98	99	90	80	72	70	63	53	49	39	31	1228	1031	872	735
26	47	99	91	92	83	74	67	65	58	50	46	38	30	1135	953	806	680
27	45	92	84	85	77	69	62	60	54	47	43	36	29	1053	883	748	630
28	44	85	78	79	72	64	58	56	50	44	41	35	28	979	822	695	566
29	43	80	73	74	67	60	54	52	47	41	38	33	27	913	766	648	546
30	41	74	68	69	62	56	50	49	44	38	35	31	25	853	716	606	510
31	40	70	64	65	58	52	47	46	41	36	33	29	24	799	670	567	478
32	38	65	60	61	55	49	44	43	38	34	31	28	23	750	629	532	449
33	37	61	56	57	52	46	41	40	36	32	29	26	22	705	591	500	422
34	35	58	53	54	49	43	39	38	34	30	28	24	21	664	557	471	397
35	34	55	50	51	46	41	37	36	32	28	26	23	20	627	526	445	375

Effective Length $K\ell$ in ft. — x-x Axis

Effective Length, $K\ell$, in feet Y-Y Axis																
0	70	423	387	345	315	280	236	182	145	104	91	66	47	3210	2934	2670	2427
1	69	419	383	340	311	276	233	179	143	102	89	65	46	3192	2917	2654	2413
2	68	414	379	334	305	272	229	176	140	101	88	64	45	3172	2899	2638	2397
3	66	409	374	328	299	266	225	172	137	99	86	62	44	3151	2879	2619	2380
4	64	403	369	321	292	260	220	167	134	97	84	60	43	3128	2858	2600	2362
5	63	397	363	312	285	253	214	162	130	94	82	58	42	3104	2835	2579	2343
6	61	390	357	304	277	246	208	156	126	92	80	56	40	3078	2811	2556	2322
7	58	383	350	294	268	238	202	150	121	89	77	54	39	3051	2786	2533	2300
8	56	376	343	284	259	230	195	143	116	86	75	51	37	3023	2760	2508	2278
9	53	368	336	273	249	221	188	136	111	83	72	48	35	2993	2732	2482	2254
10	51	359	328	262	238	211	180	129	105	79	69	46	33	2962	2703	2455	2229
11	48	350	319	250	227	201	172	121	100	76	66	43	31	2930	2673	2427	2203
12	45	341	311	237	215	191	164	113	94	72	62	39	29	2896	2641	2398	2176
13	42	331	302	224	203	180	155	104	87	68	59	36	27	2861	2609	2368	2148
14	39	321	293	210	190	168	146	95	80	64	55	33	25	2825	2575	2337	2119
15	35	311	283	196	177	156	136	85	73	60	52	29	22	2788	2541	2305	2089
16	32	300	273	181	163	144	126	75	66	55	48	25	20	2750	2505	2271	2058
17	28	289	263	165	148	130	116	67	59	51	43	22	17	2711	2468	2237	2026
18	25	278	252	149	133	117	105	60	52	46	39	20	15	2670	2430	2202	1994
19	22	266	241	133	119	105	94	53	47	41	35	18		2628	2391	2166	1960
20	20	253	229	120	108	95	85	48	42	37	32			2586	2352	2129	1926
21		241	218	109	98	86	77	44	38	34	29			2542	2311	2091	1890
22		228	205	100	89	78	70	40	35	31	26			2497	2269	2052	1854
23		214	193	91	82	72	64	36	32	28	24			2452	2226	2012	1817
24		200	180	84	75	66	59	33	29	26	22			2405	2182	1971	1780
25		186	167	77	69	61	54	31	27	24	20			2357	2138	1929	1741
26		172	154	71	64	56	50	29	25					2308	2092	1887	1702
27		160	143	66	59	52	47							2258	2045	1843	1661
28		148	133	61	55	48	43							2208	1998	1799	1620
29		138	124	57	51	45	40							2156	1949	1754	1578
30		129	116	54	48	42	38							2103	1900	1708	1535
31		121	108	50	45	39	35							2049	1849	1661	1492
32		114	102	47	42	37	33							1994	1798	1613	1447
33		107	96	44	40									1938	1745	1564	1401
34		101	90											1881	1692	1514	1355
35		95	85											1823	1637	1463	1308
36		90	80											1763	1582	1411	1260
37		85	76											1703	1525	1358	1210
38		81	72											1641	1468	1304	1160
39		76	68											1579	1409	1249	1109
40		73	65											1515	1349	1193	1057
41		69	62											1450	1288	1137	1006
42		66	59											1384	1227	1083	959
43		63	56											1320	1171	1033	915
Q_s =	0.45	1.00	1.00	1.00	1.00	0.99	0.92	0.83	0.73	0.59	0.57	0.48	0.41	1.00	1.00	1.00	1.00

TABLE 4-6 *(Continued)*

STRUCTURAL TEES—MAIN MEMBERS

Allowable Axial Compression Load, P_m (Kips)

F_y 50 ksi

Size	WT7	WT7	WT7	WT7	WT7	WT7	WT7	WT7	WT7	WT7	WT7	WT7	WT7	WT7	WT7	WT7	WT7
Weight	250	227.5	213	199	185	171	160	157	143.5	132	123	118.5	114	109.5	105.5	101	96.5
Effective Length $K\ell$ in ft. — x-x Axis																	
0	2205	2007	1878	1755	1632	1509	1413	1386	1266	1164	1086	1044	1005	966	930	891	852
1	2178	1982	1854	1732	1610	1488	1394	1366	1248	1147	1070	1028	990	951	916	877	838
2	2146	1952	1825	1704	1584	1463	1372	1343	1226	1126	1050	1009	971	933	898	860	822
3	2110	1918	1792	1673	1554	1435	1347	1316	1200	1102	1027	987	950	912	878	840	803
4	2070	1879	1755	1637	1520	1403	1319	1285	1172	1075	1001	962	925	888	855	818	782
5	2026	1838	1715	1599	1483	1368	1288	1252	1140	1045	973	934	899	862	830	794	758
6	1979	1792	1671	1557	1443	1329	1254	1215	1106	1013	942	904	869	834	803	767	732
7	1927	1743	1624	1511	1399	1288	1218	1176	1069	977	908	872	838	803	773	738	704
8	1873	1691	1573	1463	1353	1244	1179	1134	1029	940	872	837	804	770	741	708	674
9	1815	1636	1520	1411	1304	1197	1138	1089	987	900	834	800	768	735	708	675	642
10	1754	1578	1463	1357	1252	1148	1095	1042	943	858	794	761	731	698	672	640	608
11	1690	1517	1404	1300	1197	1096	1049	993	896	814	751	720	691	659	634	603	572
12	1622	1452	1342	1240	1140	1041	1001	941	847	767	707	677	648	618	595	565	535
13	1552	1385	1277	1178	1080	984	951	886	795	718	660	631	604	575	553	524	495
14	1479	1315	1209	1112	1017	924	899	829	741	666	611	583	558	529	510	481	454
15	1403	1242	1138	1044	952	862	845	769	685	613	559	533	509	482	464	437	410
16	1323	1166	1064	973	883	796	788	706	626	556	505	481	458	432	416	390	365
17	1241	1087	987	899	812	728	728	641	564	497	449	427	407	383	368	345	323
18	1155	1005	907	821	738	657	667	574	503	444	401	381	363	341	329	308	288
19	1066	919	823	741	663	590	602	515	452	398	360	342	326	306	295	276	258
20	973	832	743	669	598	532	544	465	408	359	325	309	294	277	266	249	233
21	883	755	674	607	543	483	493	422	370	326	294	280	267	251	241	226	212
22	804	687	614	553	495	440	449	384	337	297	268	255	243	229	220	206	193
23	736	629	562	506	453	402	411	351	308	272	246	233	222	209	201	189	176
24	676	578	516	464	416	370	378	323	283	250	225	214	204	192	185	173	162
25	623	532	476	428	383	341	348	297	261	230	208	198	188	177	170	160	149
26	576	492	440	396	354	315	322	275	241	213	192	183	174	164	158	148	138
27	534	456	408	367	328	292	298	255	224	197	178	169	161	152	146	137	128
28	497	424	379	341	305	271	277	237	208	183	166	158	150	141	136	127	119
29	463	396	354	318	285	253	259	221	194	171	154	147	140	132	127	119	111
30	433	370	330	297	266	236	242	207	181	160	144	137	131	123	118		
31	405	346	309	278	249	221	226	193	170	150							
32	380	325	290	261	234	208	212	182	159								
33	357	306	273	246	220	195	200										
34	337	288	257	231	207		188										
35	318	272	243				178										

Effective Length, $K\ell$, in feet — Y-Y Axis																	
0	2205	2007	1878	1755	1632	1509	1413	1386	1266	1164	1086	1044	1005	966	930	891	852
1	2192	1995	1867	1744	1622	1500	1404	1377	1258	1157	1079	1037	998	960	924	885	846
2	2177	1981	1854	1732	1611	1489	1394	1368	1249	1148	1071	1030	991	953	917	879	840
3	2162	1967	1840	1720	1599	1478	1383	1357	1239	1139	1063	1022	983	945	910	872	834
4	2145	1952	1826	1706	1586	1466	1372	1346	1229	1130	1054	1013	975	937	902	864	826
5	2127	1935	1810	1691	1572	1453	1359	1334	1218	1119	1044	1004	966	928	894	856	818
6	2108	1917	1793	1675	1557	1439	1346	1321	1206	1108	1034	994	956	919	885	847	810
7	2088	1899	1776	1659	1541	1424	1332	1307	1194	1097	1023	983	946	909	875	838	801
8	2067	1879	1757	1641	1525	1409	1317	1293	1180	1085	1011	972	936	899	865	829	792
9	2045	1859	1738	1623	1508	1393	1302	1278	1167	1072	999	961	924	888	855	819	783
10	2022	1837	1718	1604	1490	1376	1286	1263	1152	1058	987	948	913	877	844	808	772
11	1998	1815	1697	1584	1471	1359	1269	1246	1137	1044	974	936	900	865	832	797	762
12	1973	1792	1675	1563	1451	1340	1252	1229	1121	1030	960	922	888	852	820	786	751
13	1947	1768	1652	1542	1431	1322	1234	1212	1105	1015	946	909	874	840	808	774	739
14	1920	1743	1628	1520	1410	1302	1215	1193	1088	999	931	895	861	826	795	761	728
15	1892	1717	1604	1497	1389	1282	1195	1175	1071	983	916	880	847	813	782	749	715
16	1864	1690	1579	1473	1366	1261	1175	1155	1053	966	900	865	832	799	768	736	703
17	1834	1663	1553	1449	1343	1239	1155	1135	1035	949	884	849	817	784	754	722	690
18	1804	1635	1527	1423	1320	1217	1134	1115	1016	932	868	833	802	769	740	708	677
19	1773	1606	1499	1398	1295	1195	1112	1094	996	913	851	817	786	754	725	694	663
20	1741	1577	1471	1371	1270	1171	1089	1072	976	895	833	800	769	738	710	679	649
21	1708	1546	1443	1344	1245	1148	1067	1050	956	876	815	783	753	722	694	664	634
22	1675	1515	1413	1316	1219	1123	1043	1027	935	856	797	765	735	705	678	649	619
23	1641	1483	1383	1288	1192	1098	1019	1003	913	836	778	747	718	688	662	633	604
24	1606	1451	1352	1259	1164	1072	994	979	891	816	759	728	700	671	645	617	589
25	1570	1417	1321	1229	1136	1046	969	955	868	795	739	709	681	653	627	600	573
26	1533	1383	1289	1199	1108	1019	943	930	845	773	719	690	663	635	610	583	557
27	1496	1349	1256	1168	1078	992	917	904	822	751	698	670	643	616	592	566	540
28	1457	1313	1222	1136	1048	964	890	878	797	729	677	649	624	597	573	548	523
29	1419	1277	1188	1104	1018	935	863	852	773	706	655	629	604	577	555	530	506
30	1379	1240	1153	1070	987	906	834	824	748	682	633	607	583	558	535	512	488
31	1338	1202	1117	1037	955	876	806	797	722	658	611	586	562	537	516	493	470
32	1297	1164	1081	1002	922	846	776	768	696	634	588	563	541	516	496	473	451
33	1255	1124	1043	967	889	815	747	739	669	609	564	541	519	495	475	454	432
34	1211	1084	1006	931	855	783	716	710	642	584	540	518	496	474	454	433	413
35	1168	1043	967	895	821	751	685	679	614	558	516	494	474	451	433	413	393
36	1123	1001	927	857	785	718	653	649	585	531	491	470	450	429	411	392	373
37	1077	959	887	819	749	684	620	617	556	504	465	445	427	406	389	371	353
38	1030	915	846	780	712	649	588	585	527	478	441	422	404	385	369	352	335
39	983	871	804	741	676	617	558	556	500	453	419	401	384	365	350	334	318
40	935	828	764	704	643	586	531	528	476	431	398	381	365	347	333	317	302
41	890	788	727	670	612	558	505	503	453	410	379	363	347	331	317	302	287
42	848	751	693	639	583	532	481	479	431	391	361	346	331	315	302	288	274
43	809	717	661	609	556	507	459	457	412	373	345	330	316	301	288	275	261
Q_s =	1.00	1.00	1.00	1.00	1.00	1.00	1.00	1.00	1.00	1.00	1.00	1.00	1.00	1.00	1.00	1.00	1.00

TABLE 4-6 *(Continued)*

STRUCTURAL TEES—MAIN MEMBERS

Allowable Axial Compression Load, P_m (Kips)

F_y 50 ksi

Size	WT7	WT7	WT7	WT7	WT7	WT7	WT7	WT7	WT7	WT7	WT7	WT7	WT7	WT7	WT7	WT7	WT7
Weight	92	88	83.5	79	75	71	68	63.5	59.5	55.5	51.5	47.5	43.5	42	39	37	34
Effective Length $K\ell$ in ft. — x-x Axis																	
0	810	777	735	696	660	627	600	561	525	489	453	420	384	372	345	327	300
1	797	764	723	685	649	617	590	552	516	481	445	413	377	366	339	322	295
2	781	749	709	671	636	604	578	540	506	471	436	404	369	359	333	316	290
3	763	732	692	655	620	589	564	527	493	459	425	394	360	350	325	309	283
4	742	712	673	636	603	573	549	512	479	446	412	382	349	340	316	301	276
5	719	690	652	616	584	554	531	496	463	431	398	369	337	330	306	292	268
6	694	666	629	594	563	534	512	477	446	415	383	355	324	318	295	282	259
7	667	640	604	570	540	513	491	458	428	398	367	340	310	305	283	272	249
8	638	612	578	545	515	489	469	437	408	379	349	324	295	292	271	261	239
9	607	582	549	517	489	465	446	415	387	359	331	307	279	277	257	249	228
10	574	551	519	488	462	438	421	391	364	338	311	288	261	262	243	236	217
11	540	518	488	458	432	411	395	366	341	316	290	269	243	246	228	223	205
12	503	483	455	426	402	382	367	340	316	293	268	248	224	229	212	209	192
13	465	446	419	392	369	351	338	312	290	268	245	227	204	211	196	195	178
14	425	408	383	357	335	319	307	283	262	242	220	204	183	192	178	179	164
15	383	367	344	319	299	284	275	252	233	215	194	180	161	172	160	163	150
16	339	325	304	281	264	250	242	222	205	189	171	158	141	152	141	146	134
17	300	288	269	249	234	222	215	196	181	167	151	140	125	135	125	130	119
18	268	257	240	222	208	198	192	175	162	149	135	125	112	120	111	116	106
19	240	230	216	199	187	178	172	157	145	134	121	112	100	108	100	104	95
20	217	208	195	180	169	160	155	142	131	121	109	101	90	97	90	94	86
21	197	189	176	163	153	145	141	129	119	109	99	92	82	88	82	85	78
22	179	172	161	149	139	132	128	117	108	100	90	84	75	80	75	77	71
23	164	157	147	136	128	121	117	107	99	91	83	77	68	74	68	71	65
24	151	144	135	125	117	111	108	98	91	84	76	70	63	68	63	65	60
25	139	133	124	115	108	103	99	91	84	77	70	65	58	62	58	60	55
26	128	123	115	107	100	95	92	84	78	71	65	60	53	58	53	55	51
27	119	114	107	99	93	88	85	78	72	66	60	56	50	53	50	51	47
28	111	106	99	92	86	82	79	72	67	62				50	46	48	44
29	103	99	93													45	41
30																42	38
31																	
32																	
33																	
34																	
35																	

Effective Length, $K\ell$, in feet — Y-Y Axis																	
0	810	777	735	696	660	627	600	561	525	489	453	420	384	372	345	327	300
1	805	772	730	691	656	623	596	557	521	485	450	417	[illegible]	369	342	323	297
2	799	766	725	686	651	618	591	553	517	482	446	414	378	365	338	319	293
3	792	760	719	681	645	613	586	548	513	477	442	410	375	361	334	314	288
4	785	753	713	675	640	608	580	543	508	473	438	406	371	356	330	309	284
5	778	746	706	668	634	602	574	537	502	468	433	402	367	351	325	303	278
6	770	739	699	661	627	596	568	531	497	462	428	397	363	346	320	297	272
7	762	730	691	654	620	589	561	524	491	457	423	392	358	340	315	291	266
8	753	722	683	646	613	582	554	518	484	451	418	387	354	334	309	284	260
9	744	713	674	638	605	575	546	511	478	445	412	382	349	328	303	276	253
10	734	704	665	630	597	567	538	503	471	438	406	376	344	321	297	268	245
11	724	694	656	621	589	559	530	496	464	431	399	370	338	314	290	260	238
12	714	684	647	612	580	551	522	487	456	424	393	364	333	306	284	251	230
13	703	673	637	603	571	542	513	479	448	417	386	358	327	299	276	242	221
14	691	662	626	593	562	533	504	470	440	409	379	351	321	291	269	233	213
15	680	651	616	583	552	524	494	461	432	401	371	344	314	283	261	223	203
16	668	640	605	572	542	514	484	452	423	393	364	337	308	274	253	213	194
17	655	628	593	561	532	504	474	443	414	385	356	330	301	265	245	202	184
18	643	615	582	550	521	494	463	433	405	376	348	322	294	256	236	191	174
19	630	603	570	539	510	484	453	423	395	367	340	315	287	247	228	180	164
20	616	590	557	527	499	473	442	412	385	358	331	307	280	237	219	168	153
21	602	576	545	515	488	462	430	402	375	349	322	298	272	227	209	156	141
22	588	563	532	503	476	451	419	391	365	339	313	290	265	217	200	144	130
23	574	549	518	490	464	439	407	380	355	329	304	282	257	206	190	131	119
24	559	534	505	477	452	428	394	368	344	319	295	273	249	195	180	121	109
25	544	520	491	464	439	416	382	356	333	309	285	264	241	184	169	111	100
26	528	505	476	450	426	403	369	344	321	298	275	255	232	173	158	103	93
27	512	489	462	436	413	391	356	332	310	287	265	245	223	161	147	95	86
28	496	474	447	422	399	378	343	319	298	276	255	235	215	150	137	89	80
29	479	458	432	408	386	364	329	306	286	264	244	226	205	139	128	83	75
30	462	441	416	393	372	351	315	293	273	253	233	215	196	130	119	77	70
31	445	424	400	378	357	337	300	280	261	241	222	205	187	122	112	72	65
32	427	407	384	362	342	323	285	266	248	228	211	194	177	115	105	68	61
33	409	390	367	347	327	309	270	252	234	216	199	183	167	108	99	64	58
34	391	372	350	330	312	294	255	237	221	203	187	173	157	101	93	60	54
35	372	354	333	314	296	279	241	224	208	192	177	163	148	96	88	57	51
36	353	335	315	297	280	264	227	212	197	181	167	154	140	90	83	54	48
37	334	317	298	281	265	250	215	200	186	172	158	146	133	86	78	51	46
38	316	301	283	267	252	237	204	190	177	163	150	138	126	81	74	48	43
39	300	285	269	253	239	225	194	180	168	155	142	131	119	77	71	46	41
40	286	271	255	241	227	213	184	171	160	147	135	125	114	73	67	43	39
41	272	258	243	229	216	203	175	163	152	140	129	119	108	70	64	41	
42	259	246	232	218	206	194	167	155	145	133	123	113	103	66	61		
43	247	235	221	208	196	185	159	148	138	127	117	108	98	63	58		
Q_s =	1.00	1.00	1.00	1.00	1.00	1.00	1.00	1.00	1.00	1.00	1.00	1.00	1.00	1.00	1.00	1.00	1.00

TABLE 4-6 (*Continued*)

STRUCTURAL TEES—MAIN MEMBERS

Allowable Axial Compression Load, P_m (Kips)

F_y 50 ksi

Size	WT7	WT7	WT7	WT7	WT7	WT7	WT7	WT7	WT7	WT6	WT6	WT6	WT6	WT6	WT6	WT6	WT6
Weight	30.5	26.5	24	21.5	19	17	15	13	11	95	80.5	66.5	60	53	49.5	46	42.5
0	263	224	186	149	128	101	80	62	44	837	711	588	531	468	438	405	375
1	259	220	183	147	127	100	79	62	43	823	699	577	521	459	430	397	368
2	254	217	180	145	125	99	79	61	43	805	683	564	509	448	419	388	359
3	248	212	176	142	123	98	77	60	43	785	665	548	495	435	407	376	348
4	242	207	173	139	121	96	76	59	42	762	645	531	479	421	394	364	336
5	235	201	168	136	118	94	75	58	41	737	623	511	462	405	378	349	322
6	227	195	163	132	116	92	74	57	41	709	598	490	442	387	361	334	308
7	219	189	158	128	113	90	72	56	40	679	571	466	421	368	343	317	292
8	210	182	153	124	110	88	71	55	40	647	543	441	399	347	324	298	274
9	200	174	147	120	107	86	69	54	39	613	512	415	375	325	303	279	256
10	190	166	141	115	103	83	67	53	38	577	480	386	349	301	281	258	236
11	180	158	134	111	100	81	65	52	37	538	446	356	322	277	257	236	215
12	168	149	128	106	96	78	64	50	37	498	410	325	293	250	232	213	193
13	157	140	120	100	92	75	62	49	36	455	372	291	263	222	206	188	170
14	144	130	113	95	88	72	59	48	35	410	332	256	231	193	178	163	147
15	131	120	105	89	84	69	57	46	34	363	290	223	201	168	155	142	128
16	118	110	97	83	80	66	55	45	33	319	255	196	177	148	137	125	112
17	104	99	89	77	75	63	53	43	32	283	226	173	157	131	121	110	100
18	93	88	80	70	70	60	50	41	31	252	202	155	140	117	108	99	89
19	83	79	72	63	66	56	48	40	30	226	181	139	125	105	97	88	80
20	75	71	65	57	60	53	45	38	29	204	163	125	113	95	87	80	72
21	68	65	59	52	55	49	43	36	28	185	148	114	103	86	79	72	65
22	62	59	53	47	50	45	40	34	27	169	135	104	93	78	72	66	59
23	57	54	49	43	46	41	37	32	26	154	123	95	86	72	66	60	54
24	52	50	45	40	42	38	34	31	24	142	113	87	79	66	61	55	50
25	48	46	41	37	39	35	32	28	23	131	104	80	72	61	56	51	
26	45	42	38	34	36	32	29	26	22	121	97	74	67				
27	41	39	35	31	33	30	27	24	21	112	90						
28	38	36	33	29	31	28	25	23	19								
29	36	34	31	27	29	26	24	21	18								
30		32	29	25	27	24	22	20	17								
31		30	27	24	25	23	21	19	16								
32					24	21	19	17	15								
33					22	20	18	16	14								
34					21	19	17	15	13								
35								15	12								

Effective Length $K\ell$ in ft. (x–x Axis)

Effective Length, $K\ell$, in feet — Y-Y Axis																	
0	263	224	186	149	128	101	80	62	44	837	711	588	531	468	438	405	375
1	260	221	183	147	126	100	79	61	43	830	705	583	526	464	434	401	372
2	257	217	180	145	124	98	78	59	42	822	698	577	521	459	430	397	368
3	253	212	177	142	121	96	76	58	41	814	691	571	516	454	425	393	364
4	249	207	173	139	118	93	74	56	39	804	683	564	509	449	420	388	359
5	244	202	168	136	114	91	72	53	38	794	674	557	502	443	414	383	354
6	239	196	164	132	110	88	70	51	36	783	664	549	495	436	408	377	349
7	234	190	159	129	106	85	68	48	35	771	654	540	487	429	401	371	343
8	228	183	153	125	101	81	65	45	33	759	644	531	479	422	394	365	337
9	222	176	148	120	96	78	62	42	31	746	632	522	470	414	387	358	331
10	216	168	142	116	91	74	60	39	28	733	620	512	461	406	380	351	324
11	209	160	136	111	86	70	57	35	26	718	608	501	452	398	371	343	318
12	202	151	129	106	80	66	53	31	24	703	595	491	442	389	363	335	310
13	195	143	122	101	74	61	50	27	21	688	582	479	431	379	354	327	303
14	187	133	115	96	67	57	47	24	19	672	568	467	421	370	345	319	295
15	179	124	107	90	61	52	43	21	16	655	553	455	410	360	336	310	287
16	171	113	99	84	54	47	39	18	14	638	538	443	398	350	326	301	278
17	163	103	91	78	48	42	35	16	13	620	523	430	386	339	316	292	270
18	154	92	82	72	42	37	31			602	507	416	374	328	306	282	261
19	145	82	74	65	38	33	28			583	491	403	361	317	295	272	252
20	135	74	67	59	34	30	25			564	474	388	348	305	284	262	242
21	125	68	61	53	31	27	23			544	456	374	335	293	273	252	232
22	115	62	55	48	28	25	21			524	439	359	321	281	261	241	222
23	106	56	50	44	26	23	19			502	420	343	307	268	249	230	212
24	97	52	46	41	24	21	18			481	401	327	292	255	237	218	201
25	89	48	43	37	22	19				459	382	311	277	242	224	207	190
26	83	44	40	35						436	362	294	262	228	211	194	179
27	77	41	37	32						412	342	277	246	214	198	182	167
28	71	38	34	30						388	321	259	229	200	184	169	156
29	66	35	32	28						363	299	241	214	186	172	158	145
30	62	33	30	26						340	280	226	200	174	161	148	136
31	58	31	28	24						318	262	211	187	163	150	138	127
32	55									298	246	198	176	153	141	130	119
33	51									281	231	186	165	144	133	122	112
34	48									264	218	176	156	135	125	115	106
35	46									249	205	166	147	128	118	108	100
36	43									236	194	157	139	121	112	102	94
37	41									223	184	148	131	114	106	97	89
38	39									212	174	141	125	108	100	92	85
39	37									201	165	133	118	103	95	87	80
40	35									191	157	127	112	98	90	83	76
41										182	150	121	107	93	86	79	73
42										173	143	115	102	89	82	75	69
43										165	136	110	97	85	78	72	66
Q_s =	0.98	0.96	0.88	0.79	0.77	0.67	0.61	0.54	0.45	1.00	1.00	1.00	1.00	1.00	1.00	1.00	1.00

TABLE 4-6 (*Continued*)

STRUCTURAL TEES—MAIN MEMBERS

Allowable Axial Compression Load, P_m (Kips)

F_y 50 ksi

Size	WT6	WT6	WT6	WT6	WT6	WT6	WT6	WT6	WT6	WT6	WT6	WT6	WT6	WT6	WT6	WT5	WT5
Weight	39.5	36	32.5	29	26.5	25	22.5	20	18	15.5	13.5	11	9.5	8.25	7	56	50
Effective Length $K\ell$ in ft. x-x Axis																	
0	348	318	286	256	234	221	199	156	142	104	75	69	51	43	27	495	441
1	341	312	281	251	230	217	195	153	140	102	74	68	50	42	27	484	431
2	333	304	274	245	224	212	191	150	137	100	73	67	50	42	27	470	418
3	323	295	266	238	217	206	185	146	134	98	72	66	49	41	26	453	403
4	312	285	256	229	210	200	180	142	130	96	70	65	48	41	26	434	386
5	299	273	246	220	202	193	173	137	126	93	68	63	47	40	26	413	367
6	285	260	234	210	193	185	166	132	122	91	67	62	46	39	25	390	345
7	271	247	222	199	183	176	158	126	118	88	65	60	45	38	25	364	322
8	255	232	208	187	172	167	150	120	113	84	63	59	44	37	24	337	297
9	237	216	194	175	161	157	141	113	107	81	61	57	43	36	24	308	270
10	219	199	178	161	149	147	132	107	102	77	58	55	42	35	23	276	242
11	200	181	162	147	136	136	122	99	96	74	56	53	40	34	23	243	211
12	179	162	145	132	123	125	111	92	90	70	53	51	39	33	22	207	179
13	158	142	127	116	109	112	100	84	83	65	51	49	37	32	22	176	152
14	136	123	109	100	94	100	88	75	77	61	48	46	36	31	21	152	131
15	119	107	95	87	82	87	77	66	70	57	45	44	34	29	20	133	115
16	104	94	84	77	72	76	68	58	62	52	42	41	33	28	20	116	101
17	92	83	74	68	64	68	60	51	55	47	39	39	31	27	19	103	89
18	82	74	66	61	57	60	54	46	49	42	36	36	29	25	18	92	80
19	74	67	59	54	51	54	48	41	44	37	33	33	28	24	17	83	71
20	67	60	54	49	46	49	43	37	40	34	29	30	26	23	17	75	64
21	61	55	49	45	42	44	39	34	36	31	27	28	24	21	16	68	58
22	55	50	44	41	38	40	36	31	33	28	24	25	22	19	15		
23	50	46	40	37	35	37	33	28	30	26	22	23	20	18	14		
24	46	42	37	34	32	34	30	26	28	23	20	21	18	16	13		
25					30	31	28	24	25	22	19	19	17	15	13		
26						29	26		23	20	17	18	16	14	12		
27									22	19	16	17	15	13	11		
28									20	17	15	15	14	12	10		
29												14	13	11	9		
30												13	12	10	9		
31												13	11	10	8		
32														9			
33																	
34																	
35																	

Effective Length, $K\ell$, in feet Y-Y Axis																	
0	348	318	286	256	234	221	199	156	142	104	75	69	51	43	27	495	441
1	345	315	284	253	231	218	196	154	139	102	74	67	49	42	27	490	436
2	341	312	281	250	228	214	192	151	136	100	73	64	47	40	26	484	431
3	338	308	278	246	225	210	188	148	133	98	71	61	45	38	24	478	425
4	333	305	274	242	221	205	184	145	129	95	69	57	42	35	23	470	419
5	329	300	270	238	217	199	179	142	124	92	67	52	39	32	22	463	412
6	324	296	266	233	213	193	174	138	119	89	65	47	36	29	20	454	404
7	318	291	262	228	208	187	168	134	114	85	63	42	32	26	18	445	396
8	313	286	257	222	203	180	162	129	109	82	61	36	28	22	16	435	387
9	307	280	252	217	198	173	155	125	103	78	58	30	24	18	14	425	378
10	301	275	247	211	192	166	148	120	96	74	56	24	20	15	12	414	369
11	294	269	242	204	186	158	141	114	89	69	53	20	16	12	10	403	358
12	287	262	236	198	180	149	133	109	82	65	50	17	14	10	8	391	348
13	280	256	230	191	173	141	125	103	75	60	47	14	12			379	337
14	273	249	224	183	167	131	117	97	67	55	44	12				367	325
15	265	242	218	176	160	122	108	91	59	49	40					353	313
16	258	235	211	168	152	112	99	85	52	44	37					340	301
17	249	228	204	160	145	101	89	78	46	39	33					325	288
18	241	220	197	152	137	90	80	71	41	35	29					311	275
19	232	212	190	143	129	81	72	64	37	31	26					296	261
20	224	204	183	134	120	73	65	57	33	28	24					280	247
21	214	195	175	125	112	66	59	52	30	25	22					264	233
22	205	187	167	115	103	61	53	47	27	23	20					247	217
23	195	178	159	105	94	55	49	43	25	21	18					230	202
24	185	169	151	97	86	51	45	40	23	20	17					212	186
25	175	159	142	89	80	47	41	37	21	18	15					195	171
26	164	149	133	82	74	43	38	34								180	158
27	153	139	124	76	68	40	35	32								167	147
28	143	130	115	71	63	37	33	29								156	137
29	133	121	107	66	59	35	31	27								145	127
30	124	113	100	62	55	33	29	26								136	119
31	116	106	94	58	52	31	27	24								127	111
32	109	99	88	54	49	29	25	22								119	105
33	103	93	83	51	46											112	98
34	97	88	78	48	43											106	93
35	91	83	74	45	41											100	87
36	86	78	70	43	38											94	83
37	82	74	66	41	36											89	78
38	77	70	63	39	34											84	74
39	74	67	59	37	33											80	70
40	70	63	56	35	31											76	67
41	67	60	54	33	30											73	64
42	63	58	51													69	61
43	61	55	49													66	58
Q_s =	1.00	1.00	1.00	1.00	1.00	1.00	1.00	0.88	0.89	0.76	0.63	0.71	0.61	0.59	0.44	1,00	1.00

TABLE 4-6 *(Continued)*

STRUCTURAL TEES—MAIN MEMBERS

Allowable Axial Compression Load, P_m (Kips)

F_y 50 ksi

Size	WT5	WT5	WT5	WT5	WT5	WT5	WT5	WT5	WT5	WT5	WT5	WT5	WT5	WT5	WT5	WT5	WT5
Weight	44.5	38.5	36	33	30	27	24.5	22.5	19.5	16.5	14.5	12.5	10.5	9.5	8.5	7.5	5.75
Effective Length $K\ell$ in ft. — x-x Axis																	
0	393	339	318	291	265	238	216	199	172	145	128	99	81	73	63	54	27
1	384	331	310	284	258	232	211	194	168	142	125	97	79	72	62	53	27
2	372	321	301	275	250	225	203	188	163	138	122	95	77	71	61	52	26
3	359	308	289	264	240	215	195	181	157	133	118	92	75	69	59	50	26
4	343	294	276	251	229	205	185	172	150	127	114	88	73	67	57	49	25
5	325	279	261	237	216	193	174	163	142	120	109	85	70	65	55	47	25
6	306	261	244	222	201	180	162	153	133	113	103	81	67	62	53	46	24
7	285	242	226	205	186	165	149	142	123	105	97	77	63	59	51	44	24
8	262	222	207	187	169	150	135	130	113	96	91	72	60	57	49	42	23
9	237	200	186	167	151	133	120	117	102	87	84	67	56	53	46	40	22
10	211	176	164	146	132	115	103	103	90	77	76	62	52	50	44	38	21
11	183	151	140	124	111	96	86	89	77	67	68	56	48	47	41	35	20
12	155	127	117	104	93	81	72	74	65	56	60	50	43	43	38	33	19
13	132	108	100	89	79	69	62	63	55	48	51	44	39	39	35	30	19
14	114	93	86	76	68	59	53	55	47	41	44	38	34	35	31	28	18
15	99	81	75	67	60	52	46	48	41	36	39	33	29	31	28	25	17
16	87	72	66	58	52	46	41	42	36	32	34	29	26	27	25	22	15
17	77	63	58	52	46	40	36	37	32	28	30	26	23	24	22	19	14
18	69	57	52	46	41	36	32	33	29	25	27	23	20	22	19	17	13
19	62	51	47	41	37	32	29	30	26	22	24	20	18	19	17	16	12
20	56	46	42	37	34			27	23	20	22	18	16	18	16	14	11
21	50									18	20	17	15	16	14	13	10
22											18	15	14	14	13	12	9
23											16	14	12	13	12	11	8
24														12	11	10	8
25														11	10	9	7
26																8	6
27																	
28																	
29																	
30																	
31																	
32																	
33																	
34																	
35																	

Effective Length, $K\ell$, in feet — Y-Y Axis																	
0	393	339	318	291	265	238	216	199	172	145	128	99	81	73	63	54	27
1	389	335	315	288	262	236	214	196	170	143	125	97	79	71	61	52	26
2	384	331	311	284	259	233	211	193	167	141	122	95	77	68	58	49	25
3	379	327	306	280	255	229	208	189	164	138	118	92	74	64	54	46	24
4	373	322	302	276	251	226	205	184	160	135	113	88	72	59	50	42	23
5	367	316	296	271	247	222	201	180	156	131	108	85	69	54	46	38	21
6	360	310	291	266	242	217	197	175	151	127	103	80	65	48	41	34	19
7	353	304	285	260	237	213	193	169	146	123	96	76	61	42	35	29	18
8	345	297	278	254	231	208	188	163	141	118	90	71	57	35	29	23	16
9	336	289	271	248	226	203	183	157	136	114	83	66	53	27	23	18	13
10	328	282	264	241	219	197	178	150	130	109	75	61	49	22	18	15	11
11	319	274	257	234	213	191	173	143	124	103	67	55	44	18	15	12	9
12	309	265	249	227	206	185	168	136	117	98	58	49	39	15	13	10	8
13	299	257	240	220	200	179	162	128	111	92	50	42	33	13	11	9	6
14	289	248	232	212	192	173	156	120	103	86	43	37	29	11	9		
15	278	238	223	204	185	166	150	112	96	79	37	32	25				
16	267	228	214	195	177	159	143	103	88	72	33	28	22				
17	255	218	204	186	169	151	136	94	80	65	29	25	19				
18	243	208	194	177	161	144	129	85	72	58	26	22	17				
19	231	197	184	168	152	136	122	76	65	52	23	20	16				
20	218	186	173	158	143	128	115	69	58	47	21	18	14				
21	205	174	162	148	134	120	107	62	53	43	19	16	13				
22	191	162	151	137	124	111	99	57	48	39	17	15					
23	177	150	139	127	114	102	91	52	44	36							
24	163	138	128	116	105	94	84	48	41	33							
25	150	127	118	107	97	86	77	44	37	30							
26	139	117	109	99	89	80	71	41	35	28							
27	129	109	101	92	83	74	66	38	32	26							
28	120	101	94	85	77	69	61	35	30	24							
29	112	94	88	80	72	64	57	33	28	23							
30	104	88	82	74	67	60	54	31	26	21							
31	98	82	77	70	63	56	50	29	24	20							
32	92	77	72	65	59	53	47	27	23	18							
33	86	73	68	61	56	50	44	25									
34	81	69	64	58	52	47	42										
35	77	65	60	55	49	44	39										
36	73	61	57	52	47	42	37										
37	69	58	54	49	44	39	35										
38	65	55	51	46	42	37	33										
39	62	52	48	44	40	35	32										
40	59	50	46	42	38	34	30										
41	56	47	44	40	36	32	29										
42	53	45	42	38	34	31	27										
43	51	43	40														
Q_s =	1.00	1.00	1.00	1.00	1.00	1.00	1.00	1.00	1.00	1.00	1.00	0.90	0.87	0.87	0.84	0.81	0.53

TABLE 4-6 (*Continued*)

STRUCTURAL TEES—MAIN MEMBERS

Allowable Axial Compression Load, P_m (Kips)

F_y 50 ksi

Size	WT4	WT4	WT4	WT4	WT4	WT4	WT4	WT4	WT4	WT4	WT4	WT4	WT4	WT3	WT3	WT3	WT3
Weight	33.5	29	24	20	17.5	15.5	14	12	10	8.5	7.5	6.5	5	12.5	10	7.75	8
Effective Length $K\ell$ in ft. x-x Axis																	
0	295	256	212	176	154	137	123	106	88	75	67	58	33	110	88	68	71
1	287	248	205	171	150	132	120	103	86	73	65	56	32	106	85	66	68
2	276	238	197	164	143	127	115	98	83	70	63	54	31	99	79	62	64
3	262	226	186	155	135	120	109	93	79	67	60	52	30	92	73	57	60
4	246	212	174	145	126	112	102	87	75	64	58	50	29	83	66	52	55
5	229	196	160	133	116	102	94	80	70	60	54	47	28	72	57	45	49
6	209	179	144	120	104	92	85	73	65	55	51	44	26	61	48	38	42
7	187	160	127	106	91	81	76	64	59	50	47	41	25	48	37	30	35
8	164	139	109	91	77	68	65	55	53	45	43	37	23	37	28	23	27
9	138	116	89	74	62	55	54	45	46	39	38	33	21	29	22	18	21
10	113	94	72	60	51	45	43	37	38	33	33	29	20	24	18	15	17
11	93	78	59	50	42	37	36	30	32	27	28	25	18	19	15	12	14
12	78	65	50	42	35	31	30	25	27	23	24	21	15	16	13	10	12
13	67	56	42	36	30	26	26	22	23	20	20	18	13	14		9	10
14	57	48	37	31	26	23	22	19	20	17	17	15	11				
15	50	42	32	27	22	20	19	16	17	15	15	13	10				
16	44	37	28	23	20	17	17	14	15	13	13	12	9				
17	39	32							13	11	12	10	8				
18									12	10	11	9	7				
19											9	8	6				
20											9	8	6				
21																	
22																	
23																	
24																	
25																	
26																	
27																	
28																	
29																	
30																	
31																	
32																	
33																	
34																	
35																	

Effective Length, $K\ell$, in feet — Y-Y Axis																	
0	295	256	212	176	154	137	123	106	88	75	67	58	33	110	88	68	71
1	292	253	209	174	152	135	121	104	86	73	64	55	32	108	87	67	69
2	287	248	206	171	150	133	118	102	84	71	61	52	30	105	84	65	66
3	282	244	202	168	147	130	115	99	81	68	57	49	29	102	82	63	62
4	276	239	197	164	144	127	112	96	77	65	52	45	27	99	79	61	58
5	269	233	193	160	140	124	108	93	73	61	47	40	24	95	76	59	53
6	262	227	188	156	136	120	104	89	68	57	41	34	22	91	73	56	47
7	255	220	182	151	132	117	99	85	63	53	35	28	19	86	69	53	42
8	247	213	176	146	128	113	94	80	58	48	28	22	16	82	65	50	35
9	238	206	170	140	123	108	89	76	52	43	22	17	13	76	61	46	28
10	229	198	163	135	118	104	83	71	46	38	18	14	11	71	56	42	23
11	220	190	156	129	112	99	77	66	39	32	15	12	9	65	51	38	19
12	210	181	149	122	107	94	71	60	33	27	12	10	8	59	46	34	16
13	200	172	141	116	101	89	64	54	28	23	10	8	6	52	41	30	14
14	189	162	133	109	95	83	57	48	24	20	9	7		45	35	26	12
15	177	152	125	102	89	78	50	42	21	17				40	31	22	10
16	166	142	116	94	82	72	44	37	19	15				35	27	20	9
17	153	131	107	86	75	65	39	33	17	13				31	24	17	
18	141	120	97	78	68	59	35	29	15	12				27	21	16	
19	127	108	88	70	61	53	31	26	13	11				25	19	14	
20	115	98	79	63	55	48	28	24	12	10				22	17	13	
21	104	88	72	58	50	43	25	22						20	16	11	
22	95	81	65	52	45	39	23	20						18	14	10	
23	87	74	60	48	42	36	21	18						17	13	10	
24	80	68	55	44	38	33	19	16						15	12	9	
25	73	62	51	41	35	31	18	15						14	11		
26	68	58	47	38	33	28	17	14									
27	63	54	43	35	30	26											
28	59	50	40	32	28	24											
29	55	46	38	30	26	23											
30	51	43	35	28	24	21											
31	48	41	33	26	23	20											
32	45	38	31	25	21	19											
33	42	36	29	23	20	18											
34	40	34	27														
35	37																
36																	
37																	
38																	
39																	
40																	
41																	
42																	
43																	
Q_s =	1.00	1.00	1.00	1.00	1.00	1.00	1.00	1.00	1.00	1.00	1.00	1.00	0.74	1.00	1.00	1.00	1.00

TABLE 4-6 *(Continued)*

STRUCTURAL TEES—MAIN MEMBERS

Allowable Axial Compression Load, P_m (Kips)

F_y 50 ksi

Size	WT3	WT3	WT2.5	WT2.5	WT2	MT7	MT6	MT5	MT5	MT4	MT4	MT4	MT4	MT4	MT4	MT3	MT3
Weight	6	4.25	9.25	8	6.5	8.6	5.9	11.45	4.5	18.85	17.5	16.3	9.25	3.25	2.75	16.87	10
Effective Length $K\ell$ in ft. — x-x Axis																	
0	53	38	82	70	57	27	18	88	16	166	151	144	82	13	13	149	88
1	51	36	77	66	53	27	18	87	15	161	147	139	79	13	13	143	84
2	48	34	70	61	48	27	18	84	15	155	141	133	76	13	12	134	79
3	45	32	62	53	40	27	18	82	15	146	134	126	73	12	12	124	72
4	41	29	52	45	32	26	17	79	15	137	125	118	69	12	12	111	64
5	37	26	40	34	22	26	17	76	15	126	116	108	64	12	11	97	54
6	32	23	28	24	15	26	17	72	14	115	106	97	59	11	11	81	44
7	27	19	21	18	11	26	17	68	14	102	94	86	53	11	10	64	33
8	21	15	16	14	8	25	16	64	14	88	82	73	47	11	10	49	25
9	17	12	13	11		25	16	60	13	72	68	59	41	10	9	39	20
10	13	9	10			25	16	55	13	59	55	48	34	10	8	31	16
11	11	8				24	16	50	13	48	46	40	28	9	8	26	13
12	9	6				24	15	45	12	41	39	33	23	9	7	22	11
13	8	6				23	15	39	12	35	33	28	20	8	6		
14	7	5				23	15	34	11	30	28	25	17	8	5		
15						23	14	29	11	26	25	21	15	7	5		
16						22	14	26	10	23	22	19	13	6	4		
17						22	14	23	10		19		12	6	4		
18						21	13	20	10				10	5	3		
19						21	13	18	9					5			
20						20	12	17	8					4			
21						20	12	15	8					4			
22						19	11	14	7								
23						19	11		7								
24						18	11		6								
25						17	10		6								
26						17	10		5								
27						16	9										
28						16	9										
29						15	8										
30						14	8										
31						14	7										
32						13	7										
33						12											
34						12											
35						11											

Effective Length, $K\ell$, in feet — Y-Y Axis																	
0	53	38	82	70	57	27	18	88	16	166	151	144	82	13	13	149	88
1	51	36	80	69	56	27	17	86	15	164	149	142	79	12	12	146	86
2	49	34	77	67	53	26	17	84	14	161	146	139	77	11	11	142	84
3	46	32	74	64	50	25	16	81	13	157	143	136	73	10	9	138	81
4	42	30	71	61	47	24	14	77	12	154	139	133	69	9	7	133	78
5	39	27	68	58	43	22	13	74	10	149	135	129	65	7	5	128	75
6	34	24	64	55	39	21	11	69	8	145	131	125	60	5	4	122	71
7	29	20	59	51	34	19	10	65	6	140	126	121	54	4		115	67
8	24	16	54	46	29	17	8	60		135	121	116	49			108	62
9	19	13	49	42	24	16		54		129	116	111	42			101	58
10	15	10	44	37	19	14		49		123	111	106	35			93	52
11	13	8	38	32	16	11		42		117	105	100	29			84	47
12	11	7	32	27	14	10		36		110	98	95	25			75	41
13	9	6	27	23	12			31		103	92	89	21			66	35
14	8	5	24	20	10			26		96	85	82	18			57	30
15	7		21	17	9			23		89	78	76	16			49	27
16			18	15	8			20		81	70	69	14			43	23
17			16	13				18		73	63	61	12			38	21
18			14	12				16		65	56	55	11			34	18
19			13	11				14		58	50	49				31	17
20			12	10				13		52	45	44				28	15
21			10							48	41	40				25	14
22										43	37	37				23	12
23										40	34	34				21	11
24										36	31	31				19	
25										34	29	28					
26										31	27	26					
27										29	25	24					
28										27	23	23					
29										25	22	21					
30										23	20	20					
31										22		18					
32																	
33																	
34																	
35																	
36																	
37																	
38																	
39																	
Q_s =	1.00	1.00	1.00	1.00	1.00	0.36	0.35	0.88	0.40	1.00	1.00	1.00	1.00	0.46	0.54	1.00	1.00

TABLE 4-6 (*Continued*)

STRUCTURAL TEES—MAIN MEMBERS

Allowable Axial Compression Load, P_m (Kips)

F_y 50 ksi

Size	MT3	MT2.5	MT2	MT2	MT2	ST12	ST12	ST12	ST12	ST12	ST10	ST10	ST10	ST10	ST9	ST7.5	ST7.5
Weight	2.2	9.45	8.15	6.9	6.5	60	52.95	50	45	39.95	47.5	42.5	37.5	32.7	27.35	25	21.45
Effective Length $K\ell$ in ft. x-x Axis: 0	11	83	72	61	57	528	439	441	371	246	420	375	330	259	222	220	187
1	11	79	67	57	53	524	436	438	369	244	416	372	327	257	220	218	185
2	11	72	59	50	46	520	432	435	366	243	412	368	324	254	218	215	182
3	10	64	49	42	38	515	428	431	363	241	408	364	320	252	215	211	179
4	10	54	37	32	27	510	424	427	359	239	403	359	317	249	212	208	176
5	9	42	24	21	18	505	420	422	356	237	397	354	312	246	209	204	172
6	9	30	17	15	12	499	415	418	352	235	392	349	308	242	206	199	168
7	8	22	12	11	9	493	410	413	348	233	385	344	303	239	202	194	164
8	7	17	10	8		487	405	408	344	231	379	338	298	235	198	189	160
9	7	13				480	399	402	340	228	372	331	293	231	194	184	155
10	6	11				473	393	397	335	226	365	325	287	227	190	178	150
11	5					466	387	391	330	223	357	318	281	222	186	172	144
12	4					458	381	385	325	220	349	311	275	218	181	165	139
13	4					450	374	378	320	217	341	303	269	213	176	159	133
14	3					442	368	372	314	214	333	295	262	208	171	152	127
15	3					433	361	365	309	211	324	287	256	203	166	145	121
16						424	353	358	303	208	315	279	248	198	161	137	114
17						415	346	350	297	205	305	270	241	193	155	129	108
18						406	338	343	291	202	296	261	234	187	149	121	100
19						396	330	335	285	198	286	252	226	181	143	113	93
20						386	322	327	278	195	275	243	218	175	137	104	86
21						376	314	319	272	191	265	233	210	169	131	95	78
22						365	305	311	265	187	254	223	201	163	124	87	71
23						355	296	302	258	184	243	213	193	156	117	80	65
24						344	287	293	251	180	231	202	184	150	111	73	60
25						332	278	284	243	176	219	191	174	143	103	67	55
26						321	268	275	236	172	207	180	165	136	96	62	51
27						309	259	266	228	168	194	169	155	129	89	58	47
28						297	249	256	220	163	181	157	145	122	83	54	44
29						284	239	246	212	159	169	146	135	114	77	50	41
30						272	228	236	204	155	158	137	127	107	72	47	38
31						259	218	226	196	150	148	128	119	100	68	44	36
32						246	207	215	187	146	139	120	111	94	63	41	34
33						232	196	204	179	141	131	113	105	88	60	39	32
34						218	184	193	170	136	123	106	99	83	56	36	30
35						206	174	183	161	131	116	100	93	78	53	34	28

Effective Length, $K\ell$, in feet — Y-Y Axis																	
0	11	83	72	61	57	528	439	441	371	246	420	375	330	259	222	220	187
1	10	81	70	59	55	518	431	431	363	242	411	367	321	253	217	214	182
2	9	79	67	56	53	506	422	418	353	236	399	357	310	245	210	205	175
3	7	75	63	53	50	492	411	402	341	230	385	345	297	236	201	195	167
4	5	72	58	49	46	475	398	384	327	223	369	331	282	225	191	183	157
5	3	68	54	45	42	457	384	364	311	215	351	316	265	213	180	169	146
6		63	48	40	37	437	369	342	294	206	332	299	246	200	168	154	134
7		58	42	35	32	416	353	318	276	197	310	280	225	186	154	137	121
8		52	35	29	27	393	335	292	256	187	287	260	203	170	140	119	107
9		47	28	23	21	368	316	264	234	176	263	239	179	154	124	100	91
10		40	23	19	17	342	296	235	211	164	236	216	153	136	107	81	75
11		34	19	15	14	314	275	203	187	152	208	192	127	117	90	67	62
12		28	16	13	12	285	253	171	161	140	178	166	107	98	75	56	52
13		24	14	11	10	254	230	145	137	126	152	142	91	84	64	48	44
14		21	12	9	9	221	205	125	118	112	131	122	78	72	55	41	38
15		18	10	8	8	192	179	109	103	98	114	107	68	63	48	36	33
16		16	9			169	158	96	90	86	100	94	60	55	42	32	29
17		14				150	140	85	80	76	89	83	53	49	37	28	26
18		13				134	125	76	71	68	79	74	47	44	33		
19		11				120	112	68	64	61	71	66	43	39			
20						108	101	61	58	55	64	60					
21						98	92	56	52	50	58	54					
22						89	83			45	53	50					
23						82	76										
24						75	70										
25						69	65										
26							60										
27																	
28																	
29																	
30																	
31																	
32																	
33																	
34																	
35																	
36																	
37																	
38																	
39																	
Q_s =	0.58	1.00	1.00	1.00	1.00	1.00	0.94	1.00	0.94	0.69	1.00	1.00	1.00	0.90	0.92	1.00	0.99

TABLE 4-6 *(Continued)*

STRUCTURAL TEES—MAIN MEMBERS

Allowable Axial Compression Load, P_m (Kips)

F_y 50 ksi

Size	ST6	ST6	ST6	ST5	ST4	ST3.5	ST3	ST2.5	ST2	ST1.5
Weight	20.4	17.5	15.9	12.7	9.2	7.65	6.25	5	3.85	2.85
Effective Length $K\ell$ in ft. (x–x Axis)										
0	180	154	140	112	81	68	55	44	34	25
1	177	152	138	110	79	65	53	42	32	22
2	174	149	135	107	76	63	50	39	28	18
3	170	146	132	104	73	59	46	35	24	13
4	165	142	129	100	69	55	42	31	19	8
5	160	138	125	96	65	51	38	25	13	5
6	155	133	121	91	60	46	32	20	9	3
7	149	128	116	86	55	40	27	14	7	
8	142	123	111	81	49	34	21	11	5	
9	136	118	106	75	43	28	16	9		
10	128	112	100	69	36	23	13	7		
11	121	106	94	62	30	19	11	6		
12	113	99	88	55	25	16	9			
13	105	92	82	48	22	13	8			
14	96	85	75	41	19	12				
15	87	78	68	36	16	10				
16	77	70	60	32	14	9				
17	68	62	53	28	13					
18	61	55	47	25	11					
19	55	49	43	23						
20	49	45	38	20						
21	45	40	35	18						
22	41	37	32	17						
23	37	34	29	15						
24	34	31	27	14						
25	32	29	25							
26	29	26	23							
27	27	24	21							
28	25	23	20							
29	23	21	18							
30		20								
31										
32										
33										
34										
35										

Effective Length, $K\ell$, in feet — Y-Y Axis																	
0	180	154	140	112	81	68	55	44	34	25							
1	175	149	136	108	78	65	52	42	32	23							
2	168	143	130	103	74	61	49	39	29	21							
3	160	135	123	98	68	56	44	34	25	18							
4	150	126	115	91	62	50	39	29	21	14							
5	140	116	106	83	55	44	33	24	16	9							
6	128	104	96	74	48	36	26	18	11	7							
7	115	92	85	65	39	28	19	13	8	5							
8	101	78	73	54	30	21	15	10	6	4							
9	86	63	60	43	24	17	12	8	5								
10	70	51	49	35	19	14	9	6									
11	58	42	40	29	16	11	8										
12	49	36	34	24	13	10											
13	41	30	29	21	11												
14	36	26	25	18													
15	31	23	22	16													
16	27	20	19														
17	24																
18																	
19																	
20																	
21																	
22																	
23																	
24																	
25																	
26																	
27																	
28																	
29																	
30																	
31																	
32																	
33																	
34																	
35																	
36																	
37																	
38																	
39																	
Q_s =	1.00	1.00	1.00	1.00	1.00	1.00	1.00	1.00	1.00	1.00	1.00	1.00	1.00	0.90	0.93	1.00	0.99

5

CONNECTIONS

General

The latest AISC Specifications permit a wide variety of connections. The basic requirement, appropriate with the sophisticated combinations of different types of steel to be connected, different design requirements of connections, and different means of connections, is a performance requirement consistent with the overall development of the Specifications. This requirement states simply "... that the design of connections be consistent with the assumptions as to the type of construction...." (1.2). Each of the detailed requirements for the design of connections simply builds upon this basic requirement. By implicitly or explicitly requiring that the design of a particular type of connection be consistent with the design assumptions as to the type and amount of force to be transmitted, and rotation capacity (or rigidity) consistent with the rotation assumed necessary to develop the connection forces, the basic performance requirement is completed (1.2).

The Specifications explicitly recognize inelastic behavior in connections of members designed as elastic: "virtually unchanged" angles at the joints in rigid frames, "non-elastic" deformation of parts of connections in Type 2 and 3 construction, and "inelastic rotation" for wind connections with Type 2 construction (1.2). Elastic behavior in the connections of members under plastic design is implicitly recognized (2.1).

Scope

For the purposes of this Chapter, connections are most conveniently considered as classified on two bases; (1) materials used (rivets, bolts, pins, or welds), and (2) the assumed behavior of the connection (design requirements: rigid, semi-rigid, or plastic for moment; shear transmission only; tensile or compressive force only; or combinations). In addition to forming joints between two or more steel members or parts of members, connections are required to elements composed of other structural materials. For composite action with concrete elements not bonded by encasement, shear connections are required (1.11.1). Shear connections may utilize specially designed shear connectors or standard welded stud connectors (1.11.4; 1.4.6). For connection of steel column bases to transmit any direct tension or shear, anchor bolts are required (1.22).

It is not intended in this chapter to duplicate the design aids, detail data, and examples in the *AISC Handbook*, Part 4, Connections. Equally, space limitations do not permit presentation of a wide range of examples to illustrate even the recently published research in this area. (See "Selected References".) Rather, the purpose here is limited to explanations and illustrations of all applicable Specification requirements that might be overlooked or troublesome in routine work. This aim will include indication of reasonable interpretations to resolve apparent conflicts or ambiguities in the Specifications, and extension of such interpretations where the Specifications seem to have omissions. It has been desirable to extend this aim somewhat in that design aids for bearing plate and base plate connections were included as well as an extension of concrete bearing connection design to cover an apparent gap between the AISC Specifications and the ACI Building Code. During the interim between preparation and publication, AISC specifications were revised to agree with the latest ACI Building Code.

Rivets, Pins, and Bolts

Rivets and Pins. The requirements for the use of rivets and pins were established many years ago and were in many AISC Specifications. Since most of the late research has been directed toward welded, and more recently high-strength bolted connections, there has been little change in the Specifications for the use of rivets. Familiarity with the requirements for, and a sharply reduced use of, rivets in building construction results in little need for interpretations of these Specifications. Rivets of Grades 1 and 2 are available under ASTM A502 (1.4.3). Allowable stresses (for tension and bearing only) are given in Table 1.5.2.1 (1.5.2.1). Net sections for tension members must be used (1.14.2); computed as prescribed (1.14.3); and allowance of $\frac{1}{16}$ in. made plus the nominal diameter of the rivet holes (1.14.5).

The use of pin connections, originally popular in truss construction, has declined in modern steel building, and is usually encountered only for very special situations requiring special design. The general requirements for the use of pins are brief and essentially unchanged from previous Specifications (1.14.6).

Perhaps the most used application of these Specification requirements today will be in alterations or additions to existing buildings in which rivets or pins were used. For new construction, bearing-type connections can not be assumed to share stress with welds (1.15.10). If used in combination, the welds must be designed for the entire stress. In strengthening existing construction, bearing connections can be assumed to carry the in-place loads, and the new welds designed only for the additional stress (1.15.10).

Bolts. Bolts may be classified by strength as (1) low, A307, for $F_t = 20$ ksi; and (2) high, A325, for $F_t = 40$ ksi, and A490, for $F_t = 54$ ksi . . . (1.5.2.1). The ordinary low strength (A307) bolts are usable only in bearing connections (1.5.2.2). The high strength bolts may be designed for either bearing or friction connections (1.5.2.1).

The use of ordinary (A307) bolts is limited by a number of Specification requirements. The allowable stresses are low: tension $F_t = 20$ ksi on the threaded area; and shear $F_v = 10$ ksi (1.5.2.1). The slip before full bearing is achieved on a group of ordinary bolts effectively rules out the sharing of stress in a mixed connection. Holes are to be taken as $\frac{1}{16}$ in. larger than the nominal diameter (1.23.4), and the ordinary bolt does not expand to fill out the hole like a driven rivet nor can it be used for dependable friction. Stress sharing may not be assumed between ordinary bolts and rivets or welds (1.15.10; 1.15.11). In addition, low-strength bolts are not permitted in important field connections including column splices in all buildings with $H \geqslant 200$ ft., and where width/height < 0.25; also where width/height < 0.40, for $H \geqslant 100$ ft.; beam-column or column-bracing connections where $H > 125$ ft.; frames carrying cranes with more than five-ton capacity; and connections subject to vibration, impact, or stress reversal (1.15.12); nor for flange-to-web nor cover plate-to-flange connections of built-up girders (1.10.4).

High strength bolts (1.16.1) and welds are considered essentially equivalent as connections, and, for friction-type joints assembled prior to the welding, the high-strength bolts may be assumed to share stress with welds in a mixed connection (1.15.10) or with rivets (1.15.11). Gross sections may be used for the design of compression members (1.14.2), and for the flanges of both built-up and rolled-shape girders provided the area of holes is equal to or less than fifteen percent of gross flange area (1.10.1). For tension members net section area is the basis of design (1.14.2). In friction-type joints resisting direct tension, the shear stress permitted with high-strength bolts must be reduced (1.6.3).

Slotted Holes for Bolted Shear Connections. The use of short-slotted holes is permitted under 1974 AISC Specification for "Structural Joints Using ASTM A-325 or A-490 Bolts," Section 3, subject to the approval of the designer. They can be used in either friction-type or bearing-type connections, provided a washer is installed over the hole.

The normal hole size for a $\frac{3}{4}''$ ϕ bolt is $\frac{13}{16}''$, whereas a short-slotted hole is $\frac{13}{16}''$ deep by $1''$ long (or $\frac{3}{16}''$ longer in the horizontal dimension). While the Specifications state that the hole can be either vertical or horizontal, the authors suggest only the horizontal slotted method be used. End clip holes only would be slotted, not the holes in the connection beam or column. See sketch, a full scale view of the end clip holes and bolt relationship for a typical $\frac{5}{16}''$ thick web.

The advantages to this system are many, several of which are:

1. Greater erection speed with less field burning of misaligned holes.
2. The use of one size clip angle with a set gauge will accommodate web thickness from $\frac{3}{16}''$ to $\frac{9}{16}''$.
3. The reduction in sizes of clips to fabricate and stock should help reduce costs.
4. The speed of erection (and elimination of mill web thickness tolerance problems) should help reduce cost.

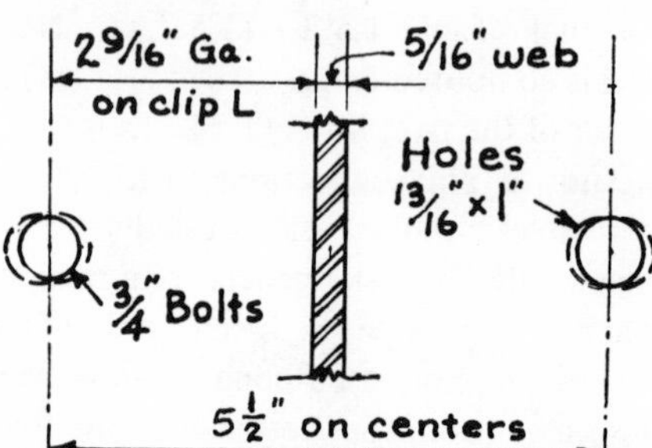

Short-slotted holes layout for clip Ls - shear connection

Welds

General. Full penetration groove-welds can be designed for full development, same stress as the base metal (1.5.2.1), by selection of the specified matching electrode and welding process (1.17.2). For all fillet, plug, and slot welds, and partial penetration groove-welds, reduced permissible stresses upon the effective throat area (1.14.7) are specified (1.5.2.1). In no case may the stresses exceed that for the base metal, or if different, the weaker base metal (1.5.2.1).

Special Considerations. A number of minor special considerations arise in the specification of welding. Generally, net sections are not a consideration except for plug and slot welds in which the gross area of the holes is deducted to check the fifteen percent maximum allowed (1.10.1; 1.14.3). The Specifications require preheating for various conditions, including all work when the temperatures are below 32°F (1.23.6). Except for single- and double-angle or similar minor members, welds are to be laid out to avoid eccentric axial force or such eccentricity must be considered in the design of the connection and the member connected (1.15.3). For the usual shear connection requiring flexibility to accommodate the necessary simple-end rotations assumed, the locations of the welds and the selection of the connection elements must be coordinated (1.15.4).

As previously noted, where welding at high-strength bolted friction-type joints is required as a mixed connection with shared stress, the final tightening of the bolts is required prior to the welding. The *sequence* of completing purely welded connections is also important, though not explicitly covered by the Specifications (1.23.6). The heat generated in the operations of welding creates intense shrinkage strains which, if restrained, leave corresponding residual stresses (1.23.6). These stresses can be relieved by local inelastic yielding, but where local inelastic yielding is also restrained or limited, warping and lamellar tearing* may result. For many welded assemblies, the simple precaution of a specified *sequence* of welding may be employed to balance the strains and to avoid warping. Even after this precaution, certain complex assemblies may be expected to retain adverse residual stresses. For cases where this condition is anticipated or suspected, stress relief by heating must be specified by the Engineer (1.23.6). (Note: this service is not provided unless it has been specified and will normally be an added cost.)

The use of a proper *sequence* to avoid creation of shrinkage stresses or to minimize same can also be specified in many connections where lamellar tearing might occur. Particularly with thicker sections, where both the direction of the shrinkage is completely restrained and the resulting stress is normal to the surface of the section, consideration should be given to the welding sequence. If the condition can not be eliminated by a practicable sequence as a first choice for a solution, it may be possible to relieve the strains without developing large stresses by use of soft wire "cushions" or by revision of the entire connection detail. At least for simple cases it should, of course, be more economical to specify a particular welding sequence. (See Examples this chapter.)

Connection Design

Classification. In addition to the general requirements previously cited for the design of connections, certain arbitrary minimum design requirements for connections have been established. All connections for members carrying calculated stress must be "designed to support not less than six kips" (except lacing, sag bars, and girts), presumably six kips shear in flexural members, six kips tension in ties, and six kips bearing in compression members, all at allowable stress levels (1.15.1). Eccentric connections of axially loaded members are to be designed to transmit the resulting moments as well as the axial force (1.15.2). These minimum requirements naturally become most significant in the design of light members such as axially loaded members in trusses. Connections for such members are required to meet an additional requirement that they transmit the design load or the minimum six kips, whichever is larger, and develop at least half of the effective strength of the member (1.15.7). Note: joists are regarded as a special very limited-size truss in which the minimum connection capacity is simply specified as twice the design load for open web steel joists (4.5); or for the longspan and deep longspan joists, as the design stress or half the allowable strength of the member (103.5b). (See Chapter 4: Joists; Examples.)

As noted previously, connection types may be classified on the basis of the connection method or the design function. Broadly, connections may be described as *flexible* (for the transmission of shear only, 1.15.4), or *rigid* (maintaining the angle between the members connected and transmitting full moment capacity of the most flexible element at the

*"Commentary on Highly Restrained Welded Connections," *Engineering Journal*, AISC, **10**, No. 3, 1973.

joint as well as the shear, 1.15.5), or *semi-rigid* (transmitting a pre-determined fraction of the full moment capacity as a rigid joint and further loads in shear as a flexible joint with corresponding angle change to supply rotation for the additional loads, 1.15.5; 1.2).

Flexible Connections. "Flexible" connections are designed to transmit shear without exceeding allowable unit stresses on the connectors as a group or the connection as a whole. The use of an average capacity for each of several connector elements sharing the total load is justified by allowing self-limiting localized stresses determined by an elastic joint analysis to exceed the yield point and create inelastic localized deformations of the connector materials, or by inelastic deformations of the connection elements (1.15.4). The simplest examples of localized deformation occur in the assembly of bearing-type bolted connections where the cumulative tolerances permitted exist on (1) out-of-round in the bolts, (2) oversize holes ($\frac{1}{16}''$), and (3) center-to-center location of the holes in the different elements connected. The extreme degree of such inelastic action occurs with a two-bolt bearing-type connection where one bolt is loosely fitted and one is very tight. Until the material of the connected element surrounding the loaded bolt or the bolt yields and deforms ($+\frac{1}{16}''$), the load is not shared and a 50 percent adjustment will be developed as the load increases. For larger (and thus more important) members, more bolts or rivets will be required and the degree of adjustment required on each will be less. Lesser adjustments are required for a long line of bolts or rivets intended to share stress equally. Even if perfectly fitted, yielding and inelastic deformations occur, maximum at and beginning at the first loaded bolt or rivet, and decreasing to a minimum at the last. (See Figs. 5-1

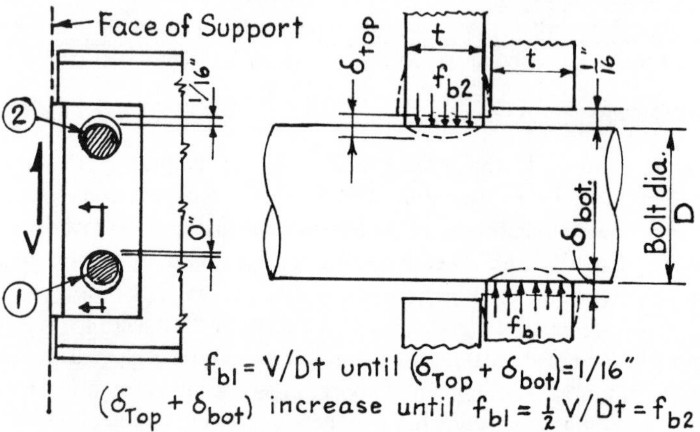

FIG. 5-1 Self-Limiting, Localized Deformations—Two Bolts.

and 5-2.) After this localized inelastic adjustment in the connectors for shear transmission, consider the inelastic adjustments that occur to reduce the "elastic theory" moments.

Inelastic deformation in the connection elements, typically angles, will occur and reduce the restraint which would transmit moment. The common double-angle shear bearing connection is extremely stiff longitudinally for the transmission of shear, and it depends upon the minor inelastic bearing deformations around each fastener to equalize the shear stresses in the fasteners. The same double-angle member is relatively flexible and will twist to permit a relatively large angular rotation reducing moment transmission. (See Fig. 5-3.)

Experience and tests confirm the practical assumptions of shear transfer only and the

Inelastic deformations occur successively in the plates at each fastener and in the fasteners, beginning and largest at the first loaded, and continuing until the elastic strains in the spaces s_1, s_2, s_3, and s_4 become proportional to equal stress in all fasteners.

FIG. 5-2 Self-Limiting Deformations—Axial Stress on Line of Separate Fasteners.

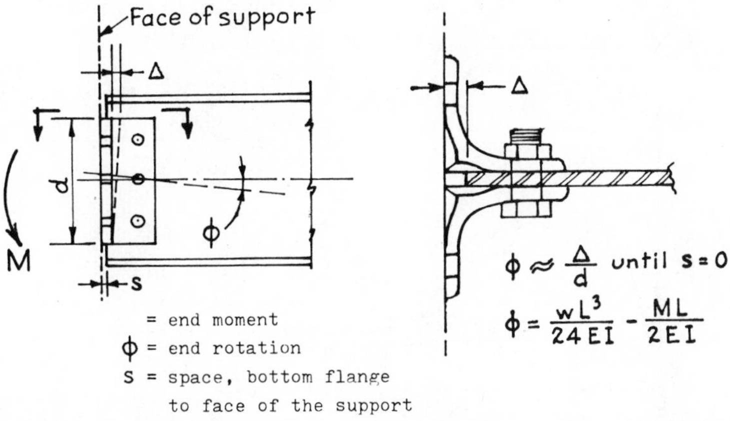

FIG. 5-3 Self-Limiting Deformation (Twist) in the Connection Elements (⌋⌊) Two Angles.

use of an average shear stress per unit weld or separate fastener. The actual rotations observed in one series range from 0.84 to 0.97 times Φ_0, the "simple beam rotation." These limits corresponding to moments ranging from three to sixteen percent were approximately the same for a single end plate connector or the common double-angle connector.[1] Figure 5-4 presents the usual device for an approximate analysis. An additional caution reported from these tests is that the moment stiffness increases abruptly when the space "*s*" (see Fig. 5-3) closes and the lower flange transmits compression to the face of the support. Coping the bottom flange where a quick analysis of the proportions of depth and required angle change show the usual clearance to be inadequate may be desirable for deep connections.

For Type 2 construction (flexible connections) all the reactions should be shown on the design drawings; alternatively, only those exceeding one half the tabulated uniform load capacity for the sections used, together with a general note that connections shall be designed for one half the capacity unless otherwise noted, should be shown.

---

[1]"Moment-Rotation Characteristics of Shear Connections," Kennedy, October, 1969, **6**, No. 4, *Engineering Journal*, AISC.

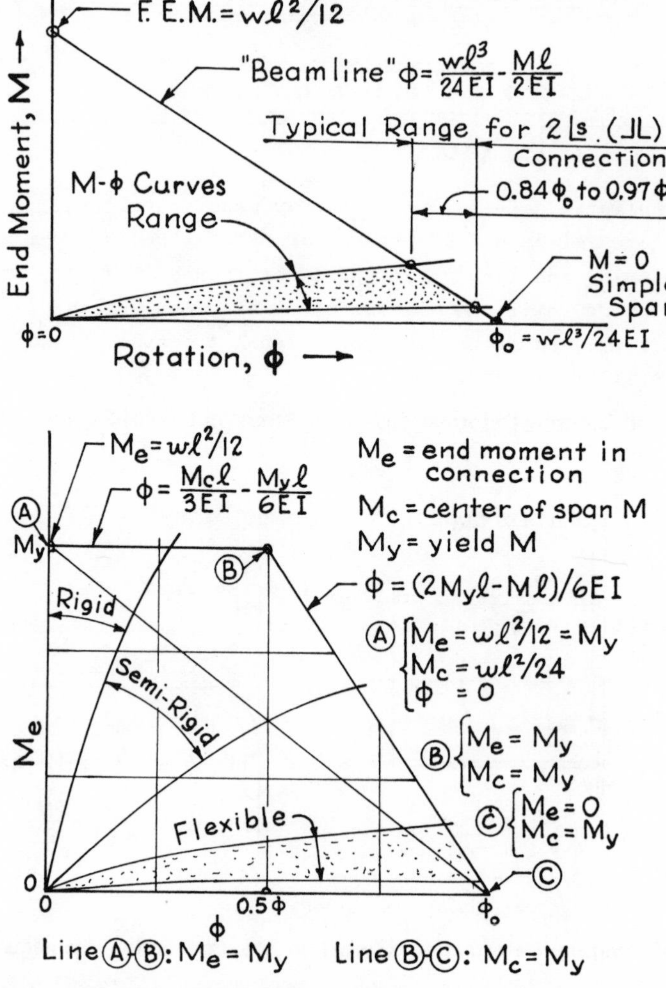

FIG 5-4 Uniformly Loaded Elastic Beam Line—Rotations at Connection.

Rigid Connections (Type 1 Construction). The AISC Specifications requirement is quite realistic: that rigid connections hold the original angles "virtually" unchanged (1.2). This requirement in elastic design is usually satisfied by connections designed to develop the full section of the flexural member or the full moment at yielding of the more flexible member connected. It will be noted from Fig. 5-4 that the rigid frame analysis ($\Phi = 0$ at the allowable stress) may be satisfied by such a connection which would have a very small rotation at 0.66 to $0.60F_y$, but would be capable of a significant rotation at yielding of the flexural member (line A-B).

A diagram similar to Fig. 5-4 but with point A representing an end moment, $M_e = M_p$, and point C, a center span moment, $M_c = M_p$, can be prepared for plastic design. Connections capable of achieving full collapse load (hinges at both ends and center of span) would be required to reach $\Phi = 0.5\Phi_0$, point B. The simpler concept of "plastic redesign," where only end hinges are required to form at the factored load, would require connections with a somewhat less rotation capacity, along line A-B. See Fig. 5-5.

Semi-Rigid Connections (Type 3 Construction). (See Fig. 5-4.) Ideally, the semi-rigid connections for Type 3 construction will behave elastically between the $\Phi = 0$ ordinate

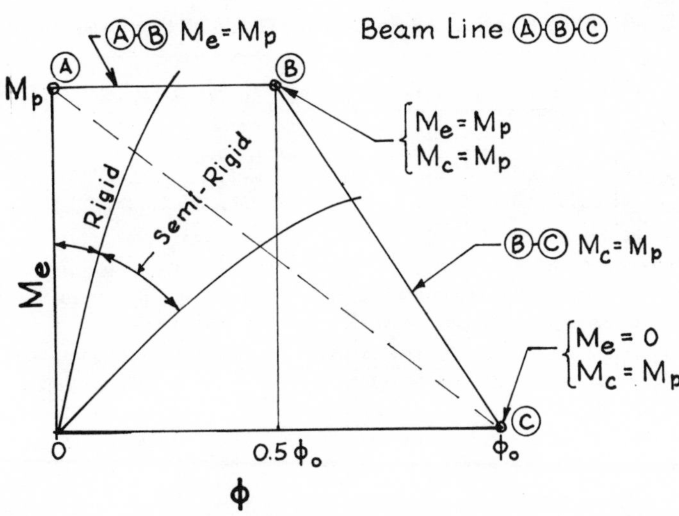

FIG. 5-5 Uniformly Loaded Plastic or Rotations at Connection.

and the "rigid" connection up to a predetermined end moment. Upon reaching this predetermined end moment (M_y for the connection), a rotation capacity sufficient to develop the yield moment at the center of the span, $M_c = M_y$, should be available. In practical cases where the nearest available rolled section will be above the design capacity required, the excess capacity will be provided at the midspan. As in plastic redesign, end hinges only will form at the full design load. Since these hinges are designed for $M_e = M_y$, more rotation capacity is required (to cross line B-C).

Masonry Bearing Connections

General. The AISC Specifications provide very conservative allowable stresses for bearing on masonry and concrete which apply in the absence of Code (statutory Building Code) regulations (1.5.5). For all masonry laid up in mortar, most statutory codes also provide low allowable stresses. Usually, codes distinguish among solid masonry units, bricks or block, and hollow units as well as among different classes of mortar and masonry materials. Values so prescribed range in general from 50 psi to 400 psi. The AISC Specification is therefore seldom applicable since it includes only stone masonry, $F_p = 0.400$ ksi, and brick laid in "cement" mortar,* $F_p = 0.250$ ksi (1.5.5). *The authors recommend use of the masonry bearing values prescribed in local Codes or those recommended by national associations dealing with masonry products.* For bearing stresses on concrete, the AISC Specifications, $F_p = 0.35 f'_c$ on the full area and $F_p = 0.35 f'_c \sqrt{A_2/A_1} \leqslant 0.70 f'_c$ on fractions of the area, utilize recent ACI Building Code (ACI 318-71) refinements for economy.** The ACI Building Code is of course usually applicable under local statutory codes. *For beam-bearing plates and column base plates on concrete, the authors recommend the use of bearing values prescribed by the ACI Building Code.*

Beam Bearing Plates. The approved design (Chapter 2, pp. 82–83, *AISC Handbook*) for beam bearing plates is the formula:

$$t = \sqrt{3 f_p (n)^2 / F_b}$$

(Continued on page 274)

---

*All mortars utilize cementitious materials and the term "cement" can be quite properly applied to all.
**Supplement No. 3, 1974.

TABLE 5-1 Allowable Bearing Pressure for Grade 36 Beam—Bearing Plates

K = 5/8"

B \ t	1/4	3/8	1/2	5/8	3/4	7/8	1	1 1/8	1 1/4	1 3/8	1 1/2	1 5/8	1 3/4
4	297	669	1190										
5	160	360	640	1000									
6	100	224	399	623	897	1221							
7	68	153	272	425	612	834	1089	1378					
8		111	197	309	444	605	790	1000	1234				
9		84	150	234	337	459	599	758	936	1133	1348		
10		66	118	184	264	360	470	595	735	889	1058	1242	1440
11		53	95	148	213	290	379	479	592	716	852	1000	1160
12			78	122	175	239	312	394	487	589	701	823	954
14			55	86	125	170	221	280	346	530	498	585	678
16				65	93	127	165	209	259	313	372	437	507
18				50	72	98	128	162	200	243	289	339	393
20					58	78	102	130	160	194	230	270	314

K = 3/4"

B \ t	1/4	3/8	1/2	5/8	3/4	7/8	1	1 1/8	1 1/4	1 3/8	1 1/2	1 5/8	1 3/4
4	360	810	1440										
5	184	413	735	1147									
6	111	250	444	694	1000	1361							
7	74	167	297	465	669	911	1190						
8	53	120	213	333	479	652	852	1078	1331				
9		90	160	250	360	490	640	810	1000	1210	1440		
10		70	125	195	280	381	498	630	779	942	1121	1316	
11		56	100	156	224	305	399	505	623	754	898	1053	1221
12			82	128	184	250	327	413	510	617	735	862	1000
14			58	90	130	176	230	292	360	436	518	608	706
16				66	96	131	171	217	268	324	385	452	524
18					74	101	132	167	207	250	298	349	405
20						80	105	133	164	199	237	278	322

K = 1"

B \ t	1/4	3/8	1/2	5/8	3/4	7/8	1	1 1/8	1 1/4	1 3/8	1 1/2	1 5/8	1 3/4
6	141	316	563	879	1266								
7	90	203	360	563	810	1103	1440						
8	62	141	250	391	563	431	1000	1266					
9		103	184	287	413	562	735	930	1148	1389			
10		79	141	220	316	431	562	712	879	1063	1266	1485	
11		63	111	174	250	340	444	563	694	840	1000	1174	1361
12		51	90	141	202	276	360	456	562	681	810	951	1103
14			62	98	141	191	250	316	391	473	562	660	766
16				72	103	141	184	232	287	347	413	485	562
18				55	79	108	141	178	220	266	316	371	431
20					62	85	111	141	174	210	250	293	340

K = 1 1/4"

B \ t	1/4	3/8	1/2	5/8	3/4	7/8	1	1 1/8	1 1/4	1 3/8	1 1/2	1 5/8	1 3/4
6	184	413	735	1148									
7	111	250	444	694	1000	1361							
8	74	167	298	465	669	911	1190						
9	53	120	213	333	479	652	852	1078	1331				
10		90	160	250	360	490	640	810	1000	1210	1440		
11		70	125	195	280	381	498	631	779	942	1121	1316	
12		56	100	156	224	305	399	505	623	754	898	1053	1222
14			68	106	153	208	272	345	425	515	612	719	834
16				77	111	151	198	250	309	373	444	522	605
18				58	84	115	150	190	234	283	337	396	459
20					66	90	118	149	184	222	264	310	360

K = 1 3/4

B \ t	1/4	3/8	1/2	5/8	3/4	7/8	1	1 1/8	1 1/4	1 3/8	1 1/2	1 5/8	1 3/4
6	360	810	1440										
7	184	413	735	1148									
8	111	250	444	694	1000	1361							
9	74	167	298	465	669	911	1190						
10	53	120	213	333	479	652	852	1078	1331				
11		90	160	250	360	490	640	810	1000	1210	1440		
12		70	125	195	280	381	498	631	779	942	1121	1316	
14			82	128	184	250	326	413	510	617	735	862	1000
16			58	90	130	176	230	292	360	436	518	608	706
18				67	96	131	171	217	268	324	385	452	524
20				52	74	101	132	167	207	250	298	349	405

K = 2"

B \ t	1/4	3/8	1/2	5/8	3/4	7/8	1	1 1/8	1 1/4	1 3/8	1 1/2	1 5/8	1 3/4
6	562	1265											
7	250	562	1000										
8	140	316	562	879	1266								
9	90	202	360	562	810	1102	1440						
10	62	141	250	391	562	766	1000	1266					
11		103	184	287	413	562	735	930	1148	1389			
12		79	141	220	316	431	562	712	879	1063	1266	1485	
14		51	90	141	202	276	360	456	562	681	810	951	1103
16			62	98	141	191	250	316	391	473	562	660	766
18				72	103	141	184	232	287	347	413	485	562
20				55	79	108	141	178	220	266	316	371	431

TABLE 5.1 (*Continued*)

$k = 7/8''$

B \ t	1/4	3/8	1/2	5/8	3/4	7/8	1	1 1/8	1 1/4	1 3/8	1 1/2	1 5/8	1 3/4
4	444	1000											
5	213	479	852	1331									
6	125	280	498	778	1121								
7	81	184	326	510	735	1000	1306						
8	58	130	230	360	518	706	922	1166	1440				
9		96	171	268	385	524	685	867	1070	1295			
10		74	132	207	298	405	529	669	826	1000	1190	1397	
11		59	105	164	237	322	421	532	657	795	947	1111	1289
12			86	134	193	262	343	434	535	648	771	905	1049
14			60	94	135	184	240	304	375	454	540	633	735
16				69	100	136	177	224	277	335	399	468	543
18				53	77	104	136	172	213	258	307	360	418
20					61	83	108	137	169	204	243	285	331

$k = 1\,1/2''$

B \ t	1/4	3/8	1/2	5/8	3/4	7/8	1	1 1/8	1 1/4	1 3/8	1 1/2	1 5/8	1 3/4
6	250	562	1000										
7	141	316	562	879	1266								
8	90	202	360	562	810	1102	1440						
9	62	141	250	391	562	766	1000	1266					
10		103	184	287	413	562	735	930	1148	1389			
11		79	141	220	316	431	562	712	879	1063	1266	1485	
12		62	111	174	250	340	444	562	694	840	1000	1174	1361
14			74	116	167	228	298	376	465	562	669	786	911
16			53	83	120	163	213	270	333	403	479	562	652
18				62	90	122	160	202	250	302	360	422	490
20					70	95	125	158	195	236	280	329	381

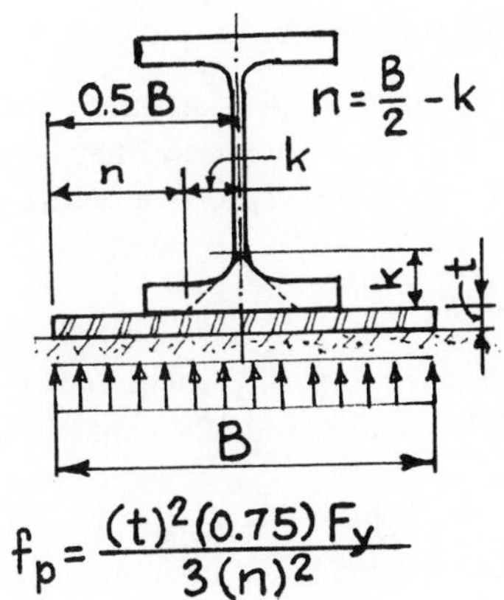

in which f_p is the bearing pressure, ksi; $F_b = 0.75\ F_y$; $n = (0.5\,B - k)$; and t = the required plate thickness, inches. See sketch, Table 5-1.

For Grade 36 steel plates (recommended by the authors to maintain minimum deflection and justify use of an average f_p), this formula reduces to:

$$f_p = \left(\frac{3t}{n}\right)^2 = \left(\frac{6t}{B - 2k}\right)^2$$

This equation has been solved for the usual design range of plate thicknesses, $\frac{1}{4}$ in. $\leqslant t \leqslant 1\frac{3}{4}$ in.; width, 6 in. $< B <$ 20 in.; and k-values, 0.625 in. $\leqslant k \leqslant$ 2.00 in. See Table 5-1 for a direct reading solution of actual pressure, f_p, computed as shown, for any values of k and t in these ranges. The solution to the selection of bearing plate size using Table 5-1 is a simple two-step operation:

Step 1: Determine the width of bearing required, $B = R/F_pN$, in which N = the length of bearing available, in., and F_p = the allowable bearing stress, ksi, prescribed by the local code. R = the total reaction, kips.

Step 2: Enter Table 5-1 with B and k (take k from the AISC Dimensions for Detailing). Select the thickness required to result in a calculated $f_p \leqslant F_p$.

It will be noted that the values of f_p are given for a range of 50 psi $< f_p <$ 1500 psi, which would apply under usual code provisions for laid-up hollow units to cast concrete with $f'_c \leqslant$ 4000 psi. It will frequently be unnecessary to use a separate bearing plate for beams bearing upon concrete. When required width of the bearing plate, B, is less than the flange width, b_f, and the required plate thickness, $t \leqslant t_f$, Table 5-1 can be utilized equally well to determine when a separate bearing plate is required. Again the solution is a quick two-step operation:

Step 1: Determine the required width, $B = R/F_pN$. When $B \leqslant b_f$, no separate bearing plate is required provided that the flange thickness is adequate in bending, $F_b \leqslant 0.75\,F_y$.

Step 2: Enter Table 5-1 with the flange thickness t_f for t; k; and flange width b_f for B. If $f_p \leqslant F_p$, no separate bearing plate is required.

Column Base Plates. The approved design formula for base plates under W-columns (Chapter 3, pp. 95-96, *AISC Handbook*) is identical with that suggested for the beam-bearing plates. The required base plate thickness, t, for concentric loads is given by the formula:

$$t = \sqrt{3f_p(m)^2/F_b} \quad \text{or} \quad t = \sqrt{3f_p(n)^2/F_b}$$

in which $m = 0.5(N - 0.95d)$ and $n = 0.5(B - 0.80b)$, where b and d are the width and depth respectively of the column section; and B and N are the width and depth respectively of the base plate; both are in inches.

Where economy is a consideration, the authors recommend use of the allowable bearing stresses on concrete as permitted in the 1971 ACI Building Code. The pertinent provisions of ACI 318-71 and the 1974 revisions thereto are:

(1) Bearing on footings–per Section 10.14 . . . . . . (15.6.2)*
(2) For bearing, $\phi = 0.70$ . . . . . . (9.2.1.4)*
(3) Allowable design** bearing stress, $= 0.85\phi f'_c$ . . . . . . (10.14.1)*

*"Building Code Requirements for Reinforced Concrete," (*ACI 318-71*) and 1975 revisions. Numbers in parentheses are Sections or Equations from "Building Code Requirements for Reinforced Concrete."
**"Design" means "factored loads" for $U = 1.4D + 1.7L$.

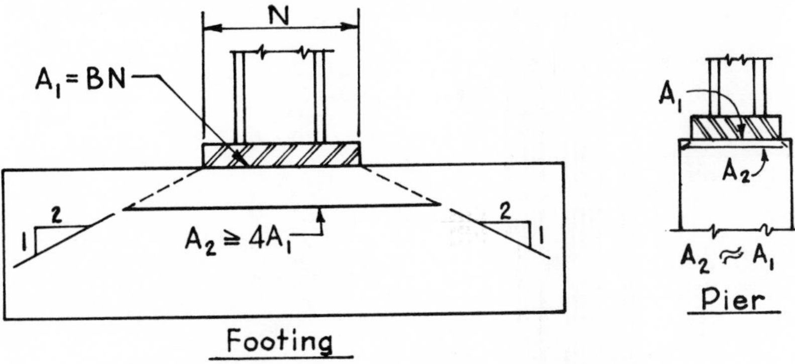

FIG. 5-6 Column Base Plates on Concrete.

(4) When the supporting surface is wider on all sides than the loaded area, the allowable stress may be multiplied by $\sqrt{A_2/A_1} \leqslant 2$ . . . . . . (10.14.2)*

(5) A_1 = the loaded area; A_2 = the area of the lower base of the largest frustrum of a right pyramid within the footing with A_1 as its top and with side slopes of 1 vertical to 2 horizontal . . . . . . (10.14.3)*

See Fig. 5-6 for the relationships between loaded area and unloaded areas. One further comparison of the codes for concrete and the AISC Specifications is needed to apply the bearing stresses of the concrete code for steel base plates. The allowable stresses of the concrete code are called the "design" stresses under "design" loads, U.

$$U = 1.4\,D + 1.7\,L \;\ldots\ldots\; \text{(Eq. 9-1)*}$$

It will be noted that the ACI allowable factored load stresses can conservatively be divided by 1.7 for the allowable stress under total dead plus live loads as used in the design of steel columns by the elastic method (Part 1); or that the factored load for design of steel columns by the plastic method can conservatively be used directly with the ACI design stresses. Applying the reduction suggested for elastic load, the allowable stresses become:

$$F_p = (0.85)\,(0.70) f_c'/1.7 = 0.35\, f_c' \;\ldots\ldots\; \text{Eq. 5-1}$$

$$F_p = (2)\,(0.35) f_c' = 0.70\, f_c' \;\ldots\ldots\; \text{Eq. 5-2}$$

The authors suggest the use of Eq. 5-1 for the design of base plates supported on reinforced concrete *piers*, and Eq. 5-2 for the design of steel base plates supported on concrete *footings* to avoid interpolation for variable ratios $\sqrt{A_2/A_1}$. It will be noted, however, that the ACI Code requirement (10.14.2)* does provide the basis for interpolation. (See Tables 5-2 for base plate designs prepared upon the basis of Eqs. 5-1 and 5-2 with $f_c' = 3000$ psi for columns of Grade 50 and base plates in Grade 36.

Anchor Bolts for Column Base Plates. The Specifications require that "Anchor bolts shall be designed to provide resistance to all conditions of tension and shear at the base of columns . . . . . ."(1.22). There are no explanations of this section in the Commen-

*"Building Code Requirements for Reinforced Concrete," (*ACI 318-71*) and 1975 revisions. Numbers in parentheses are Sections or Equations from "Building Code Requirements for Reinforced Concrete."

TABLE 5-2 Column Base Plate Design Series—(a) F_y = 36 ksi; f'_c = 3000 psi; $F_p = 0.35 f'_c$

Column Base Plate Minimum Size (Width "B" & Depth "N") and Thickness for Variable Tabulated Column Loads and Shapes

f'_c = 3000.0 F_p = 0.350 f'_c F_y (Base Plate) = 36 F_y (Column) = 50 $Kl/r \leq 150$

PLATE SIZE IN. "B" "N"	COL. LOAD KIPS	W14 730.	W14 665.	W14 605.	W14 550.	W14 500.	W14 455.	W14 426.	W14 398.	W14 370.	W14 342.	W14 320.	W14 314.	W14 287.	W14 264.	W14 246.	W14 237.	W14 228.
65 69	4709.	0.000	0.000	0.000	0.000	0.000	0.000	0.000	0.000	0.000	0.000	0.000	0.000	0.000	0.000	0.000	0.000	0.000
63 67	4432.	8.819	8.854	8.887	0.000	0.000	0.000	0.000	0.000	0.000	0.000	0.000	0.000	0.000	0.000	0.000	0.000	0.000
61 65	4163.	7.973	8.492	8.525	8.556	0.000	0.000	0.000	0.000	0.000	0.000	0.000	0.000	0.000	0.000	0.000	0.000	0.000
59 63	3902.	7.632	7.665	7.696	7.725	7.752	0.000	0.000	0.000	0.000	0.000	0.000	0.000	0.000	0.000	0.000	0.000	0.000
57 61	3650.	7.290	7.323	7.354	7.383	7.410	0.000	0.000	0.000	0.000	0.000	0.000	0.000	0.000	0.000	0.000	0.000	0.000
55 59	3407.	6.948	6.982	7.013	7.042	7.069	7.093	0.000	0.000	0.000	0.000	0.000	0.000	0.000	0.000	0.000	0.000	0.000
53 57	3172.	6.607	6.640	6.671	6.700	6.727	6.752	6.770	0.000	0.000	0.000	0.000	0.000	0.000	0.000	0.000	0.000	0.000
51 55	2945.	6.265	6.299	6.330	6.359	6.386	6.410	6.428	6.443	0.000	0.000	0.000	0.000	0.000	0.000	0.000	0.000	0.000
49 53	2726.	5.924	5.957	5.988	6.017	6.044	6.069	6.087	6.101	6.140	6.202	0.000	0.000	0.000	0.000	0.000	0.000	0.000
47 51	2516.	5.582	5.615	5.647	5.675	5.703	5.727	5.745	5.760	5.799	5.860	5.982	5.920	0.000	0.000	0.000	0.000	0.000
45 49	2315.	5.241	5.274	5.305	5.334	5.361	5.386	5.404	5.418	5.457	5.519	5.641	5.579	5.641	0.000	0.000	0.000	0.000
43 47	2122.	4.899	4.932	4.963	4.992	5.019	5.044	5.062	5.077	5.116	5.177	5.299	5.237	5.299	5.349	0.000	0.000	0.000
41 45	1937.	4.557	4.591	4.622	4.651	4.678	4.702	4.721	4.735	4.774	4.836	4.957	4.896	4.957	5.008	5.048	5.069	0.000
39 43	1760.	4.216	4.249	4.280	4.309	4.336	4.361	4.379	4.393	4.433	4.494	4.616	4.554	4.616	4.666	4.707	4.728	4.747
37 41	1592.	3.874	3.908	3.939	3.968	3.995	4.019	4.037	4.052	4.091	4.153	4.274	4.213	4.274	4.325	4.365	4.386	4.406
36 40	1511.	3.704	3.737	3.768	3.797	3.824	3.849	3.867	3.881	3.920	3.982	4.103	4.042	4.103	4.154	4.194	4.215	4.235
35 39	1433.	3.533	3.566	3.597	3.626	3.653	3.678	3.696	3.710	3.749	3.811	3.933	3.871	3.933	3.983	4.024	4.045	4.064
34 38	1356.	0.000	3.395	3.426	3.455	3.482	3.507	3.525	3.539	3.579	3.640	3.762	3.700	3.762	3.812	3.853	3.874	3.893
33 37	1282.	0.000	0.000	3.256	3.285	3.312	3.336	3.354	3.369	3.408	3.469	3.591	3.529	3.591	3.641	3.682	3.703	3.723
32 36	1209.	0.000	0.000	3.085	3.114	3.141	3.165	3.184	3.198	3.237	3.299	3.420	3.359	3.420	3.471	3.511	3.532	3.552
31 35	1139.	0.000	0.000	0.000	2.943	2.970	2.995	3.013	3.027	3.066	3.128	3.250	3.188	3.250	3.300	3.340	3.362	3.381
30 34	1070.	0.000	0.000	0.000	0.000	2.799	2.824	2.842	2.856	2.895	2.957	3.079	3.017	3.079	3.129	3.170	3.191	3.210
29 33	1004.	0.000	0.000	0.000	0.000	2.628	2.653	2.671	2.686	2.725	2.786	2.908	2.846	2.908	2.958	2.999	3.020	3.039
28 32	940.	0.000	0.000	0.000	0.000	0.000	2.482	2.500	2.515	2.554	2.616	2.737	2.676	2.737	2.788	2.828	2.849	2.869
27 31	878.	0.000	0.000	0.000	0.000	0.000	0.000	2.330	2.344	2.383	2.445	2.566	2.505	2.566	2.617	2.657	2.678	2.698
26 30	818.	0.000	0.000	0.000	0.000	0.000	0.000	0.000	2.173	2.212	2.274	2.396	2.334	2.396	2.446	2.487	2.508	2.527
25 29	761.	0.000	0.000	0.000	0.000	0.000	0.000	0.000	0.000	2.042	2.103	2.225	2.163	2.225	2.275	2.316	2.337	2.356
24 28	705.	0.000	0.000	0.000	0.000	0.000	0.000	0.000	0.000	0.000	1.932	2.054	1.992	2.054	2.104	2.145	2.166	2.186
23 27	652.	0.000	0.000	0.000	0.000	0.000	0.000	0.000	0.000	0.000	0.000	1.883	1.822	1.883	1.934	1.974	1.995	2.015
22 26	600.	0.000	0.000	0.000	0.000	0.000	0.000	0.000	0.000	0.000	0.000	0.000	0.000	1.713	1.763	1.803	1.824	1.844
21 25	551.	0.000	0.000	0.000	0.000	0.000	0.000	0.000	0.000	0.000	0.000	0.000	0.000	0.000	1.592	1.633	1.654	1.673
20 24	503.	0.000	0.000	0.000	0.000	0.000	0.000	0.000	0.000	0.000	0.000	0.000	0.000	0.000	0.000	1.462	1.483	1.502
19 23	458.	0.000	0.000	0.000	0.000	0.000	0.000	0.000	0.000	0.000	0.000	0.000	0.000	0.000	0.000	0.000	0.000	1.332

		W14 219.	W14 211.	W14 202.	W14 193.	W14 184.	W14 176.	W14 167.	W14 158.	W14 150.	W14 142.	W14 136.	W14 127.	W14 119.	W14 111.	W14 103.	W14 95.	W14 87.
40 44	1847.	0.000	0.000	0.000	0.000	0.000	0.000	0.000	0.000	0.000	0.000	0.000	0.000	0.000	0.000	0.000	0.000	0.000
38 42	1675.	4.598	4.617	4.637	0.000	0.000	0.000	0.000	0.000	0.000	0.000	0.000	0.000	0.000	0.000	0.000	0.000	0.000
36 40	1511.	4.256	4.275	4.295	4.316	4.335	0.000	0.000	0.000	0.000	0.000	0.000	0.000	0.000	0.000	0.000	0.000	0.000
35 39	1433.	4.085	4.105	4.124	4.145	4.165	4.186	0.000	0.000	0.000	0.000	0.000	0.000	0.000	0.000	0.000	0.000	0.000
34 38	1356.	3.914	3.934	3.953	3.974	3.994	4.015	4.036	0.000	0.000	0.000	0.000	0.000	0.000	0.000	0.000	0.000	0.000
33 37	1282.	3.744	3.763	3.783	3.804	3.823	3.844	3.865	3.895	0.000	0.000	0.000	0.000	0.000	0.000	0.000	0.000	0.000
32 36	1209.	3.573	3.592	3.612	3.633	3.652	3.673	3.695	3.714	3.733	0.000	0.000	0.000	0.000	0.000	0.000	0.000	0.000
31 35	1139.	3.402	3.422	3.441	3.462	3.482	3.503	3.524	3.543	3.563	3.584	0.000	0.000	0.000	0.000	0.000	0.000	0.000
30 34	1070.	3.231	3.251	3.270	3.291	3.311	3.332	3.353	3.372	3.392	3.413	3.413	0.000	0.000	0.000	0.000	0.000	0.000
29 33	1004.	3.061	3.080	3.099	3.121	3.140	3.161	3.182	3.202	3.221	3.242	3.242	3.263	0.000	0.000	0.000	0.000	0.000
28 32	940.	2.890	2.909	2.929	2.950	2.969	2.990	3.011	3.031	3.050	3.071	3.071	3.093	3.112	0.000	0.000	0.000	0.000
27 31	878.	2.719	2.738	2.758	2.779	2.798	2.820	2.841	2.860	2.880	2.901	2.901	2.922	2.941	2.962	0.000	0.000	0.000
26 30	818.	2.548	2.568	2.587	2.608	2.628	2.649	2.670	2.689	2.709	2.730	2.730	2.751	2.770	2.792	2.811	0.000	0.000
25 29	761.	2.377	2.397	2.416	2.437	2.457	2.478	2.499	2.519	2.539	2.559	2.559	2.580	2.600	2.621	2.640	2.661	0.000
24 28	705.	2.207	2.226	2.246	2.267	2.286	2.307	2.328	2.348	2.367	2.388	2.388	2.409	2.429	2.450	2.469	2.491	2.510
23 27	652.	2.036	2.055	2.075	2.096	2.115	2.136	2.158	2.177	2.196	2.218	2.218	2.239	2.258	2.279	2.299	2.320	2.339
22 26	600.	1.865	1.885	1.904	1.925	1.945	1.966	1.987	2.006	2.026	2.047	2.047	2.068	2.087	2.108	2.128	2.149	2.168
21 25	551.	1.694	1.714	1.733	1.754	1.774	1.795	1.816	1.835	1.855	1.876	1.876	1.897	1.917	1.938	1.957	1.979	1.998
20 24	503.	1.523	1.543	1.562	1.584	1.603	1.624	1.645	1.665	1.684	1.705	1.705	1.726	1.746	1.767	1.786	1.807	1.827
19 23	458.	1.353	1.372	1.392	1.413	1.432	1.453	1.474	1.494	1.513	1.534	1.534	1.555	1.575	1.596	1.616	1.637	1.656
18 22	415.	0.000	1.201	1.221	1.242	1.261	1.283	1.304	1.323	1.343	1.364	1.364	1.385	1.404	1.425	1.445	1.466	1.485
17 21	374.	0.000	0.000	0.000	0.000	1.091	1.112	1.133	1.152	1.172	1.193	1.193	1.214	1.233	1.254	1.274	1.295	1.315
16 20	335.	0.000	0.000	0.000	0.000	0.000	0.000	0.962	0.981	1.001	1.022	1.022	1.043	1.063	1.084	1.103	1.124	1.144
15 19	299.	0.000	0.000	0.000	0.000	0.000	0.000	0.000	0.000	0.000	0.000	0.851	0.872	0.892	0.913	0.932	0.953	0.973

TABLE 5.2 (*Continued*)

Column Base Plate Minimum Size (Width "B" & Depth "N") and Thickness for Variable Tabulated Column Loads and Shapes

f'_c = 3000.0 F_p = 0.350f'_c F_y (Base Plate) = 36 F_y (Column) = 50 $Kl/r \leq 150$

PLATE SIZE IN. "B" "N"	COL. LOAD KIPS	W14 84.	W14 78.	W14 74.	W14 68.	W14 61.	W14 53.	W14 48.	W14 43.
25 27	708.	0.000	0.000	0.000	0.000	0.000	0.000	0.000	0.000
24 26	655.	2.456	0.000	0.000	0.000	0.000	0.000	0.000	0.000
23 25	603.	2.285	2.288	0.000	0.000	0.000	0.000	0.000	0.000
22 24	554.	2.114	2.117	2.381	0.000	0.000	0.000	0.000	0.000
21 23	507.	1.943	1.946	2.210	2.214	0.000	0.000	0.000	0.000
20 22	461.	1.772	1.776	2.039	2.043	2.049	0.000	0.000	0.000
19 21	418.	1.602	1.605	1.868	1.873	1.878	0.000	0.000	0.000
18 20	377.	1.431	1.434	1.697	1.702	1.707	1.972	0.000	0.000
17 19	339.	1.260	1.263	1.527	1.531	1.537	1.801	1.806	0.000
16 18	302.	1.089	1.093	1.356	1.360	1.366	1.631	1.635	1.639
15 17	267.	0.919	0.922	1.185	1.190	1.195	1.460	1.464	1.468
14 16	235.	0.748	0.751	1.014	1.019	1.024	1.289	1.293	1.297
13 15	204.	0.577	0.580	0.844	0.848	0.853	1.118	1.122	1.127
12 14	176.	0.000	0.000	0.000	0.000	0.000	0.000	0.000	0.000

PLATE SIZE IN. "B" "N"	COL. LOAD KIPS	W12 190.	W12 161.	W12 133.	W12 120.	W12 106.	W12 99.	W12 92.	W12 85.	W12 79.	W12 72.	W12 65.	W12 58.	W12 53.	W12 50.	W12 45.	W12 40.
38 40	1595.	0.000	0.000	0.000	0.000	0.000	0.000	0.000	0.000	0.000	0.000	0.000	0.000	0.000	0.000	0.000	0.000
37 39	1515.	4.587	0.000	0.000	0.000	0.000	0.000	0.000	0.000	0.000	0.000	0.000	0.000	0.000	0.000	0.000	0.000
36 38	1436.	4.417	0.000	0.000	0.000	0.000	0.000	0.000	0.000	0.000	0.000	0.000	0.000	0.000	0.000	0.000	0.000
35 37	1359.	4.246	0.000	0.000	0.000	0.000	0.000	0.000	0.000	0.000	0.000	0.000	0.000	0.000	0.000	0.000	0.000
34 36	1285.	4.075	4.096	0.000	0.000	0.000	0.000	0.000	0.000	0.000	0.000	0.000	0.000	0.000	0.000	0.000	0.000
33 35	1212.	3.904	3.925	0.000	0.000	0.000	0.000	0.000	0.000	0.000	0.000	0.000	0.000	0.000	0.000	0.000	0.000
32 34	1142.	3.733	3.755	0.000	0.000	0.000	0.000	0.000	0.000	0.000	0.000	0.000	0.000	0.000	0.000	0.000	0.000
31 33	1074.	3.563	3.584	3.604	0.000	0.000	0.000	0.000	0.000	0.000	0.000	0.000	0.000	0.000	0.000	0.000	0.000
30 32	1007.	3.392	3.413	3.434	0.000	0.000	0.000	0.000	0.000	0.000	0.000	0.000	0.000	0.000	0.000	0.000	0.000
29 31	943.	3.221	3.242	3.263	3.269	0.000	0.000	0.000	0.000	0.000	0.000	0.000	0.000	0.000	0.000	0.000	0.000
28 30	881.	3.050	3.072	3.092	3.098	0.000	0.000	0.000	0.000	0.000	0.000	0.000	0.000	0.000	0.000	0.000	0.000
27 29	822.	2.880	2.901	2.921	2.927	2.940	0.000	0.000	0.000	0.000	0.000	0.000	0.000	0.000	0.000	0.000	0.000
26 28	764.	2.709	2.730	2.750	2.757	2.769	2.774	0.000	0.000	0.000	0.000	0.000	0.000	0.000	0.000	0.000	0.000
25 27	708.	2.538	2.559	2.580	2.586	2.598	2.604	2.608	0.000	0.000	0.000	0.000	0.000	0.000	0.000	0.000	0.000
24 26	655.	2.367	2.388	2.409	2.415	2.427	2.433	2.438	2.444	0.000	0.000	0.000	0.000	0.000	0.000	0.000	0.000
23 25	603.	2.196	2.218	2.238	2.244	2.257	2.262	2.267	2.274	2.277	0.000	0.000	0.000	0.000	0.000	0.000	0.000
22 24	554.	2.026	2.047	2.067	2.073	2.086	2.091	2.096	2.103	2.106	2.112	0.000	0.000	0.000	0.000	0.000	0.000
21 23	507.	1.855	1.876	1.897	1.903	1.915	1.920	1.925	1.932	1.935	1.941	1.961	0.000	0.000	0.000	0.000	0.000
20 22	461.	1.684	1.705	1.726	1.732	1.744	1.750	1.754	1.761	1.765	1.770	1.790	2.047	0.000	0.000	0.000	0.000
19 21	418.	1.513	1.534	1.555	1.561	1.573	1.579	1.584	1.591	1.594	1.599	1.620	1.876	1.878	0.000	0.000	0.000
18 20	377.	1.343	1.364	1.384	1.390	1.403	1.408	1.413	1.420	1.423	1.429	1.449	1.705	1.707	1.970	0.000	0.000
17 19	339.	0.000	1.193	1.213	1.220	1.232	1.237	1.242	1.249	1.252	1.258	1.278	1.535	1.537	1.799	1.804	0.000
16 18	302.	0.000	0.000	1.043	1.049	1.061	1.067	1.071	1.078	1.082	1.087	1.107	1.364	1.366	1.628	1.633	1.639
15 17	267.	0.000	0.000	0.872	0.878	0.890	0.896	0.901	0.907	0.911	0.916	0.936	1.193	1.195	1.458	1.462	1.468
14 16	235.	0.000	0.000	0.000	0.707	0.720	0.725	0.730	0.737	0.740	0.745	0.766	1.022	1.024	1.287	1.292	1.297
13 15	204.	0.000	0.000	0.000	0.000	0.000	0.554	0.559	0.566	0.569	0.575	0.595	0.851	0.853	1.116	1.121	1.127
12 14	176.	0.000	0.000	0.000	0.000	0.000	0.000	0.000	0.000	0.000	0.000	0.424	0.661	0.683	0.945	0.950	0.956
11 13	150.	0.000	0.000	0.000	0.000	0.000	0.000	0.000	0.000	0.000	0.000	0.000	0.510	0.512	0.775	0.779	0.785

PLATE SIZE IN. "B" "N"	COL. LOAD KIPS	W10 112.	W10 100.	W10 89.	W10 77.	W10 72.	W10 66.	W10 60.	W10 54.	W10 49.	W10 45.	W10 39.	W10 33.
29 31	943.	0.000	0.000	0.000	0.000	0.000	0.000	0.000	0.000	0.000	0.000	0.000	0.000
28 30	881.	3.358	0.000	0.000	0.000	0.000	0.000	0.000	0.000	0.000	0.000	0.000	0.000
27 29	822.	3.188	0.000	0.000	0.000	0.000	0.000	0.000	0.000	0.000	0.000	0.000	0.000
26 28	764.	3.017	3.026	0.000	0.000	0.000	0.000	0.000	0.000	0.000	0.000	0.000	0.000
25 27	708.	2.846	2.856	2.865	0.000	0.000	0.000	0.000	0.000	0.000	0.000	0.000	0.000
24 26	655.	2.675	2.685	2.694	0.000	0.000	0.000	0.000	0.000	0.000	0.000	0.000	0.000
23 25	603.	2.505	2.514	2.524	2.546	0.000	0.000	0.000	0.000	0.000	0.000	0.000	0.000
22 24	554.	2.334	2.343	2.353	2.375	2.395	0.000	0.000	0.000	0.000	0.000	0.000	0.000
21 23	507.	2.163	2.173	2.182	2.204	2.224	2.243	0.000	0.000	0.000	0.000	0.000	0.000
20 22	461.	1.992	2.002	2.011	2.034	2.053	2.073	2.094	0.000	0.000	0.000	0.000	0.000
19 21	418.	1.821	1.631	1.841	1.863	1.882	1.902	1.923	1.944	0.000	0.000	0.000	0.000
18 20	377.	1.651	1.660	1.670	1.692	1.712	1.731	1.752	1.773	1.793	0.000	0.000	0.000
17 19	339.	1.480	1.489	1.499	1.521	1.541	1.560	1.581	1.602	1.622	1.807	0.000	0.000
16 18	302.	1.309	1.319	1.328	1.351	1.370	1.389	1.411	1.432	1.451	1.636	0.000	0.000
15 17	267.	1.138	1.148	1.157	1.180	1.199	1.219	1.240	1.261	1.280	1.465	1.470	0.000
14 16	235.	0.967	0.977	0.987	1.009	1.028	1.048	1.069	1.090	1.110	1.294	1.299	1.302
13 15	204.	0.000	0.806	0.816	0.838	0.858	0.877	0.898	0.919	0.939	1.124	1.128	1.132
12 14	176.	0.000	0.000	0.645	0.667	0.687	0.706	0.727	0.749	0.768	0.953	0.957	0.961
11 13	150.	0.000	0.000	0.000	0.000	0.516	0.536	0.557	0.578	0.597	0.782	0.786	0.790
10 12	125.	0.000	0.000	0.000	0.000	0.000	0.000	0.000	0.000	0.426	0.611	0.616	0.619
9 11	103.	0.000	0.000	0.000	0.000	0.000	0.000	0.000	0.000	0.000	0.441	0.445	0.448
8 10	83.	0.000	0.000	0.000	0.000	0.000	0.000	0.000	0.000	0.000	0.000	0.000	0.000

TABLE 5.2 (*Continued*)

Column Base Plate Minimum Size (Width "B" & Depth "N") and Thickness for Variable Tabulated Column Loads and Shapes

f'_c = 3000.0 F_p = 0.350 f'_c F_y (Base Plate) = 36 F_y (Column) = 50 $Kl/r \leq 150$

PLATE SIZE IN. "B" "N"	COL. LOAD KIPS	W8 67.	W8 58.	W8 48.	W8 40.	W8 35.	W8 31.	W8 28.	W8 24.	W8 20.	W8 17.
22 24	554.	0.000	0.000	0.000	0.000	0.000	0.000	0.000	0.000	0.000	0.000
21 23	507.	2.467	0.000	0.000	0.000	0.000	0.000	0.000	0.000	0.000	0.000
20 22	461.	2.297	0.000	0.000	0.000	0.000	0.000	0.000	0.000	0.000	0.000
19 21	418.	2.126	2.166	0.000	0.000	0.000	0.000	0.000	0.000	0.000	0.000
18 20	377.	1.955	1.996	0.000	0.000	0.000	0.000	0.000	0.000	0.000	0.000
17 19	339.	1.784	1.825	1.865	0.000	0.000	0.000	0.000	0.000	0.000	0.000
16 18	302.	1.613	1.654	1.695	1.735	0.000	0.000	0.000	0.000	0.000	0.000
15 17	267.	1.443	1.483	1.524	1.564	1.585	0.000	0.000	0.000	0.000	0.000
14 16	235.	1.272	1.312	1.353	1.394	1.415	1.434	1.497	0.000	0.000	0.000
13 15	204.	1.101	1.142	1.182	1.223	1.244	1.263	1.326	0.000	0.000	0.000
12 14	176.	0.930	0.971	1.011	1.052	1.073	1.093	1.155	1.161	0.000	0.000
11 13	150.	0.759	0.800	0.841	0.881	0.902	0.922	0.985	0.990	1.158	0.000
10 12	125.	0.000	0.629	0.670	0.710	0.731	0.751	0.814	0.819	0.988	0.990
9 11	103.	0.000	0.000	0.499	0.540	0.561	0.580	0.643	0.648	0.817	0.819
8 10	83.	0.000	0.000	0.000	0.000	0.000	0.409	0.472	0.478	0.646	0.648
7 9	66.	0.000	0.000	0.000	0.000	0.000	0.000	0.301	0.307	0.475	0.478

$F_p = 0.700 f'_c$ (Other design values same as top of page)

		W14 730.	W14 665.	W14 605.	W14 550.	W14 500.	W14 455.	W14 426.	W14 398.	W14 370.	W14 342.	W14 320.	W14 314.	W14 287.	W14 264.	W14 246.	W14 237.	W14 2[illegible]
51 55	5890.	0.000	0.000	0.000	0.000	0.000	0.000	0.000	0.000	0.000	0.000	0.000	0.000	0.000	0.000	0.000	0.000	0.000
49 53	5453.	8.886	0.000	0.000	0.000	0.000	0.000	0.000	0.000	0.000	0.000	0.000	0.000	0.000	0.000	0.000	0.000	0.000
47 51	5033.	7.895	7.942	0.000	0.000	0.000	0.000	0.000	0.000	0.000	0.000	0.000	0.000	0.000	0.000	0.000	0.000	0.000
45 49	4630.	7.412	7.459	7.503	0.000	0.000	0.000	0.000	0.000	0.000	0.000	0.000	0.000	0.000	0.000	0.000	0.000	0.000
43 47	4244.	6.929	6.975	7.020	7.060	0.000	0.000	0.000	0.000	0.000	0.000	0.000	0.000	0.000	0.000	0.000	0.000	0.000
41 45	3874.	6.445	6.492	6.536	6.577	6.616	0.000	0.000	0.000	0.000	0.000	0.000	0.000	0.000	0.000	0.000	0.000	0.000
39 43	3521.	5.962	6.009	6.053	6.094	6.133	6.167	0.000	0.000	0.000	0.000	0.000	0.000	0.000	0.000	0.000	0.000	0.000
37 41	3185.	5.479	5.526	5.570	5.611	5.650	5.684	5.710	0.000	0.000	0.000	0.000	0.000	0.000	0.000	0.000	0.000	0.000
36 40	3023.	5.238	5.285	5.329	5.370	5.408	5.443	5.469	5.489	0.000	0.000	0.000	0.000	0.000	0.000	0.000	0.000	0.000
35 39	2866.	4.996	5.043	5.087	5.128	5.167	5.201	5.227	5.247	5.303	5.390	0.000	0.000	0.000	0.000	0.000	0.000	0.000
34 38	2713.	4.755	4.802	4.846	4.887	4.925	4.960	4.986	5.006	5.061	5.148	0.000	0.000	0.000	0.000	0.000	0.000	0.000
33 37	2564.	4.513	4.560	4.604	4.645	4.683	4.718	4.744	4.764	4.820	4.907	5.079	4.992	0.000	0.000	0.000	0.000	0.000
32 36	2419.	4.272	4.319	4.363	4.404	4.442	4.477	4.502	4.523	4.578	4.665	4.837	4.750	0.000	0.000	0.000	0.000	0.000
31 35	2278.	4.030	4.077	4.121	4.162	4.200	4.235	4.261	4.281	4.337	4.424	4.596	4.509	4.596	0.000	0.000	0.000	0.000
30 34	2141.	3.789	3.836	3.880	3.921	3.959	3.994	4.019	4.040	4.095	4.182	4.354	4.267	4.354	4.425	0.000	0.000	0.000
29 33	2009.	3.547	3.594	3.638	3.679	3.717	3.752	3.778	3.798	3.853	3.941	4.113	4.026	4.113	4.184	4.241	0.000	0.000
28 32	1881.	3.306	3.353	3.397	3.438	3.476	3.511	3.536	3.557	3.612	3.699	3.871	3.784	3.871	3.942	4.000	4.030	4.057
27 31	1757.	3.064	3.111	3.155	3.196	3.234	3.269	3.295	3.315	3.370	3.458	3.630	3.543	3.630	3.701	3.758	3.788	3.816
26 30	1637.	2.823	2.870	2.914	2.955	2.993	3.028	3.053	3.074	3.129	3.216	3.388	3.301	3.388	3.459	3.517	3.547	3.574
25 29	1522.	2.581	2.628	2.672	2.713	2.751	2.786	2.812	2.832	2.887	2.975	3.147	3.059	3.147	3.218	3.275	3.305	3.333
24 28	1411.	0.000	2.387	2.431	2.472	2.510	2.545	2.570	2.591	2.646	2.733	2.905	2.818	2.905	2.976	3.034	3.063	3.091
23 27	1304.	0.000	2.145	2.189	2.230	2.268	2.303	2.329	2.349	2.404	2.492	2.664	2.576	2.664	2.735	2.792	2.822	2.849
22 26	1201.	0.000	0.000	1.948	1.988	2.027	2.062	2.087	2.108	2.163	2.250	2.422	2.335	2.422	2.493	2.551	2.580	2.608
21 25	1102.	0.000	0.000	0.000	1.747	1.785	1.820	1.846	1.866	1.921	2.008	2.181	2.093	2.181	2.252	2.309	2.339	2.366
20 24	1007.	0.000	0.000	0.000	0.000	1.544	1.578	1.604	1.624	1.680	1.767	1.939	1.852	1.939	2.010	2.068	2.097	2.125
19 23	917.	0.000	0.000	0.000	0.000	0.000	1.337	1.363	1.383	1.438	1.525	1.698	1.610	1.698	1.769	1.826	1.856	1.883
18 22	831.	0.000	0.000	0.000	0.000	0.000	0.000	1.121	1.141	1.197	1.284	1.456	1.369	1.456	1.527	1.584	1.614	1.642
17 21	749.	0.000	0.000	0.000	0.000	0.000	0.000	0.000	0.000	0.955	1.042	1.214	1.127	1.214	1.286	1.343	1.373	1.400
16 20	671.	0.000	0.000	0.000	0.000	0.000	0.000	0.000	0.000	0.000	0.000	0.000	0.000	0.000	0.000	1.101	1.131	1.159
15 19	598.	0.000	0.000	0.000	0.000	0.000	0.000	0.000	0.000	0.000	0.000	0.000	0.000	0.000	0.000	0.000	0.000	0.000

		W14 219.	W14 211.	W14 202.	W14 193.	W14 184.	W14 176.	W14 167.	W14 158.	W14 150.	W14 142.	W14 13[illegible].	W14 [illegible]7.	W14 119.	W14 111.	W14 103.	W14 95.	W14 87.
29 31	1887.	0.000	0.000	0.000	0.000	0.000	0.000	0.000	0.000	0.000	0.000	0.000	0.000	0.000	0.000	0.000	0.000	0.000
28 30	1763.	3.704	3.709	0.000	0.000	0.000	0.000	0.000	0.000	0.000	0.000	0.000	0.000	0.000	0.000	0.000	0.000	0.000
27 29	1644.	3.463	3.468	3.477	0.000	0.000	0.000	0.000	0.000	0.000	0.000	0.000	0.000	0.000	0.000	0.000	0.000	0.000
26 28	1528.	3.221	3.226	3.236	3.244	3.253	0.000	0.000	0.000	0.000	0.000	0.000	0.000	0.000	0.000	0.000	0.000	0.000
25 27	1417.	2.980	2.985	2.994	3.002	3.012	3.022	0.000	0.000	0.000	0.000	0.000	0.000	0.000	0.000	0.000	0.000	0.000
24 26	1310.	2.738	2.743	2.753	2.761	2.770	2.780	2.810	2.837	0.000	0.000	0.000	0.000	0.000	0.000	0.000	0.000	0.000
23 25	1207.	2.497	2.502	2.511	2.519	2.529	2.539	2.568	2.596	2.623	0.000	0.000	0.000	0.000	0.000	0.000	0.000	0.000
22 24	1108.	2.255	2.260	2.270	2.278	2.287	2.297	2.327	2.354	2.382	2.412	2.465	0.000	0.000	0.000	0.000	0.000	0.000
21 23	1014.	2.014	2.019	2.028	2.036	2.046	2.055	2.085	2.113	2.140	2.170	2.223	2.233	0.000	0.000	0.000	0.000	0.000
20 22	923.	1.772	1.777	1.787	1.794	1.804	1.814	1.844	1.871	1.899	1.929	1.982	1.992	1.999	2.016	0.000	0.000	0.000
19 21	837.	1.531	1.536	1.545	1.553	1.563	1.572	1.602	1.630	1.657	1.687	1.740	1.750	1.758	1.774	1.802	0.000	0.000
18 20	755.	1.289	1.294	1.304	1.311	1.321	1.331	1.361	1.388	1.416	1.446	1.499	1.509	1.516	1.533	1.560	1.590	0.000
17 19	678.	1.048	1.053	1.062	1.070	1.080	1.089	1.119	1.147	1.174	1.204	1.257	1.267	1.275	1.291	1.319	1.349	1.376
16 18	604.	0.806	0.811	0.821	0.828	0.838	0.848	0.878	0.905	0.933	0.963	1.016	1.025	1.033	1.050	1.077	1.107	1.135
15 17	535.	0.000	0.000	0.000	0.000	0.000	0.000	0.000	0.000	0.000	0.000	0.774	0.784	0.792	0.808	0.836	0.866	0.893
14 16	470.	0.000	0.000	0.000	0.000	0.000	0.000	0.000	0.000	0.000	0.000	0.000	0.000	0.000	0.000	0.000	0.000	0.000

Column Base Plate Minimum Size (Width B & Depth N) and Thicknes for Variable Tabulated Column Loads and Shapes

f'_c = 3000.0 F_p = 0.700 f'_c F_y (Base Plate) = 36 F_y (Column) = 50 $Kl/r \leq 150$

PLATE SIZE IN. "B" "N"	COL. LOAD KIPS	W14 84.	W14 78.	W14 74.	W14 68.	W14 61.	W14 53.	W14 48.	W14 43.
18 20	755.	0.000	0.000	0.000	0.000	0.000	0.000	0.000	0.000
17 19	678.	1.762	0.000	0.000	0.000	0.000	0.000	0.000	0.000
16 18	604.	1.541	1.545	0.000	0.000	0.000	0.000	0.000	0.000
15 17	535.	1.299	1.304	1.676	1.682	0.000	0.000	0.000	0.000
14 16	470.	1.058	1.062	1.435	1.441	1.449	0.000	0.000	0.000
13 15	409.	0.816	0.821	1.193	1.199	1.207	0.000	0.000	0.000

PLATE SIZE IN. "B" "N"	COL. LOAD KIPS	W12 190.	W12 161.	W12 133.	W12 120.	W12 106.	W12 99.	W12 92.	W12 85.	W12 79.	W12 72.	W12 65.	W12 58.	W12 53.	W12 50.	W12 45.	W12 40.
27 29	1644.	0.000	0.000	0.000	0.000	0.000	0.000	0.000	0.000	0.000	0.000	0.000	0.000	0.000	0.000	0.000	0.000
26 28	1528.	3.831	0.000	0.000	0.000	0.000	0.000	0.000	0.000	0.000	0.000	0.000	0.000	0.000	0.000	0.000	0.000
25 27	1417.	3.589	0.000	0.000	0.000	0.000	0.000	0.000	0.000	0.000	0.000	0.000	0.000	0.000	0.000	0.000	0.000
24 26	1310.	3.348	3.378	0.000	0.000	0.000	0.000	0.000	0.000	0.000	0.000	0.000	0.000	0.000	0.000	0.000	0.000
23 25	1207.	3.106	3.136	0.000	0.000	0.000	0.000	0.000	0.000	0.000	0.000	0.000	0.000	0.000	0.000	0.000	0.000
22 24	1108.	2.865	2.895	0.000	0.000	0.000	0.000	0.000	0.000	0.000	0.000	0.000	0.000	0.000	0.000	0.000	0.000
21 23	1014.	2.623	2.653	2.682	0.000	0.000	0.000	0.000	0.000	0.000	0.000	0.000	0.000	0.000	0.000	0.000	0.000
20 22	923.	2.382	2.412	2.441	2.450	0.000	0.000	0.000	0.000	0.000	0.000	0.000	0.000	0.000	0.000	0.000	0.000
19 21	837.	2.140	2.170	2.199	2.208	2.225	0.000	0.000	0.000	0.000	0.000	0.000	0.000	0.000	0.000	0.000	0.000
18 20	755.	1.899	1.929	1.958	1.966	1.984	1.992	0.000	0.000	0.000	0.000	0.000	0.000	0.000	0.000	0.000	0.000
17 19	678.	1.657	1.687	1.716	1.725	1.742	1.750	1.757	1.766	0.000	0.000	0.000	0.000	0.000	0.000	0.000	0.000
16 18	604.	1.416	1.446	1.475	1.483	1.501	1.509	1.515	1.525	1.530	0.000	0.000	0.000	0.000	0.000	0.000	0.000
15 17	535.	1.174	1.204	1.233	1.242	1.259	1.267	1.274	1.283	1.288	1.296	0.000	0.000	0.000	0.000	0.000	0.000
14 16	470.	0.933	0.963	0.992	1.000	1.018	1.025	1.032	1.042	1.047	1.054	1.083	0.000	0.000	0.000	0.000	0.000
13 15	409.	0.691	0.721	0.750	0.759	0.776	0.784	0.791	0.800	0.805	0.813	0.841	1.204	1.207	0.000	0.000	0.000
12 14	352.	0.000	0.000	0.000	0.000	0.000	0.000	0.000	0.000	0.000	0.000	0.600	0.963	0.966	1.337	0.000	0.000
11 13	300.	0.000	0.000	0.000	0.000	0.000	0.000	0.000	0.000	0.000	0.000	0.000	0.721	0.724	1.096	1.102	1.111
10 12	251.	0.000	0.000	0.000	0.000	0.000	0.000	0.000	0.000	0.000	0.000	0.000	0.000	0.000	0.000	0.000	0.000

PLATE SIZE IN. "B" "N"	COL. LOAD KIPS	W10 112.	W10 100.	W10 89.	W10 77.	W10 72.	W10 66.	W10 60.	W10 54.	W10 49.	W10 45.	W10 39.	W10 33.
20 22	923.	0.000	0.000	0.000	0.000	0.000	0.000	0.000	0.000	0.000	0.000	0.000	0.000
19 21	837.	2.576	0.000	0.000	0.000	0.000	0.000	0.000	0.000	0.000	0.000	0.000	0.000
18 20	755.	2.335	2.348	0.000	0.000	0.000	0.000	0.000	0.000	0.000	0.000	0.000	0.000
17 19	678.	2.093	2.107	2.120	0.000	0.000	0.000	0.000	0.000	0.000	0.000	0.000	0.000
16 18	604.	1.851	1.865	1.879	1.910	0.000	0.000	0.000	0.000	0.000	0.000	0.000	0.000
15 17	535.	1.610	1.624	1.637	1.669	1.696	0.000	0.000	0.000	0.000	0.000	0.000	0.000
14 16	470.	1.368	1.382	1.396	1.427	1.455	1.482	1.512	0.000	0.000	0.000	0.000	0.000
13 15	409.	1.127	1.140	1.154	1.186	1.213	1.241	1.271	1.300	0.000	0.000	0.000	0.000
12 14	352.	0.885	0.899	0.912	0.944	0.972	0.999	1.029	1.059	1.086	0.000	0.000	0.000
11 13	300.	0.644	0.657	0.671	0.703	0.730	0.758	0.787	0.817	0.845	1.106	1.112	0.000
10 12	251.	0.000	0.000	0.000	0.000	0.000	0.000	0.000	0.000	0.603	0.865	0.871	0.876
9 11	207.	0.000	0.000	0.000	0.000	0.000	0.000	0.000	0.000	0.000	0.623	0.629	0.634

PLATE SIZE IN. "B" "N"	COL. LOAD KIPS	W8 67.	W8 58.	W8 48.	W8 40.	W8 35.	W8 31.	W8 28.	W8 24.	W8 20.	W8 17.
15 17	535.	0.000	0.000	0.000	0.000	0.000	0.000	0.000	0.000	0.000	0.000
14 16	470.	1.799	0.000	0.000	0.000	0.000	0.000	0.000	0.000	0.000	0.000
13 15	409.	1.557	1.615	0.000	0.000	0.000	0.000	0.000	0.000	0.000	0.000
12 14	352.	1.316	1.373	1.431	0.000	0.000	0.000	0.000	0.000	0.000	0.000
11 13	300.	1.074	1.132	1.189	1.246	0.000	0.000	0.000	0.000	0.000	0.000
10 12	251.	0.833	0.890	0.947	1.005	1.035	0.000	0.000	0.000	0.000	0.000
9 11	207.	0.591	0.649	0.706	0.763	0.793	0.821	0.910	0.000	0.000	0.000
8 10	167.	0.000	0.000	0.000	0.000	0.000	0.579	0.668	0.676	0.000	0.000
7 9	132.	0.000	0.000	0.000	0.000	0.000	0.000	0.427	0.434	0.672	0.676
6 8	100.	0.000	0.000	0.000	0.000	0.000	0.000	0.000	0.000	0.000	0.000

tary. The AISC Specifications provide no allowable concrete stresses for the development of such anchor bolts in tension or in bearing upon concrete. The 1971 ACI Building Code (318-71) and the 1975 revision also fail to provide allowable stresses for bolts (plain bars) in bearing or for their development in tension. Older ACI codes, however, traditionally allowed for plain bars 50 percent of the bond stress allowed for deformed bars, but not to exceed 160 psi in tension or compression for any strength of concrete (ACI 318-63, Section 1301 (c) (4)). The authors suggest that this value be used for all plain anchor bolts using the full length of the bolt whether hooked or straight. For important footings or where uplift tension may involve large forces, it will usually be preferable to use an embedded base plate at the lower end of the bolts as a positive anchorage. This positive anchorage should be provided where bolt diameter $\geqslant$ 1.5 in.

Since both steel and concrete code requirements for bolts in bearing are lacking, provision of anchor bolts for shear resistance becomes a matter of judgment. For very large shears as at the base of tall buildings, a positive provision for the transfer of shear to the foundations at the first floor level will avoid the uncertainty. The column continuing to the footings below the basement level can then be designed readily for concentric compression. Another device to provide positive shear transfer is the use of lugs welded to the base plate for embedment in the concrete. For lesser shears or where most of the shear can be considered to be resisted by friction and where no other means are available, the authors suggest using shear dowel forces upon the anchor bolts as assumed in the design of pavements for transfer of loads across a formed joint.

Dia. Bolt	Net Shear to Bolt (lbs.)	Total Shear; μ = 0.2 on Base Bolts Tightened, 50 ksi
$\frac{1}{2}''$	–	1,200 lbs.
$\frac{3}{4}''$	2,500	3,200 lbs.
1″	3,000	5,000 lbs.
$1\frac{1}{2}''$	4,000	12,000 lbs.
2″	4,500	20,000 lbs.
$2\frac{1}{2}''$	4,700	30,000 lbs.
3″	4,900	45,000 lbs.

"Load Carrying Capacity of Dowels at Transverse Pavement Joints," H. Marcus; ACI Proc. V. 48, No. 13. Disc. P. C. Disario; Closure, Marcus re "Anchor Bolts."

Specifications

General. Efficiency in design time, fabrication, and erection for routine conditions, suggests that the selection of connections be considered part of the detailer's function. The designer must, of course, provide the design requirements necessary for another to complete the details of the connections. For special conditions, any special design requirements or limitations on the types of connector materials, connection fasteners, etc. must also be provided either as specification requirements or details and general notes on the design drawings (1.1.1).

Flexible Connections (Type 2 Construction). A general note to indicate the type of construction on design drawings should be standard practice, although Type 2 construction is often taken for granted unless otherwise specified. To permit maximum economy in detailing connections, it is also preferable to show all beam reactions on the design drawings. Otherwise it is customary to detail these connections for half the uniform load capacity for the section, span, and grade of steel used. Showing all reactions not only permits some economy in those less than half the uniform load capacity, but also aids the designer's memory not to omit showing those which exceed this amount and therefore must be shown.

On his own initiative, the detailer should not be expected to select connections to create nil end moments or to provide the required rotation capacity. The designer should indicate the typical materials, type, and details of the connections desired as well as any special details such as coped bottom flanges for more rotation capacity where they are critical to the design. For eccentric connections such as brackets, the design drawings should either show the connection desired in detail and/or give the reaction and the moment for same.

Rigid Connections (Type 1 Construction). Again a general note that the design is based upon Type 1 construction is required. The portions of the structure which are to be of Type 1, such as "E-W beam-column frames;" joints, such as "E-W beam-column connections only;" fixed column bases, if any, and in which direction, if only one; etc. should be so identified. Since Type 1 construction requires virtually full moment development of members at each end connection, it is unnecessary to show design moments unless less than full development is required as when connected to a more flexible support. Shears should be shown.

Semi-Rigid Connections (Type 3 Construction). Since the use of this type of construction is less common, the general note becomes more important. For these connections, the design must indicate both moment and shear reaction since the moment assumed in design is known only to the designer.

Web Stiffeners

Design Requirements. Rigid and semi-rigid joints consisting of beam-column connections with the beams framed to the flanges of the columns must be investigated to determine whether stiffeners on the column web opposite the beam flanges are required (1.15.5). Four conditions for such analyses are prescribed:

(1) column web thickness, t_w, limited by the allowable shear stress within the joint (C 1.5.1.2);
(2) t_w limited by buckling opposite the compressive flange of the beam (1.15.5, Eq. 1.15-2);
(3) t_w limited by inelastic yielding opposite the compression flange of the beam (1.15.5, Eq. 1.15-1); and
(4) column flange thickness, t_f, limited by inelastic yielding opposite the tension flange of the beam (1.15.5, Eq. 1.15-3).

Finally, if stiffeners are required opposite either or both flanges of the beam, the minimum area of the stiffeners, A_{st}, must be calculated (1.15.5, Eq. 1.15-4). A_{st} is the total area of the pair of stiffeners, one on each side of the web.

For important connections, involving built-up columns and different grades of steel, a complete analysis beyond the scope of the AISC formulas is desirable. For important connections to rolled section columns where mixed grades other than $F_y = 36$ or 50 ksi are involved, the user is referred to Section 1.15.5 of the Specifications. For unimportant connections (not repeated sufficiently to justify design time) a conservative solution is to provide stiffeners equal in thickness to the beam flange delivering the concentrated loads, located opposite each beam flange, and extending to the column centerline for beams on one side only or full depth for beams on opposite sides.

For the common cases of rolled section columns and beams, involving Grades 36 and 50 ksi only, the complete five step design procedure can be simplified to three steps:

(1) Check column web thickness for buckling and allowable shear stress. See Tables 5-3 and 5-4 for solutions using Grades 36 and 50 ksi respectively.
(2) Check the required area of stiffeners, A_{st}. If the result is negative none are required. If positive, stiffeners are required opposite the compression flange of the beam.
(3) If stiffener area, $A_{st} > 0$, check whether stiffeners are required opposite the tension flange of the beam. If required, area $= A_{st}$. See Table 5-5 for solutions for columns of Grades 36 or 50 ksi.

TABLE 5-3 Minimum Thickness of Column Web in Connection—Grade 36—for Buckling and Allowable Shear

Col. "T" $= d_c$	Minimum Web Thickness, t_w, inches										
	Buckling (1.15-2) (in.)	Allowable shear = 0.4 F_y = 14.4 ksi (C1.5.1.2) For Total Flange Force, F, from Beam (kips)									
		25	50	75	100	125	150	175	200	225	250
4"	0.14	0.46	0.91								
5"	0.17	0.37	0.73	1.10							
6"	0.20	0.31	0.61	0.91							
7"	0.24	0.26	0.52	0.78	1.04						
8"	0.27	0.23	0.46	0.69	0.91	1.14					
9"	0.30	0.20	0.41	0.61	0.81	1.02	1.21				
10"	0.34	0.18	0.37	0.55	0.73	0.91	1.10				
11"	0.37	0.17	0.33	0.50	0.66	0.83	1.00	1.16			
12"	0.40	–	0.31	0.46	0.61	0.76	0.91	1.07			
13"	0.44	–	0.28	0.42	0.56	0.70	0.84	0.98	1.13		
14"	0.47	–	0.26	0.39	0.52	0.65	0.78	0.91	1.04	1.17	
15"	0.50	–	0.25	0.37	0.49	0.61	0.73	0.85	0.98	1.10	1.07
16"	0.54	–	0.23	0.34	0.46	0.57	0.69	0.80	0.92	1.03	1.00

TABLE 5-4 Minimum Thickness of Column Web in Connection—Grade 50—for Buckling and Allowable Shear

Col. "T" $= d_c$	Minimum Web Thickness, t_w, inches										
	Buckling (Equation 1.15-2) (in.)	Allowable shear = 0.4 F_y = 20.0 ksi (C1.5.1.2) For Total Flange Force, F, from Beam (kips)									
		25	50	75	100	125	150	175	200	225	250
4"	0.16	0.33	0.66								
5"	0.20	0.26	0.52	0.79							
6"	0.24	0.22	0.44	0.66	0.88	1.09					
7"	0.28	0.19	0.38	0.56	0.75	0.94	1.12				
8"	0.32	0.17	0.33	0.50	0.66	0.82	0.99	1.15			
9"	0.36	–	0.29	0.44	0.58	0.73	0.88	1.02	1.16		
10"	0.40	–	0.26	0.40	0.53	0.66	0.79	0.92	1.05	1.18	1.32
11"	0.44	–	0.24	0.36	0.48	0.60	0.72	0.84	0.96	1.08	1.20
12"	0.48	–	0.22	0.33	0.44	0.55	0.66	0.77	0.88	0.99	1.10
13"	0.52	–	0.20	0.30	0.40	0.51	0.61	0.71	0.81	0.91	1.01
14"	0.56	–	0.19	0.28	0.38	0.47	0.56	0.66	0.75	0.85	0.94
15"	0.59	–	0.18	0.26	0.35	0.44	0.53	0.61	0.70	0.79	0.88
16"	0.63	–	0.17	0.25	0.33	0.41	0.50	0.58	0.66	0.74	0.82

Derivation of Design Aids (Tables 5-3, 5-4, and 5-5). The three-step procedure is developed as follows:

Step 1.

Buckling. Column web, $t_w > d_c\sqrt{F_y}/180$ . . . . . (Eq. 1.15-2) For d_c, use detail dimension "T" for the column section. See Tables 5-3 and 5-4 for solutions for F_y = 36 and 50 ksi.

$$\text{Allowable shear} = 0.4\,F_y = \frac{12\,M}{0.95\,A_{bc}\,t_w} \;\ldots\ldots \text{(C1.5.1.2)}$$

Substitute $d_b F \approx 12\,M$, where d_b = beam depth, in.; and F = flange force, kips.

For

$$F_y = 36 \text{ ksi}, t_w \geqslant F/13.68\, d_c \ldots\ldots \text{(See Table 5-3.)}$$

For

$$F_y = 50 \text{ ksi}, t_w \geqslant F/19.00\, d_c \ldots\ldots \text{(See Table 5-4.)}$$

Step 2.

For stiffeners of the same grade as the columns, the required area, $A_{st} \geqslant C_1 A_f - t_w(t_b + 5k)$ . . . . . (Eq. 1.15-4). t_w = thickness of column web, in.; t_b = thickness of beam flange, in.; k is detail dimension for column fillet; C_1 = the ratio of beam-to-column yield stress; and A_f = area of beam flange.

Step 3.

Column flange thickness, t_f, opposite tension flange of the beam.

$$t_f \geqslant 0.4\sqrt{C_1 A_f} = 0.4 \sqrt{\frac{F_{yb} F}{F_{yc}(0.66)F_{yb}}} \ldots\ldots \text{(Eq. 1.15-3)}$$

in which F_{yc} and F_{yb} are yield points for the column and the beam respectively. See Table 5-5 for solutions of the minimum flange thickness for use without stiffeners for columns of Grades 36 and 50 ksi. $t_f = 0.49\sqrt{F/F_y}$.

TABLE 5-5 Minimum Thickness of Unstiffened Column Flange Opposite Tension Flange of Beam (Eq. 1.15-3)

Column F_y (ksi)	Minimum Unstiffened Col. Flange Thickness, t_f, (in.) For Total Flange Force from Beam, F, (kips)									
	25	50	75	100	125	150	175	200	225	250
36	0.41	0.58	0.71	0.82	0.91	1.00	1.08	1.15	1.23	1.29
50	0.35	0.49	0.60	0.69	0.78	0.85	0.92	0.98	1.04	1.10

JOINT DESIGN EXAMPLES

General

As previously noted, typical details as shown on the design drawings should show sufficient design data for the completion of the details by others. In addition, the type of detail desired must be indicated whenever there are design limitations involved that preclude the use of certain details. It is unnecessary, and even undesirable in terms of design time, for the designer to proceed further than these objectives. For most shear connections, the design shear only in terms of $(D + L)$ allowable design stress is sufficient; for some the designer should also indicate the acceptable types of standard shear connection as where use of bolts would reduce gross flange area unacceptably, a welded type should be shown. More detail must be shown for moment connections, of course, in addition to the design shear and moment, particularly where a minimum rotation capacity is required.

The following joint design examples are presented with these objectives. Calculations which must be considered by the designer for each are included. Note that design values

(M, V, etc.) shown for detailers' use on the typical details have been simplified and include in some cases allowances for partial loads and erection conditions not included in the analyses of Chapter 2. Also, all have been adjusted for use with allowable stresses for $(D + L)$ loads.

EXAMPLE 1 (See Fig. 5-7.)

Interior beam-column; one-story building–Type 1 Construction. (See Fig. 2-5(a) for the analysis.)

Design Shears. For this beam-over-column, no shear data needed.

Design Moments. Beam moments need not be shown. The column moment at the center of the joint is 92.5 k-in. for $(D + L)$ and 192 k-in. for $(D + L + W)$. The latter is critical since $192 > (\frac{4}{3})(92.5)$. Reducing moment to the face of the beam,

$$\frac{(192 + 150)}{180}\frac{(18)}{2} = 17 \text{ k-in.};\ 192 - 17 = 175 \text{ k-in.}$$

Moment for detailing at allowable $(D + L)$ stress,

$$M = (192 - 17)(0.75) = 131 \text{ k-in.}$$

$50.6/7.06 \pm 175/20.8 = 7.17 \pm 8.41$; tension = 1.24 ksi, compression = 15.58 ksi
Tensile force on welds to column flange = $(1.24)(6.5 \times 0.398) = 3.24$ ksi; Cap plate as a cantilever: $L = 1$ in.; $M = 3.24$ k-in.; $S = (8)(t)^2/6$. For Grade 36 (since the column is Grade 36), $M = 0.6 \times 36\,(S)$.

$$F_b = 0.75\,F_y = 27.0 \text{ ksi} \ldots\ldots (1.5.1.4.3).$$

Required thickness, $t = \sqrt{\frac{(6)}{1.33 \times 21.6}\frac{(16.4)}{(8)}} = 0.3$ in. Use $\frac{3}{8}''$ PL. Note that erection

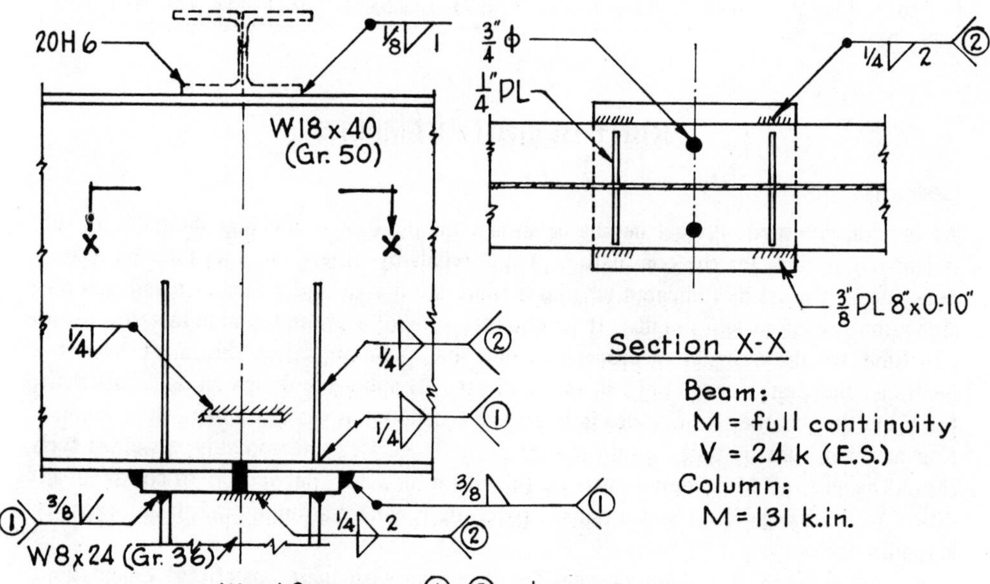

FIG. 5-7 Typical Interior Beam–Column.

bolts are located inside the column depth so as to leave full flanges gross section for complete capacity at the face of the supports.

Web Stiffeners. Web stiffeners, if required, can be half-depth since the column is not continued above (1.15.5). Enter Tables 5-4 and 5-5 with detail dimensions: $W18 \times 40$; $t_w = 0.316$ in.; $T = 15.75$ in.; $t_f = 0.524$ in.; and $W8 \times 24$; $t_f = 0.398$ in.; $k = 0.875$ in. Note that the terms "beam" and "column" in these tables are reversed for the beam-on-column connection.

Buckling. Enter Table 5-4 with $T = 15.75$ in. The minimum web thickness without stiffeners, $t_w = 0.62$ in. > 0.316 in. and so stiffeners are required.

Shear. Enter Table 5-4 with $T = 15.75$ in. and $F = (15.58)(6.5 \times 0.398) = 40$ k. Minimum web thickness for allowable shear, $t_w = 0.17$ in. < 0.316 in. OK.

Minimum flange thickness for beam flange opposite the tension flange of the column. Enter Table 5-5 with $F = 3.24$ for Grade 50, read $t_f = (3.24/25)\ (0.35) = 0.04$ in. $<$ 0.524 in. OK for tension; required only for compression.

Minimum area of A36 stiffeners, $A_{st} = C_1 A_f - t_w(t_b + 5k)$ . . . (1.15.5; Eq. 1.15-4)

$$A_{st} = \frac{(36)}{(50)}(6.5 \times 0.398) - 0.316\,(0.398 + 5 \times 0.875)$$

$$= 0.36 \text{ in.}^2 \text{ for the pair. Use } \tfrac{1}{4} \text{ PL.} \times 2.5 \text{ in.}$$

$$b/t = 2.5/0.25 = 10;\ 95\sqrt{36} = 15.83 > 10, \text{ OK}$$

The desired sequence of shop weld for the stiffeners is shown as (1) to flange and (2) to web to minimize lamellar tensions.

Joist connection. Field-welded connections to the beam are shown for both top and bottom chords (Specification J- and H- Series, 5.6.-b).

EXAMPLE 2 (See Fig. 5-8, *Exterior Column, One-Story Building—Type 1 Construction.* See Fig. 2-5(a), Chapter 2, for analysis.)

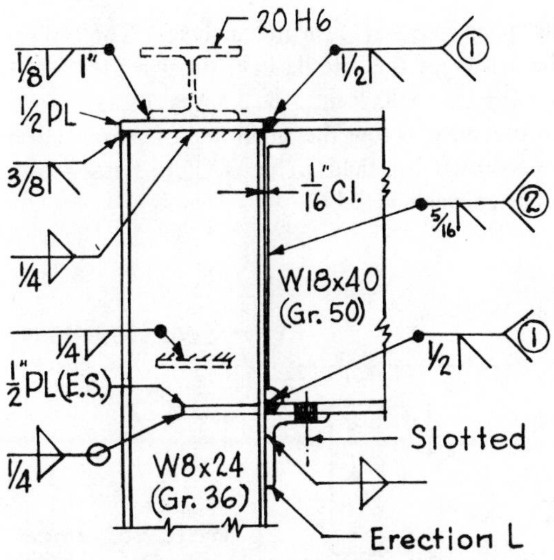

FIG. 5-8 Typical Exterior Beam-Column Connection—Example 2.

Design Shear = 15 k. Design $(D + L)$ moment = 336 kip-in. Flange forces = $\frac{336}{17}$ = 20 k.

Detail Dimensions: $W8 \times 24$, Grade 36, t_w = 0.245 in.; t_f = 0.398 in.; k = 0.875; T = 6.12 in.

$W18 \times 40$, Grade 50, t_w = 0.316 in.; t_f = 0.524 in.; k = 1.16; T = 15.75 in.

Fillet welds–use 21 ksi; other welds, 21 ksi . . . . . (1.5.3).

Cap plate. Use $t = \frac{1}{2}$ in. for entire flange weld. Use minimum fillet welds $\frac{1}{4} \times 6''$ E.S. to column web and far edge, 6.5 in.

Moment. Use single bevel full penetration welds for each flange.

Shear. Use fillet welds at web.

Web Stiffeners.

1. Buckling. Enter Table 5-3 (Grade 36) with T = 6.12 in. Read minimum t_w for buckling = 0.205 in. < 0.245 in. OK.
2. Shear. Enter Table 5-3 with $F = 20k$; T = 6.12. Interpolate minimum t_w = (20/25) (0.30) = 0.24 in. ≈ 0.245 in. OK.
3. Minimum column flange t_f opposite beam tension, F. Enter Table 5-5 with F = 20 for Grade 36, interpolate. Minimum t_f = (20/25) (0.41) = 0.328 < 0.398 in. OK.
4. Area of stiffener opposite compression flange

$$\text{Minimum } A_{st} = C_1 A_f - t_w(t_b + 5k)$$

$$= \frac{50}{36}(0.524 \times 6) - 0.245\ [0.524 + 5(1)]$$

$$= 3.01 \text{ in.}^2 \text{ Stiffeners are required.}$$

Use half-depth stiffener, PL $\frac{1}{2} \times 0'$-3″ both sides. Field weld sequence to minimize lamellar tension ① ② Flanges; ③ web (1.23.6).

EXAMPLE 3 (See Fig. 5-9, *Moment Splice in Beam*.)

One Story Building, Type 1 Construction. (See Fig. 2-5(a) for the analysis.) The typical detail will be designed for location at the center of the middle bay. Design shear $(D + L)$ = 12 k for erection. Design moment $(D + L)$ = 600 k-in. > 512. A web plate shear connection is provided for erection; since the shear is low the small loss of gross section for bolt holes is negligible. The sequence shown for field welds is ① to flanges and ② to the web to mimimize tension in the flange welds (1.23.6).

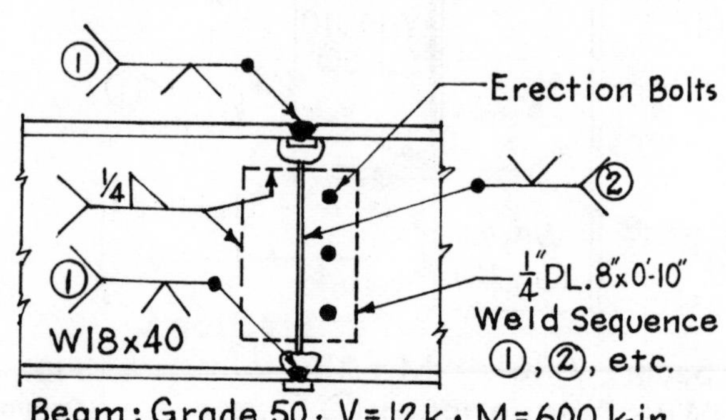

FIG 5-9 Typical Moment Splice in Beams–Example 3.

EXAMPLE 4 (See Fig. 5-10, *Column Base Plate for Fixed Base.*)

See Fig. 2-5(a) for the analysis.

	$(D \& L \& W)(0.75)$	$D \& L$
Exterior	$P = 20.3k(0.75) = 15.2k$ $M = 262k''(0.75) = 197k''$	$P = 20k$ $M = 167k''$
Interior	$P = :51(0.75) = 38k$ $M = 150(0.75) = 113k$	$P = 50k$ $M = 46.2k''$

Plate size to accommodate the anchor bolts required is 7 in. X 14 in. (1.16.5) for use of the same detail at both interior and exterior columns.

Note that ordinary bolts (A307) with positive bottom anchorage to the concrete are shown. In computing the bearing pressure upon the concrete to determine the required base plate thickness, initial tension is assumed zero. This assumption is conservative for design of the base plate, and realistic even if bolts are hand tightened since creep in the concrete and grout is likely to dissipate an initial tension. Anchorage into the concrete, however, should be adequate for full bolt tension up to yielding. Bearing pressure capacity of the concrete should be adequate for both loading conditions; (1) $(D + L)$ vertical load plus wind moment reduced to 0.75 for allowable stress design and (2) $(D + L)$ concentric vertical load plus full yield point tension in the bolts, to include the possibility of fully tightened bolts.

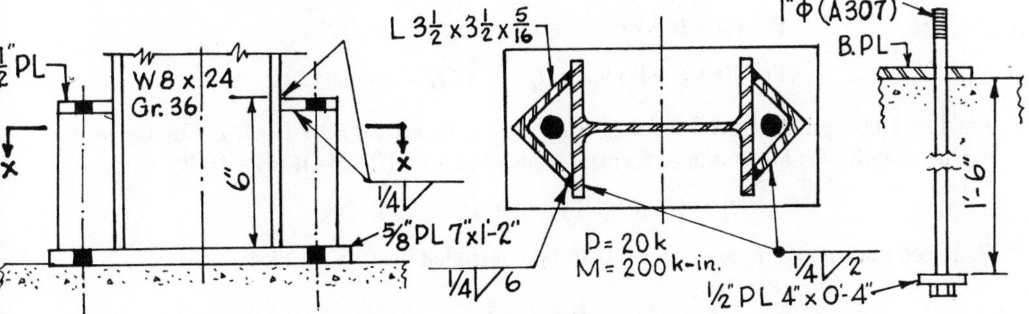

FIG. 5-10 Typical Column-Base Plate Connection—Example 4.

Interior Column: $e = \frac{112.5}{37.5} = 3.0$ in. $> \frac{14}{6}$ (kern). Compute elastic stress for zero initial bolt tension. Assume base plate is perfectly rigid and concrete is elastic.*

Bearing pressure, $f_p = 0.89$ ksi maximum. See sketch (a).

Exterior Column: $e = \frac{197}{15.2} = 13.0$ in. $> \frac{14}{6}$. Compute concrete bearing stress for uplift causing bolt tension (elastic analysis for cracked section). See sketch (b). $f_p = 1.12$ ksi. On the triangular corner projection of the base plate as a cantilever, 3 in. X 3.5 in. $F_b = 0.75\,F_y = 27.0$ ksi . . . . . . (1.5.1.4.3).

$$M = \frac{(3 \times 3.5)}{2}(1.12)\frac{(0.707 \times 3.5)}{3} = 4.9 \text{ k-in.}$$

Required thickness of the base plate, t

$$t = \sqrt{\frac{6M}{bf}} = \sqrt{\frac{(6)(4.9)}{(5)(27)}} = 0.46 \text{ in.}$$

**Design of Welded Steel Structures*, O. W. Blodgett, The James F. Lincoln Arc Welding Foundation, 1966. Pages 3.3-8, -10.

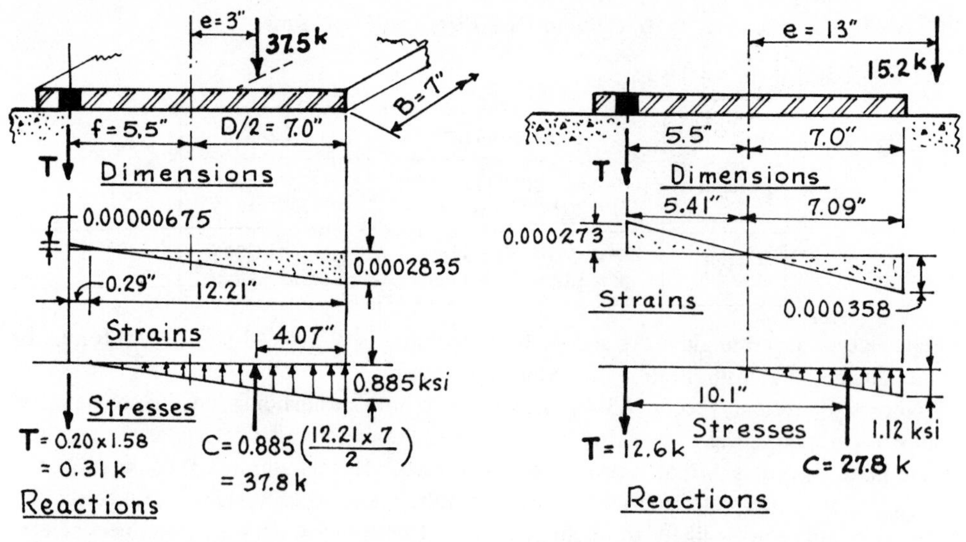

The ACI Building Code (ACI 318-71), Section 10.14, allows a design (factored load) bearing pressure = $(2)(0.85)(0.7) f_c' = 1.19 f_c'$ for the condition here, a footing area more than four times the area of the base plate.

$$\text{For } (D + L) \text{ loads}, f_p = 1.19 f_c'/1.7 = 0.700 f_c'$$

$$\text{For } (D + L + W) \text{ loads}, f_p = 1.19 f_c'/1.3 = 0.915 f_c'$$

For $f_c' = 3{,}000$ psi, $f_p = 2.1$ ksi for $(D + L)$; $f_p = 2.75$ ksi for $(D + L + W)$. OK for both loading conditions: Maximum concentric load = $50k + (2)(0.7854)(36) = 107k$.

$$f_p = 107/(7 \times 14) = 1.1 \text{ ksi} < 2.75. \text{ OK}$$

Required thickness of base plate, $t = 0.46$ in. controlled by $f_p = 1.12$ ksi.

Use $\frac{5}{8}$" PL

Anchor bolt tension $(D + L)$ total = 12.6 k. 2-1" ϕ ($T = 12.11k$ at 20 ksi) provides some allowance for initial tension within elastic limits. For the connection to the column, allow a height of 6 in. for strain in the bolts. The length available for fillet weld = 2 X 6 in. = 12 in. Use minimum size $\frac{1}{4}$" fillets. Field weld column to base plate merely as a precaution against displacement after remainder of structure is in place.

Tension embedment values for plain bolts are not provided under the 1971 Building Code. For large tensile forces the designer has four options: (1) use of standard (A307 or A325 Grade) plain bolts with positive anchorage as shown; (2) threaded deformed reinforcing bars used with straight embedment where footing depth is unlimited, or standard 90° or 180° end hooks in shallow footings; (3) proprietary anchor devices with guaranteed anchorage established by test; or (4) designing his own anchorage. Caution: threaded reinforcing bars require machining off the deformations and net area is based on actual size of *threaded* area. Calculate ultimate anchorage based conservatively upon a 60° angle frustrum of a cone about a single bolt or frustrum of a group of bolts using an ultimate unit shear value, $v_u = (4)(0.85)\sqrt{f_c'}$ (ACI 11.10.3 for two-way shear). For light columns as in this example, use of plain bolts is suggested to save extra machining and threading cost for reinforcing bars. Grout and leveling details are not shown.

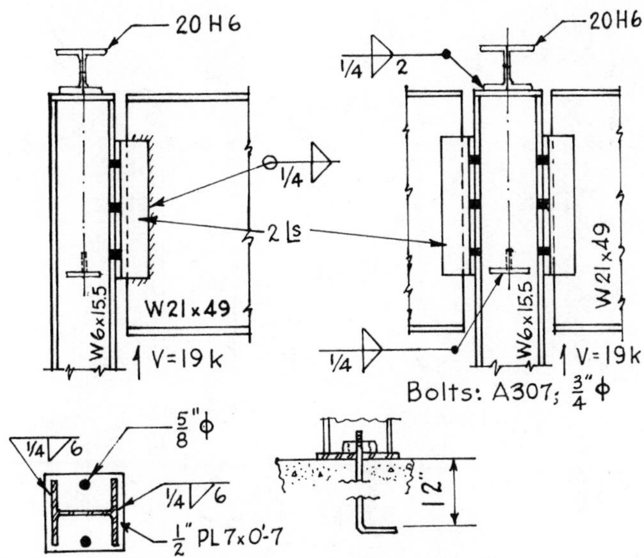

FIG 5-11 Typical Details—Type 2 Construction—Example 5.

EXAMPLE 5

(See Fig. 5-11, *Typical Details for One-Story Building*, Type 2 Braced Construction.) For analysis, see Fig. 2-5(b). The main decision required of the designer here is whether to show typical details. Member sizes and end reactions are shown on plans and elevations. Typical details are required only if a particular erection sequence or method of connection is required. If the designer needed or wanted shop-welded, field-bolted assembly, the typical details of Fig. 5-11 are more specific than a "General Note" to this effect, but the cost of design time (of showing these details) would be justified only if these particular arrangements were required.

EXAMPLE 6

(See Fig. 5-12. *Typical Details for One-Story Building, Type 2 with Wind Bracing at Top of Columns.*) For analysis, see Fig. 2-5(c). Columns, $W8 \times 24$; beams, $W21 \times 49$. Exterior column, $P = 25k$; $M = 185$ k-in.; interior column, $P = 48k$; $M = 200$ at center of joint. Beam shears, $V = 20k$. For use with allowable $(D + L)$ stress.

Detail dimensions, $W8 \times 24$; $t_f = \frac{3}{8}''$; $b = 6\frac{1}{2}''$;
$W21 \times 49$, $t_w = \frac{3}{8}''$; $t_f = \frac{9}{16}''$; $b = 6\frac{1}{2}''$.
Reduced moment, $M = 200 \times \frac{3}{4} = 150$ k-in.
Flange forces in beams $= T = \frac{150}{20} = 7.5k$

For this design, flexible connections are required to permit a rotation capacity for wind moment only. This rotation is best supplied by use of bolted angles which will deform easily to the necessary ϕ-rotation. The shear requirement could be satisfied by either web or seat connections. A seat connection will thus serve for shear, to transmit moment, and for easy erection while it can deform to the end rotation angle. Use A307 bolts.

Seat Angle. See AISC Table V-A. For beam web, $t_w = \frac{3}{8}$ in. The 4 in. OSL requires $t = \frac{5}{8}$ in. for $V = 24.3k$ with a 6 in. long seat. Select Type C connection, Table V-C for 6-$\frac{3}{4}''$ ϕ bolts; capacity $V = 26.5k$. From Table V-D, use $L8 \times 4 \times \frac{5}{8} \times 0'\text{-}6''$. Field-weld bottom flange to seat angle.

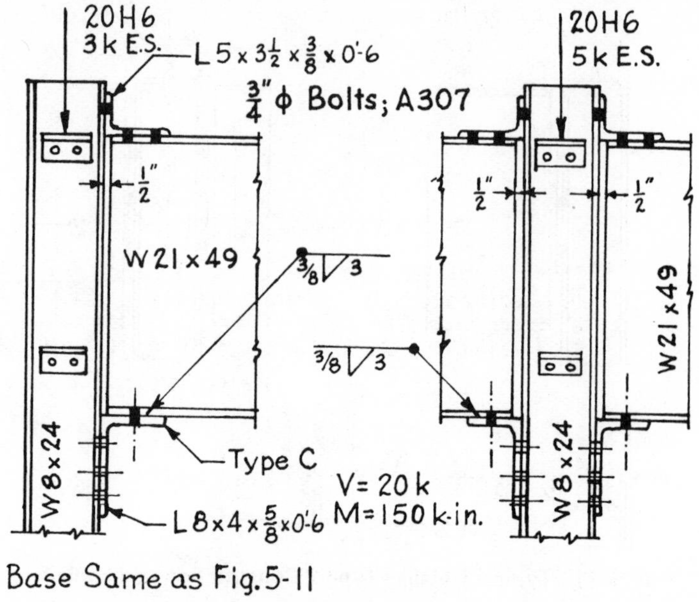

Base Same as Fig. 5-11

FIG. 5-12 Type 2 Construction--Wind Bracing at Top—Example 6.

Top Angle. Tension = 7.5*k*. $\frac{3}{4}''$ ϕ bolt tension capacity = 6.69*k*; single shear = 4.42*k*. For flexibility, check that angle will yield in bending at less than the flange tension. Try *L*5 × 3$\frac{1}{2}$ × $\frac{3}{8}$ × 0′-6. Assume reverse curvature for maximum stiffness.

$$\text{Yield: } M = F_y S = \frac{(36)(6)(0.375)^2}{6} = 5 \text{ k-in.}$$

Gauge 2$\frac{1}{2}''$; dia. bolt head = 1$\frac{1}{8}''$; *k* for $\frac{3}{8}$ *L* = 0.88 in. (1.16.4). $M \approx (7.5)(\frac{1}{2})(2.00 - 0.38 - 0.06) = 6$ k-in.

EXAMPLE 7. *Typical Details, One-Story Building, Type 2 Construction, Braced, Modified for Cantilever-Suspended Span System.*

(See Fig. 2-5(e) for the analysis. Columns are Grade 36, *W*6 × 15.5; beams, Grade 50, *W*14 × 34 in outer spans; and *W*14 × 26 in the center span.)

Beam splices designed for shear only are assumed to behave as hinged, and thus are required to have maximum flexibility. The end joints are designed as hinged, but unavoidable small negative moments are acceptable. The center joints are designed for full continuity of the beams. With all spans loaded the balancing cantilever and end moments leave zero moment to the interior columns. Under partial loadings, minimum column moments will develop, and are acceptable. Since this framing is for a roof system, it is feasible to use beam-on-column joints to keep unavoidable connection moments small and for simplicity in details and erection procedures. (See Fig. 5-13.)

Exterior Column Joint. $V = 19k$; (allow) $M = 0.5dV = 48$ k-in. for bearing on inner flange.

*W*6 × 15.5: $b_f = 6$ in.; $t_f = 0.269$ in.; $t_w = 0.235$ in.; $k = 0.75$.; $T = 4.5$ in.
*W*14 × 34: $b_f = 6.75$ in.; $t_f = 0.453$ in.; $t_w = 0.287$ in.; $k = 1.06$ in.; $T = 11.88$ in.

The transfer of shear from beams to columns is through direct bearing and no calculations are required. Connections are required only for transfer of horizontal shear and for erec-

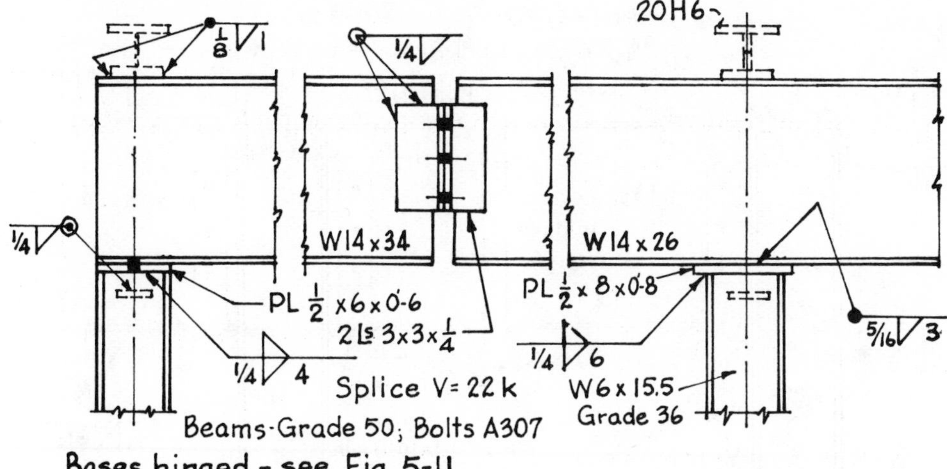

FIG. 5-13 Type 2 Construction—Cantilever-Suspended Span, Braced—Example 7.

tion. The designer must determine the need for web stiffeners. Note that this joint is unrestrained (1.15.4) and that the only need for web stiffeners is to prevent web crippling opposite a compressive flange reaction (1.10.10).

Applying the maximum interior column reaction on one flange

Interior Column:

$$\frac{R}{t + (N + 2k)} \leqslant 0.75\, F_y \;\ldots\ldots \text{(Eq. 1.10-8)}$$

$$\frac{49.6}{0.287\ [8 + 2\,(1.06)]} = 15.6 < 0.75\,(36) = 27.0$$

and so end bearing stiffeners are not required for either column. Use two $\frac{3}{4}$ in. ϕ bolts for erection and a minimum $\frac{1}{2}$ in. cap plate (1.16.4; 1.16.5).

Interior Column Joint. At this joint the design requires full continuity, and the use of erection bolts would reduce the net section more than 0.15 times the area of the compressive flange (1.10.1; 1.14.3; 1.14.4). For this reason, field welds will be utilized sufficient to develop the nominal moment allowed in-plane of the beam-column. These welds will be located opposite the flanges so as to provide also some lateral bracing during erection till the top is braced by the roof joists. Weld required: $F = \frac{48}{6} = 8k$. Use $\frac{1}{4}$ in. fillets 3 in. long at the ends of an 8 in. cap plate. Use a $\frac{1}{2}$ in. cap plate with $\frac{1}{4}$ in. welds top and bottom.

Splice. The most flexible shear connection is two pairs of angles as short as possible. From AISC Table 1-A3, read for 3-$\frac{3}{4}$ in. bolts, capacity = 26k in shear; $(\frac{109}{0.25}) = 27k$ in bearing on $\frac{1}{4}$ in. thick material.

Column Bases. Same as in Example 5.

EXAMPLE 8. *Typical Details, One Story Building, Type 3 Construction, Sidesway Uninhibited.*

(See Fig. 2-5(f) for the analysis. See Fig. 5-14 for the details.) The columns are Grade 36, $W8 \times 24$; beams, Grade 50, $W18 \times 40$.

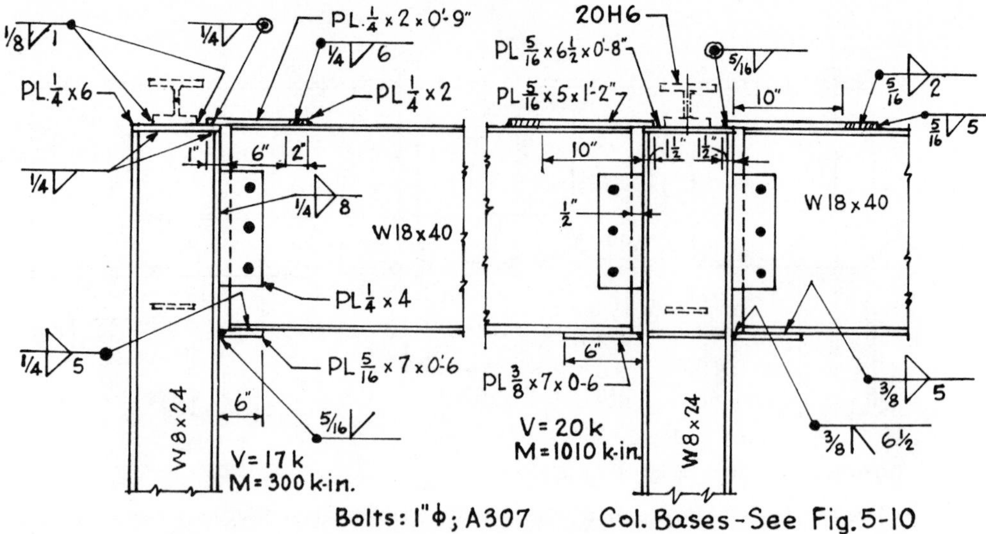

FIG. 5-14 Typical Details—Type 3 Construction—Example 8.

Column: $b_f = 6.5$ in.; $t_f = 0.398$ in.; $t_w = 0.245$ in.; $T = 6.12$ in.; $k = 0.88$ in.
Beam: $b_f = 6.00$ in.; $t_f = 0.524$ in.; $t_w = 0.316$ in.; $T = 15.75$ in.; $k = 1.06$ in.

Exterior Column-Beam Joint. $V = 17k; M = 300$ k-in. Flange forces, $F = \frac{300}{18} = 17k$. The flange plate connections are intended to yield at the design moment. Since welding to the columns is limited to Grade 36, connection materials will also be computed for the use of Grade 36. The required area for the flange plate at the top, $A = \frac{17}{36} = 0.47$ in.^{2}. Use $\frac{1}{4}$ *PL* 2 in. wide. A rule of thumb for free length is 1.5 times the width for tensile strain. The bottom flange plate should have more areas as the strain is more easily accommodated in tension. In this case (roof beam) the strain length available is ample since the top plate can extend over the column cap plate sufficiently for connection (1 in.). Use $\frac{1}{4}$ *PL* 2″ × 0′-10″, and provide ample welds so that the plate yields not the welds. To avoid yielding of the bottom plate in compression or buckling, $A = \frac{0.47}{0.6} = 0.78$ in.^{2} Use $\frac{5}{16}$ *PL* 3″ × 0′-6″. The design assumes all moment transferred through the flange plates, and all shear through a vertical side plate, field bolted. Thus, the bolts do not "share" stress with the welds (1.15.10), and maximum flexibility is provided with some inelastic deformation (1.2). Three $\frac{3}{4}$″ A307 bolts in $\frac{1}{4}$ in. thick material are adequate; capacity = $26.5k$.

Interior Column-Beam Joint. $V = 20k$; $M = 1010$ k-in. The same sizes of members are used as at the exterior joint. Flange forces, $F = \frac{1010}{18} = 56k$. Plate area for the top flange plate, $A = \frac{56}{36} = 1.6$ in.^{2}. Use $\frac{5}{16}$ *PL* 5 in. wide extending over the cap plate 1 in. for ample end welds. Allow a free length (free of welds) 10 in. long. For the lower flange plate, $A = \frac{1.6}{0.6} = 2.7$ in.^{2}. Use $\frac{3}{8}$ *PL* 7 in. wide. Use same web plates for shear as at the end columns.

Web Stiffeners.

Buckling. Enter Table 5-3 with $T = 6.12$ in. Read minimum $t_w = 0.20$ in. < 0.245 in. OK.

Shear. Enter Table 5-3 with $F = 17k$ for the exterior column. Read minimum $t_w = 0.31 \times \frac{17}{25} = 0.21 < 0.245$ in. OK. For the interior column, estimate unbalanced load

$F = 0.6 \times 56 = 33.6k$. Read minimum $t_w = 0.31 \times \frac{33.6}{25} = 0.42 > 0.245$. Stiffeners are required on the interior columns only. The area of stiffeners,

$$A_{st} \geqslant \tfrac{36}{36}(2.7) - 0.245(0.375 + 5 \times 0.88) = 1.5 \text{ in.}^2$$

Two stiffeners, one on each side of the column web, 3 in. in depth, will require minimum thickness, $t = \frac{1.5}{6} = 0.25$ in. Use $\frac{1}{4}$ in. $\times$ 3 in. stiffeners.

EXAMPLE 9. *2nd Floor Exterior Column-Beam Connection, 25 Story Building, Type I Construction.*

(See Fig. 5-15. For the analysis see Fig. 2-11(b) Wind and Fig. 2A-43 for $(D + L)$.) The critical loading here is $0.75(D + L + W)$. Beam shear, $V = 42k$; $M = 3922$ k-in. Beams are Grade 36; columns are Grade 50. Detail dimensions of the sections used are:

Column: $W14 \times 202$; $b_f = 15.75$ in.; $t_f = 1.5$ in.; $t_w = 0.93$ in.; $k = 2.19$ in.; $T = 11.25$ in.
Beam: $W24 \times 84$; $b_f = 9.00$ in.; $t_f = 0.772$ in.; $t_w = 0.47$ in.; $T = 21$ in.; $k = 1.56$ in.

For full development of the beam flanges, use full depth, single-bevel welds ($\frac{3}{4}$ in.) direct to the column flanges. For maximum web development in moment (shear is minor) use a single-web plate, $t = t_w$ of beam $= \frac{1}{2}$ in. Shop-weld the web plate to the column; use slotted holes for two $\frac{3}{4}''$ ϕ erection bolts; complete the web connection by field-welded fillet full depth ($T_e = \frac{7}{16}$ in.).

Web Stiffeners.

Shear. Enter Table 5-4 with $T = 11.25$ in.; $F = \frac{3922}{24} = 163\,k$. Read minimum shear thickness = 0.72 in. < 0.93 in. OK.

Buckling. Read minimum thickness = 0.45 in. < 0.93 in. OK. Opposite the Tension Flange. Enter Table 5-5 with $F = 163k$; for Grade 50, read minimum thickness = 0.87

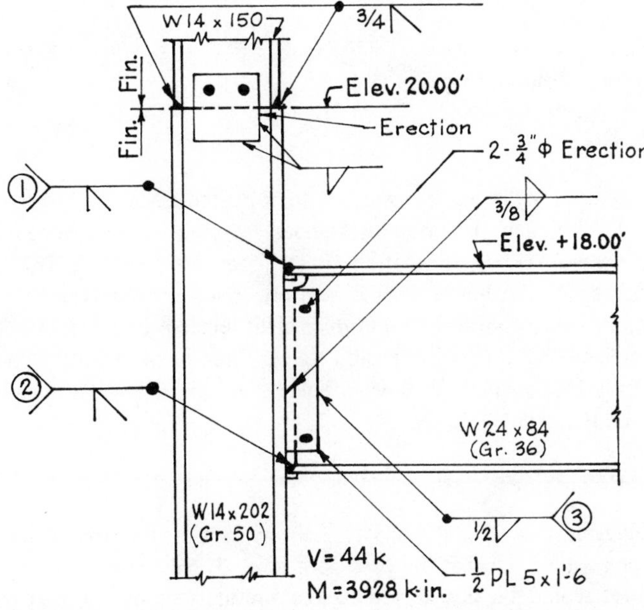

FIG. 5-15 Exterior Column-Beam Connection at Second Floor—Example 9.

in. < 0.93 OK. Opposite the Compression Flange. Min. $A_{st} = C_1 A_f - t_w(t_b + 5k)$

$$A_{st} = \frac{36}{50}(9 \times 0.772) - 0.93(0.772 + 5 \times 2.19) < 0 \text{ OK.}$$

No web stiffeners are required.

Column Splices. Location-Elevation 20.00 (2′-0″ above the top of the second floor beam). Upper section $W14 \times 150$; depth = 14.87 in.; $t_f = 1.12$ in.; lower section, $W14 \times 202$; depth = 15.62 in.; $t_f = 1.5$ in. Finish both bearing surfaces. Use single bevel $\frac{3}{4}$ in. weld for each flange. Welded area is greater than 0.50 X gross area (1.15.8).

EXAMPLE 10. *2nd Floor, Interior Column-Beam Connection, 25 Story Building, Type 1 Construction.*

(See Fig. 5-16. For the analysis see Fig. 2-11(b) Wind, and Fig. 2A-43 for $(D + L)$.) Detail dimensions of the sections used are:

Column: $W14 \times 264$; Grade 50; $t_f = 1.938$ in.; $t_w = 1.205$ in.;
Beam: $W24 \times 84$, Grade 36; same as Ex. 9.

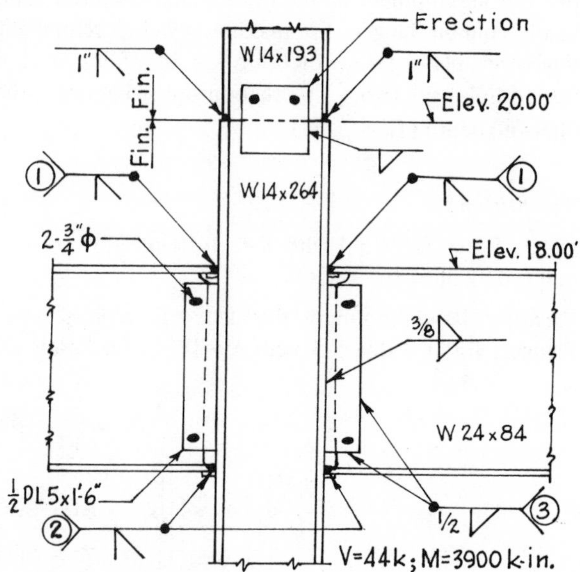

FIG. 5-16 Interior Column-to-Beam Connection at Second Floor—Example 10.

Note that the calculations of Ex. 9 need not be repeated since the interior column is a heavier section; both flange and web thicknesses are greater; unbalanced flange forces from the beams are smaller. By inspection, no stiffeners are required. The same size web plates will suffice here. With the thicker sections involved, observance of the welding sequence becomes more important to minimize weld tensions and potential lamellar tearing. The general sequence is to weld tension flanges first, bottom flanges second, and the web last. The fully restrained weld is on the web, and any tension there is balanced by compression in the flanges (1.23.6).

EXAMPLE 11. *Column Base Details.*

The 25 story building example of Chapter 2 showed an assumption of fixed bases (for purposes of frame analysis only) at the first floor level. It was assumed that all wind shear forces at this level would be transferred to the foundations by the first floor framing.

The same column section shown for first floor to second floor in practice would be carried in one piece through to base plates on footings just below the basement floor level. These base plates are to be designed for concentric load, compression only. Assume concrete strength of $f'_c = 3000$ psi for the footing; also assume that the area of the footing, pile cap, or caisson supporting the base plate will be at least four times that of the base plate itself.

The added column loads from the first floor $(D + L)$ are $D = 115$ psf and $L = 100$ psf. (See Appendix 2A for analysis with $(D + L) = 215$ psf. See Table 5-2(b) for base plates.)*

Interior Columns	*Exterior Columns*
$P = 1{,}651k$	$P = 962k$
$W14 \times 264$	$W14 \times 202$
Read: Min. $t = 3.7$ in.	Read: Min. $t = 2.0$ in.
size $27'' \times 31''$	size $21'' \times 23''$
for $P = 1757k$	for $P = 1014k$

Computed stress = 0 for both shear and tension. (Section 1.15.8 is inoperative for this application.) The requirements for column bases and anchor bolts are likewise inoperative for this condition except as they are satisfied by bearing of the finished base of the column section upon the base plate also finished as specified (1.21; 1.22).

In order to provide a minimum of connection capability for practicable construction and a rational approach to the design thereof, consider that the sections referred to are inoperative for the finished structure only. Consider that it is quite conceivable in construction that it will not be convenient nor economical to complete the first floor framing to carry all shears to completed foundations until the steel framing reaches the second column splice (four floors). It would not only be prudent to design the base plate connections for this condition but would be required by the specifications (1.21; 1.22; and 1.15.8). If it is merely considered reasonable that the construction loads plus wind on the first four incomplete floors will not exceed those in the actual building, no new analysis need be made. See Appendix 2A, analyses for $(W + D + L)$ and $(D + L)$ loadings; use the moments and loads for the 21st story level.

Exterior Columns: $(D + L)$: $P = 158k; M = 836$ k-in.
$(W + D + L)$: $P = 166k; M = 1056$ k-in.
Interior Columns: $(D + L)$: $P = 278k; M = 202$ k-in.
$(W + D + L)$: $P = 285k; M = 746$ k-in.

Exterior Columns	*Interior Columns*
$P = 158k; M = 836$ k-in.	$P = 278k$;
	$M = 0.75 \times 746 = 560$ k-in.
Tension in anchor bolts	Tension in anchor bolts
$T \approx \frac{836}{19} = 44k$	$T \approx \frac{560}{27} = 21k$

Anchor Bolts A307
Exterior Columns. Use $2\text{—}1\frac{3}{8}''$ ϕ (capacity $= 2 \times 23 = 46k$)
Interior Columns. Use $2\text{—}1''$ ϕ (capacity $= 2 \times 12 = 24k$)

---

*For bearing pressure $f_p = 0.700 f'_c$; applicable where the column is concentric over an area larger than four times the base plate.

Check Plate Thickness for Bending on Tensile Side
Exterior, $M = 44 \times 3.5 = 154$ k-in. $S = bt^2/6 = (21/6)\,t^2$

$$\text{Required, } t = \sqrt{\frac{6M}{(0.75F_y)(21)}} = \sqrt{\frac{(6)(154)}{(27)(21)}} = 1.27 < 2.0 \text{ in. OK.}$$

Interior, $M = (21)(5.25) = 110$ k-in. $S = (27)\,t^2/6$

$$\text{Required } t = \sqrt{\frac{(6)(110)}{(27)(27)}} = 1'' < 4.0'' \text{ OK.}$$

Welds (to flange)
Exterior. Minimum size fillet is $\frac{3}{8}$ in. to 2 in. plate (1.17.5)

$$\frac{M}{d} = \frac{836}{15.6} = 54k.$$

$$\text{Required length of weld} = \frac{54}{0.375 \times 18} = 8''$$

Interior. Minimum size fillet is $\frac{1}{2}$ in. to 4 in. plate (1.17.5)

$$\frac{M}{d} = \frac{560}{16.5} = 34k$$

$$\text{Required length of weld} = \frac{34}{0.5 \times 18} = 4''$$

(See Fig. 5-17 for detail and dimensions.)

The requirements for column bases and anchor bolts are satisfied (1.21; 1.22).

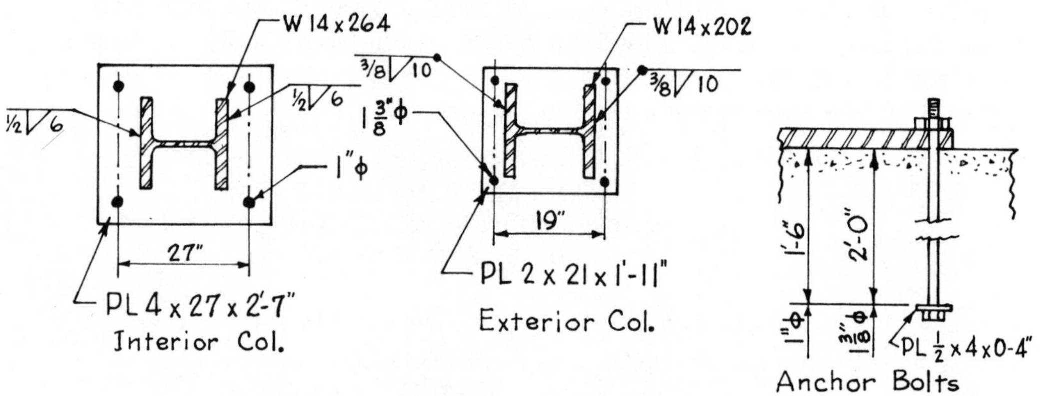

FIG. 5-17 Typical Column Base Details—Example 11.

SELECTED REFERENCES—CONNECTIONS

1. "Wind Connections with Simple Framing," Disque; July, 1964; **1**, No. 3; *Engineering Journal*, AISC.
2. "Plastic Design of Eccentrically Loaded Fasteners," Abolitz; July, 1966; **3**, No. 3; *Engineering Journal*, AISC.
3. "High Strength Bolting," Munse; January, 1967; **4**, No. 1; *Engineering Journal*, AISC.

4. "Moment-Rotation Characteristics of Shear Connections," Kennedy; October, 1969, **6**, No. 4; *Engineering Journal*, AISC.
5. "Ultimate Capacity of Single Bolted Angle Connections," J. B. Kennedy & G. R. Sinclair, **95**, No. ST8, Aug. 1969, *ASCE*.
6. "Design of Steel Bearing Plates," Fling; April, 1970; *Engineering Journal*, AISC.
7. "Maximum A325 and A490 Bolt Loads at a Glance," Stetina; July, 1970; **7**, No. 3; *Engineering Journal*, AISC.
8. "Analysis of Frames with Partial Connection Rigidity," Karl M. Romstad and Chittoor V. Subramanian, **96**, No. ST11, Nov. 1970, *ASCE*.
9. "Further Studies of Inelastic Beam-Column Problem," W. F. Chen, **97**, No. ST2, Feb. 1971, *ASCE*.
10. "Plastic Behavior of Eccentrically Loaded Connections," Shermer; April, 1971, **8**, No. 2; *Engineering Journal*, AISC.
11. "Treatment of Eccentrically Loaded Connections in the AISC Manual," Higgins; April, 1971, **8**, No. 2, *Engineering Journal*, AISC.
12. "Cyclic Yield Reversal in Steel Building Connections," Popov and Pinkney: July, 1971; **8**, No. 3; *Engineering Journal*, AISC.
13. "Eccentrically Loaded Welded Connections," Lorne J. Butler, Shubendu Pal, and Geoffrey L. Kulak, **98**, No. ST5, May, 1972, *ASCE*.
14. "Moment-Rotation Curves for Locally Buckling Beams," J. Jay Climenhaga and R. Paul Johnson, **99**, No. ST6, June, 1972, *ASCE*.
15. "Analysis of T-Stub Flange-to-Column Connections," Ben Kato and William McGuire, **99**, No. ST5, May, 1973, *ASCE*.
16. "Cyclic Loading of Steel Beams and Connections," E. P. Popov and V. V. Bertero, **99**, No. ST6, June, 1973, *ASCE*.
17. "Ultimate Strength Design of Tubular Joints," J. Blair Reber, Jr., **99**, No. ST6, June, 1973, *ASCE*.
18. "Sustained and Fluctuating Loading on Bolted Joints," G. D. Base, L. C. Schmidt, D. S. Mansell, and L. K. Stevens; **99**, No. ST10, Oct. 1973, *ASCE*.
19. "Building Code Requirements for Reinforced Concrete," (*ACI 318-71*) as revised Jan. 1973.
20. "Inclusions and Susceptibility to Lamellar Tearing of Welded Structural Steels," J. C. M. Farrar, AWS, *Welding Journal*, August, 1974. (Note: Metallurgical Studies.)

6

COLUMNS

Definitions

A number of common terms used loosely among the AISC Specifications, Commentary, and technical literature on structural steel design and research add to the practitioners' confusion. The authors have no pretensions toward, nor intent to attempt, standardization of definitions. Merely for the purposes of this chapter and intelligible communication of the concepts herein, the following definitions were adopted and followed:

Beams. Principal solid-web flexural members without axial compressive load.

Beam-Columns. Principal solid-web members which are designed to resist combinations of axial compression and bending in which the axial compressive stress $f_a < 0.15\ F_a$.

Braced. A condition in which *external* restraint prevents relative displacement of the ends of the column at right angles to the axial load and in the direction of the bending under consideration.

Buckling. Lateral deflection of the centroidal axis between braced ends which increases from the elastic state through an inelastic state to failure. (Similar failure of a portion of a column cross-section such as an unsupported flange or thin web is herein referred to as local buckling.)

Columns. Principal compression members used primarily for the support of axial compression loads. *Ends* of columns refer to points of support or points where load conditions change, including the restraints that provide bracing. *Length* of columns is the distance between ends (measured center-to-center of joints for analysis of frames, and between the faces of lateral restraints for determining *unbraced length*).

Drift. *Relative* lateral displacement of the *ends* of unbraced columns; customarily the result of gravity and lateral load primary effects computed separately and combined by superposition. When the secondary effects of shear deformations and differential changes in the axial lengths of parallel members are to be considered, any resulting lateral displacements are included in drift by superposition.

Lateral deflection. Displacement of the centroidal axis *between* braced or unbraced ends. If excessive, this displacement becomes buckling.

Lateral displacement. *Relative* movement of the centroidal axis at the ends of a column at right angles to the direction of the axial load and in the direction of bending under consideration. Drift, sidesway, or the sum of drift plus sidesway.

Sidesway. An increase in the *relative* lateral displacement of the ends of a column (an increase in the drift). Sidesway under an axial load ($P\Delta$) is a secondary effect of lateral displacement, and as such it is not usually computed. When sidesway is to be considered, it is computed as a series of increments, all proportional to the axial load; the first, a function of the drift; the second, drift plus the first increment of sidesway; etc.

Unbraced. A condition in which lateral displacement will occur. Restraint against such movement may consist of the column itself (cantilever) or a frame including the column, and externally any flexible bracing element, such restraints being insufficient to hold the ends relatively immovable.

M_s. Secondary moment magnifier. A variable expression derived from the AISC Specifications, Eq. (1.6-1a), to provide for the effect of secondary moments, ($P\Delta$), V, and PL/AE, not usually computed in column design.

$$M_s = C_m/(1 - f_a/F_e') \ldots\ldots \text{(See Fig. 6-7.)}$$

Δ. The sum of the lateral deflection and lateral displacement at any point in the length of the column resulting from behavior of the column under any loading considered.

δ. Any increment in Δ originally computed for only the primary effects of unsymmetrical gravity or lateral loads.

COLUMN DESIGN—CONDITIONS

General

The applications of the AISC Specification requirements for column design are simple, but the selection of the requirements appropriate to each shape or built-up section is not simple, except for the special case of direct compression without moment or lateral displacement.

For convenience, the problem of steel column design can be subdivided into two general categories. Direct compression is the simplest case, but, unfortunately for the designer, this case has limited application (1.5.1.3). A column or strut may be considered in direct compression only when it is laterally braced, loaded concentrically with connections of sufficient flexibility so that end moments may be assumed zero (Type 2 Construction), and not subjected to transverse loads of any kind between end supports. The second category, axial compression plus bending, is the general case encountered in most design (1.6.1).

Slenderness Effects in Axially Loaded Columns of Braced Frames

In both categories of column design, the effect of column slenderness (expressed as Kl/r) must be considered as a reduction factor applied to allowable stresses.[1] There are no "short columns" for which slenderness may be entirely disregarded where F_a can be taken as a constant (Eqs. 1.5-1, 1.5-2 and 1.5.3) and F_b can be taken as a constant

[1] "Effective Column Length–Tier Buildings," T. R. Higgins, *AISC Journal*, January 1964.

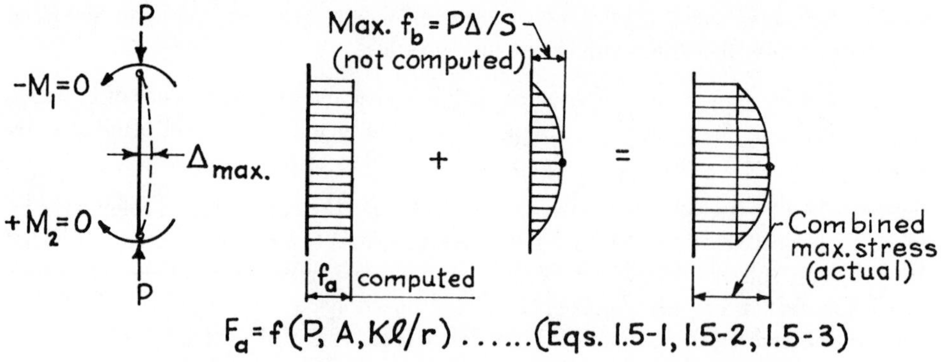

P

$-M_1 = 0$

$\Delta_{max.}$

$+M_2 = 0$

P

Max. $f_b = P\Delta/S$ (not computed)

+

=

f_a computed

Combined max. stress (actual)

$F_a = f(P, A, K\ell/r)$ ......(Eqs. 1.5-1, 1.5-2, 1.5-3)

FIG. 6-1 Special Case—Axial Load Only, Laterally Braced.

(1.5.1.4), although for columns with very light axial loads and large moments, the slenderness effect is related to parameters other than (Kl/r) for allowable bending stress compression (Eq. 1.6-2).

The reduction in allowable axial compression stress to allow for slenderness is proportional to modified Euler expressions for values of Kl/r larger than C_c (Eq. 1.5-2). For $F_y = 36$ ksi, $C_c = 126.1$; for $F_y = 50$ ksi, $C_c = 107.0$. For Kl/r values smaller than C_c, the reduction is based upon an equation developed by the Column Research Council. For Kl/r values equal to C_c, both expressions are equivalent (Eqs. 1.5-1 and 1.5-2). It may be helpful to regard these equations as providing for the secondary bending compression stress due to the lateral deflection of a column between end supports. (See Fig. 6-1.)

For the usual routine design, it is unnecessary to compute Δ and $P\Delta$, and the corresponding increase in f_b as the three allowable axial compression formulas (for F_a) automatically provide reasonable safety factors against buckling (slenderness effects) in braced main members, varying from $\frac{5}{3} = 1.667$ for $Kl/r = 0$, to $\frac{23}{12} = 1.92$ at $Kl/r \geqslant C_c$ with $l/r \leqslant 200$. Although the AISC Specification does not explicitly define columns as "primary" or "main" compression members, the provision for a larger allowable axial compression (F_{as}) on "axially loaded bracing and secondary" members should *not* be applied to columns (Eq. 1.5-3). The Nomenclature defines F_{as} as applicable to "bracing and *other* secondary members." The authors suggest revising the section containing Eq. 1.5-3 to read ". . . axially loaded bracing and *other* secondary members . . . ." and interpretation thereof in the Commentary to agree with the Nomenclature of the Specifications.

Axial Compression and Bending (1.6.1)

General. In the general case of column design, bi-axial bending is combined with axial compression. The sum of the ratios of axial stress to allowable axial stress and bending stress in each direction to allowable bending stress in each direction is limited to 1.0 (1.6.1). Allowable bending stresses are established separately in each direction to provide against local buckling (see Chapter 3, Beams). Since the allowable axial stress is established by the maximum slenderness ratio (Kl/r), it is additive in only one direction, that of maximum Kl/r. The process of superimposing the three effects, therefore, does not produce a simple uniform "safety factor," although it is conservative, particularly where bending occurs about only one axis not in the direction of the axial load slenderness reduction.

Braced Frames. Where lateral bracing is provided, the Specifications permit the use of $K = 1.0$, ". . . unless analysis shows that a smaller value may be used . . ." (Table

C1.8.1).[1] The use of $K = 1.0$ is recommended for preliminary design. Where bending is present, lateral deflection will occur in the plane of bending. This lateral deflection (Δ) displaces the centroid of the member and a "secondary" additional moment ($P\Delta$) is developed. In practical applications, it is not economically feasible to include this moment in computing f_b. The Specifications provide for this effect by requiring use of a moment magnification factor for secondary moments. (See "M_s" in Definitions.) The maximum effect of this secondary moment will add to the maximum original moment directly in only one special case. Where single curvature develops, $M_1 = M_2 = 0$, or $M_1 = -M_2$, the maximum deflection Δ and the accompanying maximum moment $P\Delta$ occur at the mid-height, and stresses f_a, f_b, and $P\Delta/S$ are directly additive for the maximum combined moment. (See Fig. 6-2(a).) In the general case, the location of the primary moment and the maximum combined moment will occur somewhere in the half length of the column, beginning at the support with the larger end moment, M_2. See all cases of loading, Fig. 6-2(a), (b), (c), and (d). It becomes necessary to consider two possible critical sections designated (1) and (2) in Fig. 6-2.

At critical sections (1) the maximum moment is M_2, calculated from an elastic frame analysis of primary moments for gravity loads only. No increase in stress due to lateral displacement of the column can occur at this section as displacement here is prevented by the lateral bracing. At this section, therefore, the allowable stresses are $F_a = 0.60\,F_y$, and F_b is reduced only by local buckling considerations. The allowable stress, F_b, here is a function of (F_y), $(F_y, \sqrt{F_y}, b_f/2t_f)$, or $[(l/r_T)^2, F_y, M_1/M_2]$ as for bending in flexure only (1.5.1.4, 1.6.1, 1.6.2, and Eq. 1.6-1b).

The second possible critical section (2) develops in the length of the column between supports where lateral deflection caused by primary bending moments, $\Delta \neq 0$. For this condition, the allowable stress must be reduced from that at critical section (1) to maintain a reasonably uniform factor of safety against buckling (reduction of F_a) and to allow for the increase, $P\Delta$, in primary bending moment above that computed. The equation applicable for this critical section (Eq. 1.6-1a) requires the use of variable reduction in F_a and provides a moment magnification factor applicable to f_b which approximates the effect of the added (but *not* computed) secondary moment, $P\Delta$.

The stresses plotted in Fig. 6-2 are intended to represent the distribution of compressive stress along the flange in which the maximum combined stress occurs. This flange is located on the left side of the column in sketches (a), (b), and (d). In sketch (c), depending upon the relative magnitudes of the wind moment and end moments, the maximum stress could occur either above or below the midheight. Sketch (c) shows the usual pattern where wind moment is low and the maximum combined stress is in the right side flange. Note also that, simply for convenience, all sketches show M_2, the larger end moment, at the bottom, which may require inversion to match a particular real condition encountered.

See Fig. 6-5 (or enter Table 1, *AISC Manual of Steel Construction*, with the effective length ratio Kl/r, based upon the minimum r, usually r_y) for values of F_a. See Figs. 6-6 and 6-7 for graphic solutions of the secondary moment factor, M_s, applicable over the full range of design, and Table 3-1 for values of M_{Rx} and M_{Ry}. These values are applicable for use in both design equations for all laterally braced columns. ($f_{bx}/F_{bx} = M_x/M_{Rx}$ and $f_{by}/F_{by} = M_y/M_{Ry}$). See also column design examples herein for illustration of these shortcut solutions for combined bending.

---

[1] "Design of Beam-Columns," Ira Hooper; *AISC Journal*, April 1967. Nomograph for the effective length factor, K, where columns are laterally braced. For final design where more accurate calculation than the assumed values of Table C1.8.1 is desirable.

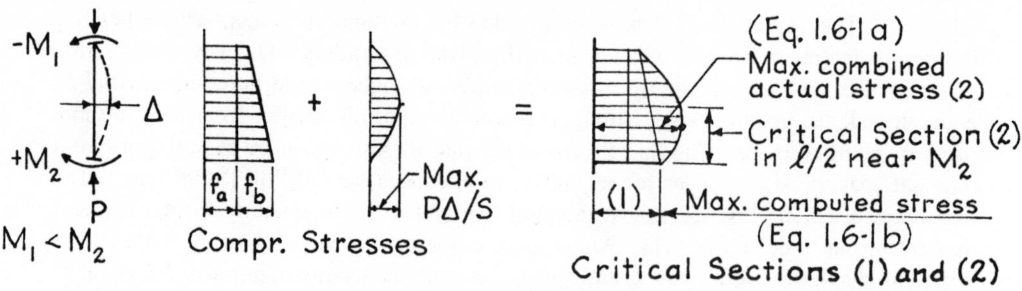

(a) Single Curvature, M_1/M_2 negative; No Lateral Loads Between End Supports; $f_a > 0.15F_a$

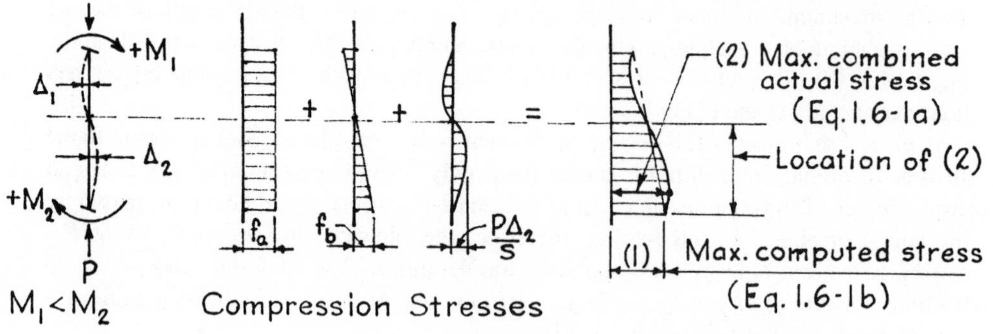

(b) Double Curvature, M_1/M_2 positive; No Lateral Loads Between End Supports; $f_a > 0.15F_a$

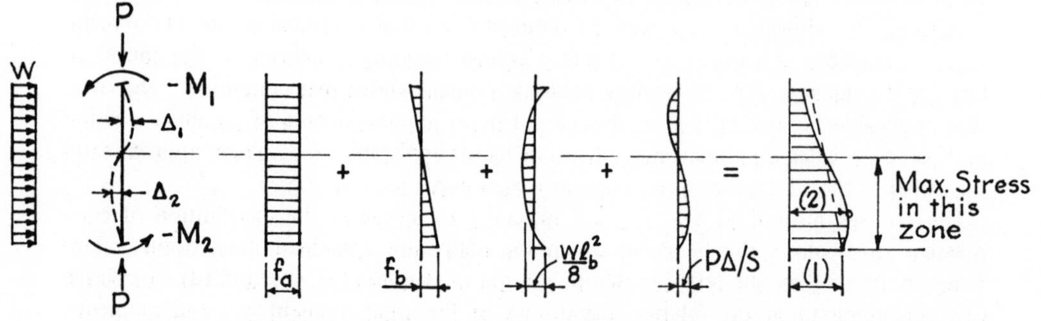

(c) Double Curvature, M_1/M_2 positive; Lateral Load Between End Supports; ($f_a > 0.15\ F_a$)

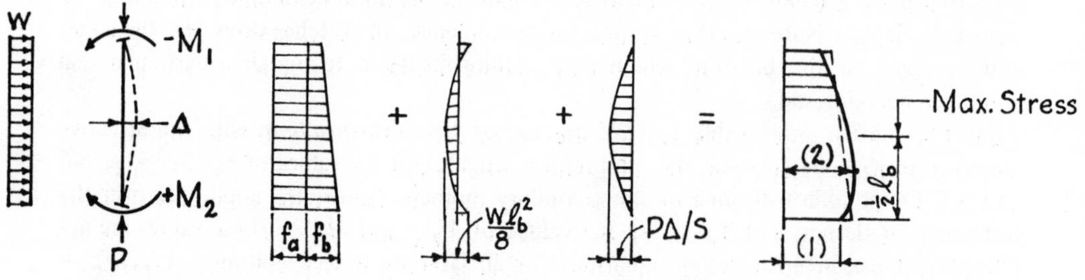

(d) Single Curvature, M_1/M_2 negative; Lateral Load Between End Supports; ($f_a > 0.15\ F_a$)

FIG. 6-2 General Case—Columns Laterally Braced. (a) Single Curvature, M_1/M_2 Negative; No Lateral Loads Between End Supports; $f_a > 0.15\ F_a$. (b) Double Curvature, M_1/M_2 Positive; No Lateral Loads Between End Supports; $f_a > 0.15\ F_a$. (c) Double Curvature, M_1/M_2 Positive; Lateral Load Between End Supports; $f_a > 0.15\ F_a$. (d) Single Curvature, M_1/M_2 Negative; Lateral Load Between End Supports; $f_a > 0.15\ F_a$.

Beam–Columns. As the axial load becomes smaller, the effect of the lateral deflection, the $P\Delta$ moment, becomes less. To provide a smooth transition between the general loading case for columns with relatively large axial load stresses, to that for beams with relatively small axial load stresses compared to bending moment stresses, the Specifications permit neglect of the moment magnification factor when $f_a \leqslant 0.15 F_a$ (Eq. 1.6-2). Practical design for beams with some axial load and columns supporting principally lateral forces is thereby simplified. (See Fig. 6-3.)

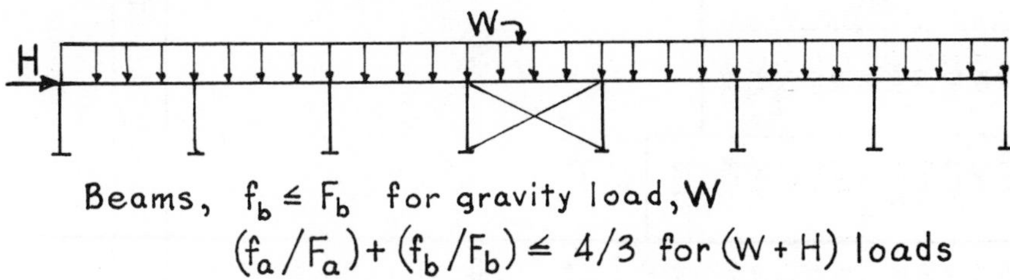

W
H
Beams, $f_b \leq F_b$ for gravity load, W
$(f_a/F_a) + (f_b/F_b) \leq 4/3$ for (W + H) loads

FIG. 6-3 Special Case–Small Axial Loads; $f_a \leqslant 0.15 F_a$.

Axial Compression and Bending with Lateral Forces–Columns in Unbraced Frames

General–K, C_m, and $P\Delta$. Up to this point, only braced construction, in which relative lateral displacement at each story is prevented, has been considered. When resistance to lateral displacement is provided only by the beam-column frame itself, relative lateral displacements at each end of the columns develop, both under pattern gravity loads alone and due to lateral forces, wind or seismic. The designer is confronted by three additional factors associated with this relative lateral displacement (drift).

The effective length factor, K, can no longer be conservatively assumed 1.0 or estimated at lower values. For special limited cases it may be assumed equal to 2.0; but, since it may greatly exceed even this value, in the general case it must be computed (1.8.3). A very convenient nomograph for the determination of K is available (Commentary 1.8).[1] (See Fig. 6-4.)

The second factor affected is the coefficient, C_m, used to modify the secondary moment factor, M_s (see Fig. 6-7). For conditions where drift occurs, C_m becomes a constant 0.85 (1.6.1), but can conservatively be taken as $(1 - 0.18\, f_a/F_e')$; (Commentary 1.6.1). See Table 6-1.

The third factor is the magnitude of the drift plus sidesway, the absolute value of relative displacement between ends of the column in the story under consideration. This factor cannot be estimated. When separate elastic analyses are made for gravity pattern loadings and wind loading and combined by superposition for maximum effect, usually only the primary value of drift, Δ, has been determined. An additional increase in Δ, "sidesway," or $\delta_1 \Delta$, occurs proportional to the moment, $P\Delta$, increasing this moment by $P\delta_1 \Delta$, which in turn increases the displacement by $\delta_2 \Delta$, etc. For columns with heavy axial loads, this secondary moment due to "sidesway" can be significant. Typical analyses for highrise buildings with total lateral displacement, Δ, near the generally accepted limit $0.0025 \times$ height show the total secondary sidesway about $0.10\ \Delta$. The Specifications require the use of a semi-arbitrary magnification factor for the primary moments, $C_m/(1 - f_a/F_e')$ to provide for these secondary effects (Eq. 1.6-1a).

---

[1]Commentary Fig. C1.8.2, K for "Uninhibited Sidesway."

TABLE 6-1 Reduction Factor, C_m, with Sidesway (Comm. 1.6.1)

Kl/r	F'_e	Allowable Axial Stress, f_a															
		2	4	6	8	10	12	14	16	18	20	22	24	26	28	30	32
0	∞	1.000	1.000	1.000	1.000	1.000	1.000	1.000	1.000	1.000	1.000	1.000	1.000	1.000	1.000	1.000	1.000
21	339	.999	.998	.997	.996	.995	.994	.993	.992	.991	.989	.988	.987	.986	.985	.984	.983
25	239	.998	.997	.995	.994	.992	.991	.989	.988	.986	.985	.983	.982	.980	.979	.977	.976
30	166	.998	.996	.993	.991	.989	.987	.985	.983	.980	.978	.976	.974	.972	.970	.967	.965
35	122	.997	.994	.991	.988	.985	.982	.979	.976	.973	.970	.968	.965	.962	.959	.956	.953
40	93.33	.996	.992	.988	.985	.981	.977	.973	.969	.965	.961	.958	.954	.950	.946	.942	.938
45	73.74	.995	.990	.985	.980	.976	.971	.966	.961	.956	.951	.946	.941	.937	.932	.927	.922
50	59.73	.994	.988	.982	.976	.970	.964	.958	.952	.946	.940	.934	.928	.922	.916	.910	.903
55	49.37	.993	.985	.978	.971	.964	.956	.949	.942	.934	.927	.920	.912	.905	.898	.891	.883
60	41.48	.991	.983	.974	.965	.957	.948	.939	.931	.922	.913	.904	.896	.887	.878	.870	.861
65	35.34	.990	.980	.969	.959	.949	.939	.929	.919	.908	.898	.888	.878	.868	.857		
70	30.48	.988	.976	.964	.953	.941	.929	.917	.906	.894	.882	.870	.858				
75	26.55	.986	.973	.959	.946	.932	.919	.905	.892	.878	.864	.850					
80	23.33	.985	.969	.954	.938	.923	.907	.892	.877	.861							
85	20.67	.983	.965	.948	.930	.913	.896	.878	.861								
90	18.44	.980	.961	.941	.922	.902	.882	.863									
95	16.55	.978	.956	.935	.913	.891	.869										
100	14.93	.976	.952	.928	.904	.879	.855										
105	13.54	.973	.945	.920	.894	.867											
110	12.34	.971	.942	.912	.883	.854											
115	11.29	.968	.936	.904	.872												
120	10.37	.965	.931	.896	.861												
125	9.56	.926	.925	.887													
130	8.84	.959	.919	.878													
140	7.62	.953	.906	.858													
145	7.10	.949	.899														
150	6.64	.946	.892														
155	6.22	.942	.884														
160	5.83	.938	.877														
170	5.17	.930	.861														
180	4.61	.922															
190	4.14	.913															
200	3.73	.903															

Below line, use $C_m = 0.85$

REDUCTION FACTOR

$C_m = 1 - 0.18 f_a/F'_e \leqslant 0.85$
with Sidesway

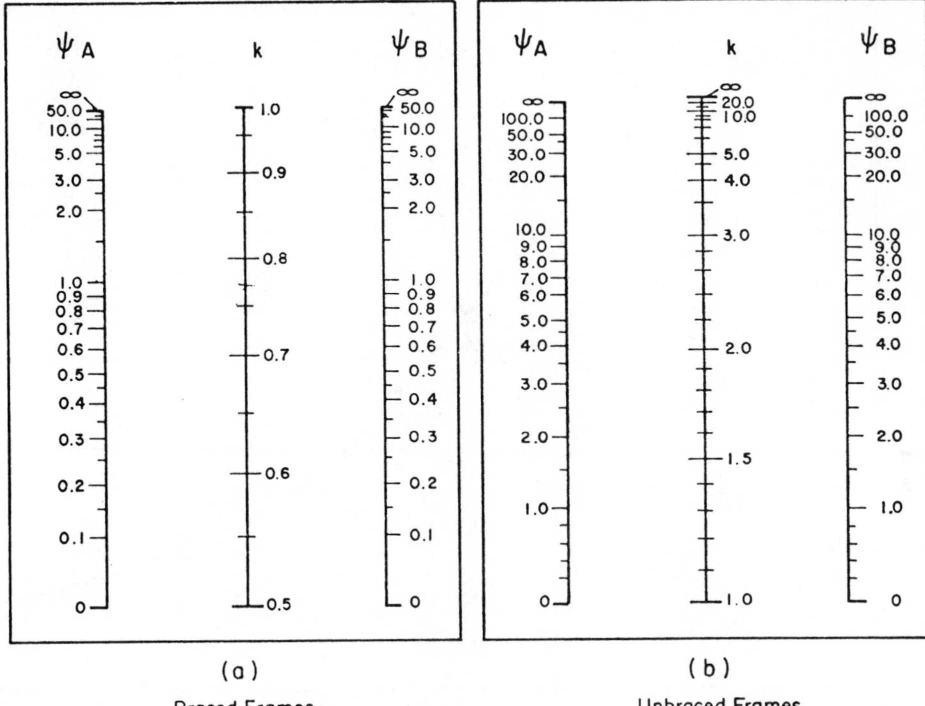

ψ = Ratio of Σ (EI/ℓ_c) of compression members to Σ (EI/ℓ) of flexural members in a plane at one end of a compression member

k = Effective length factor

FIG. 6-4 Effective Length Factors Nomograph*

***From "Commentary on Building Code Requirements for Reinforced Concrete" (ACI 318-71) with 1973 Supplement, Courtesy American Concrete Institute.**

Arbitrary Allowance by Specifications

General. For routine design the structural engineer cannot afford "special analyses" described at length in research reports or technical papers, for the analysis of all the complex conditions encountered in practical applications. For routine design of lowrise buildings, simple conservative design requirements that can be carried out manually are essential. Similarly, a manual procedure for workable preliminary selection of sections is essential for the input required where the final analysis and design is to be performed by computer. The Specifications provide complete freedom for the exercise of the designer's ingenuity to achieve economy both in design time and in construction cost for frames which must provide lateral stability. The Specifications are nearly pure "performance requirements." Where separate lateral bracing is not to be provided, the Engineer must:

1. Provide for general stability of the structure and each compression element thereof (1.8.1).
2. For slenderness ratios, use the effective length unbraced, Kl, and the radius of gyration corresponding (1.8.1).
3. Kl must be determined by a rational method; $K \geqslant 1.0$; and $Kl \geqslant$ actual unbraced length (1.8.3; see Commentary Fig. C1.8.2).

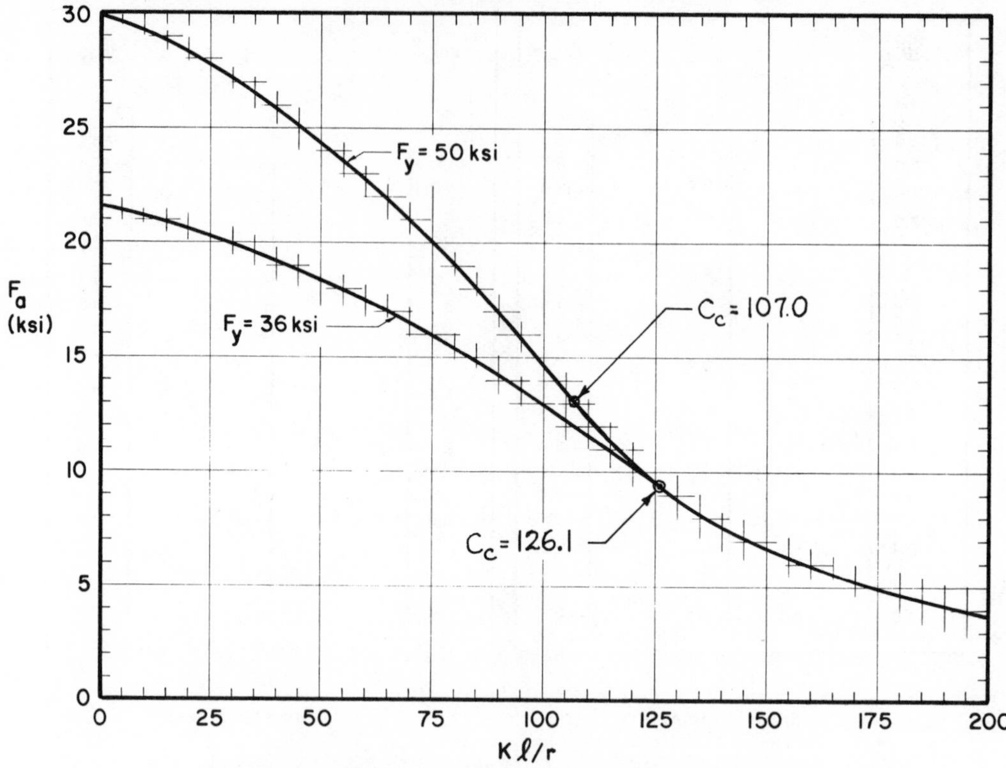

FIG. 6-5 Allowable Axial Stress—Columns Without Moment.

The specifications define three constants required for column design, two of which must be determined in advance by the Engineer before applying the short-cut procedures of this chapter for the solution of the column design formulas prescribed:

1. An axial compression constant, $C_c = \sqrt{\dfrac{2\pi^2 E}{F_y}}$. For F_y = 36 ksi, C_c = 126; for F_y = 50 ksi, C_c = 107. For other values of F_y see Appendix A, Table 1, *AISC Manual of Steel Construction*. Values of F_a for the full range of Kl/r are conveniently available in Table 1. The use of Table 1 obviates the need of computing C_c or solving Eqs. 1.5-1 and 1.5-2 for F_a. (See also Fig. 6-5.)
2. A bending compression (or tension) constant, C_b.
   a. In Eq. 1.6-a, for checking the critical section (2) as shown in Fig. 6-2, C_b = 1.0 in braced frames. Enter Table 3-1, Chapter 3, with this value for solutions of F_b.
   b. In Eqs. 1.6-1, for checking the two critical sections of Fig. 6-2 in unbraced frames,

   $$C_b = 1.75 + 1.05(M_1/M_2) + 0.3(M_1/M_2)^2 \leqslant 2.3.$$

   (See Table 6-2.)
   c. *Recommendation:* For all preliminary designs, routine design of lowrise structures, and columns with large bending moments between the ends, take C_b = 1.0 for use in both Eqs. 1.6-1a and 1.6-1b in braced or unbraced frames.
3. A "moment reduction" factor constant, C_m, for use with the secondary moment factors of Fig. 6-7. The Specifications permit use of C_m = 0.85 for all columns in

TABLE 6-2 Bending Coefficient; C_b

M_1/M_2	C_b	M_1/M_2	C_b
−1.0	1.000	+1.0	2.300
−0.9	1.048	↑	↑
−0.8	1.102		
−0.7	1.162	↓	↓
−0.6	1.228	+0.5	2.300
−0.5	1.300	+0.4	2.218
−0.4	1.378	+0.3	2.092
−0.3	1.462	+0.2	1.972
−0.2	1.552	+0.1	1.858
−0.1	1.648	0	1.750

$C_b = 1.75 + 1.05(M_1/M_2) + 0.3(M_1/M_2)^2 \leqslant 2.3$

unbraced frames subject to drift, or conservatively, the values in Table 6-1. For columns in braced frames, see Fig. 6-6.

Individual Column Design. At this point, the designer will have completed preliminary elastic analyses using assumed relative stiffnesses of members (see Chapter 2). The designer will possess the following data for the preliminary design of each column:

1. Axial gravity loads.
2. End conditions (at least one end restrained) and the sign and approximate magnitude of end moments, M_1 and M_2, under (a) gravity loads, and (b) wind, seismic, or other transverse loads.

Where lateral bracing is not provided, unless the designer can neglect lateral displacements on the basis of experience or intuition based thereon, the analysis should include the sum of the primary wind, or seismic moments, and pattern live load moments, and the primary drift resulting. The sum of the moments will be magnified for the $(P\Delta)$ effect. (The drift should be computed for comparison to accepted limits, even though it is not required to complete the column design using the arbitrary allowance for the $(P\Delta)$ effect.)

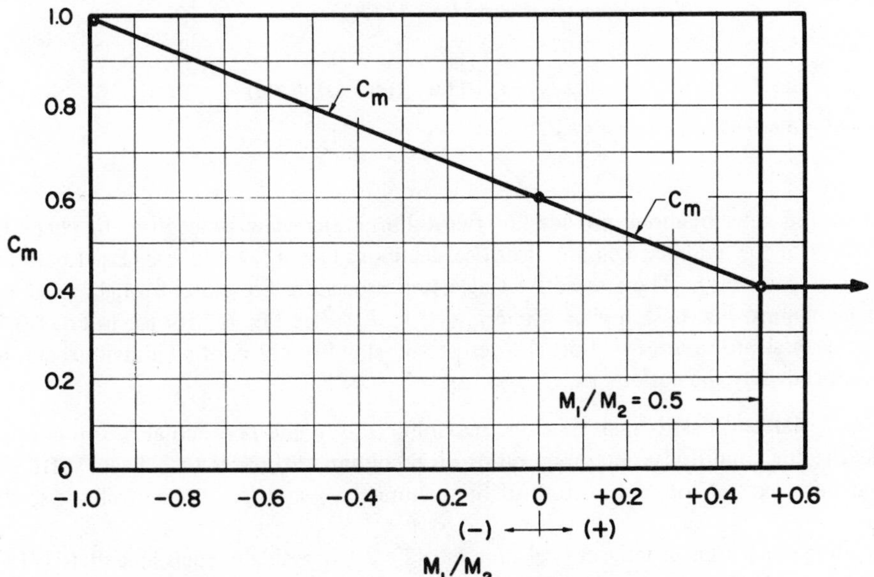

FIG. 6-6 Reduction Factor, C_m—Sidesway Prevented.

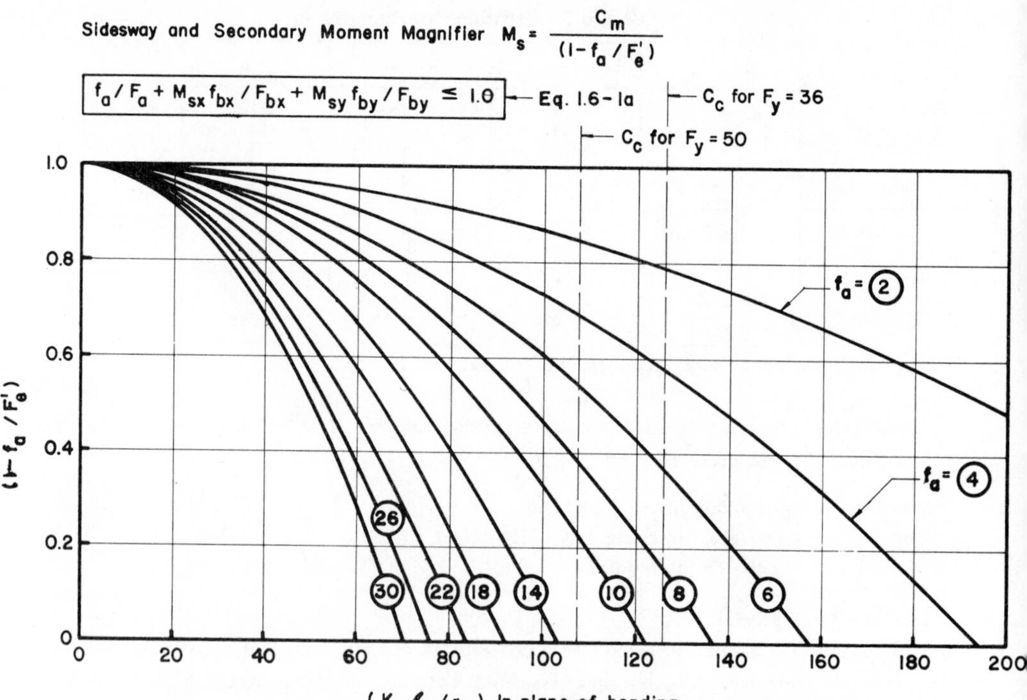

FIG. 6-7 Secondary Moment Magnification Factor, M_s.*

*For load combinations where allowable stresses may be increased one-third (1.5.6), F'_e may also be increased one-third (1.6.1).

Let

S_c = sidesway coefficient for F'_e as read from Table 6-7.

S'_c = sidesway coefficient for 1.333 F'_e

Then

$$M'_s = C_m/S'_c = C_m/[1.0 - (1.0 - S_c)\ (0.75)]$$

where M'_s = moment magnifier for $4F'_e/3$.

The AISC Specifications provide for two (alternate) review procedures to check the adequacy of the assumed column at critical sections (1) and (2), Fig. 6-2 (Eq. 1.6-1b and 1.6-1a, respectively). The alternative single step procedure Eq. 1.6-2, for light axial load (beam-columns, Fig. 6-3) applies where $f_a \leqslant 0.15\ F_a$. See Fig. 6-7 for use in Eq. 1.6.1a.

The general procedure for short-cut design and stability check of an individual column recommended by the authors is:

Step 1. Collect elastic analysis data: maximum P; M_1 and M_2, actual l_b for all applicable loading conditions; determine ratios of Σ column stiffnesses to Σ beam stiffnesses about both axes at top and bottom of the columns, G_T and G_B, respectively. See Fig. 6-4.

Loading conditions usually critical may be: (1) $D + L$ with all panels loaded, (2) $D + L$ with pattern loading for live loading. For frames with lateral movement possible, either combination (1) or (2) separately or combined with wind may be critical, and Δ for primary drift with wind plus Δ for pattern live loading should be included.

Step 2. Record design data for assumed section. $A, S, r, r_T, d/A_f, b_f/2t_f, d/t_w$ and b_f (included in Table 3-1).

Step 3. Determine effective length factor, K. Use the stiffness ratios, G_T and G_B, in the nomograph applicable, for braced or sidesway permitted, from Fig. 6-4. Calculate $K_x l_x / r_x$ and $K_y l_y / r_y$.

For load condition (D + L)

Step 4. Read F_a from Fig. 6-5 or AISC Table 1, appropriate with F_y used. Use the larger value, $(K_x l_x / r_x)$ or $(K_y l_y / r_y)$. Calculate f_a / F_a.

If $f_a/F_a > 0.15$:

Step 5. For bending in a braced plane, use $C_b = 1.0$. Otherwise, calculate M_1/M_2 (negative for reverse curvature), and read C_b from Table 6-2. Enter Beam Selection Table 3-1 or 3-2. For trial column size, read or calculate resisting moments, M_{Rx} and M_{Ry}. For M_{Ry}, compare the tabulated value of $b_f/2t_f$ to footnote constants and use the appropriate value, Columns (7), (8), or (9). For M_{Rx}, when criteria in footnotes (1) or (2) apply, use the value in column (2) when a value is shown in column (2). When columns (4) and (5) or (5) and (6) are applicable, use the larger value of the applicable pair, noting that the value used may not exceed that in column (3).

Step 6. Calculate or select C_m. See Fig. 6-6 for C_m Sidesway Prevented or Table 6-1 for C_m. Use $(K_b l_b / r_b)$. Read the "moment magnifier" in Fig. 6-7. Calculate factors, M_{sx} and M_{sy}. ($M_s = C_m$/moment magnifier.) Where sidesway occurs, compare $M_x(M_{sx} - 1)$ and $M_y(M_{sy} - 1)$ to $P\Delta_y$ and $P\Delta_x$. The conservative approach recommended, when one end of the column is hinged or not considered to be resisting moment, is to use $(M_x + P\Delta_y)$ and $(M_y + P\Delta_x)$ whenever $P\Delta > (M)(M_s - 1)$. The Specifications do not specifically require this conservative approach, but, of course, do require "stability" (1.8.1).

Step 7. Solve Eq. 1.6-1a. With values from steps 4 through 6, this equation reduces to

$$\frac{f_a}{F_a} + M_{sx}\left(\frac{M_{2x}}{M_{rx}}\right) + M_{sy}\left(\frac{M_{2y}}{M_{ry}}\right) \leqslant 1.0^*$$

Step 8. Calculate C_b. Using calculated C_b, repeat step 5, and solve Eq. 1.6-1b (with values of M_r appropriate for calculated C_b).

$$\frac{1}{0.6}\left(\frac{f_a}{F_a}\right) + \frac{M_{2x}}{M_{rx}} + \frac{M_{2y}}{M_{ry}} \leqslant 1.0$$

Step 9. If the assumed section is adequate for all $(D + L)$ loadings, steps 7 and 8, repeat steps 4 through 8 for loading conditions $(D + L + W)$ increasing limits in steps 7 and 8 to 1.333. Neglect the increase in F_e' unless equations near limit 1.333.

For beam-columns, $f_a/F_a \leqslant 0.15$:

Step 5. For loads $(D + L)$ with calculated C_b, enter Beam Selection Table, Chapter 3; determine M_{rx} and M_{ry}.

Step 6. Solve Eq. 1.6-2. With values from the Beam Selection Table, this equation becomes:

$$\frac{f_a}{F_a} + \frac{M_{2x}}{M_{rx}} + \frac{M_{2y}}{M_{ry}} \leqslant 1.0$$

Step 7. For loads $(D + L + W)$. Repeat step 5 and step 6 with the limit 1.333, again neglecting the increase in F_e'.

---

*Use for M_{2x} and M_{2y} values at the same end of the column; if necessary, check both ends of the column to determine the controlling combination.

DESIGN EXAMPLES USING SPECIFICATION FORMULAS WITH PRIMARY ANALYSIS DATA

In the following examples, the sketches show simplified summaries of analyses (from Chapter 2 and Appendix 2A). Some values have been rounded off. Moment values to columns have been increased from zero from analyses for shear connections at the column faces to include shear times one half column depth. For simplicity where the frame considered is braced against joint translation at right angles to the plane of bending under consideration, the effective length factor, $K_{y-y} = 1.0$ has been used. This approach conservatively neglects incidental moment restraints due to lateral bracing connections and base restraints (such as anchor bolts outside the plane of bending considered). Where successive examples illustrate different selection of sections or connections for the same condition, values from the preceding examples are freely used as a designer would to save time. Also in successive examples, calculations obviously not critical in preceding examples are not repeated. These short cuts not only save space here, but make the examples more representative of design procedures to save design time.

EXAMPLE 1

Check preliminary selections of sizes for the exterior columns of Design Example 1, Chapter 2. (See Fig. 2-5(a).) The structure is Type 1 Construction (rigid) with no lateral bracing in the plane of the frame. It is laterally braced at right angles. The column length, l_{bx}, for bending about its x-x axis is (180 - 18) in. = 162 in. The unbraced length about the y-y axis is 100 in.

Fig. 2-5(a) shows the maximum column loads and moments for design, including secondary effects. For this example, the primary analyses based upon moment only will be used. (See Sketch A; data from Chapter 2—Appendix). Use $F_y = 36$ ksi.

Dead Load + Live Load

Step 1. $P = 19.9$ k; $M_{1x} = +167$ k-in.; $M_{2x} = +336$ k-in. $l_{bx} = 162$ in.; $l_{by} = 100$ in.; $M_{1y} = 0$; M_{2y} = (negligible); Beam: W 18 × 40.

Step 2. Assumed: W 8 × 24. $A = 7.06$ in.^{2}; $S = 20.8$ in.^{3}; $d/t_w = 32.4$; $r_y = 1.61$; $r_x = 3.42$; $r_T = 1.78$; $d/A_f = 3.07$; $b_f/2t_f = 8.17$; $b_f = 6.5$ in.

Step 3. In the direction of bending about axis x-x, take $G_{bot} = 1.0$; calculate $G_{top} = \dfrac{82.5/162}{612/352} = 0.293$

From Fig. 6-5, Uninhibited Sidesway, $K = 1.2$; braced direction, take $K = 0.8$.

$$Kl_{bx}/r_{bx} = (1.2)(162)/3.42 = 56.8;$$

$$Kl_y/r_{by} = (0.8)(100)/1.61 = 49.7.$$

Step 4. From Fig. 6-6, read $F_a = 17.73$ ksi for $Kl/r = 56.8$. $f_a = 19.9/7.06 = 2.82$ ksi. $f_a/F_a = 0.159 > 0.15$.

Step 5. $M_1/M_2 = 0.5$. See Table 6-2; $C_b = 2.3$;
$l_b/r_T = 162/1.78 = 91$. Enter Table 3-1 for $F_y = 36$ ksi.

$53.3\sqrt{C_b} = 81 < 91 < 119\sqrt{C_b} = 180$. Use larger value of columns (4) and (5) in Table 3-1. Read 4.633 for column (4), and 6.78 for column (5).

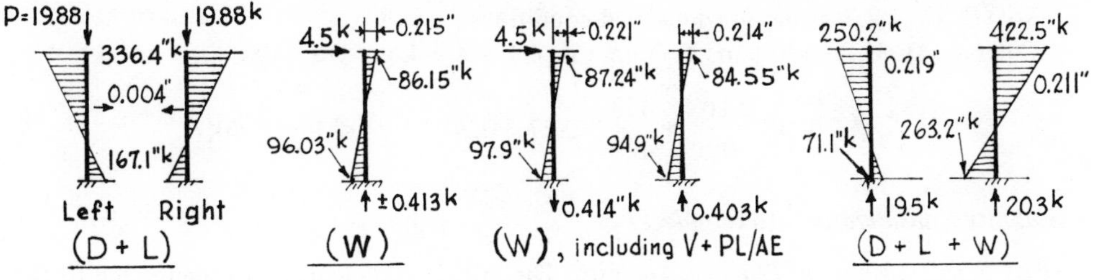

Exterior Columns, One Story Building - Example 1

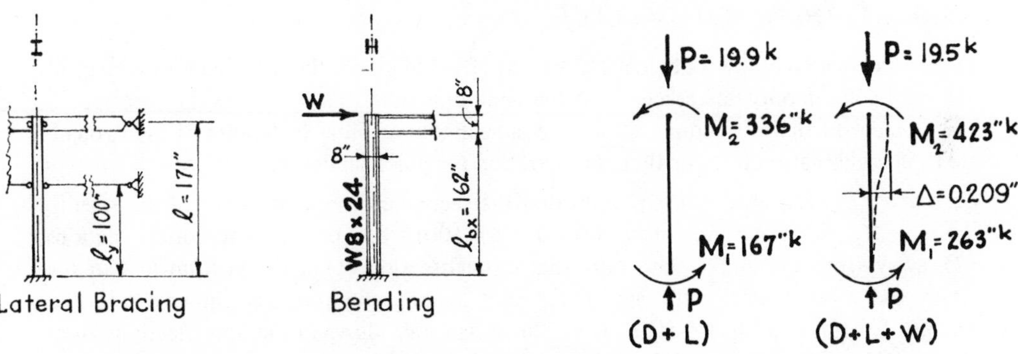

SKETCH A Exterior Column Dimensions—Primary Analysis.

Column (4): Read 4.633×10^{-4}

$$M_R = (41.1) - \frac{(162)^2\ (4.633)}{(2.3)\ (10{,}000)} = 35.8 \text{ k-in.} \leqslant 37.4$$

$$= 430 \text{ k-in.}$$

Column (5): (1000) (6.78) (2.3)/162 = 96.3 k-ft. > 37.4. Use 37.4 × 12 = 449 k-in.

Step 6. $C_m = 0.85$ (sidesway). Fig. 6-7, read 0.93.

$M_s = 0.85/0.93 = 0.914$

Step 7. Eq. 1.6-1a; $f_a/F_a + M_{sx} f_{bx}/F_{bx} + M_{sy} f_{by}/F_{by} \leqslant 1.0$. 0.159 + (0.914) (336/449) = 0.843 < 1.0 OK.

Step 8. Use $C_b = 2.3$. $l_b/r_T = 91$. Columns (4) and (5) apply. Use resisting moment from Step 5.

$$\text{Eq. 1.6-1b.}\quad f_a/0.6\,F_a + f_{bx}/F_{bx} + 0 \leqslant 1.0$$

$$\frac{0.159}{0.6} + \frac{336}{449} = 1.01 \approx 1.0 \text{ OK.}$$

Dead Load + Live Load + Wind Load

Step 1. $P = 20.3\ k; M_{1x} = 263$ k-in; $M_{2x} = 423$ k-in; $\Delta_{yy} = 0.209$ in.

Step 4. $F_a = 17.73$ ksi. $f_a = \frac{20.3}{7.06} = 2.88$ ksi. $f_a/F_a = 0.162 > 0.15$. $M_1/M_2 = 0.62$.

Steps 5, 6, and 7. $C_b = 1.0$. Table 3-1 from previous load conditions, Column (4); $M_R = 347$ k-in; Column (5): $M_R = 449$ k-in. Use 449 k-in.

Eq. 1.6-1a: 0.162 + (0.914) (423/449) = 1.02 < 1.333 OK.

Step 8. $C_b = 2.3$ From previous load conditions,
Column (4): $M_R = 430$ k-in. Column (5): $M_R = 449$ k-in. Use 449 k-in.

$$\text{Eq. 1.6-1b: } \frac{0.162}{0.6} + \frac{423}{449} = 0.27 + 0.94 = 1.21 < 1.333 \text{ OK.}$$

Stability Considerations (Example 1)

(PΔ) Effect Using the Primary Analysis. The initial increments in moments due to the effect of drift, Δ, under vertical load, *P*, are:

Left col.: (19.9) (−0.215) = −4.3 k-in.
Right col.: (20.3) (+0.215) = +4.4 k-in.

Due to the relatively low axial loads, the (*P*Δ) effect increases the top column moment of the right column only one percent. At the same time note that it reduces the corresponding moment in the left column about the same amount. Thus, instability of the two columns in combination is no further consideration for practical design.

Secondary Effects. Interestingly, note that the secondary moment factor of the Specifications, M_s, reduced moments (1.00 − 0.914) (100) = 8.6 percent for critical sections (2), away from the ends. Also note that critical section (1) at the bottom face of the beam was most critically stressed under both loading conditions as shown by formulas, Eq. 1.6-1(a) and Eq. 1.6-1(b). Thus, the design procedures of the Specifications automatically provided ample stability in this example.

See also Fig. 2A-1, showing that the inclusion of the effects of shear and differential axial length change deformations reduced both moment and Δ at the critical right column. The moment was decreased and Δ increased at the left column, reducing somewhat the effect of (*P*Δ) on both as a combination.

Left col.: (−87.7 + 86.25) + (19.9) (0.225) = −3.38 k-in.
Right col.: (84.55 − 86.15) + (20.3) (0.210) = +2.66 k-in.

Overall Structure. The first increment of the (*P*Δ) moment on the entire structure may be compared to the total wind moment.

$$\frac{P\Delta}{M_w} = \frac{(140.4)(0.215)}{(4.5)(171)} = \frac{30.2}{769.5} = 0.04$$

This comparison shows that the (*P*Δ) effect is comparable to an increase in wind of about 4 percent.

EXAMPLE 2

Check the preliminary selections of sizes for the interior columns of Design Example 1, Chapter 2. As in Example 1, use $F_y = 36$ ksi. (Use the data in Sketch B.) The unbraced length in both directions here is 162 in.

EXAMPLE 2. INTERIOR COLUMNS, Fig. 2-5(a).

Dead Load + Live Load

Step 1. $P = 50.3\ k; M_1 = 46$ k-in.; $M_2 = 92.5$ k-in.; $l_{bx} = 162$ in.; $l_{by} = 162$ in.

$$G_{top} = \frac{82.5/162}{(2)(612)/(352)} = 0.146; \text{ take } G_{bot} = 1.0$$

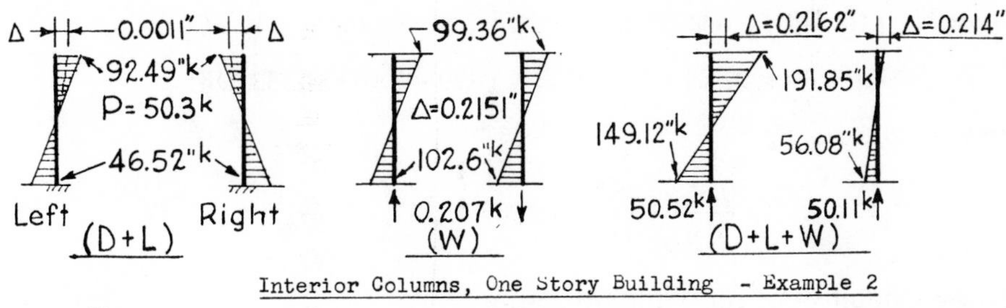

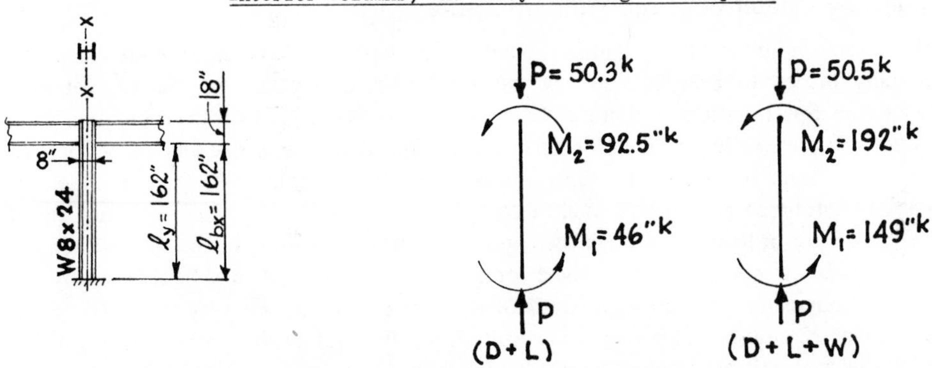

SKETCH B Interior Column Dimensions—Primary Analysis.

Step 2. $W8 \times 24$ (Same as Ex. 1)
Step 3. From Fig. 6-5, $K_x = 1.18$. Take $K_y = 0.8$ (Table C1.8.1-b).

$$Kl_b/r_b = (1.18)\,(162)/3.42 = 55.8;$$

$$Kl_y/r_y = (0.8)\,(162)/1.61 = 80.5$$

Step 4. From Fig. 6-6, $F_a = 15.30$ ksi (for bending about axis y-y). $f_a = 50.3/7.06 = 7.12$ ksi; $f_a/F_a = 0.466 > 0.15$.
Step 5. $C_b = 2.3$. $l_b/r_T = 91$. Enter Table 3-1. Columns (4) and (5) apply.

$$M_R = 449 \text{ k-in. (from Ex. 1)}$$

Step 6. $C_m = 0.85$. In Fig. 6-7, read 0.83.

$$M_s = 0.85/0.83 = 1.024$$

Step 7. Eq. 1.6-1a.

$$0.466 + (1.024)\,(92.5)/449 = 0.677 < 1.00 \text{ OK.}$$

Step 8. $C_b = 1.75 + 1.05\,(0.5) + 0.3\,(0.5)^2 = 2.35 \approx 2.3$

$$M_R = 449 \text{ k-in. (From Ex. 1)}$$

$$\text{Eq. 1.6-1b: } \frac{0.466}{0.6} + \frac{92.5}{449} = 0.98 < 1.0 \text{ OK.}$$

Dead Load + Live Load + Wind Load

Step 1. $P = 50.5$ k; $M_1 = 149$ k-in; $M_2 = 192$ k-in; $\Delta = 0.21$ in.
Step 4. $f_a = 50.5/7.06 = 7.15$ ksi; $f_a/F_a = 7.15/15.30 = 0.467 > 0.15$
Steps 5, 6, and 7. $C_b = 2.3$; Column (4): $M_R = 347$ k-in.; Column (5): $M_R = 449$ k-in.

Use same M_s as for $D + L$. Neglect increase in F'_e. See footnote to Fig. 6-7.

$$\text{Eq. 1.6-1a: } 0.467 + (1.024)(192)/499 = 0.90 < 1.333 \text{ OK.}$$

Step 8. Eq. 1.6-1b. $C_b = 2.3$

$$\frac{0.468}{0.6} + \frac{192}{449} = 1.208 < 1.333 \text{ OK.}$$

Sidesway—Stability of the Total Structure

The lateral stability of the structure, as well as that of individual columns when not laterally braced to prevent sidesway, must be assured (1.8.1). For the case of sidesway, the Specifications require a rational method (1.8.3). For the preliminary selection of sizes, or where all columns in a story are of the same size and carry approximately the same loading for the same unbraced lengths, the column-by-column design for stability can be an economical final design. For the general case, where columns are of different sizes or loading conditions are quite different, exterior columns and interior columns, and few columns are involved; or where some columns in a bent of many are oriented oppositely; a stability analysis of the entire structure story-by-story will permit significant economy. This economy results from distributing the surplus bracing effect of the heavier, stiffer, or more lightly loaded (than the average of all) columns to change the condition of the others from unbraced, $K > 1.0$, to braced $K \leqslant 1.0$.[1]

The secondary moment magnifiers, M_s, can be computed for the entire unbraced structure in each direction as:

$$M_s = \frac{0.85}{1 - \Sigma P/\Sigma(AF'_e)}$$

In the design of the individual columns, M_s can then be taken as the larger value computed for the entire story as above or for the individual column assumed braced.

A very conservative approach to the determination of individual column M_s factors is to regard these columns as braced only for the determination of the effective length factors, K, and simply to use $K = 1.0$. This procedure conforms entirely to the letter of the AISC Specifications (1.8.3). It will, of course, permit an increase in the factor, F'_e, reducing M_s correspondingly (C1.6.1). Slightly less conservative, but still conforming to the letter of the AISC Specifications is to calculate K for a braced column (Fig. 6-4) and to use $C_m = 0.85$ as in the computation of the story M_s (1.6.1). The authors would not consider it conservative to compute C_m as for a column immovably braced against drift, $\Delta = 0$. It is, of course, proper to use the reduced value of $K = 1.0$ also for the calculation of F_a.

For all cases, the value of K for a braced column will be less than one-half that for an unbraced column (by inspection of Fig. 6-4). A quick estimate comparing the term, M_s, for each braced individual column to the story term, M_s, becomes possible since $F'_e \sim (1/K)^2$. The individual column need not be investigated further as a braced column if the story term $M_s > \left(\dfrac{C_m}{1 - f_a/4F'_e}\right)$ where C_m, f_a, and F'_e are taken for the individual column. Take the same value for F'_e unbraced as used in calculating the story term. Read C_m from Table 6-1. Where all columns in a story are the same section, the expression

[1] "The Effect of Length of Columns in Unbraced Frames," J. A. Yura, *AISC Journal*, April, 1971; and Discussion of "Stability under Elastic Support," A. Zweig, *AISC Journal*, July, 1965.

for a story moment magnifier reduces to

$$M_s = \frac{0.85}{1 - \dfrac{\Sigma P}{A \Sigma F'_e}}$$

where ΣP = the total story load and A = the area of one column.

If the columns are oriented in the same direction so that Kl_b/r_b can be taken as equal for all, neglecting the effect of different ratios of joint stiffness, G, the equation reduces further to

$$M_s = \frac{0.85}{1 - \Sigma P/nAF'_e} \quad \text{where } n = \text{the number of columns}$$

Substituting for the total load divided by the total area, $f_a = (\Sigma P/nA)$

$$M_s = \frac{0.85}{1 - f_a/F'_e}$$

The solution for this term, M_s, may be read from Fig. 6-7, entering with f_a, where f_a = the average for all the columns. The braced condition for individual columns need not be investigated further if 0.25 f_a for the individual column is less than the average f_a for all of the columns in the story.

In this short-cut approach, all column moments in a story are conservatively considered to *increase* with sidesway. If individual columns are to be investigated, it will be noted (as in Examples 1 and 2) that the gravity load moments on approximately half the columns are *reduced* by wind and sidesway effects. The device in Examples 1 and 2 of "pairing" such columns may be applied for a more accurate determination of stability.

Rational Method (Secondary ($P\Delta$) Sidesway Moments)

General Approach. As the use of computers for structural analysis and design has become more prevalent, the economics of design time change. With the aid of computer solutions, use of the arbitrary formulas to allow for drift and sidesway achieves no great economy of design time. The requirement that general stability be provided for the structure as a whole and for each compression element can be satisfied directly (1.8.1). The secondary moment magnifier term can be replaced by a direct solution for the secondary moment.

Let $M'_s = P\ (\Delta + \delta_1 + \delta_2 + \cdots \delta_n)$ where M'_s is the computed sum of the secondary moments due to the vertical load, P, on an individual column and Δ = primary story drift due to the sum of the pattern live loading, axial shortening, and wind effects; δ_1 = the increase in the drift due to the added moment, $P\Delta$, (sidesway); δ_2 = increase due to $P\delta_1$; etc. The authors recommend the use of a modified Eq. 1.6-1a as follows:

$$\frac{f_a}{F_a} + \frac{C_{mx}(M_{2x} + M'_{sx})}{S_x F_{bx}} + \frac{C_{my}(M_{2y} + M'_{sy})}{S_y F_{by}} \leqslant 1.0$$

Note that the moment reduction factor, C_m, is retained to allow for the variable location of the critical section (2) in the length of the column; see Fig. 6-2. A conservative value for C_m is recommended; see Table 6-1.

Advantages. One advantage of this rational approach is that the solution for stability of the whole structure and each compression element is combined. The surplus stiffness or capacity of lightly loaded columns or columns oriented in different directions is distributed directly for maximum efficiency. It is unnecessary to compute capacity for each column with K-braced and K-sidesway. One solution for Eq. 1.6-1a and Eq. 1.6-1b will suffice for each critical loading condition.

The disadvantage of this approach is that, at the present writing, most computer programs available solve for Δ, including the effect of axial shortening, but not the effect of $P\Delta$; or solve for Δ, including the effect of $P\Delta$, but not axial shortening. The axial shortening effect becomes more significant with increase in length stressed or level of stress, and with differential shortening proportional to the actual average compressive stress, f_a, throughout the height of columns.

Since the steel frame is customarily erected as a unit complete to the top before the application of most of the gravity loads, most of the axial shortening adjustments of the frame will result in changes in the moments. Similarly, for a low wide structure, the frame will customarily be completed before full lateral load (wind or seismic) causes differential axial shortening in the horizontal members which will result in a variable Δ across the frame. It is, however, not difficult to combine the results of the two analyses or to estimate the effects of such axial shortening and superimpose same upon an analysis manually, nor to program the combination. See Examples 1 and 12 for comparisons of the rational and arbitrary solutions for columns in one and 25-story building structures.

EXAMPLE 3. See Fig. 2-5(b) *One-Story Structure. Type 2, Braced.*

Columns $W6 \times 15.5$; beams $W21 \times 49$. $F_y = 36$ ksi.

Interior Columns: (D + L) loading condition only;

$W6 \times 15.5$. $P = 46.8\ k$; $M_1 = M_2 = 0$; $\Delta_x = 0$; $\Delta_y = 0$; unbraced length, $l = 15 \times 12 - (0.5 \times 21) = 169.5''$. $K = 1.0$. $A = 4.56$ in.^{2}. $r_{xx} = 2.57''$; $r_{yy} = 1.46''$. Enter Fig. 6-5 (or AISC Table 1-36) with $Kl/r = 116$; read $F_a = 10.8$ ksi; $f_a = 46.8/4.56 = 10.3$ ksi OK

Exterior Columns: (D + L) loading condition only;

$W6 \times 15.5$. See sketch (C). $P = 23.4\ k$.

$$M_1 = -M_2 = V \times 0.5 \text{ depth}$$

$$= (18.7)(0.5 \times 6) = 56.1''\ k$$

$$M_1/-M_2 = -1.0 \text{ Single curvature (1.6.1)}$$

$$C_b = 1.75 + 1.05(-1.0) + 0.3(-1.0)^2 = 1.0$$

$$l_b = 169.5;\ r_T = 1.53;\ l_b/r_T = 111;\ S = 10$$

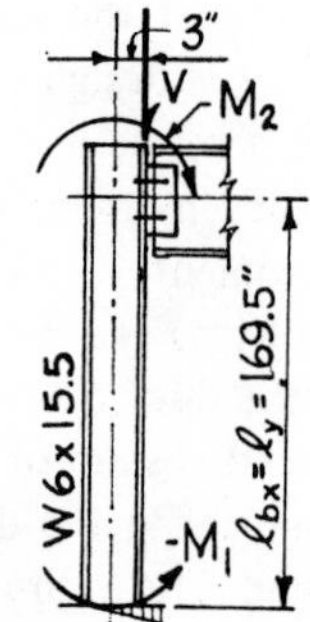

SKETCH C Exterior Column—Type 2—Braced.

Enter Table 3-1. $L_c = 6.32$ ft. $< L_b$

$$53\sqrt{C_b} < 104 < 119\sqrt{C_b}$$

Column (4): Read 2.656; Col. (1): 19.7; Col. (3): 18.0

$$M_R = 19.7 - \frac{(2.65)(169.5)^2}{10{,}000\,(1.0)} \leqslant 18.0 \text{ k-ft.}$$

$$M_R = 12.1 < 18.0 \text{ k-ft.}$$

Column (5): $M_R = \dfrac{(1{,}000)\,(2.68)\,(1.0)}{169.5} = 15.8 < 18$ k-ft.

$M_R = 15.8 \times 12 = 190$ k-in.

$f_a = 23.4/4.56 = 5.13$ ksi; $F_a = 10.6$ ksi.

Enter Fig. 6-6 with $M_1/M_2 = -1.0$, read
$C_m = 0.6 - 0.4\,(-1.0) = 1.0$
$K = 1.0$; $Kl_b/r_b = 169.5/2.57 = 66$. Enter Fig. 6-7, read 0.85;
$M_s = 1.0/0.85 = 1.18$

Eq. 1.6-1a: $5.13/10.6 + (1.18)\,(56.1)/190 = 0.82 < 1.0$ OK.

Eq. 1.6-1b: $5.13/21.6 + (56.1)/190 = 0.533 < 1.0$ OK.

EXAMPLE 4. (See Fig. 2-5(c), *Wind Connections at Top of Columns only.*)

Column $W8 \times 24$; beam $W21 \times 49$; $F_y = 36$ ksi.

Exterior Column. Shear connection, $e = 0.5 \times d = 4''$; $V = 18.7\ k$. Assume $M_1 = 0$

$(D + L)$ loading. $P = 23.4\ k$; $-M_2 = (18.7)\,(4) = 74.8$ k-in.
Single curvature, $M_1/M_2 = 0$; $C_b = 1.75$; $K = 2.0$ $l_b = 180 - 0.5\ (21) = 169.5''$; $r_T = 1.78$; $r_x = 3.42$; $r_y = 1.61$

$l_b/r_T = 169.5/1.78 = 95$; $A = 7.06$; $S_{xx} = 20.8$

Enter Table 3-1. $53\sqrt{C_b} < 95 < 119\sqrt{C_b}$; use columns (4) and (5). Read Col. (1): 41.1; Col. (3): 37.4; Col. (4): 4.633; and Col. (5): 6.78.

Col. (4): $M_R = 41.1 - \dfrac{(4.633)\,(169.5)^2}{10{,}000\,(1.75)} = 33.5 < 37.4$

Col. (5): $M_R = (1{,}000)\,(6.78)\,(1.75)/169.5 = 70 > 37.4$

Use $37.4 \times 12 = 449$ k-in.

$K_y = 1.0$; $K_x = 2.0$; $(K_x l_x/r_x) = 99$; $(K_y l_y/r_y) = 105$

Enter Fig. 6-5 with $Kl/r = 169.5/1.61 = 105$; read $F_a = 12.3$ ksi. (From Example 3, $C_m = 1.0$.) Calculate $f_a = 23.4/7.06 = 3.31$ ksi. Enter Fig. 6-7, with $f_a = 3.31$ and $Kl_b/r_b = 169.5/3.42 = 50$, read 0.95; calculate $M_s = 1.0/0.95 = 1.05$.

$$Eq.\ 1.6\text{-}1a:\ f_a/F_a + M_s(M_2/M_R) = \frac{3.31}{12.3} + (1.05)\,\frac{74.8}{449} \leqslant 1.0$$

$$= 0.443 < 1.0 \text{ OK}$$

$$Eq.\ 1.6\text{-}1b:\ \frac{3.31}{(0.6)\,(36)} + \frac{74.8}{449} = 0.153 + 0.167 \leqslant 1.0$$

$$= 0.320 < 1.0 \text{ OK.}$$

$(D + L + W)$ loading. $P = 24.3\ k$; $M_2 = 184$ k-in; $M_1 = 0$ k-in. $\Delta = 0.84''$;

Single curvature $M_1/M_2 = 0$

$C_b = 1.75$

Enter Table 3-1. $53\sqrt{C_b} = 70 < 95 < 119\sqrt{C_b} = 157$; use (4) and (5).

Col. (4): $M_R = 41.4 - \dfrac{(4.633)(169.5)^2}{10{,}000\,(1.75)} = 33.8 < 37.4$ k-ft.

Col. (5): $M_R = (1{,}000)(6.78)(1.75)/(169.5)$

$= 70 > 37.4$ k-ft.

Use $37.4 \times 12 = 449$ k-in.

Eq. 1.6-1a:

Also, $C_m = 0.85$; $F_a = 12.3$ ksi.

Calculate $f_a = 24.3/7.06 = 3.44$. Enter Fig. 6-7 with $f_a = 3.44$ and $Kl_b/r_b = 99$, read 0.77.*

Calculate $M_s = 0.85/0.77 = 1.10$

$$\frac{3.44}{12.3} + (1.10)\frac{184}{449} = 0.730 < 1.333 \text{ OK} \ldots\ldots \text{(Eq. 1.6-1a)}$$

$$\frac{3.44}{(0.6)(36)} + \frac{184}{449} = 0.159 + 0.410 = 0.569 < 1.333 \text{ OK} \ldots\ldots \text{(Eq. 1.6-1b)}$$

Stability: $P(\Delta) = (24.3)(0.84) = 20$ k-in. - negligible.

Interior Column. Check $(D + L + W)$ loading condition only. $W8 \times 24$; $P = 47.3\ k$; $M_1 = 0$; $M_2 = 200$ k-in; $\Delta = 0.84''$.

$C_b = 1.75 + 0 + 0 = 1.75$; $f_a = 47.3/7.06 = 6.70$ ksi;
$F_a = 12.3$ ksi; for $C_b = 1.0, M_{Rx} = 449$ k-in;
$C_m = 0.85$.

Enter Fig. 6-7, $Kl_b/r_b = 99$; $f_a = 6.70$ ksi, read 0.56.** Calculate $M_{sx} = 0.85/0.56 = 1.52$

Eq. 1.6-1a: $\dfrac{6.70}{12.3} + (1.52)\dfrac{200}{449} = 1.22 < 1.333 \ldots\ldots$ OK.

Eq. 1.6-1b: $C_b = 1.75$; use columns (4) and (5) in Table 3-1; from the previous calculation, $M_{Rx} = 449$ k-in.

$$\frac{6.70}{(0.6)(36)} + \frac{200}{449} = 0.310 + 0.445 = 0.755 < 1.333 \text{ OK.}$$

Stability: $P(\Delta) = (47.3)(0.84) = 40$ - negligible.

EXAMPLE 5. (see Fig. 2-5(d) *Wind Connections Top and Bottom.*)

By inspection of Example (4), the wind moment connections for the base can be eliminated. It is also evident that the wind moment connections at the top could be eliminated if moment-resisting anchor-bolted bases are provided. The approximate upper-bound capacity required can be taken as

$$(0.25)(4.5\ k)(180'') = 203 \text{ k-in.}$$

---

*Conservatively neglect the minor correction to include a one-third increase in F_e' for $(W + D + L)\cdot$ $1.00 - 0.77 = 0.23$; $(0.75)(0.23) = 0.17$. The corrected value becomes $(1.00 - 0.17) = 0.83$; corrected $M_s = 1.02$. To save time, use the correction only when results are over the limit, 1.333.

**Conservatively neglect the minor correction to include a one-third increase in F_e' for $(W + D + L)$. $1.00 - 0.56 = 0.44$; $(0.75)(0.44) = 0.33$. The corrected value becomes $(1.00 - 0.33) = 0.67$; corrected $M_s = 1.233$. To save time, use the correction only when results are over the limit, 1.333.

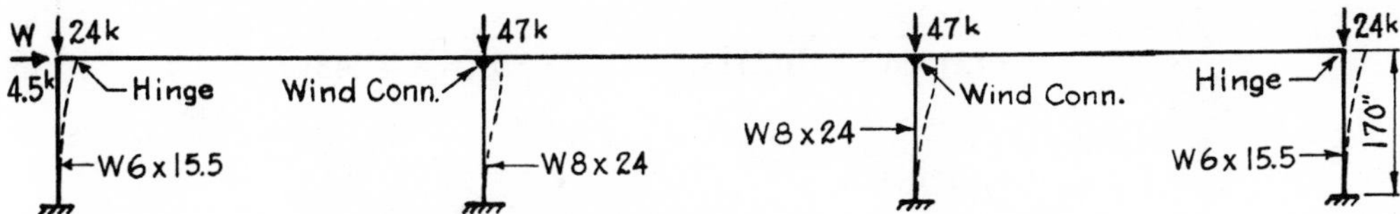

SKETCH D Mixed Sizes—Type 2 + Wind Connections.

The fixed base instead of the moment-connections at the top is structurally preferable for the improved stability and reduction in the temporary bracing required, during erection.

Additional Trials. Another possible refinement of the preliminary design is to reduce the size of the exterior columns for economy. Try $W6 \times 15.5$ for the exterior columns, fixed at the base and using simple shear connections at the top. The moment magnification factor will be computed for the entire structure so as to allow the surplus bracing capacity of the interior columns with fixed bases and wind moment top connections to brace the smaller exterior columns acting as cantilevers. A simplified manual analysis will serve to illustrate this approach. (See sketch (D).)

(2) Exterior columns: $I = 30.1$; $A = 4.56$; $r_b = 2.57$; $r_y = 1.46$

(2) Interior columns: $I = 82.5$; $A = 7.06$; $r_b = 3.42$; $r_y = 1.61$

Sum of the relative stiffnesses:

$$\Sigma(I/L) = 2 \times (30.1)/(170)\,(4.0) + 2 \times (82.5)/(170)\,(1.0)$$
$$= 1.06$$

Exterior columns:

$$\text{Wind moment}, M_2 = \frac{0.045}{1.06}(4.5\ k)\,(170'') = 32.4 \text{ k-in.}$$

Total $D + L + W$: $M_2 = 32.4 + 56.1 = 88.5$ k-in; $Kl_b/r_T = 111$; $K_{y\text{-}y} = 1.0$; $Kl/r = (1.0) \cdot (170)/1.46 = 116$; braced $F_a = 10.85$ ksi. (If base were fixed about y-y, $K_{y\text{-}y} = 0.8$)

$K_{x\text{-}x} = 2.1$; $Kl/r = (2.1)\,(170)/2.57 = 139$; for the story average with sidesway, $F'_e = 7.73$; as a braced column, $Kl_b/r_b = (0.8)\,(170)/2.57 = 53$, $F'_e = 53.16$.

Interior columns:

$$\text{Wind moment}, M_1 = M_2 = \frac{0.485}{1.06}(4.5)\,(0.5 \times 170)$$
$$= 175 \text{ k-in.}$$

Total $D + L + W$: $M_2 = 175$ k-in.

$K_{y\text{-}y} = 1.0$; $Kl/r = 170/1.61 = 106$; $F_a = 12.20$ ksi
$K_{x\text{-}x} = 1.2$, $Kl_b/r_b = (1.2)\,(170)/3.42 = 60$; $F_a = 17.43$;
$F'_e = 41.48$.

Secondary moment magnifier for the story: sidesway $C_m = 0.85$.

$$M_s = \frac{C_m}{1 - \Sigma P/\Sigma(AF'_e)}$$

$$= \frac{0.85}{1 - (2)(24 + 47)/(2)(4.56 \times 7.73 + 7.06 \times 41.48)}$$

$$= \frac{0.85}{1 - 142/656} = 1.085$$

Check for loading condition $(D + L + W)$ only.

Exterior column: $P = 24$ k; $M_1 = 56.1$ k-in.; $M_2 = 88.5$ k-in.; $f_a = 24/4.56 = 5.26$ ksi; $F_a = 10.85$ ksi; $C_b = 1.0$; from Example 2, $M_R = 190$ k-in.

Eq. 1.6-1a: $5.26/10.85 + (1.085)(88.5)/190 = 0.99$

< 1.33 OK.

$$C_b = 1.75 + 1.05(-56.1/88.5) + 0.3(-56.1/88.5)^2 \leqslant 2.3$$

$$= 1.20$$

Enter Table 3-1. $53.3\sqrt{C_b} = (58.3)\sqrt{1.2} = 58.4 < Kl_b/r_T = 111$

Use columns (4) and (5). From Column (3), read $M_R = 18$ k-ft. $= 216$ k-in.

Column (5), read: $M_R = (1000)(2.68)(1.2)/170 = 18.9$

> 18.0

Column (4), $M_R < 18.0$ k-ft. (Example 2). Use 216 k-in.

Eq. 1.6-1b: $\dfrac{5.26}{(0.6)(36)} + \dfrac{88.5}{216} = 0.243 + 0.410 = 0.653 < 1.33$ ...... OK.

Interior Column. W8 × 24; $f_a = 47/7.06 = 6.65$ ksi; from the previous calculation, $F_a = 12.33$ ksi; $M_1 = M_2 = 175$ k-in.

Eq. 1.6-1a: Previous calculations; $C_b = 1.0$;

$M_R = 449$ k-in.

$$\frac{6.65}{12.20} + (1.085)\frac{175}{449} = 0.968 < 1.33 \text{ ...... OK.}$$

Eq. 1.6-1b: $C_b = 1.75 + 1.05 + 0.3 = 2.83 > 2.3$.

Use 2.3.

Enter Table 3-1; $53.3\sqrt{2.3} = 81 < Kl_b/r_T = 96$; use columns (4) and (5). Read in column (3), $M_R = 37.4 \times 12 = 449$ k-in.

Column (4): $M_R = 41.1 - \dfrac{(4.633)(170)^2}{10{,}000(2.3)} = 35.2 < 37.4$ k-ft.

Column (5): $M_R = (1000)(6.78)(2.3)/170$

$= 91.7 > 37.4$ k-ft. Use 449 k-in.

$$\frac{6.65}{(0.6)(36)} + \frac{175}{449} = 0.698 < 1.33 \text{ ...... OK.}$$

It will be noted that additional trials become less time-consuming as values from earlier trials can be picked up and utilized, usually requiring merely a check of the footnotes to Table 3-1 (or Table 3-2 for $F_y = 50$ ksi) to ensure that the same formulas apply.

EXAMPLE 6. (See Fig. 2-5(e).) *Braced Cantilever-Suspended Slab System.*

Unbraced length = 180" - 10" = 170". All columns are $W6 \times 15.5$; $F_y = 36$ ksi; shear-connected top and at base.

$(D + L)$ loading condition only; see Table 2-2(b)

Exterior Columns. $P = 20.7k$; $V_1 = (1.7)(9.4) = 16k$; single curvature.

$$M_1 = -M_2 = (16)(0.5 \times 6") = 48"k; A = 4.56; I_x = 30.1;$$

$$S_x = 10; K_{yy} = 1.0; r_y = 1.46; Kl/r = (1.0)(170)/1.46 = 116.$$

$$K_{xx} = 1.0; r_x = 2.57; Kl_b/r_b = (1)(170)/2.57 = 66;$$

$$l_b/r_T = 111.$$

Enter Fig. 6-5, $F_a = 10.85$; $f_a = 20.7/4.56 = 4.54$ ksi. $f_a/F_a = 4.54/10.85 = 0.418 \geqslant 0.15$; Use Eqs. 1.6-1a and 1.6-1b . . . . . . (1.6.1)

Enter Fig. 6-6, $C_m = 1.0$; enter Fig. 6-7; for $f_a = 4.54$ ksi; read 0.86. $M_s = C_m/0.86 = 1.16$.

Eq. 1.6-1a: Enter Table 3-1: use $C_b = 1.0$. $53.3\sqrt{C_b} = 53.3 < l_b/r_b = 111 < 119\sqrt{C_b} = 119$. Use columns (4) and (5), the larger value of M_R but not to exceed column (3) = $18 \times 12 = 216$ k-in.

$$\text{Column (4): } M_R = 19.7 - \frac{(2.656)(170)^2}{(10{,}000)(1.0)} = 12.0$$

$$< 18.0 \text{ k-ft.}$$

$$\text{Column (5): } M_R = (1000)(2.68)(1.0)/170 = 15.76$$

$$< 18.0 \text{ k-ft.}$$

Use $15.76 \times 12 = 189$ k-in.

$$\frac{4.54}{10.85} + \frac{(1.16)(48)}{189} = 0.418 + 0.295 = 0.713 < 1.00 \text{ . . . . . . OK.}$$

Eq. 1.6-1b: Enter Table 3-1:

$$C_b = 1.75 + 1.05(-1) + 0.3(-1)^2 = 1.0$$

$$\frac{4.54}{(0.6)(36)} + \frac{48}{189} = 0.210 + 0.254 = 0.464 < 1.00 \text{ . . . . . . OK.}$$

Interior Columns. $P = 49.6\,k$; $M_1 = M_2 = 0$; max. $Kl/r = 116$; $F_a = 10.85$ ksi; $f_a = 49.6/4.56 = 10.88$ ksi; same M_R as exterior columns.

$$\textit{Eq. 1.6-1a:} \quad \frac{10.88}{10.85} = 1.002 \approx 1.00 \text{ . . . . . . OK.}$$

EXAMPLE 7. (See Fig. 2-5(f), *Semi-rigid, Type 3, with Sidesway.*)

All columns are $W8 \times 24$; $F_y = 36$ ksi; with unbraced length $l_{yy} = 170"$; $l_{xx} = 170"$. $A = 7.06$; $I_x = 82.5$; $S_x = 20.8$; $r_x = 3.42$; $r_y = 1.61$; $r_T = 1.78$.

Exterior Columns. $(D + L)$ loading condition
$P = 21.4k$; $M_1 = +150$ k-in.; $M_2 = +300$ k-in.; $M_1/M_2 = +0.5$, double curvature. For K_{yy}: see Fig. 6-4(a), estimate

$$G_{bot} = 1.0, \text{ for fixed base;}$$

$$G_{top} = \frac{\Sigma k \text{ for } W8 \times 24/14.1}{\Sigma k \text{ for 20H36 joist}/25} = 1/2; \text{ use } K_{yy} = 0.7$$

for bending about minor axis, $Kl/r = (0.7)(170)/1.61 = 74$. K_{xx}: see Fig. 6-4(b), $G_{bot} = 1.0$; $G_{top} = 300/220 = 1.4$, estimated from the ratio of the moments; use $K_{xx} = 1.4$.

$$Kl_b/r_b = (1.4)(170)/3.42 = 70; \; l_b/r_T = 96$$

Enter Fig. 6-5 with $Kl/r = 74$, $F_a = 15.9$ ksi. Calculate $f_a = 21.4/7.06 = 3.03$. $f_a/F_a = 0.19 > 0.15$. Use Eqs. 1.6-1a and 1.6-1b. Enter Fig. 6-7, read 0.9; take $C_m = 0.85$; $M_s = 0.85/0.9 = 0.94$. Enter Table 6-2 with $M_1/M_2 = 0.5$, $C_b = 2.3$.

Enter Table 3-1. $53.3\sqrt{2.3} = 81 < 96 < 119\sqrt{2.3} = 181$, use columns (4) and (5).

$$\text{Column (4): } M_R = 41.1 - \frac{(4.633)(170)^2}{(10{,}000)(2.3)} = 35.3 < 37.4 \text{ k-ft.}$$

$$\text{Column (5): } M_R = (1000)(6.78)(2.3)/170 = 91.7 > 37.4.$$

Use $37.4 \times 12 = 449$ k-in.

Eq. 1.6-1b:

$$\frac{3.03}{(0.6)(36)} + \frac{300}{449} = 0.140 + 0.668 = 0.81 < 1.0 \ldots\ldots \text{OK.}$$

Eq. 1.6-1a: $C_b = 1.0$; use columns (4) and (5).

$$\text{Column (4): } M_R = 41.1 - \frac{(4.633)(170)^2}{(10{,}000)(1.0)} = 27.7 < 37.4 \text{ k-ft.}$$

$$\text{Column (5): } M_R = (1000)(6.78)(1.0)/170 = 39.8 > 37.4 \text{ k-ft.}$$

Use 449 k-in.

$$\frac{3.03}{15.9} + \frac{(0.94)(300)}{449} = 0.191 + 0.628 = 0.819 < 1.0 \ldots\ldots \text{OK.}$$

$(D + L + W)$ loading condition

$P = 21.4k$; $M_1 = 300$ k-in.; $M_2 = +306$ k-in.; $M_1/M_2 = 1$. See Table 6-2. $C_b = 2.3$. Enter Table 3-1.

$$53.3\sqrt{2.3} = 81 < l_b/r_T = 96 < 119\sqrt{2.3} = 181.$$

Eq. 1.6-1a: Use $C_b = 1.0$; use columns (4) and (5), previously calculated.

$$0.191 + (0.94)(306)/449 = 0.191 + 0.641 = 0.832 < 1.33 \ldots\ldots \text{OK.}$$

Eq. 1.6-1b: Use $C_b = 2.3$; columns (4) and (5); use previous values. $0.140 + 306/449 = 0.822 < 1.33$ OK.

Interior Column. $(D + L)$ loading condition

$P = 39.4\,k; M_1 = +23$ k-in.; $M_2 = +45$ k-in.;
$M_1/M_2 = +0.5$, same as exterior column with $(D + L)$;
$C_b = 2.3$; use same $F_a = 15.9$ ksi; $f_a = 39.4/7.06 = 5.58$ ksi;
K_{xx} = see Fig. 6-4(b), $G_A = 1.0$; take $G_B = 20$; $K_{xx} = 2.1$;
$Kl_b/r_b = (2.1)(171)/3.42 = 105$; $l_b/r_T = 96$; $C_m = 0.85$;
$M_s = 0.94$.

Eq. 1.6-1a: $C_b = 1.0$

$$\frac{5.58}{15.9} + \frac{(0.94)(45)}{449} = 0.351 + 0.094 = 0.445 < 1.0 \ldots\ldots \text{OK.}$$

Eq. 1.6-1b: $C_b = 2.3$

$$\frac{5.58}{(0.6)(36)} + \frac{45}{449} = 0.258 + 0.100 = 0.358 < 1.0 \ldots\ldots \text{OK.}$$

$(D + L + W)$ loading condition

$P = 48.8\,k; M_1 = +80$ k-in.; $M_2 = +123$ k-in.;
$f_a = 48.8/7.06 = 6.91$ ksi

Eq. 1.6-1a:

$$\frac{6.91}{15.9} + \frac{(0.94)(123)}{449} = 0.435 + 0.257 = 0.692 < 1.33 \ldots\ldots \text{OK.}$$

Final Condition of Loading: $P = 48.8\,k$; $M_1 = M_2 = 0$. OK by inspection of preceding calculations.

Stability of Structure

After hinges develop (see Fig. 2-5f), the structure as a whole is still in the elastic range. (See Sketch (E).) An approximate upper-bound calculation of the drift under wind load only can be based upon the deflection of the leeward column as a cantilever.

$$\Delta = \frac{P(l)^3}{12EI} = \frac{(0.9)(171)^3}{(12)(29 \times 10^3)(82.5)} = 0.154''$$

The sidesway moment, $P\Delta = (49)(0.154) = 7.5$ k-in. is negligible compared to the wind moments.

An increase in the lateral force will continue deflection in the elastic range up to the connection capacity at the tops of the exterior columns; $(300 - 80) = 220$ k-in. at the windward column and 300 k-in. at the leeward column. The approximate surplus elastic resistance to lateral force can be estimated as $(220 + 300)/170 = 3\,k$. This force represents an additional 66.6 percent times the specified wind force.

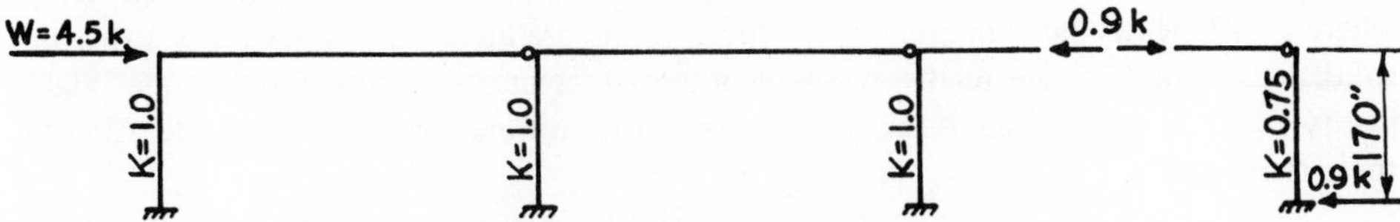

SKETCH E Type 3–Stability After Hinges Form.

The distribution of the surplus stability to brace the lighter or shear-connected columns or to reduce column size for some of the columns with Type 3 design is complicated by the changes in the effective length factors, K, that develop as the behavior at a connection changes from elastic to plastic.* It is feasible for estimates to compute average moment magnification factors for the structure with no hinges or with the maximum number of hinges before instability, as upper and lower bound stability factors. In this example, consider all columns cantilevered from fixed based (hinged at the top). $G_{top} = \infty$; $G_{bot} = 1.0$; (see Fig. 6-4(b)), $K = 2.4$. $Kl_b/r_b = (2.4)(171)/3.42 = 120$; $F'_e = 10.37$ ksi.

$$M_s = \frac{0.85}{1 - \Sigma P/\Sigma AF'_e} = \frac{0.85}{1 - f_a/F'_e}, \text{ where } f_a = \Sigma P/nA$$

$$f_a = (2)(21.4 + 48.8)/(4)(7.06) = 4.97 \text{ ksi}$$

$$M_s = \frac{0.85}{1 - 4.97/10.37} = 1.63 \text{ (upper bound)}$$

By inspection, the exterior columns are the more nearly loaded to the allowable limit than the interior columns. Substituting the story factor into Eq. 1.6-1a for the exterior columns:

$$0.191 + (1.63)(0.641) = 1.24 < 1.33 \ldots\ldots \text{OK}$$

Thus, the structure is shown to possess surplus lateral bracing capacity at the specified loads and stability under higher lateral loads even after the top connections become hinges. An upper bound solution for the drift, Δ, can likewise be simply estimated under 1.66 times the specified lateral loading as:

$$\Delta = \frac{(1.66)(4.5)}{(4)(0.9)} (0.154'') = 0.32''$$

$$0.32/170 = 0.0019 < 0.0025 \text{ considered acceptable.}$$

SUMMARY—ONE STORY COLUMN DESIGN EXAMPLES

General

Examples one through six illustrate applications of allowable stress design for columns in a one story structure. It has been demonstrated for this typical one story structure that refinements of analysis and design for the consideration of effects other than moment upon deformation are unnecessary to ensure safety. The allowances for safety built into the AISC Specifications to provide for secondary effects (shear, differential axial length changes, and drift) are generally more than adequate for safety. Where a more refined analysis and design are attempted in an effort for economy, tonnage savings, particularly in the light column and beam sizes utilized in these examples, are frequently limited by the available rolled shapes in a given depth series. The depth, of course, may have to be maintained for stiffness and the next smaller depth as from 8 in. to 6 in. or 6 in. to 5 in. is too large a reduction to enable achievement of a weight saving. Only where a sufficiently large number of identical frames are involved in a one story structure so that the total of the relatively small potential savings in tonnage per frame might justify the time for an elaborate analysis, would the authors recommend consideration of these refinements.

---

*References (7) and (9).

Pattern Live Loading

Although pattern live loading on the roof is a matter of concern for the design of the flexural members (ponding), it will be noted that generally the load conditions critical for the column design were full live and dead load or full live, dead, and wind loads. See Chapter 3—Beams, for examples of the roof beam design.

MULTI-STORY COLUMN DESIGN

General Considerations

Splices. Several design considerations absent or not significant in one or two-story construction become important for multi-story columns. Shipping limitations and erection convenience limit the practical length of column elements. The cost of column splices includes material and expensive field labor. In the 25 story structure for Design Example 2, Chapter 2, these conflicting requirements were satisfied by the use of two-story lengths with field splices assumed just above alternate floor levels.

Size. Another, less important consideration is the continuity of the size selected. Column sections with matching depths are somewhat more conveniently spliced than if the depths are greatly unequal. Architectural details for the fit of other materials are also simplified if the depth of a multi-story column can be maintained. For this example, this consideration was solved by the use of $W14$ shapes for which the widest choice of weights is available. The following design examples reviewing the preliminary selections used for the analyses show a definite excess capacity for the upper story columns. Possible savings in steel weight here, however, tend to be offset by the added cost of splices, fit of other materials, etc. The decision to retain the $W14$ size or to seek savings from tonnage reduction is partly a matter of judgment, as well as availability of lighter shapes with a lesser depth and sufficient stiffness for the unbraced lengths used.

Lateral Forces. The lateral force loading condition, $(W + D + L)$, becomes far more important, of course, as the height increases. This effect raises many considerations not significant in the one story examples where the one-third increase in the allowable stress usually provides sufficient capacity (1.5.6). The requirements for stability (1.8.1, 1.8.2, and 1.8.3) though brief, require far more consideration than in the one story examples.

The multi-story example herein is intended to illustrate the cost (in added tonnage) of a frame with sidesway not prevented (1.8.3) where drift is limited to $0.0025H$. If no walls can be utilized for bracing, the cost of this added tonnage may be compared to the cost of diagonal bracing. If walls were to be employed only for shear bracing, the cost of such walls including the time added to the construction schedule usually is excessive compared to either diagonal bracing or added tonnage and connection cost in a rigid frame.

Secondary Stress Effects. As the total height, story height, and number of stories increase, the secondary stress effects become more significant. When considering lateral displacement, the problem is complicated by the number of variables involving both structural properties and human response. The absolute magnitude of lateral displacement is too important to rely upon the $0.0025H$ computed limit applied to a "standard analysis" regardless of height. A prudent approach is to evaluate secondary stress effects with increasing refinements as height increases. The authors believe that, *for the particular conditions of the example selected*, the secondary stress effects are significant, having reached a magnitude above which dependence upon the $0.0025H$ rule-of-thumb with "standard analyses" is not justifiable.

MULTI-STORY COLUMN DESIGN

EXAMPLE 8.

Check the design of the columns shown for the 24th and 25th story levels in analysis Example 2 (Chapter 2). (See Fig. 2-6.) Use the analysis which includes the effects of shear and differential axial length changes of columns. Neglect the secondary ($P\Delta$) effects, drift and sidesway. For analysis data see Fig. 2-9. For the $(D + L)$ loading condition deduct moments for (W) only from those for $(W + L + D)$; for axial loads, see the $(D + L)$ output for columns.

Step 1. Design data

	Exterior Column	*Interior Column*
Size	$W14 \times 43$, Grade 50	$W14 \times 43$, Grade 50
l_b	144" - 21" = 123"	123"
25th	$P = 33.3\ k;\ f_a = 2.63$ ksi	$P = 56.1\ k;\ f_a = 4.43$ ksi
24th	$P = 70.6\ k;\ f_a = 5.58$ ksi	$P = 117.3\ k;\ f_a = 9.27$ ksi
M_y	0 k-ft.	0 k-ft.
K_y	1.0	1.0
M_x (25)	M_2 = 109 k-ft.; M_1 = 82.5 k-ft.	M_2 = 19.2 k-ft.; M_1 = 18.4 k-ft.
(24)	M_2 = 67.0 k-ft.; M_1 = 61.8 k-ft.	M_2 = 18.7 k-ft.; M_1 = 17.4 k-ft.

Step 2. $W14 \times 43;\ r_y = 1.89;\ I_x = 429;\ r_x = 5.82$

	Exterior Columns	*Interior Columns*
Beam	$W21 \times 44$	$W21 \times 44$
I_x	843	843
span	25 ft.	25 ft.
$k = I/L$	843/25 = 33.72	(2) (843)/25 = 67.44

Step 3. Effective Length of Column

$$G_{top} = G_{bot} = \frac{(2)(429)/12}{33.72} = 2.12 \text{ for the exterior column (24th floor)}$$

$G_{top} = 1.06; G_{bot} = 2.12$ for the exterior column (25th floor)

$G_{top} = G_{bot} = 1.06$ for the interior column (24th floor)

$G_{top} = 0.53; G_{bot} = 1.06$ for the interior column (25th floor)

(See Fig. 6-4.) Read for above data unbraced frames, $K_x = 1.6$ for the exterior column and $K_x = 1.33$ for the interior column (24th floor). $K_x = 1.5$ for the exterior column and 1.25 for the interior column (25th floor). Use $K_x = 1.6$ and 1.33.

Step 4. $K_b l_b / r_b$

K_y.	1.0	1.0
	$K_y l_b / r_b = (1.0)(123)/1.89 = 65.1$	
$K_x l_b / r_b$	$(1.6)(123)/5.82 = 33.8$ $< K_y$	$(1.33)(123)/5.82 = 28.1$ $< K_y$

Enter Fig. 6-5 with $K_b l_b / r_b = 65.1$; read $F_a = 21.8$ ksi

f_a / F_a	$5.6/21.8 = 0.257 > 0.15$	$9.3/21.8 = 0.426 > 0.15$

Step 5. (Eq. 1.6-1a)

Calculate M_1 / M_2 to determine C_b. (See Table 6-2.) Read for C_b

M_1 / M_2	$82.5/109 = 0.76$	$17.4/18.7 = 0.93$

Note that both cases, exterior and interior columns, are for reverse curvature bending, and so both top and bottom moments are taken as positive. From Table 6-2,

C_b	2.3	2.3

Enter Beam Selection Table 3-2, Grade 50 For $W14 \times 43$, $l_b = 123''$; $L_b = 10.25'$; and $C_b = 2.3$, read as follows: $L_c = 7.16' < L_b$; columns (1) and (2) do not apply. Compute $l_b / r_T = 123/2.14 = 57.5$.

For $C_b = 2.3, 45\sqrt{C_b} = 68.2 > 57.5$; for M_{Rx} use column (5) $\leqslant$ Col. (3).

Col. (5). $M_{Rx} = (1000)(19.36)(2.3)/123 \leqslant 156.7$ k-ft.
$= 362 \geqslant 156.7$ Use 156.7 k-ft.

Step 6. Moment Magnification Factor.

Enter Table 6-1 with f_a and $K_x l_{bx} / r_{bx}$ values and read C_m.

C_m	= 0.992 for ext.	= 0.991 for int.

Enter Fig. 6-7 with above values, and read 0.96 and 0.95 for exterior and interior columns. Compute the moment magnification factors for secondary moments

M_{sx}	$= 0.992/0.96 = 1.04$	$= 0.991/0.95 = 1.04$

Step 7. Solve AISC Eq. 1.6-1a $M_y = 0$

$$f_a / F_a + M_{sx}(M_2 / M_R) = 1.00$$

Exterior column: $0.257 + (1.04)(67.0/156.7) = 0.70 \leqslant 1.00$ OK.
Interior column: $0.426 + (1.04)(18.7/156.7) = 0.55 \leqslant 1.00$ OK.

Solve AISC Eq. 1.6.2 for exterior column (25th floor) since $f_a / F_a = 2.63/21.8 = 0.121 < 0.15$

Exterior column: $0.121 + (109)/156.7 = 0.817 < 1.00$. OK.

Step 8. Solve AISC Eq. 1.6-1b $M_y = 0$. Check both columns at 24th floor.

$45\sqrt{2.3} = 68.2 > l_b/r_T = 57.5$. Enter Table 3-2 for resisting moments. The resisting moment will be calculated from the moment factor (*MF*) of column (5), but not to exceed M_{Rx} of column (3).

$$M_{Rx} = 1000\,(MF)C_b/l_b \leqslant 156.7 \text{ k-ft}$$

$$= (1000)\,(19.36)\,(2.3)/123$$

$$= 362 \text{ k-ft.} \geqslant 156.7 \text{ k-ft. Use } 156.7 \text{ k-ft.}$$

Eq. 1.6-1b $f_a/0.6\,F_y + M_2/M_{Rx} \leqslant 1.0\ (M_y = 0)$

	$5.6/30 + 67.0/156.7$ $= 0.614 \leqslant 1.0$ OK.	$9.3/30 + 18.5/156.7$ $= 0.428 \leqslant 1.0$ OK.

The capacity for both columns is adequate for $(D + L)$ loads.

(D + L + W) Loading

Step 1. Design Data

24th	$P = 70.6 + 1.4 = 72\ k$	$P = 117.3 + 0.3 = 117.6\ k$
25th	$M_2 = 109$ k-ft.; $M_1 = 80.3$ k-ft.	$M_2 = 28.8$ k-ft.; $M_1 = 26.1$ k-ft.
24th	$M_2 = 70.6$ k-ft.; $M_1 = 70.6$ k-ft.	$M_2 = 37.3$ k-ft.; $M_1 = 32.5$ k-ft.

Steps 2 through 8.

By inspection the increase in both *P* and *M* for both columns due to the wind load effect are less than the increase in allowable stress, 33 percent, applied to Eqs. 1.6.

EXAMPLE 9.

Note that the next two-story section of the columns of Example 7 are the same size, *W*14 X 43, in Fig. 2-6. Repeat the review of Example 7 at the 22nd floor level to determine whether the preliminary selections are adequate.

Step 1. Design data.

	Exterior Columns	*Interior Columns*
$(D + L)$	$M_y = 0$ $P = 140\ k$	$M_y = 0$ $P = 238.5\ k$
22nd fl.	$f_a = 11.11$ ksi $M_2 = 66.5$ k-ft., $M_1 = 64.9$ k-ft.	$f_a = 18.93$ ksi $M_2 = 15.1$ k-ft.; $M_1 = 13.7$ k-ft.
$D + L + W$	$M_2 = 86.8$ k-ft.; $M_1 = 81.4$ k-ft. $P = 145\ k; f_a = 11.51$ ksi	$M_2 = 52.9$ k-ft.; $M_1 = 46.7$ k-ft. $P = 245\ k; f_a = 19.45$ ksi

Steps 2, 3, and 4. From Example 7

	$l_{by} = l_{bx} = 123''$	$l_{by} = l_{bx} = 123''$
x-x	$K_b l_b/r_b = 33.8$	$K_b l_b/r_b = 28.1$
y-y	$K_b l_b/r_b = 65.1$	$K_b l_b/r_b = 65.1$
	$F_a = 21.8$ ksi	$F_a = 21.8$ ksi

Step 4.

f_a/F_a $(D+L)$	$11.11/21.8 = 0.510$	$18.93/21.8 = 0.868$
f_a/F_a $(D+L+W)$	$11.51/21.8 = 0.528$	$19.45/21.8 = 0.892$

Step 5. (Eq. 1.6-1a) Enter Table 3-2, with $C_b = 2.3$, and read as in Example 7, that M_{Rx} is the value computed from column (5), but not larger than that in column (3)

	$M_{Rx} = 156.7$ k-ft.	$M_{Rx} = 156.7$ k-ft.

Step 6. Enter Table 6-1 with f_a and Kl/r about x-x; read

	$C_m = 0.985$	$C_m = 0.974$

Enter Fig. 6-7 with these values; read 0.92 and 0.86.

M_{sx}	$= 0.985/0.92 = 1.07$	$= 0.974/0.86 = 1.13$

Step 7. (Eq. 1.6-1b) $f_a/F_a + M_{sx}(M_2/M_{Rx}) + M_{sy}(M_2/M_{Ry}) \leqslant 1.00$

$(D+L)$	$0.510 + 1.07\,(66.5/156.7)$ $= 0.967 \leqslant 1.00$ OK.	$0.868 + 1.13\,(15.1/156.7)$ $= 0.980 \leqslant 1.00$ OK.
$(D+L+W)$	$0.538 + 1.07\,(86.8/156.7)$ $= 1.131 \leqslant 1.333$ OK.	$0.892 + 1.13\,(52.9/156.7)$ $= 1.273 \leqslant 1.333$ OK.

Step 8. (Eq. 1.6-1b) $C_b = 2.3$ and $M_{Rx} = 156.7$ k-ft. from Example 7. $f_a/0.6F_y + (M_2/M_{Rx}) + (M_2/M_{Ry}) \leqslant 1.00$

$(D+L)$	$11.11/30 + 66.5/156.7$ $= 0.795 \leqslant 1.00$ OK.	$18.93/30 + 15.1/156.7$ $= 0.728 \leqslant 1.00$ OK.
$(D+L+W)$	$11.51/30 + 86.8/156.7$ $= 0.937 \leqslant 1.333$ OK.	$19.45/30 + 52.9/156.7$ $= 0.986 \leqslant 1.333$ OK.

The preliminary selection, $W14 \times 43$, is adequate for both the exterior and interior columns down to the 22nd floor level.

EXAMPLE 10.

Check the design for the section shown as the exterior column in the first story of the high-rise example (Chapter 2). (See sketch (F).) Refer also to Fig. 2-6.

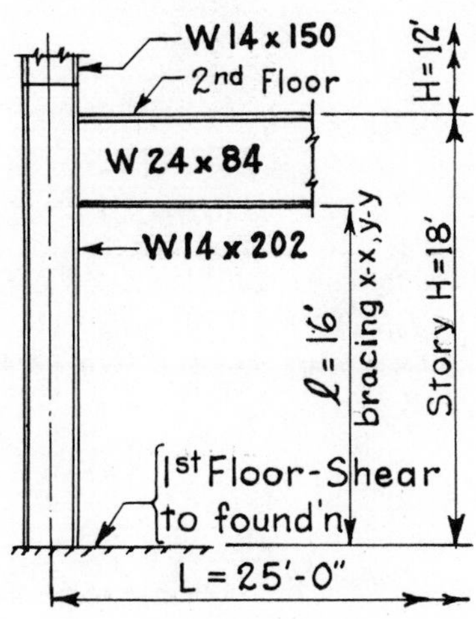

SKETCH F Exterior Column—1st Floor.

Use $C_m = 0.85$ and the AISC moment magnification factor, M_s (Fig. 6-6); otherwise neglect $P\Delta$ and axial shortening effects. Analysis is similar to Fig. 3-20.

Analyses data: $(D + L)\ P = 902\ k; M_2 = 44$ k-ft.; $M_1 = 24$ k-ft.
$(D + L + W)\ P = 1199\ k; M_2 = 358$ k-ft.; $M_1 = 197$ k-ft.
$M_1/M_2 > 0.5$

Sections and section properties:

Column. $W14 \times 202; A = 59.4; I_x = 2540; r_x = 6.54;$
$I_y = 980; r_y = 4.06; r_T = 4.44$
Beam. $W24 \times 84; I_x = 2370.$

Column above. $W14 \times 150; I_x = 1790$

Effective length factor, k; see Fig. 6-4:

$$\text{Top. } G_A = \frac{\dfrac{2540}{17 \times 12} + \dfrac{1790}{12 \times 12}}{\dfrac{2370}{(25 \times 12)}} = 3.1$$

Bottom. $G_B = 1.0$ for fixed base (Comm. 1.8, Fig. C1.8.2) Read from Fig. 6-4, $k_x = 1.59$; take $k_y = 1.00$ for braced but free to rotate)

Effective lengths, $k_x l_{bx}/r_x = (1.59)(16 \times 12)/6.54 = 46.7$
$k_y l_{by}/r_y = (1.00)(16 \times 12)/4.06 = 47.3$

Loading Condition $(D + L)$: $f_a = 902/59.4 = 15.18$ ksi; read Fig. 6-5, $F_a = 24.5$ ksi. For $M_1/M_2 > 0.55$, in Table 6-2 read $C_b = 2.3$.

See Table 3-2. For $l_b/r_T = (16 \times 12)/4.44 = 43.2$, $45.2\sqrt{C_b} > 43.2$. Enter column (3), read $M_R = 812.4$ k-ft.

$$f_a/F_a + C_m f_b/(1 - f_a/F_e')(F_b) \leqslant 1.00 \ldots\ldots \text{(Eq. 1.6-1a)}$$

See Fig. 6-7, for $kl/r = 46.7$ and $f_a = 15.18$ ksi, read

$$M_s = C_m/0.78 = 0.85/0.78 = 1.09 \text{ for } (D + L)$$

$$\frac{f_a}{F_a} + M_s \frac{M}{M_R} = \frac{15.18}{24.5} + \frac{(1.09)(44)}{812.4} = 0.62 + 0.06$$

$$= 0.68 < 1.00; \text{ OK.}$$

$$f_a/(0.6F_y) + f_b/F_b \leqslant 1.00 \ldots\ldots \text{(Eq. 1.6-1b)}$$

$$\frac{15.18}{(0.6)(50)} + \frac{44}{812.4} = 0.51 + 0.06 = 0.57 < 1.00; \text{ OK}$$

Loading condition $(D + L + W)$: $f_a = 1199/59.4 = 20.2$ ksi; in Fig. 6-7 read 0.70 for $(1 - f_a/F_e')$. $(1 - 0.70) = 0.30$; $(1 - 0.30 \times 0.75) = 0.775$ with $F_e'\ (\frac{4}{3})$. $M_s = 0.85/0.775 = 1.10$ secondary moment factor $(W + D + L)$

$$\frac{20.2}{24.5} + (1.10)\frac{358}{812.4} = 0.824 + 0.485 = 1.31 < 1.33 \ldots\ldots \text{OK.}$$

$$\frac{20.2}{(0.6)(50)} + \frac{358}{812.4} = 0.673 + 0.441 = 1.114 < 1.33; \text{ OK.}$$

EXAMPLE 11

Check the column of Example 10 using the variable value of C_m suggested as more accurate (Comm. 1.6.1). See Table 6-1. For the more critical loading condition, $(D + L + W)$, $f_a = 20.2$ ksi, read $C_m = 0.946$ by interpolation.

The revised Eq. 1.6-1a becomes:

$$0.824 + \frac{0.946}{0.85}(0.485) = 0.824 + 0.540 = 1.36 > 1.33$$

This conservative solution indicates that the column selected may be over-stressed approximately three percent, although the design has satisfied the arbitrary moment magnification factors provided by the Specifications.

EXAMPLE 12

Continue the investigation of the column section of Examples 10 and 11. In this example, the actual $P\Delta$ effect expected will be computed for comparison to the results of Examples 10 and 11. See the analyses data for the loading condition $(D + L + W)$ in Chapter 2. See pages 56 and 60.

$$P = 1199 + 332 - 297 = 1234\ k$$

$$M = 358 + 370 - 334 = 394 \text{ k-ft.}$$

It will be noted that both the vertical load and the moment increase with the effect of $P\Delta$.

$$f_a = 1234/59.4 = 20.8 \text{ ksi}$$

Eq. 1.6-1a for sections between supports, omitting the arbitrary M_s factor, and using the conservative C_m, becomes

$$\frac{20.8}{(24.5)} + \frac{(0.946)(394)}{812.4} = 0.849 + 0.459 = 1.31 < 1.33; \text{ OK.}$$

Eq. 1.6-1b becomes

$$\frac{20.8}{0.6\,(50)} + \frac{394}{812.4} = 0.693 + 0.485 = 1.17 < 1.33; \text{ OK.}$$

This comparison indicates that the arbitrary allowance for moment magnification provided in the AISC Specifications as applied in Example 10 was impressively realistic, yielding exactly the same answer as the computer solution including the $(P\Delta)$ effect. $(1.31 = 1.31)$. At least for this example and structures similar to, and within the limits of, this example the use of the variable factor for C_m $(C_m = 1 - 0.18 f_a/F_e')$ tabulated in Table 6-1 in the arbitrary moment magnification factor, M_s, of Fig. 6-7 seems to provide a convenient conservative solution to avoid the need of computing $P\Delta$ effects. Where computer solutions are not available or to select sections for input into computer analysis solutions, this procedure is recommended.

EXAMPLE 13

The effect of differential axial shortening can usually be neglected safely except for very highly stressed long columns with very stiff girders. For the longest columns, $F_y = 50$ ksi, of the 25-story structure example, the maximum effect of differential axial shorten-

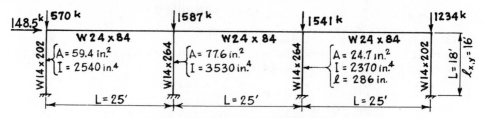

SKETCH G Axial Forces at First Floor

ing in the columns would be expected in the first story where columns are 18 ft. in height and the floor beams are the heaviest. This maximum effect will be evaluated by an approximate computation. (See sketch (G).)

Shortening of columns in the first story is

$$\Delta = \frac{P\,L}{A\,E} = f_a \frac{(18 \times 12)}{(29{,}000)} = (0.007448276)\, f_a$$

Exterior column: $f_a = 9.60$ ksi
Interior column: $f_a = 20.45$ ksi

Maximum differential shortening, in the exterior bay,

$$\Delta_m = (0.007448276)\,(20.45 - 9.60) = 0.081 \text{ in.}$$

Added shear in the second floor beam, V_Δ

$$V_\Delta = \Delta \frac{12\,EI}{(L)^3} = 0.081 \frac{(12)\,(29{,}000)\,(2370)}{(286)^3}$$

$$V_\Delta = 2.86\ k$$

Added fixed-end moment to second floor beam,

$$M_\Delta = \Delta \frac{6\,EI}{(L)^2} = \frac{(0.081)}{(12)} \frac{(6)\,(29{,}000 \times 2370)}{(286)^2}$$

$$M_\Delta = 34.0 \text{ k-ft.}$$

See Chapter 2, Analysis; minimum end moments in the beam due to wind are 336 k-ft. Thus, the effect of differential shortening increased moment 34.0/336 = 10 percent.

The effect of axial shortening is thus demonstrated to be negligible for this building, being about the same order of magnitude as a differential temperature of 50°F. The effect of the differential axial shortening in the beams causing a variable Δ for each column is likewise customarily neglected since the axial forces (due to the horizontal shears; i.e., wind) are small and are resisted in part by the diaphram effect of the usual floor system. (See Chapter 3, Beams. See also Chapter 2, Analysis.)

SELECTED REFERENCES—COLUMNS

1. "Effective Tier Length–Tier Buildings," T. R. Higgins; January, 1964; **1**, No. 1; *Engineering Journal*, AISC.
2. "Column Stability Under Elastic Support," T. R. Higgins, April, 1965; **2**, No. 2; *Engineering Journal*, AISC.

3. "Column Stability in Type 2 Construction," F. DeFalco and F. J. Marino; April, 1966, **3**, No. 2; *Engineering Journal*, AISC.
4. "The Effective Length of Columns in Unbraced Frames," J. A. Yura; January, 1971; **8**, No. 1; *Engineering Journal*, AISC.
5. "Aspects of Column Design in Tall Steel Buildings," J. Springfield and P. F. Adams; **98**, *No. ST5*; May, 1972, ASCE.
6. "Calculation of Effective Lengths and Effective Slenderness Ratios of Stepped Columns," J. P. Anderson and J. H. Woodward; October, 1972; **9**, No. 4; *Engineering Journal*, AISC.
7. "Inelastic *K*-Factor for Column Design," R. O. Disque; April, 1973; **10**, No. 2; *Engineering Journal*, AISC.
8. "Experiments on Laterally Loaded Steel Beam-Columns," G. W. English and P. F. Adams; **99**, *No. ST7*; July, 1973; ASCE.
9. "Stability of Steel Structures," J. A. Yura, paper presented at AISC Chicago meeting, No. 41, Sept. 6, 1974.

7

PLASTIC DESIGN

INTRODUCTION

Students and practitioners of structural design in steel have always been intellectually annoyed by the "wasted" material at points away from midspan in a shear-connected, rolled-section, simple beam. Inherent in production of steel shapes is the uniform bending capacity throughout the length for negative or positive bending. Similarly, the shape of the bending moment diagram for simple beams is inherently maximum near the midspan and zero at each end. This intellectual dissatisfaction led, first, to the use of continuity in steel design to reduce the midspan moment and to distribute moment to the ends. While the resultant tonnage reductions were gratifying, some residual dissatisfaction remained. As long as an elastic analysis is employed, both for the structure and the design of all sections, it is generally impossible to equalize the positive and negative moments, and the difference is still regarded as "wasted" (capacity). Type 2 Construction with "Wind Connections" and Type 3 Construction with "Semi-rigid" provide partial utilization, at least, with slight modifications of the elastic analysis (at joints only). (See Chapter 2, Design Example 1.)

The concept of plastic stress analysis (of sections) and plastic structural analysis of members in frames probably evolved from a desire for full utilization of the cross-section (including the web) and equalizing the positive and negative moments at the critical points in a member. The plastic structural analysis offers not only a simple solution, but the only realistic evaluation of the real safety factor against total collapse of a continuous frame. Since yield point stresses are involved in plastic behavior, the use of real or specified loads magnified by suitable factors becomes necessary to replace the "safety factor" applied to reduce the yield point for allowable stress design with the real loads. This device then logically allows application of the "factored" loads under which the structure will develop plastic behavior. One of the advantages of using the system of factored loads (strength design) is that separate load factors can be established to suit the probability of overload. The Specifications provide factors of 1.7 for gravity loads (dead plus live) and 1.3 for combined gravity plus wind or seismic load (2.1).

The concept of plastic design is not yet fully developed from research through the AISC

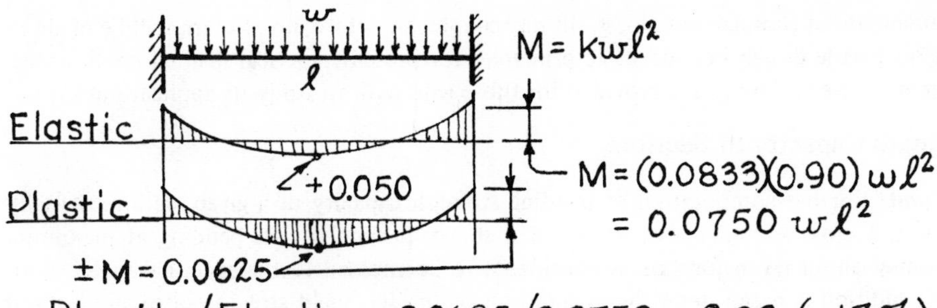

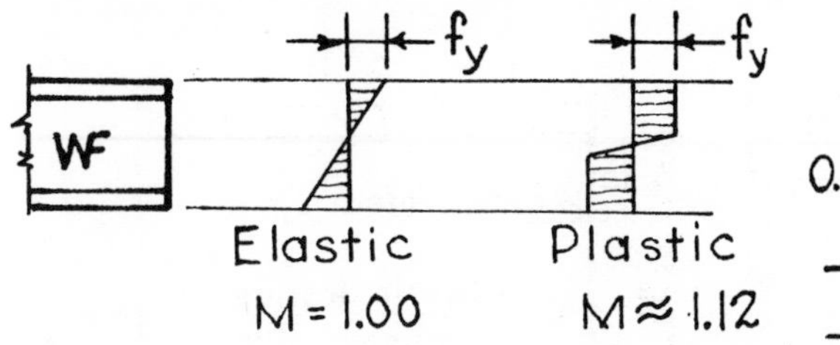

Service stress = $F_y/1.7$

Plastic = $0.59\ F_y$

$0.66 F_y \geqq$ elastic $\geqq 0.60\ F_y$

$-0.12 + (0.66 - 0.59) = -5\%$

$-0.12 + (0.60 - 0.59) = -11\%$

FIG. 7-1 Potential Weight Savings—Plastic vs. Elastic Design.

Specifications to computer analysis and design programs for general use. In considering the use of plastic design, its potential benefits are best limited to comparisons with elastic design for Type 1 Construction (rigid) where the connection requirements (and cost) are nearly identical. Where plastic design is permitted, the maximum potential economy in tonnage savings may be seen to approach seventeen percent by equalization of fully fixed elastic end moments reduced ten percent (Section 1.5.1.4.1), plus perhaps ten percent by utilization of the full cross-section of rolled shapes at full stress. (See Fig. 7-1.)

It will be noted that the comparisons in Fig. 7-1 are very crude and the full "potential" tonnage savings indicated will seldom be achieved since (1) fully fixed end supports for negative moment are seldom developed, and (2) most sections used would qualify as compact for $F_b = 0.66 F_y$.

Scope

Until 1974, the AISC Specifications limited plastic design applications to continuous beams, one and two story rigid frames, and *braced* multi-story frames over two stories (2.1). Under Supplement No. 3 (1974), the restrictions upon the use of plastic design were removed (2.1). Following general practice in elastic analysis, a primary plastic analysis (neglecting $P\Delta$) is permitted for all one and two story structures (2.3). For the stability analyses of *braced* multi-story structures (over two stories) axial deformations and lateral deflection in the individual members of the vertical bracing system must be included (2.3.1). Similarly, the stability analyses of *unbraced* multi-story frames must include axial deformation and ($P\Delta$) effects (2.3.2). These analyses must be applied for two load conditions in general: (1) factored gravity loads and (2) the sum of factored gravity and horizontal loads.

In spite of its promise for tonnage savings and other advantages, the use of plastic design could not become general until statutory codes permitted its use without restrictions and included specific requirements for its application. The use of plastic design can now be expected to become general gradually as statutory codes are up-dated to include Supplement No. 3, and as plastic analysis computer programs are developed to implement the requirements of Supplement No. 3. In this chapter therefore, no attempt will be made to develop plastic design beyond basic principles for present practical manual applications. A list of selected references is provided for those who wish to study its application further.

Strength Capacity of Sections

Flexural. For the computation of bending strength capacity of a given section, virtually all of the cross-sectional area of a rolled shape, particularly in bending at maximum efficiency about its major axis, is considered to be stressed to the specified yield point. This condition is based upon the assumption of the flat yield stress level of an idealized stress-strain curve. See Fig. 7-2(a). Other simplifying assumptions commonly employed

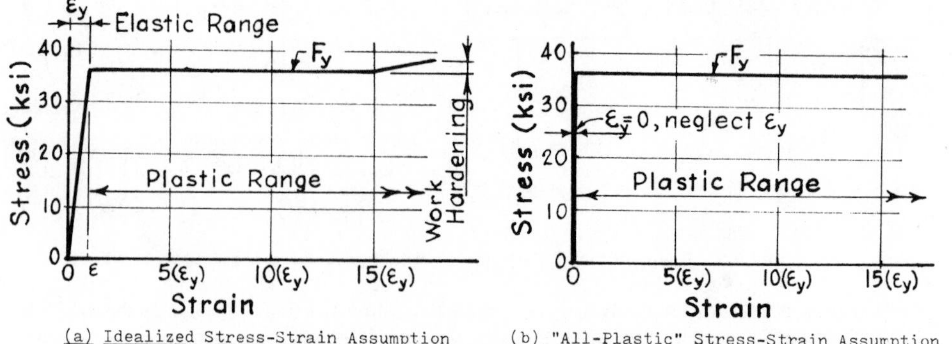

FIG. 7-2 Stress—Strain Relationships; Simplifying Assumptions.

are to neglect the small area of the web near the neutral axis where stress is elastic, varying from nil to F_y. This assumption is convenient for computations and sufficiently accurate since the moment of these stresses is small. In effect, this procedure assumes zero strain from zero stress to the stress, F_y. (See Fig. 7-2(b).) In practical applications for standard rolled shapes, tabulated values of Z_x and Z_y are provided, where Z is defined as the plastic section modulus.[1] The plastic bending moment capacity is computed simply as

$$M_p = F_y Z_x, \text{ or } M_p = F_y Z_y$$

about the major axis, x-x, or minor axis, y-y, respectively. Values of M_p for $F_y = 36$ ksi and $F_y = 50$ ksi are also conveniently tabulated.[2]

Compact sections are most suitable for plastic design because of their ability to develop the yield point stress very nearly through the full depth before local buckling. The "shape factor" (see Fig. 7-1) is greater than 1.0, averaging 1.12 (Comm. 2.1). The non-compact section "shape factor" is 1.0 since local buckling will occur shortly after the outer fibers yield. It will be noted that the tabulated plastic section properties for some sections are limited to $F_y = 36$ ksi for this reason.

Shear. Properly speaking, there is no real plastic analysis required for the shear capacity. The ultimate shear capacity as such is not computed. Assumed shear capacity under the factored loading conditions required for the plastic design is limited safely to

$$V_u \leqslant 0.55 F_y \, td \ldots\ldots (2.5)$$

where t = web thickness and d = web depth.

[1]"Properties for Designing," AISC *Manual of Steel Construction*.
[2]"Plastic Design Selection Table," *Manual of Steel Construction*.

Compression. Width-thickness ratios are limited for members in which a possible plastic hinge under compression may form. These limits are a function primarily of $(1/\sqrt{F_y})$ but also of other variables (Comm. 2.7). A tabulation of these limits for rolled shapes at various levels of F_y is provided (2.7). For box or built-up section flanges, the limit is $190/\sqrt{F_y}$.

Similarly, limits are provided upon the depth-thickness ratio of the webs in such members:

$$d/t \leqslant (412)(1 - 1.4P/P_y)/\sqrt{F_y} \quad \text{for } P/P_y \leqslant 0.27 \text{ and}$$

$$d/t \leqslant 257/F_y \quad \text{for } P/P_y > 0.27 \ldots\ldots \text{(Eqs. 2.7-1a and 2.7-1b respectively).}$$

Shapes for which these limits must be investigated when subject to axial load plus bending are so marked in the "Plastic Design Selection Table."

Columns. Two major design formulas for plastic design of columns in which a hinge may occur parallel the two formulas for elastic design (2.4). As for elastic design, these two formulas provide for safety at two critical sections; (1) at the face of the end joint and (2) at intermediate points. See Chapter 6, Fig. 6-2. The slenderness ratios for such columns are also limited, $l/r < C_c$ (2.4). The two formulas are:

$$P/P_{cr} + C_m M/(1 - P/P_e)M_m \leqslant 1.00 \ldots\ldots \text{(Eq. 2.4-2)}$$

$$P/P_y + M/(1.18M_p) \leqslant 1.00; M \leqslant M_p \ldots\ldots \text{(Eq. 2.4-3).}$$

The new terms introduced for plastic design at this point are:

$$P_{cr} = 1.7AF_a \ldots\ldots \text{(Eq. 2.4-1)}$$

$$P_e = (23/12)AF'_e \ldots\ldots (2.4)$$

for columns braced in the weak direction, $M_M = M_p$ and, for columns unbraced in the weak direction, $M_m = [1.07 - (l/r)\sqrt{F_y}/3160]M_p \leqslant M_p$, where M_m is the maximum moment capacity with no axial load.

Structural Analysis

Continuous Beams. The structural analysis for plastic design can be very simple and is one of the inducements to use of plastic design. The usual idealized example for a continuous beam is based on the ideal case in Fig. 7-1. See Fig. 7-3. All critical moments

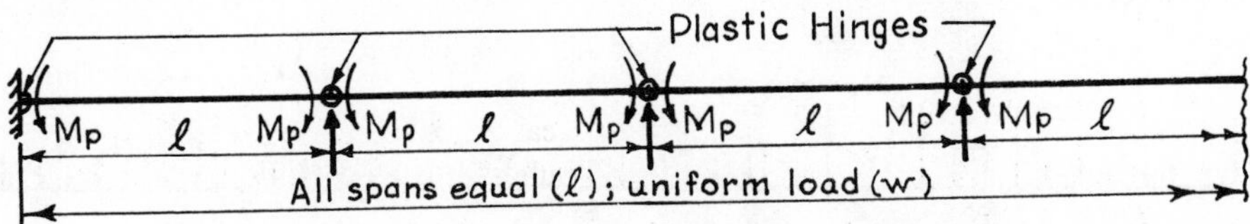

FIG. 7-3 Special Case—Idealized Example.

are made equal to M_P. Hinges form first at the negative moment areas. The final collapse mechanism develops when $+M = M_P$. The analysis is simply $+M = 0.0625\,wl^2$, and the design consists of selecting a uniform beam section with M_P as little as possible above the required value to achieve the potential saving of 17%.

In most practical applications this special case is seldom encountered. Most ends will be either simply supported or framed into a support with less than fixed conditions. Even in

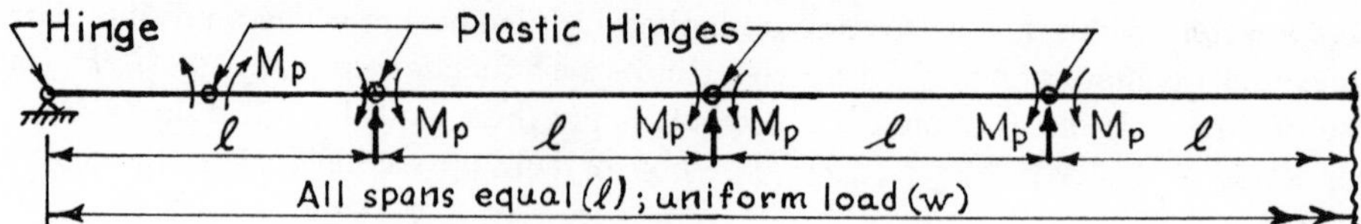

FIG. 7-4 Special Case—Common Example.

the common case where all spans are equal, the end restraints encountered, being variable from (zero up to M_P), complicate the analysis and reduce potential savings in tonnage. (See Fig. 7-4.)

For the somewhat common case of Fig. 7-4, the economy of the plastic analysis is limited to the end spans if the beam section is to be uniform throughout. Where splices are required because of length, maximum economy will require heavier sections in the end spans. With this design, the heavier section can be extended across the first support and spliced near the inflection point or the splice can be located in the first bay. For the case in Fig. 7-4, the difference in weight for these two splice locations is negligible. See Table 2-2(b) for a quick guide to splice locations.

For the completely general case where spans are not equal, loads are not equal or uniform, and some degree of restraint exists at the end, the achievement of maximum economy is much more complex. Maximum economy for the final design is dependent upon selection of sections limited to available shapes and variable live-to-dead load ratios considered in pattern loadings.

The designer has two basic approaches open at this point. The theoretical maximum economy can be achieved only by postulating every conceivable "collapse mechanism," investigating the effect of every loading condition upon each mechanism, and selecting sections with plastic moment capacity as little as possible above that required for the worse cases, "lower bound" solutions. For the simple special case of Fig. 7-3, this approach not only achieves maximum economy of tonnage, but is actually the quickest analysis. The second approach to analysis preferred by the authors for the practical reason of minimum design time yields slightly less capacity than any collapse mechanism and is, therefore, safely below the "lower bound" load capacity without sacrifice of significant economy. Much of the real saving will depend not on the mere maximum design moments from analysis, but the art of selecting sections from the limited list of rolled shapes most appropriate for the structure and connecting same for minimum fabrication and field time. It will be noted that the Specifications permit any rational analysis (2.1).

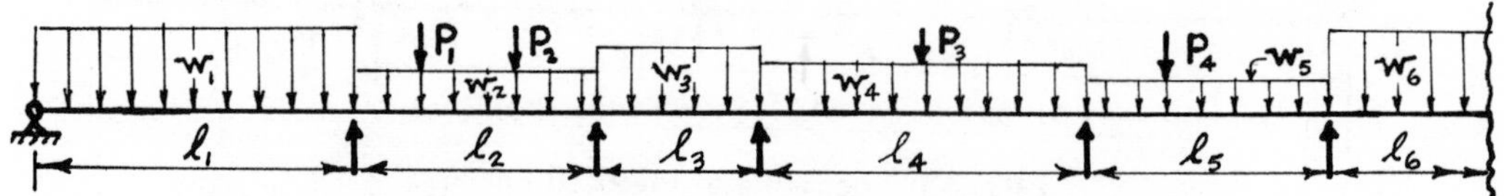

FIG. 7-5 General Case—Variable Spans, Loads, Number of Spans, etc.

Collapse Mechanism. To illustrate the two approaches consider the general case of a beam continuous over supports. The collapse mechanism approach involves the following permutations of load condition and variable effects from adjoining spans repeated for each span with no foreseeable certainty that maxima moments will be compatible with available rolled shape properties. (See Fig. 7-5 and 7-6.)

Interior Span, i

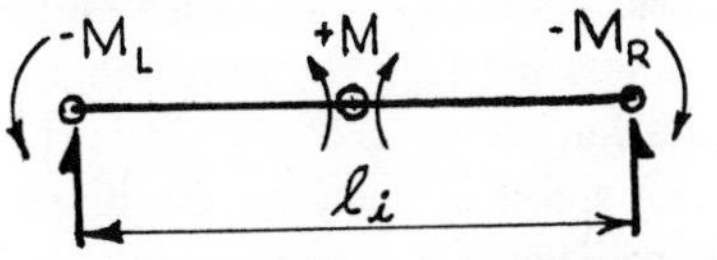

FIG. 7-6 Collapse Mechanism for an Interior Span.

(1) $1.7\,(D + L)$ all spans
(2) $1.7\,(D + L)$ spans h and i; other spans $1.0(D)$
(3) $1.7\,(D + L)$ spans i and j; other spans $1.0(D)$
(4) $1.7\,(D + L)$ spans h and j; $1.0(D)$ on span i. (Center hinge upward)
(5) $1.7\,(D + L)$ span i; $1.0(D)$ other spans

Repeat all five conditions for $1.3\,(D + L + W)$ if wind or seismic forces are included.

Repeat all ten conditions as often as trial section shapes result in incompatible results in adjacent spans.

Plastic Re-design. The second approach might be termed a "plastic re-design." It is based upon use of an elastic analysis which can be secured from conveniently available computer programs or manually by approximate methods for complex structures or even ready-made coefficients for simple structures. This approach is best illustrated also by an example. See Fig. 7-7.

Interior Span. (Data from beam diagrams 37, 38, 39, from *AISC Manual of Steel Construction*, pages 2–211, for four equal spans.) Assuming that available shapes provided section moduli for the same overall system (splice locations, etc.) equally closely fitting elastic design moments, the heavier section in the outer bays was reduced to 0.0900/(0.9 × 0.1205) = 0.93 (seven percent reduction).

It will be noted that maximum deflections even at the factored loads cannot extend far into the plastic range. Under each of the three extreme loading conditions shown (dead

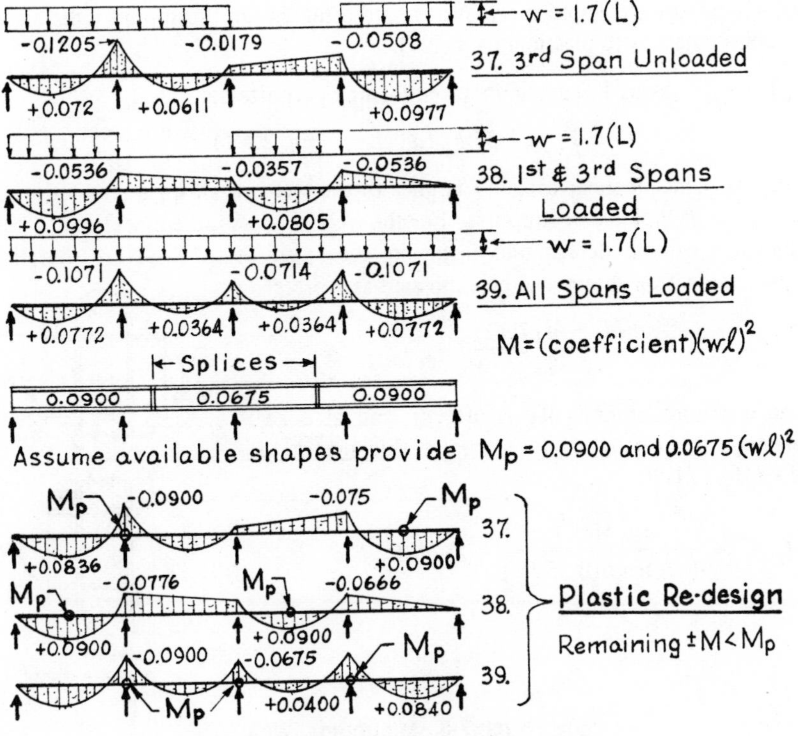

FIG. 7-7 Plastic Redesign—4 Span Beam.

load neglected for pattern load effect) most of the structure is still entirely in the elastic range. A minimum of hinges are postulated to reduce peak moments only. The reserve capacity existing for further loads to form additional hinges for even the lower bound collapse mechanism is conservatively neglected. In addition to most of the potential economy in tonnage saving, in controllable deflection, and speed of analyses, this direct method of plastic re-design is also attractive in that it will yield converging results by additional trials if further refinement is desired.

Design

General. The objective of the plastic design requirements for the final selection of sections is to ensure that the plastic moment capacity, M_p, can, in fact, be developed and maintained through the hinge rotation necessary to develop the other factored load moments from the plastic analysis. The design requirements are intended to assure sufficient strength and rotation capacity at a final hinge moment, M_p, without premature failure due to (1) shear, (2) local buckling of webs or flanges, (3) general buckling of the entire section, or (4) connections.

Flexure. Flexural strength capacity (for compact sections) may be assumed as M_p (Comm. 2.1). No design moments can exceed M_p for either beams or columns (2.4-3). See Table 7-1. For columns under combined flexure and bending of the prescribed factored loads, interaction equations as in elastic design are used to establish limits on axial load, P, and moment, M (Eqs. 2.4-2; 2.4-3). As in elastic design, the two equations are intended to provide limits for critical sections at the end (face of joint) and intermediate points. See Chapter 6, Columns. See also Design Examples, Example 1, Chapter 7. Lateral bracing in members with plastic hinges must meet special requirements near hinges which include bracing at the locations of all plastic hinges (except the last one to form which creates a collapse mechanism), and a maximum unsupported distance to the bracing point adjacent to plastic hinges (2.9).

Shear. The total shear, V_u, at the factored loading permitted is

$$V_u \leqslant 0.55\, F_y\, td \;\ldots\ldots\; \text{(Eq. 2.5-1)}.$$

This limiting value of shear capacity applies to the webs of all members within connections as well as at the face of supports. For the web areas within connections, a minimum web thickness can be determined from Eq. 2.5-1; see Fig. 7-8. Note that the shear within the connection shown can be expressed as

$$V = \frac{(M_2 - M_1)(12)}{0.95\, d_b}$$

where an opposing moment M_1 is present, and at edge columns $M_1 = 0$. Solving for the minimum thickness, t_w, with $M = (M_2 - M_1)$:

$$t_w = \frac{12M}{(0.95\, d_b)(d_c)(0.55\, F_y)} = 23M/A_{bc}F_y$$

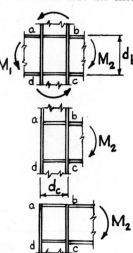

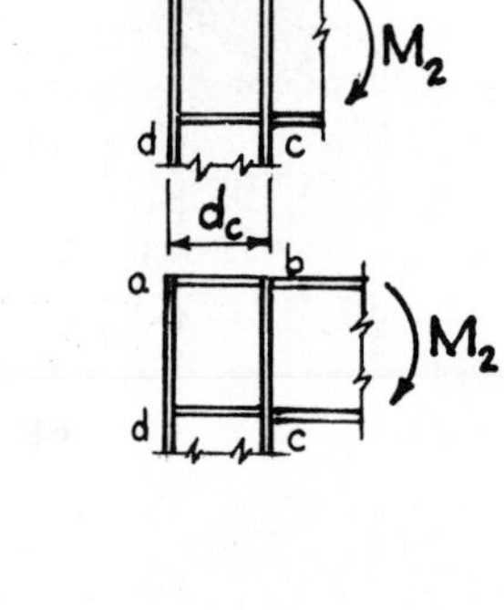

FIG. 7-8 Minimum Web Thickness—Excess Shear.

TABLE 7-1 Ratio of Maximum Moment with $P = 0$ on Columns Unbraced in the Weak Direction to Plastic Moment, M_m/M_p

Ratio of Maximum Moment with $P = 0$ to Plastic Moment, M_m/M_p							
$(C_c = \underline{\ *\ })$ l/r_y	F_y						
	36	42	45	50	55	60	65
126.1*	0.831	–	–	–	–	–	–
125.0	0.833	–	–	–	–	–	–
120.0	0.842	–	–	–	–	–	–
116.7*	–	0.831	–	–	–	–	–
115.0	0.852	0.834	–	–	–	–	–
112.8*	–	–	0.831	–	–	–	–
110.0	0.861	0.844	0.836	–	–	–	–
107.0*	–	–	–	0.831	–	–	–
105.0	0.871	0.855	0.847	0.835	–	–	–
102.0*	–	–	–	–	0.831	–	–
100.0	0.880	0.865	0.858	0.846	0.835	–	–
97.7*	–	–	–	–	–	0.831	–
95.0	0.890	0.875	0.868	0.857	0.847	0.837	–
93.8*	–	–	–	–	–	–	0.831
90.0	0.899	0.885	0.879	0.869	0.859	0.849	0.840
85.0	0.909	0.896	0.890	0.880	0.871	0.862	0.853
80.0	0.918	0.906	0.900	0.891	0.882	0.874	0.866
75.0	0.928	0.916	0.911	0.902	0.894	0.886	0.879
70.0	0.937	0.926	0.921	0.913	0.906	0.898	0.891
65.0	0.947	0.937	0.932	0.925	0.917	0.911	0.904
60.0	0.956	0.947	0.943	0.936	0.929	0.923	0.917
55.0	0.966	0.957	0.953	0.947	0.941	0.935	0.930
50.0	0.975	0.967	0.964	0.958	0.953	0.947	0.942
45.0	0.985	0.978	0.974	0.969	0.964	0.960	0.955
40.0	0.994	0.988	0.985	0.980	0.976	0.972	0.968
35.0	1.0	0.998	0.996	0.992	0.988	0.984	0.981
30.0	–	1.0	1.0	1.0	1.0	0.996	0.993
25.0	–	–	–	–	–	1.0	1.0

where shear in the connection exceeds the value, V_u, or where the web thickness is less than the minimum as computed above, the web must be reinforced for the excess shear by diagonal stiffeners or provided with a web plate adding to the thickness in the area a-b-c-d. See Fig. 7-8. (See AISC Commentary 2.5; 1.5.1.2.)

Local Buckling. For members containing plastic hinges, the *width/thickness* ratios of flanges in compression are limited (2.7):

(1) flanges of rolled shapes or single-web builtup members in compression should not exceed

F_y	36	42	45	50	55	60	65
$b_f/2t_f$	8.5	8.0	7.4	7.0	6.6	6.3	6.0

(2) for box sections and cover plates, the width/thickness ratio should not exceed $190/\sqrt{F_y}$

Depth/thickness ratios are also limited for members containing plastic hinges depending upon the compressive stress level and the yield stress, as follows:

For beams ($P = 0$) . . . . . . $d/t_w \leqslant 412/\sqrt{F_y}$

For columns where $0 < P/P_y \leqslant 0.27$,

$$d/t_w \leqslant 412(1 - 1.4\,P/P_y)/\sqrt{F_y} \;\ldots\ldots \text{(Eq. 2.7-1a)*}$$

For columns where $P/P_y > 0.27$,

$$d/t_w \leqslant 257/\sqrt{F_y} \;\ldots\ldots \text{(Eq. 2.7-1b)}$$

(See Table 7-2.)

TABLE 7-2 Maximum Depth/Thickness Ratios Under Plastic Bending (In the Vicinity of Plastic Hinges)—Webs Inside Connections or Near Connections

		Maximum Depth/Thickness Ratios						
	P/P_y	F_y (ksi)						
		36	42	45	**50**	55	60	65
For General Use in Columns	> 0.27	**42.8**	39.7	38.3	**36.3**	34.7	33.2	31.9
↑	0.26	**43.7**	40.4	39.1	**37.1**	35.3	33.8	32.5
	0.24	**45.6**	42.2	40.8	**38.7**	36.9	35.3	33.9
	0.22	**47.5**	44.0	42.5	**40.3**	38.4	36.8	35.4
	0.20	**49.4**	45.8	44.2	**42.0**	40.0	38.3	36.8
	0.18	**51.4**	47.6	45.9	**43.6**	41.6	39.8	38.2
	0.16	**53.3**	49.3	47.7	**45.2**	43.1	41.3	39.7
Columns or Beam-Columns	0.14	**55.2**	51.1	49.4	**46.8**	44.7	42.8	41.1
	0.12	**57.1**	52.9	51.1	**48.5**	46.2	44.3	42.5
	0.10	**59.1**	54.7	52.8	**50.1**	47.8	45.7	43.9
	0.08	**61.0**	56.5	54.5	**51.7**	49.3	47.2	45.4
	0.06	**62.9**	58.2	56.3	**53.4**	50.9	48.7	46.8
	0.04	**64.8**	60.0	58.0	**55.0**	52.4	50.2	48.2
↓	0.02	**66.7**	61.8	59.7	**56.6**	54.0	51.7	49.7
For General Use in Beams $P = 0$	≈ 0	**68.7**	63.6	61.4	**58.3**	55.6	53.2	51.1

*Note that Eq. 2.7-1b is no longer a plastic design equivalent to Eq. 1.5-4a since Supplement No. 3 changed Eq. 1.5-4a. Commentary section 2.7 should also have been changed.

Maximum Unsupported Lengths. Special limitations are provided upon the maximum spacing of supports and lateral bracing for members in which plastic hinges will occur at the (factored) design loadings.

(1) *In the plane of bending*—The slenderness ratio, l_b/r_b, of columns in which plastic hinges will develop at the ultimate loading shall not exceed C_c (2.9; 1.5.1.3). In this requirement, it would seem logical to consider "ultimate loading" to be the *factored loadings* which are the minimum strength requirement (2.9). Thus, if hinges in the columns will not develop under any factored loading required, the limitations on the unbraced length in the plane of loading would simply be reflected in the interaction equation (2.4-2) as reduced factors in P_{cr} (reduced F_a) and in P_e (reduced F_e').

(2) *Lateral Bracing*. The maximum unsupported lengths, L_b, for beams in the region of moments less than the plastic moment, M_p, are identical to those for allowable stress design (Eqs. 1.5-6a and 1.5-6b); and for columns (Eqs. 1.6-1a and 1.6-1b); except adjacent to plastic hinges (where $M = M_p$) bending about the major axis of rolled sections (2.9). The maximum unbraced length from the bracing required at the plastic hinges bending about the major axis to adjacent bracing points l_{cr} is limited as follows:

	F_y = 36 ksi	F_y = 50 ksi	Eq.
where $+1.0 > M/M_p > -0.5$, l_{cr}/r_y =	63.2	52.5	(2.9-1a)
where $-0.5 \geqslant M/M_p > -1.0$, l_{cr}/r_y =	38.2	27.5	(2.9-1b)

See also Fig. 7-9.

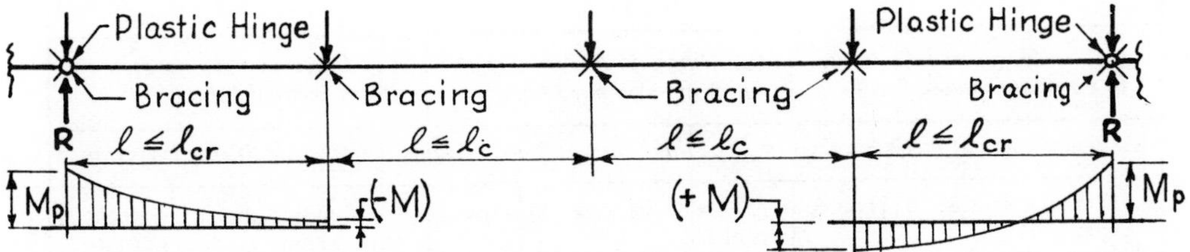

FIG. 7-9 Maximum Spacing of Lateral Bracing.

Note that the critical unbraced length, l_{cr}, is a severe limitation for light columns where bracing points may well be eight to fifteen ft. apart. For this reason location of plastic hinges in the columns of light sections used as columns in one and two story frames should be avoided unless lateral bracing is available continuously or at very close spacings. See table above. The plastic re-design approach to analysis simply avoids hinges in such columns. See Example 1. See Table 7-3 for various conditions.

Connections. All connections must be designed for M_p or such lesser maximum design moments as required by the analyses for factored loads (2.8). Web stiffnesses must be provided at plastic hinges to prevent premature failure by local web crippling (2.6). Such stiffeners are required when web thicknesses are less than the minimum required for elastic design of Type 1 Construction (1.15.5). See Chapter 5, Connections, Examples.

TABLE 7-3 Column Design and Bracing Requirements with Plastic Hinges

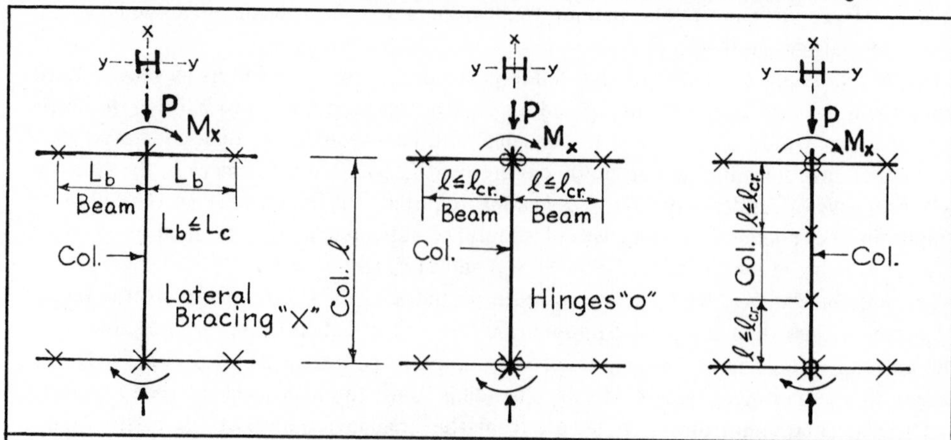

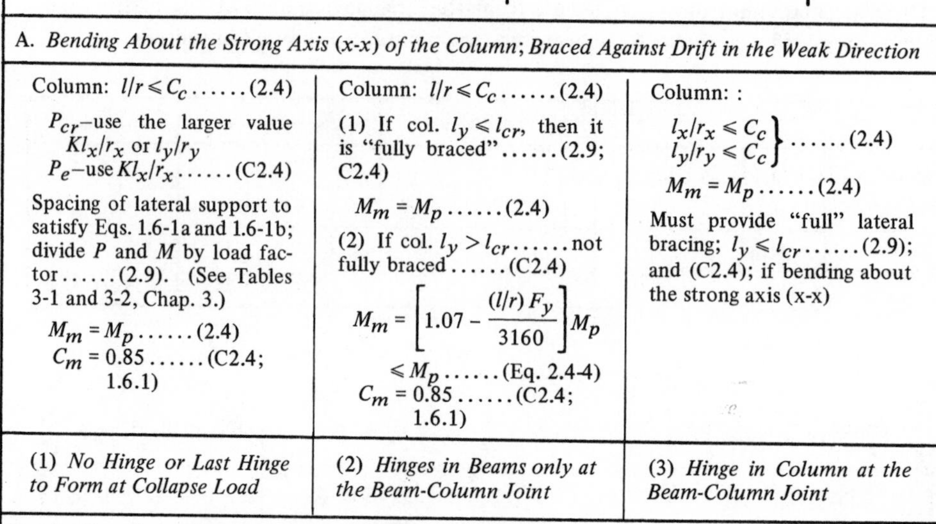

A. *Bending About the Strong Axis (x-x) of the Column; Braced Against Drift in the Weak Direction*

Column: $l/r \leqslant C_c$ . . . . . . (2.4) P_{cr}–use the larger value Kl_x/r_x or l_y/r_y P_e–use Kl_x/r_x . . . . . . (C2.4) Spacing of lateral support to satisfy Eqs. 1.6-1a and 1.6-1b; divide P and M by load factor . . . . . . (2.9). (See Tables 3-1 and 3-2, Chap. 3.) $M_m = M_p$ . . . . . . (2.4) $C_m = 0.85$ . . . . . . (C2.4; 1.6.1)	Column: $l/r \leqslant C_c$ . . . . . . (2.4) (1) If col. $l_y \leqslant l_{cr}$, then it is "fully braced" . . . . . . (2.9; C2.4) $M_m = M_p$ . . . . . . (2.4) (2) If col. $l_y > l_{cr}$ . . . . . . not fully braced . . . . . . (C2.4) $M_m = \left[1.07 - \frac{(l/r)\,F_y}{3160}\right] M_p$ $\leqslant M_p$ . . . . . . (Eq. 2.4-4) $C_m = 0.85$ . . . . . . (C2.4; 1.6.1)	Column: : $l_x/r_x \leqslant C_c$, $l_y/r_y \leqslant C_c$ . . . . . . (2.4) $M_m = M_p$ . . . . . . (2.4) Must provide "full" lateral bracing; $l_y \leqslant l_{cr}$ . . . . . . (2.9); and (C2.4); if bending about the strong axis (x-x)
(1) *No Hinge or Last Hinge to Form at Collapse Load*	(2) *Hinges in Beams only at the Beam-Column Joint*	(3) *Hinge in Column at the Beam-Column Joint*

B. *Bending About the Weak Axis (y-y) of the Column; Braced Against Drift in the Strong Direction*

(1) Same as case A(1) except	(2) Same as case A(2) except	(3) Bending about the weak axis the limit l_{cr} does not apply. Otherwise same as case A(1) except . . . . . . (M_m)
$M_m = \left[1.07 - \frac{(l/r)\,F_y}{3160}\right] M_p \leqslant M_p$ . . . . . . (Eq. 2.4-4)		

C. *Columns in Multi-Story Structures Braced Against Drift in Both Directions*–The requirements in cases A(1), (2), and (3) apply to bending in either direction or both except that the "full" lateral bracing (maximum spacing, l_{cr}) applies only for M_x.

DESIGN EXAMPLES

EXAMPLE 1. *One Story Rigid Frame Analysis*

Use the structure and elastic analyses summarized in Chapter 2, Fig. 2-5(a). As a first trial select the next lighter-weight sections in each series for both the columns and beams. The "upper bound" (collapse mechanism) solutions in plastic design involving columns as in this frame are complicated by the column formulas for combined axial load and moment. In order to consider a final "collapse mechanism" for this frame hinges must be

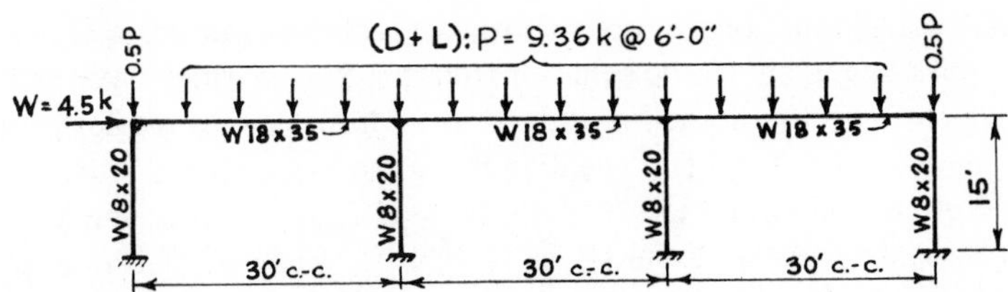

FIG. 7-10 One Story Building with Loads, Dimensions, and First Trial Member Sizes—Design Example 1.

located in the columns for a predicted plastic moment capacity. This value determines and is dependent upon the axial load, and might require several trials. The simpler approach of assuming hinges in the beams only (where axial loads are negligible) will be employed here. This simpler approach is particularly advantageous for plastic "redesign" of a completed elastic design as in this example. (See Table 7-2.) The use of lighter sections of the same depth throughout will not change the relative stiffnesses significantly, and so the available elastic analyses will be satisfactory for practical purposes. The elastic analyses will be applied in direct proportion to the factored loads required in plastic design (2.1). See Fig. 7-10 and the first sketch in Fig. 7-11 of the first trial structure for service (unfactored) loads and analysis.

Record plastic section data:

Beam, $W18 \times 35 – P_y = {}^*371\,k; M_p = 200 \times 12 = 2400$ k-in.
Column, $W8 \times 20 – P_y = 212\,k; M_p = 57.3 \times 12 = 688$ k-in.

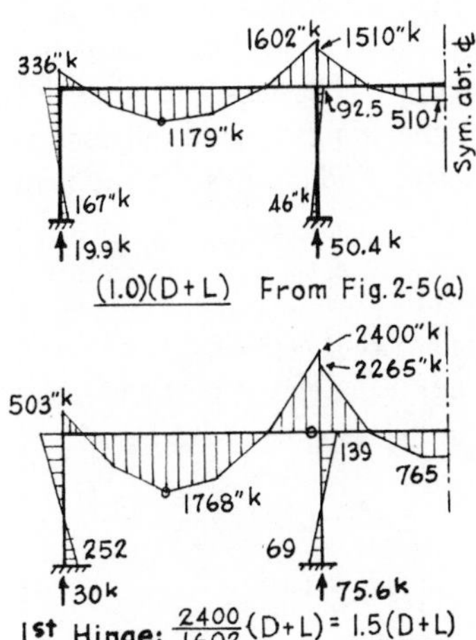

FIG. 7-11 (Elastic) Analysis to Formation of First Hinges.

Note that the assumed structure will behave elastically as loads increase from $1.0\,(D + L)$ or $1.0\,(D + L + W)$ to $1.7\,(D + L)$ and $1.3\,(D + L + W)$ respectively. (See the sequence of loading sketches, Fig. 7-11 for $(D + L)$ to $1.7\,(D + L)$ loading. Note limits of Table 7-2.)

The first hinge forms in the beam at the outer face of the interior column when $M =$ 2400 k-in. The structure is assumed elastic throughout up to this point, and so the load level is $(2400/1602)(D + L) = 1.5(D + L)$. Select points of maximum elastic moment ($M_{max.}$) for the first hinges. Apply M_p at these points to the member in which the hinge

*See Plastic Design Selection Table in AISC Manual.

forms, and increase all other (elastic) moments and reactions in the ratio $M_P/M_{max.}$. This procedure assumes essentially elastic behavior from the load at which outer fibers yield to the load at which the fully plastic hinge develops, a load increment approximately proportional to the moment at the first yielding times the shape factor. (See Fig. 7-1.) It neglects elasto-plastic behavior. Use of the ratio M_p/M_{max}, permits simple extension of the elastic analysis data to the useful stages of "plastic" analysis. This ratio is important as the limit of the purely elastic behavior since the deflections increase at a higher order of magnitude after the hinges form. The elastic deflection overload ratio of 1.5 is eminently reassuring.

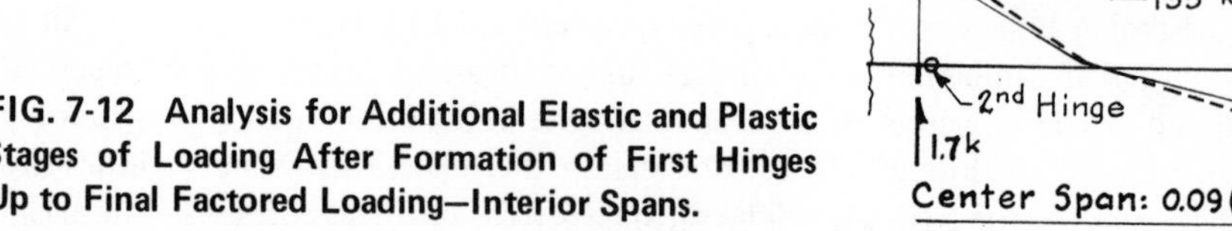

FIG. 7-12 Analysis for Additional Elastic and Plastic Stages of Loading After Formation of First Hinges Up to Final Factored Loading—Interior Spans.

The remaining required overload capacity is only $0.2\,(D + L)$. It will be noted that part of this increment is still in the elastic range of behavior for the center span. See Fig. 7-12. For the center span, add loading of

$$\frac{(2400 - 2265)}{2265}(1.5)(D + L) = 0.09\,(D + L),$$

which will result in the formation of the second hinges. Up to this point assume elastic behavior. For loading beyond this point the plastic hinges at each end of the center span create the equivalent of an elastic simply supported single span. Until the maximum positive moment reaches M_p causing a third hinge, and creating a collapse mechanism, the deflection is still limited elastically. The remainder of the required factored load is $(1.7 - 1.59)(D + L) = 0.11\,(D + L)$. This additional load creates added positive moment $+M = (0.11)(18 \times 12)(P) = 222$ k-in. The total is

$$+M = 765 + 46 + 222 = 1033 \text{ k-in.} < M_p = 2400 \text{ k-in.}$$

The assumed section of the center span is adequate for strength at $1.7\,(D + L)$.

The outer spans behave for the $0.2\,(D + L)$ load increment as an elastic post and beam rigidly connected with the far end of the beam hinged at its support. See Fig. 7-13. The relative stiffnesses become

$$\text{Col. } K_c = I_x/l_x = 69.4/171 = 0.408$$
$$\text{Bm. } K_b = (0.75)(I_x/l_x), \text{ hinged} = (0.75)(513/360) = 1.06$$

Distribution factors are

$$\text{Col. D. F.} = 0.28$$
$$\text{Bm. D. F.} + 0.72$$
$$\text{Use } P = 0.2(9.36)k = 1.872\ k$$
$$\text{F.E.M.} = 0.6\ PL$$
$$\text{Fixed end moment} = (18 \times 12)P = +404$$

See Fig. 7-13 for distribution (one-cycle). $H = 56.5/(0.33 \times 171) = 0.99\,k$.

$$V_2 = \frac{-113 + 1.872(6 + 12 + 18 + 24)(12)}{30 \times 12} = 3.43\,k$$

$$V_1 = (4)(1.872) - 3.43 = 4.05\,k$$

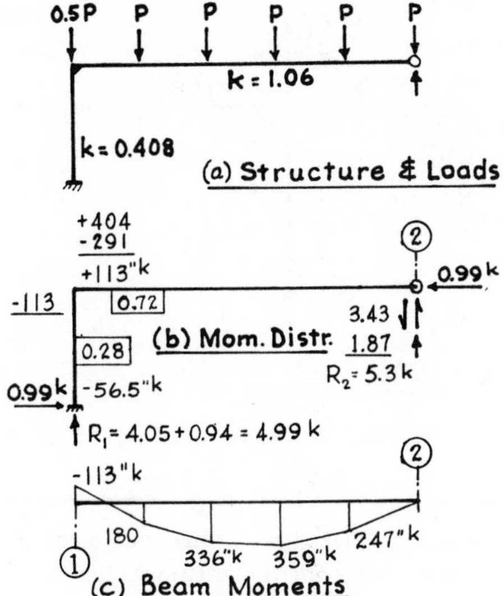

FIG. 7-13 Elastic Structure Remaining in Exterior Spans after First Hinges Up to Final Factored Loading.

Under the final full load in all spans, the plastic hinges in the beams at the interior columns provide balancing moments. Columns ② and ③, however, carry unbalanced moment at the intermediate stages of loading as well as unbalanced moments due to unbalanced shears on each face of the columns at the plastic design load. See Fig. 7-14.

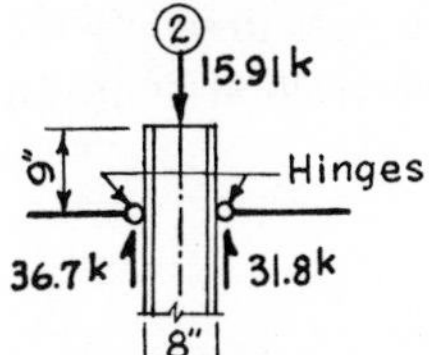

FIG. 7-14 Final Direct Loads and Shears at Interior Columns.

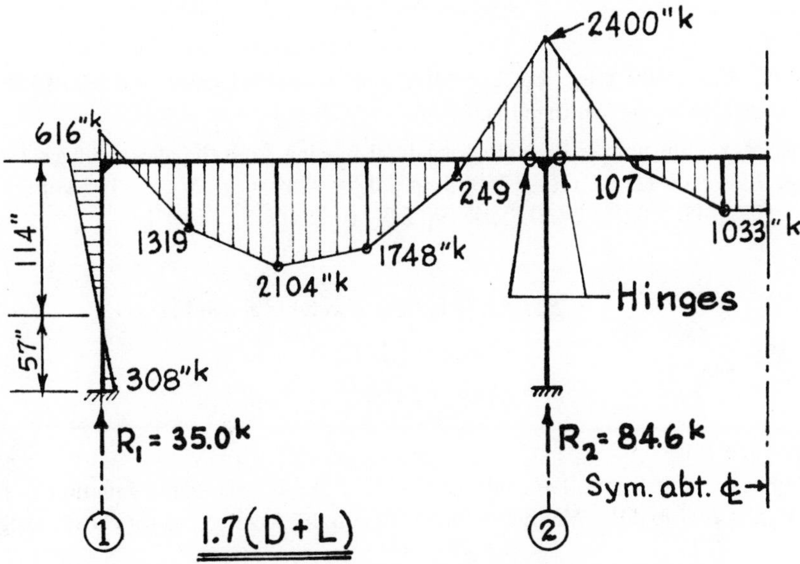

FIG. 7-15 Superimposed Analyses Data for Design at 1.7 $(D + L)$.

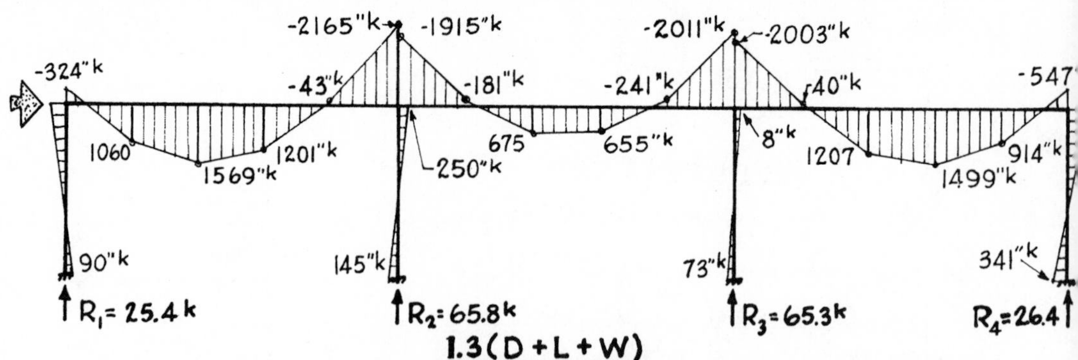

FIG. 7-16 One Story Building—Moments and Reactions for Design Loading 1.3 (*D* + *L* + *W*)—Example 1.

Final moments and column reactions by superposition of each loading stage increment are shown in Fig. 7-15, which represents the final analyses data for the plastic design under loading condition $1.7\,(D + L)$.

For this example, the second prescribed loading condition, $1.3\,(D + L + W)$, for plastic design becomes a simple proportion of 1.3 times the elastic analysis for $1.0\,(D + L + W)$ since no hinges form under this condition. All moments and reactions are elastic ($<M_p$ or P_y) and all (except moments on the interior columns) are less than those for loading condition $1.7\,(D + L)$. See Fig. 7-16.

Figs. 7-15 and 7-16 represent a simple solution for the plastic analyses satisfying Section 2 of the Specifications as a basis for the final selection of member sizes to satisfy the plastic strength requirements. (See also "Ponding," Chapter 3, for deflection limits on roof members.)

STRENGTH OF MEMBERS

EXAMPLE 1. *One Story Rigid Frame—Design Check*

Beams.

$W18 \times 35$; $M_p = 2400$ k-in.; $M_{max.} = 2400$ k-in. maximum shear, $V = 36.7\,k$; all other moments and shears are less than maximum.

Flexure: Maximum unbraced length for lateral bracing from the plastic hinges, $l = 72$ in. See Fig. 7-9. $(1.7)\,(D + L)M_p = 2400$ k-in.; $M_L = +249$ k-in.; $M_R = +107$ k-in. Critical $M/M_p = 107/2400 = 0.045 > -0.5$; use Eq. 2.9-1a

$$\frac{l_{cr}}{r_y} = \frac{1375}{F_y} + 25 = 63.19 \text{ in. See also table above Fig. 7-9.}$$

$$l_{cr} = (1.23)\,(63.19) = 77.7 >= 72 \text{ in. OK}$$

$(1.3)\,(D + L + W)$.

Maximum $M = 2165$ k-in. M_p ratio $= 181/2165 > -\ 0.5$ use same equation as for $1.7\,(D + L)$; $l_{cr} = 77.7$ in. OK. Maximum unbraced length elsewhere, as for elastic design

$$L_c = 6.3 > 6.0 \text{ ft. OK.}$$

Local buckling

$$b_f/2t_f = 6.99 \leqslant 8.5 \ldots\ldots (2.7) \text{ for } F_y = 36 \text{ ksi; OK.}$$

$$d/t_w = 59.4 \leqslant 412/\sqrt{F_y} = 68.7 \ldots\ldots (\text{Eq. 2.7-1a}). \text{ OK.}$$

Shear outside the plastic hinges–

$$\text{Allowable } V_u = 0.55\,F_y\,td \ldots\ldots (2.5)$$

$$= (0.55)(36)(0.298)(17.71)$$

$$= 104.5\,k > 36.7\,k; \text{ OK.}$$

Shear buckling inside the connection at the plastic hinges

(1) at interior columns. (See Fig. 7-14.) $M = 250$ k-in.

$$\text{Min. web, } t_w \geqslant \frac{(23)(250/12)}{A_{bc}F_y} \ldots\ldots (\text{Comm. 2.5})$$

$$t_w = \frac{(23)(20.8)}{(8 \times 18)(36)} = 0.092 \text{ in.}$$

Beam $W18 \times 35$, $t_w = 0.298$ in. OK; column $W8 \times 20$, $t_w = 0.248$ in. OK. The connection may be detailed with the beam either over the column or framing into it. (See Fig. 7-8.)

(2) at the exterior columns. (See Fig. 7-15); $M = 616$ k-in. (not a plastic hinge)

$$\text{Min. web, } t_w = (23)(616/12)/(8 \times 18)(36) = 0.23 \text{ in. OK.}$$

Exterior Columns (① and ④). $W8 \times 20$; $P_y = 212\,k$; $M_p = 57.3 \times 12 = 688$ k-in.; A 5.89 sq. in.; $l_{bx} = 162$ in.; $l_{by} = 100$ in.; $r_x = 3.43$ in.; $r_y = 1.25$ in. From Fig. 6-4, $K_x = 1.2$; $K_y = 0.8$ ...... (C 1.8.1-b); and $C_m = 0.85$ ...... (1.6.1); $P = 35\,k$, $M = 616$ k-in.; 1.7 $(D + L)$

$$K_x\,l_{bx}/r_x = (1.2)(162)/3.43 = 56.7; F_e' = 46.46 \text{ ksi}$$

From AISC Table 2; or Chap. 6, Table 6-1.

$$P_e = (23/12)A\,F_e' = (23/12)(5.89)(46.46) = 524.5\,k \ldots\ldots (2.4)$$

$$K_y\,l_y/r_y = 0.8(100)/1.25 = 64 > 56.7$$

For $Kl/r = 64$, $F_a = 17.04$ ksi; (see Fig. 6-5).

$$P_{cr} = 1.7\,A\,F_a = 1.7(5.89)(17.04) = 170.6 \text{ k} \ldots\ldots (\text{Eq. 2.4-1, 2.4})$$

$$P/P_{cr} + C_m M/(1 - P/P_e)M_m \leqslant 1.00 \ldots\ldots (\text{Eq. 2.4-2})$$

Moment at bot. of beam, $M_f = 616(1 - 9/114) = 567$ k-in.

$$\frac{35.0}{170.6} + \frac{(0.85)(567)}{(1 - 35/524)688} = 0.205 + 0.751 = 0.956 \leqslant 1.0 \text{ OK.}$$

$$P/P_y + M/(1.18)(688) \leqslant 1.00; M \leqslant M_p \ldots\ldots (\text{Eq. 2.4-3})$$

$$\frac{35}{212} + \frac{567}{1.18(688)} = 0.165 + 0.698 = 0.863 \leqslant 1.00 \ldots\ldots \text{OK.}$$

Interior Columns (② and ③) $W8 \times 20$; $l_{bx} = l_{by}$ 162 in.; $K_x = 1.2$; $K_y = 0.8$; section properties same as those for the exterior columns.

1.7(D + L): Maximum $P = 84.6\ k$; $M = (36.7 - 31.8)(4) = 19.8$ k-in., see Fig. 7-11; maximum $M = 159$ k-in. at the centerline, $M_f = 159(1 - 9/114) = 146.4$ k-in. at face with $P = 75.6k$.

$$K_x l_{bx}/r_x = 56.7; K_y l_y/r_y = (0.8)(162)/1.25 = 103.7$$

Enter Fig. 6-5 with $Kl/r = 103.7$, read $F_a = 12.51$ ksi. $P_{cr} = (1.7)(5.89)(12.51) = 125.3\,k$; $P_e = 524.5\,k$, same as for the exterior column.

$$\frac{84.6}{125.3} + \frac{(0.85)(19.8)}{(1 - 84.6/524.5)(688)} = 0.675 + 0.029 = 0.704$$

$$\leqslant 1.00 \ldots\ldots \text{OK} \ldots\ldots \text{(Eq. 2.4-2)}$$

$$\frac{75.6}{125.3} + \frac{(0.85)(146.4)}{(1 - 75.6/524.5)(688)} = 0.603 + 0.211 = 0.814$$

$$\leqslant 1.00 \ldots\ldots \text{OK} \ldots\ldots \text{(Eq. 2.4-2)}$$

$$\frac{75.6}{212} + \frac{146.4}{(1.18)(688)} = 0.357 + 0.180 = 0.537 \leqslant 1.0 \ldots\ldots \text{OK} \ldots\ldots \text{(Eq. 2.4-3)}$$

$$\frac{84.6}{212} + \frac{19.8}{(1.18)(688)} = 0.399 + 0.024 = 0.423 \leqslant 1.0 \text{ OK.}$$

1.3 (D + L + W). Design $P = 65.8\,k$; $M = 250$ k-in. at the centerline; moment at the face, M_f. Inflection point is $(250)(162)/250 + 145 = 102.5$ in. from top.

$$M_f = 250(1 - 9/102.5) = 228 \text{ k-in.}$$

$$\frac{65.8}{125.3} + \frac{(228)(0.85)}{(1 - 65.8/524.5)(688)} = 0.525 + 0.322 = 0.847$$

$$\leqslant 1.00 \ldots\ldots \text{OK} \ldots\ldots \text{(Eq. 2.4-2)}$$

$$\frac{65.8}{212} + \frac{228}{(1.18)(688)} = 0.310 + 0.281 = 0.591 \leqslant 1.00 \text{ OK} \ldots\ldots \text{(Eq. 2.4-3)}$$

All Columns. Maximum shear, $V = 1.3 \times 4 = 5.2\,k$. $V_u = 0.55(36)(0.248)(8.14) = 40.0\,k$ ...... OK. See Chapter 5 for bearing value of anchor bolts on concrete.

Review of these results for possible further savings shows that the exterior columns and beams in the exterior spans offer little further reduction. The results of the two column strength criteria (Eqs. 2.4-2 and 2.4-3) are 0.932/1.00 and 0.924/1.00 respectively. The beams show $+M = 2104$ k-in. versus the plastic moment capacity used for $-M = 2400$ k-in. In the center span both the beam and the interior columns show more unused capacity. Achievement of further weight savings for these members will depend in practice upon availability of lighter sections sufficiently stiff to control deflections. Ideally the choices would be a lighter weight section of the same depth in each case (not possible for the beams).

Interior Columns. Try $W8 \times 17$. $r_y = 1.22$ in.; $r_x = 3.36$ in.; $A = 5.01$ sq. in.; $P_y = 180\,k$; $M_p = 47.7x\ 12 = 572$ k-in.

$$K_y l_y/r_y = (0.8)(162)/1.22 = 106; F_a = 12.20 \text{ ksi}$$
$$K_x l_x/r_x = (1.2)(162)/3.36 = 57.9; F_e' = 44.5 \text{ ksi}$$
$$P_{cr} = (1.7)(5.01)(12.20) = 103.9\,k$$
$$P_e = (23/12)(44.5)(5.01) = 427.3\,k$$

$$\frac{84.6}{103.9} + \frac{(0.85)(19.8)}{(1 - 84.6/427.3)(572)} = 0.814 + 0.037 = 0.851$$

$\leqslant 1.00 \ldots\ldots$ OK $\ldots\ldots$ (Eq. 2.4-2)

$$\frac{75.6}{103.9} + \frac{(0.85)(146.4)}{(1 - 75.6/427.3)(572)} = 0.7276 + 0.2643 = 0.992$$

$\leqslant 1.00 \ldots\ldots$ OK $\ldots\ldots$ (Eq. 2.4-2)

$$\frac{65.8}{103.9} + \frac{(0.85)(228)}{(1 - 65.8/427.3)(572)} = 0.633 + 0.400 = 1.033$$

$\geqslant 1.000 \ldots\ldots$ N.G. $\ldots\ldots$ (Eq. 2.4-2)

Further trials for smaller columns ② and ③ will not be attempted because the plastic section moduli tabulated for available rolled shapes between that for the *W*8 X 20 and the *W*8 X 17 which might provide the required strength are for heavier sections.

Interior Span Beams. Try *W*16 X 26. M_p = 132 X 12 = 1584 = k-in. which is about half way between required $+M$ = 1033 k-in. and $-M$ = 2400 k-in.

Without repeating the analyses for the intermediate (elastic) stages of loading, note that the *W*16 X 26 will provide the required elastic bending capacity for the positive moment to avoid formation of an additional hinge. If the *W*18 X 35 is extended over the interior columns into the center span so as to locate the splices near the inflection points of the center span, the *W*16 X 26 would satisfy the strength requirements without changing the plastic analysis significantly. See Chapter 6, Connections for considerations of splicing.

Weight Savings. The basic plastic design using the *W*18 X 35 for all beams and the *W*8 X 20 for all columns would provide the following reduction in weight of steel per sq. ft.:

Elastic design

Joists, 20H6: 9.6 X 16 X 30	= 4608 lb.
Beams, *W*18 X 40: 40 X 3 X 30	= 3600
Columns, *W*8 X 24: 24 X 4 X 13.5	= 1296
Total per 30 ft. wide bay	= 9504 lb.

Plastic Design (F_y = 36 ksi) Savings in Steel

Beams, *W*18 X 35: 5 X 3 X 30	= 450 lb.	= 12.5%
Columns, *W*8 X 20: 4 X 4 X 13.5	= 216	= 16.7%
Total savings per bay	= 666 lb.	= 7.0%

SELECTED REFERENCES—PLASTIC DESIGN

1. "Research in Plastic Design of Multi-Story Framing," G. C. Driscoll and L. S. Beedle; *AISC Jour.*, July, 1964.
2. "A Method of Combining Mechanisms in Plastic Analysis," G. A. Pincus; *AISC Jour.*, July, 1965.
3. "A Plastic Method for Unbraced Frame Design," J. H. Daniels; *AISC Jour.*, Oct., 1966.
4. "Lehigh Conference on Plastic Design of Multistory Frames—A Summary," G. C. Driscoll, Jr.; *AISC Jour.*, Apr. 1966.
5. "Plastic Design by Moment Balancing," E. H. Gaylord; *AISC Jour.*, Oct., 1967.
6. "Generalized Superposition Method in Plastic Analysis," G. Pincus; *AISC Jour.*, Oct., 1967.
7. "Deformation Analysis of Structures Near Collapse Load," K. P. Lavingia; *AISC Jour.*, July 1968.

8. "Probability of Plastic Collapse Failure," J. L. Jorgenson and J. E. Goldberg; **95**, *No. ST8*; Aug., 1969; ASCE.
9. "A Modification to the Subassemblage Method of Designing Unbraced Multistory Frames," O. de Buen; *AISC Jour.*, Oct., 1969.
10. "Elastic-Plastic Analysis of Frameworks," G. A. Morris and S. J. Fenves; **96**, *No. ST5*; May, 1970; ASCE.
11. "Deflection Analysis for Shakedown," D. G. Eyre and T. V. Galambos; **96**, *No. ST7*; July, 1970; ASCE.
12. "Further Studies of Inelastic Beam-Column Problem," W. F. Chen; **97**, *No. ST2*; Feb., 1971; ASCE.
13. "Applied Plastic Design of Unbraced Multistory Frames," R. O. Disque; *AISC Jour.*, Oct., 1971.
14. "Storywise Plastic Design for Multistory Steel Frames," L. Z. Emkin and W. A. Little; **98**, *No. ST1*; Jan., 1972; ASCE.
15. "Inelastic Multistory Frame Buckling," B. M. McNamee and Le-Wu Yu; **98**, *No. ST7*; July, 1972; ASCE.
16. "Deformation Analysis of Elastic-Plastic Frames," D. E. Grierson; **98**, *No. ST10*; Oct. 1972; ASCE.
17. "Static and Dynamic Cyclic Yielding of Steel Beams," A. M. Almuti and R. D. Hanson; **99**, *No. ST6*; June, 1973, ASCE.
18. "Second-Order Collapse Load Analysis: LP Approach," L. H. Martin and P. J. Wainwright; **99**, *No. ST11*; Nov., 1973; ASCE.
19. *Plastic Design in Steel* (Book), AISC.
20. *Plastic Design of Braced Multistory Steel Frames* (Book), AISC.

SUBJECT INDEX

SPECIFICATION SECTION INDEX

"Specification for the Design, Fabrication, and Erection of Structural Steel for Buildings" and "Commentary," AISC.